Recent Advances in Gel Chemistry: Properties and Applications

Recent Advances in Gel Chemistry: Properties and Applications

Edited by:

Maria Emilova Velinova

ARCLER
PRESS

www.arclerpress.com

Recent Advances in Gel Chemistry: Properties and Applications

Maria Emilova Velinova

Arler Press
224 Shoreacres Road
Burlington, ON L7L 2H2
Canada
www.arclerpress.com
Email: orders@arclereducation.com

ISBN: 978-1-77407-772-6 (Hardcover)

DECLARATION

ABOUT THE EDITOR

Maria Velinova is Ph.D. holder in Quantum chemistry at the University of Sofia since April 2012. Her major research experience is in the field of Computational Chemistry, especially in statistical mechanics methods applied to different sorts of biomolecules. Member of the Laboratory of Quantum and Computational Chemistry at the University of Sofia.

TABLE OF CONTENTS

LIST OF CONTRIBUTORS

Emily R. Draper
School of Chemistry, University of Glasgow, Glasgow, G12 8QQ, UK.

Dave J. Adams
School of Chemistry, University of Glasgow, Glasgow, G12 8QQ, UK.

Ty Christoff-Tempesta
Department of Materials Science and Engineering, Massachusetts Institute of Technology, Cambridge, MA 02139, USA

Andrew J. Lew
Department of Chemistry, Massachusetts Institute of Technology, Cambridge, MA 02139, USA

Julia H. Ortony
Department of Materials Science and Engineering, Massachusetts Institute of Technology, Cambridge, MA 02139, USA

Emily R.Draper
School of Chemistry, University of Glasgow, Glasgow G12 8QQ, UK

Dave J.Adams
School of Chemistry, University of Glasgow, Glasgow G12 8QQ, UK

Dipankar Ghosh
Department of Chemistry, Science Institute, University of Iceland, Dunhagi 3, 107 Reykjavík, Iceland

Matthew T. Mulvee
Department of Chemistry, Durham University, South Road, Durham DH1 3LE, UK

Krishna K. Damodaran
Department of Chemistry, Science Institute, University of Iceland, Dunhagi 3, 107 Reykjavík, Iceland

Xuenan Feng
Beijing Key Laboratory for Science and Application of Functional Molecular and

Crystalline Materials, Department of Chemistry, University of Science and Technology Beijing, Beijing, China

Chenxi Liu
Beijing Key Laboratory for Science and Application of Functional Molecular and Crystalline Materials, Department of Chemistry, University of Science and Technology Beijing, Beijing, China

Xiqian Wang
Beijing Key Laboratory for Science and Application of Functional Molecular and Crystalline Materials, Department of Chemistry, University of Science and Technology Beijing, Beijing, China

Yuying Jiang
Beijing Key Laboratory for Science and Application of Functional Molecular and Crystalline Materials, Department of Chemistry, University of Science and Technology Beijing, Beijing, China

Gengxiang Yang
Beijing Key Laboratory for Science and Application of Functional Molecular and Crystalline Materials, Department of Chemistry, University of Science and Technology Beijing, Beijing, China

Rong Wang
Beijing Key Laboratory for Science and Application of Functional Molecular and Crystalline Materials, Department of Chemistry, University of Science and Technology Beijing, Beijing, China

Kaishun Zheng
Beijing Key Laboratory for Science and Application of Functional Molecular and Crystalline Materials, Department of Chemistry, University of Science and Technology Beijing, Beijing, China

Weixiao Zhang
Beijing Key Laboratory for Science and Application of Functional Molecular and Crystalline Materials, Department of Chemistry, University of Science and Technology Beijing, Beijing, China

Tianyu Wang
Beijing Key Laboratory for Science and Application of Functional Molecular and Crystalline Materials, Department of Chemistry, University of Science and Technology Beijing, Beijing, China

Jianzhuang Jiang
Beijing Key Laboratory for Science and Application of Functional Molecular and Crystalline Materials, Department of Chemistry, University of Science and Technology Beijing, Beijing, China

Juan V. Alegre-Requena
Laboratorio de Organocatálisis Asimétrica. Departamento de Química Orgánica Instituto de Síntesis Química y Catálisis Homogénea (ISQCH) CSIC-Universidad de Zaragoza C/ Pedro Cerbuna 12, 50009 Zaragoza, Spain
Institut für Organische Chemie, Universität Regensburg, Universitätsstr. 31, 93053 Regensburg, Germany

Marleen Häring
Institut für Organische Chemie, Universität Regensburg, Universitätsstr. 31, 93053 Regensburg, Germany

Isaac G. Sonsona
Laboratorio de Organocatálisis Asimétrica. Departamento de Química Orgánica Instituto de Síntesis Química y Catálisis Homogénea (ISQCH) CSIC-Universidad de Zaragoza C/ Pedro Cerbuna 12, 50009 Zaragoza, Spain
Institut für Organische Chemie, Universität Regensburg, Universitätsstr. 31, 93053 Regensburg, Germany

Alex Abramov
Institut für Organische Chemie, Universität Regensburg, Universitätsstr. 31, 93053 Regensburg, Germany

Eugenia Marqués-López
Laboratorio de Organocatálisis Asimétrica. Departamento de Química Orgánica Instituto de Síntesis Química y Catálisis Homogénea (ISQCH) CSIC-Universidad de Zaragoza C/ Pedro Cerbuna 12, 50009 Zaragoza, Spain

Raquel P. Herrera
Laboratorio de Organocatálisis Asimétrica. Departamento de Química Orgánica Instituto de Síntesis Química y Catálisis Homogénea (ISQCH) CSIC-Universidad de Zaragoza C/ Pedro Cerbuna 12, 50009 Zaragoza, Spain

David Díaz Díaz
Institut für Organische Chemie, Universität Regensburg, Universitätsstr. 31, 93053 Regensburg, Germany
Institute of Advanced Chemistry of Catalonia – Spanish National Research Council (IQAC–CSIC), Jordi Girona 18–26, 08034 Barcelona, Spain

Isobel Tibbetts
Department of Chemistry, School of Life Sciences, University of Sussex, Brighton BN1 9QJ, UK

George E. Kostakis
Department of Chemistry, School of Life Sciences, University of Sussex, Brighton BN1 9QJ, UK

Asif Mahmood
Department of Materials Science and Engineering College of Engineering, Peking University, Beijing 100871

Wei Xia
Department of Materials Science and Engineering College of Engineering, Peking University, Beijing 100871

Nasir Mahmood
Department of Materials Science and Engineering College of Engineering, Peking University, Beijing 100871

Qingfei Wang
Department of Materials Science and Engineering College of Engineering, Peking University, Beijing 100871

Ruqiang Zou
Department of Materials Science and Engineering College of Engineering, Peking University, Beijing 100871

Abhijeet K. Chaudhari
Multifunctional Materials & Composites (MMC) Laboratory, Department of Engineering Science, University of Oxford, Parks Road, Oxford, OX1 3PJ, UK.

Jin-Chong Tan
Multifunctional Materials & Composites (MMC) Laboratory, Department of Engineering Science, University of Oxford, Parks Road, Oxford, OX1 3PJ, UK.

Li-Chun Zhao
College of Pharmacy, Guangxi University of Chinese Medicine, Nanning 530200, China

Mei Tang
College of Pharmacy, Guangxi University of Chinese Medicine, Nanning 530200, China

Qian-Hua Zhang
College of Pharmacy, Guangxi University of Chinese Medicine, Nanning 530200, China

Zhi-Yi Hu
College of Pharmacy, Guangxi University of Chinese Medicine, Nanning 530200, China

Hong-Wei Gao
College of Pharmacy, Guangxi University of Chinese Medicine, Nanning 530200, China

Xia-Yun Liao
College of Pharmacy, Guangxi University of Chinese Medicine, Nanning 530200, China

Gang Wang
College of Pharmacy, Guangxi University of Chinese Medicine, Nanning 530200, China

Jing Leng
College of Pharmacy, Guangxi University of Chinese Medicine, Nanning 530200, China

Shirin Rafieian
Department of Chemical Engineering, Islamic Azad University, North Tehran Branch, Tehran, Iran

Hamid Mirzadeh
Polymer and Color Engineering Department, Amirkabir University of Technology, Tehran, Iran

Hamid Mahdavi
Department of Novel Drug Delivery Systems, Iran Polymer and Petrochemical Institute, 15 km Tehran-Karaj Highway, Pajuhesh Science and Technology Park, Pajuhesh Boulevard, Tehran, Iran.

Mir Esmaeil Masoumi
Department of Chemical Engineering, Islamic Azad University, North Tehran Branch, Tehran, Iran

Francesco Picchioni
Department of Chemical Engineering, Engineering and Technology Institute

Groningen (ENTEG), University of Groningen, Nijenborgh 4, 9747 AG Groningen, The Netherlands

Henky Muljana
Department of Chemical Engineering, Engineering and Technology Institute Groningen (ENTEG), University of Groningen, Nijenborgh 4, 9747 AG Groningen, The Netherlands
Department of Chemical Engineering, Parahyangan Catholic University, Ciumbuleuit 94, Bandung 40141, West Java, Indonesia

B.Marco-Dufort
Macromolecular Engineering Laboratory, Department of Mechanical and Process Engineering, ETH Zürich, Zürich 8092, Switzerland

M.W.Tibbitt
Macromolecular Engineering Laboratory, Department of Mechanical and Process Engineering, ETH Zürich, Zürich 8092, Switzerland

Keiichi Imato
Graduate School of Engineering, Kyushu University, 744 Motooka, Nishi-ku, Fukuoka 819-0395, Japan;
Department of Organic and Polymeric Materials, Tokyo Institute of Technology, 2-12-1 Ookayama, Meguro-ku, Tokyo 152-8550, Japan

Masamichi Nishihara
Institute for Materials Chemistry and Engineering, Kyushu University, 744 Motooka, Nishi-ku, Fukuoka 819-0395, Japan

Atsushi Irie
Graduate School of Engineering, Kyushu University, 744 Motooka, Nishi-ku, Fukuoka 819-0395, Japan;

Atsushi Takahara
Graduate School of Engineering, Kyushu University, 744 Motooka, Nishi-ku, Fukuoka 819-0395, Japan;
Institute for Materials Chemistry and Engineering, Kyushu University, 744 Motooka, Nishi-ku, Fukuoka 819-0395, Japan

Hideyuki Otsuka
Graduate School of Engineering, Kyushu University, 744 Motooka, Nishi-ku, Fukuoka 819-0395, Japan;
Department of Organic and Polymeric Materials, Tokyo Institute of Technology, 2-12-1 Ookayama, Meguro-ku, Tokyo 152-8550, Japan
Institute for Materials Chemistry and Engineering, Kyushu University, 744 Motooka,

Nishi-ku, Fukuoka 819-0395, Japan

Christian Weinberger
Department of Chemistry—Inorganic Functional Materials, Paderborn University, 33098 Paderborn, Germany

Dirk Kuckling
Department of Chemistry—Organic and Macromolecular Chemistry, Paderborn University, 33098 Paderborn, Germany

Michael Tiemann
Department of Chemistry—Inorganic Functional Materials, Paderborn University, 33098 Paderborn, Germany

Hong He
College of Petroleum Engineering, China University of Petroleum (East China), No.66, West Changjiang Road, Qingdao 266580, China

Yefei Wang
College of Petroleum Engineering, China University of Petroleum (East China), No.66, West Changjiang Road, Qingdao 266580, China

Xiaojie Sun
College of Petroleum Engineering, China University of Petroleum (East China), No.66, West Changjiang Road, Qingdao 266580, China

Peng Zhang
College of Chemistry and Chemical Engineering, Chongqing University of Science and Technology, Chongqing 401331, China

Dandan Li
College of Petroleum Engineering, China University of Petroleum (East China), No.66, West Changjiang Road, Qingdao 266580, China

LIST OF ABBREVIATIONS

LMWG	Low molecular weight gelators
MMFF	Merck Molecular Force Field
NMR	Nuclear Magnetic Resonance
SEM	scanning electron microscopy
SANS	Small angle neutron scattering
SAXS	Small angle X-ray scattering
STED	Stimulated emission depletion
TEM	Transmission electron microscopy
USANS	Ultra-small angle neutron scattering
XRD	X-Ray Diffraction
AMP	Adenosine monophosphate
ATP	Adenosine triphosphate
RGD	Arginylglycylaspartic acid
BDT	Boc-*O*-dodecyl-l-tyrosine
CNT	Carbon nanotubes
DCP	Diethyl chlorophosphate
DMMP	Dimethyl methylphosphonate
ECM	Extracellular matrix
TTA	triplet-triplet annihilation
LMWGs	Low-molecular-weight gels
CMC	Critical micelle concentration
NMR	Nuclear magnetic resonance
SEM	Scanning electron microscopy
SANS	Small-angle neutron scattering
TEM	Transmission electron microscopy
INO	Isonicotinic acid end
MGC	Minimum gel concentration
MWGs	molecular weight gelators

SCXRD	Single-crystal X-ray diffraction
TEM	Transmission electron microscopy
CCMV	Complex of cowpea chlorotic mottle virus
FRET	Förster resonance energy transfer
HPC	Hydroxylpropyl cellulose
PDT	Photodynamic therapy
VS-PANI	Vinyl substituted polyaniline
ATR	Attenuated total reflection
CGC	Critical gelation concentration
DFS	Dynamic frequency sweep
FESEM	Field emission scanning electron microscopy
TLC	Thin-layer chromatography
LMW	Low-molecular-weight
DSC	Differential scanning calorimetry
NMR	nuclear magnetic resonance spectroscopy
ADF	Annular dark field
CTAB	Cetyltrimethylammonium bromide
DOE	Department of Energy
HFGO	Highly fluorinated graphene oxide
EXAFS	Extended X-ray absorption fine structure
HFGO	Ghly fluorinated graphene oxide
MOFs	Metal–organic frameworks
MOG	Metal–organic gel
PDF	Pair distribution function
PXRD	Powder X-ray diffraction
CuBTC	copper benzene tricarboxylate
COFs	Covalent organic frameworks
DFT	Density functional theory
GC-MS	Gas chromatography-mass spectrometry
PSD	Pore size distribution
SBUs	Secondary building units
TGA	Thermogravimetric analysis
TGA	Thermogravimetric analysis
XPS	X-ray photoelectron spectroscopy

AFM	Atomic force microscopy
HCR	High concentration reaction
LMWG	Low-molecular-weight gels
ZIF	Zeolitic imidazolate framework
BET	Brunauer–Emmett–Teller
TGA	Thermal gravity analysis
AAS	Atomic Absorption Spectrometer
AFM	Atomic force microscopy
CLF	Chitosan–lactidefibrinogen
DMA	Dynamic Mechanical Analysis
ECM	Extracellular matrix
HMM	High molecular mass
SM	Stereo microscopy
OM	Optical microscopy
STM	Tunneling microscopy
ATRP	Atom transfer radical polymerization
HPC	Hydroxypropyle cellulose
HPE	Hyper-branched polyester
DDSs	Drug delivery systems
BAEC	Bovine aortic endothelial cells
FDA	Food and Drug Administration
SHA	Salicylhydroxamic acid
CD	Circular dichroism
ICP	inductively coupled plasma
XPS	X-ray photoelectron spectroscopy
GPC	Gel-permeation chromatographic
PSB	Phosphate saline buffer
TPA	Texture profile analysis
EGDMA	Ethylene glycol dimethacrylate
EISA	Evaporation-induced self-assembly
HEMA	Hydroxyl ethyl methacrylate
DS	Degree of syneresis

PREFACE

Gels are dilute cross-linked systems, which manifest no flow when in the steady-state. In the last few years, an extended range of applications from medicine to material science has arisen a renewed interest in gels, commonly classified in supramolecular gels and polymer gels. The first ones differ from permanently covalently cross-linked polymer gels because the cross-linking can be reversed by the input of energy, for example, by heating, and are assembled from distinct constituent molecules. This book reviews the recent advances in gel chemistry, considering their properties and applications.

Section 1 treats the properties and applications of supramolecular gels, considering how to tune gels features by functional group modification. Moreover, it presents the supramolecular gels with high strength by tuning of calix[4]arene-derived networks.

Section 2 focuses on metal-organic gels and their biomedical applications. Moreover, it discusses a mechano-responsive supramolecular metal-organic framework (MOF) gel material rich in ZIF-8 nanoplates, and the fabrication of a porous MOF for delivery and inhibiting human osteosarcoma cells.

Section 3 begins with treating the history and applications of hydrogels. In particular, it reviews hydrogels based on dynamic covalent and noncovalent bonds. Lastly, it presents the design of moldable hydrogels for biomedical applications and a catalytic chiral gel microfluidic reactor assembled via dynamic covalent chemistry.

Section 4 focuses on diarylbibenzofuranone-based dynamic covalent polymer gels prepared via radical polymerization and subsequent polymer reaction, and on the effect of method type and instrument variables in assessing mucoadhesion in polymer gels.

Finally, the last Section 5 deals with hydrogels as porogens for nanoporous inorganic materials, and with the development and evaluation of organic/inorganic combined gel for conformance control in high temperature and high salinity reservoirs.

SECTION 1
SUPRAMOLECULAR GELS

CHAPTER 1

How Should Multicomponent Supramolecular Gels Be Characterised?

Emily R. Draper[a] **and Dave J. Adams** [a]

School of Chemistry, University of Glasgow, Glasgow, G12 8QQ, UK.

Citation: Emily R. Draper and Dave J. Adams. "How should multicomponent supramolecular gels be characterised?" Chem. Soc. Rev., 2018, 47, 3395 https://doi.org/10.1039/C7CS00804J

ABSTRACT

Low molecular weight gels, or supramolecular gels, are formed when small molecules self-assemble into fibrous structures. Above a critical concentration, the entanglement and cross-linking of these structures leads to the formation of a self-supporting gel. There are many examples where a single component is used to form such gels. There is however an ever-increasing interest in using multiple components. Here, if each component is able to form a gel by itself, a range of fibre types are possible, formed by either random or specific associations between the low molecular weight gelators (LMWG). The properties of the networks will depend on how the LMWG assemble into the primary fibrous structures and then how these primary structures entangle. As such, to understand these gels, it is necessary to understand the networks across multiple length scales. Here, we discuss the current state of the art, the effectiveness of the different techniques that have been used, and hopefully provide the impetus for the field to move away from the cartoon-level discussion of assembly.

KEY LEARNING POINTS

1. Multicomponent supramolecular gels can be formed by the self-assembly of two or more independent gelators into fibrous structures that entangle.
2. On self-assembly, different possibilities exist: the gelators can interact specifically to lead to ordered fibres containing both gelators or randomly to form fibres that contain random amounts of both gelator. Self-sorted fibres may also be formed, where each fibre contains only one of the gelators. Whilst often not stressed, mixtures of these possibilities are conceptually possible.
3. Once these primary structures are formed, entanglement happens over longer length scales and the fibres can interact in different ways.

4. The properties of the gels will not only be a result of this primary assembly, but also to how the fibers entangle, cross-link and interact.
5. To understand these systems fully therefore, it is necessary to characterize the gels across multiple length-scales, from molecular interactions to bulk properties.

INTRODUCTION

Gels can be formed by the self-assembly of small molecules called low molecular weight gelators (LMWGs).[1–3] To form the gels, the LMWG assemble into long fibers, which cross-link in some manner, either by entanglement or branching for example, to form a network. The networks are held together only by the non-covalent forces that induce the self-assembly of the LMWG and so the gels tend to be reversible. These LMWG have been investigated for a wide range of applications, for example for controlled release, cell culturing, optoelectronics and directing crystallisation.[1,4–6]

For the majority of cases, a single LMWG is used to form the gels. There are also a number of examples where more than one component is needed to form a gel; essentially here different components come together in situ to form a LMWG.[7] Recently, there has been more and more interest in the idea of gels formed from multiple LMWG. In these cases, each of the components is a LMWG in its own right.[7,8] There are a number of reasons why mixing different LMWG is interesting. For example, one could imagine a case where one LMWG provides the structural elements of the gel, whilst another drives specific interactions with a desired target.[9] There are examples where the different components have different electronic properties, allowing the formation of an analogy of a bulk heterojunction.[10] It may also be possible to access different properties by mixing components than can be achieved with a single component, such as a temporal trigger for gelation,[11] or this approach can allow gels that can be photo-patterned to be formed.[12,13]

Whilst there is an ever-increasing number of papers on multicomponent gels, there are very few detailed discussions as to how best to characterize these systems. Characterizing single component systems fully across all length-scales can be challenging. Moving to multicomponent systems increases the complexity dramatically. Conceptually, when two LMWG are mixed, there are a range of possibilities. Assuming that the overall self-assembly process still leads to the formation of fibers, the LMWG can interact specifically to lead to ordered fibers containing both gelators (e.g.Fig. 1a), or randomly to form fibers that contain random amounts of both LMWG (Fig. 1b).

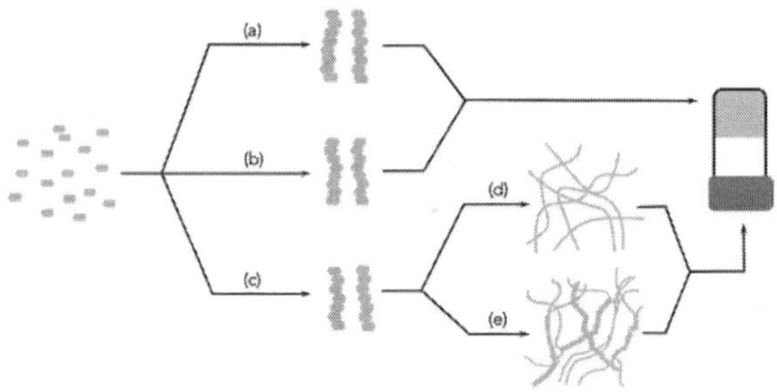

Figure. 1 In a mixture of two LMWG, the primary assembled fibres can be formed by (a) specific interactions leading to ordered fibres; (b) random incorporation of the two gelators; (c) self-sorted fibres. Self-sorted fibres can go on to form (d) homo- or (e) hetero-aggregates.

Self-sorted fibres may also be formed, where each fiber contains only one of the LMWG (Fig. 1c). Of course, it is possible that the assembly leads to a mixture of these possibilities, or mainly but not exclusively one type of assembly. If there are more than two LMWG, the number of scenarios becomes increasingly complex, but all of the examples reported so far are (to the best of our knowledge) only formed by two components.

Importantly, the properties of the gels will not only be a result of this primary assembly, but also to how the fibers entangle, cross-link and interact. Hence, in general, it is important to understand not just the primary assembled structures, but also the structure on a longer length-scale. This is often the most difficult aspect to understand.

Here, we will discuss how these multicomponent gels can be characterized and understood. Currently, the literature has a number of examples, but there are cases where the assembly is assumed, or very little data used to prove the assembly type. In some cases, of course, full characterization may not be needed, but our contention is that the understanding is necessary such that the systems can be designed. As a single example, if we assume that self-sorted fibers are formed and one network is formed before the second, the properties of the gel will be a result of the two networks, how those networks are interacting (if at all), whether the first network was modified by the presence of the second (at the time un-gelled) component, and whether the second network was affected by growing in the presence of the first. This is important; it has been shown that gel networks can be significantly affected by growing in the presence of a non-gelling additive.[14] To understand how to carry this out reproducibly, the gel would need to be characterized at many length-scales. Similarly, if we wish one component to interact specifically with something (e.g. with a cell), we would need this component to be placed in a position where this interaction can occur; again, characterization using a range of techniques will be necessary to understand if this is the case.

In this Tutorial Review, we discuss the current state of the art. We do not attempt to cover how multicomponent systems can be designed to co-assemble or self-sort, but rather we focus on how we can tell the difference post-gelation. For the interested reader, recent reviews cover the design elements of multicomponent systems.[7,8] As this is a Tutorial Review, we are restricted in the overall number of references. We have therefore chosen a range of papers to discuss that cover the concepts. We are in no way suggesting that these are the only, nor necessarily the most important multicomponent gel papers, only that these allow us to cover the widest concepts within the restricted reference list. We have also chosen to focus our discussion around gels formed primarily by assembly of the LMWG into fibrous structures; assembly into other structures such as vesicles and lamellar structures is also known to be possible to form gels.[1,2] As far as we know, there are no examples where mixed systems have been prepared from such LMWG however. Finally, many of the examples presented here are for where at least one of the components is a LMWG based around an amino acid or oligopeptide. However, this is not always the case as we have chosen to show the widest range of techniques.

MULTICOMPONENT ASSEMBLY

There are now a number of papers that describe the use of multicomponent LMWG systems.[7,8] A number of these are based on functionalized amino acids or peptides. As noted, whilst there is conceptually no limit to the number of components that could be mixed, the examples reported so far focus on a mixture of two LMWGs. It is also most common to focus on a single ratio and absolute concentration of each LMWG, but of course this does not need to be the case. As such, whilst the potential complexity is very high, most reports limit this where possible.

It is unsurprisingly most common to mix LMWG whose self-assembly as a single component has also been examined. As such, in many cases, the two-component system is compared to that of the single component and differences are used to infer information about the two-component system.

However, it is clear that using this approach, using only a single technique is sometimes perhaps not as informative as might be assumed. For example, both self-sorted and co-assembled gels (as proven by a range of techniques) were found to have rheology data higher for the two-component gel than for gels formed from either of the two components alone;[11] only by using a number of complementary techniques could the absolute assembly type be assigned. It is also surprising that data are often presented without any comments on reproducibility. It is also very common for a single type of assembly to be assigned, as in 100% self-sorted. It is however, not clear that it would be possible to tell if a mixture of assembly types had occurred and to what degree.

With these caveats in mind, we believe that it is necessary to characterise the gels across multiple length-scales. Here, we mean that it is necessary to understand how the system has self-assembled from the molecular level to the multi-micron length-scale. The interactions at the molecular scale essentially will inform us as to the assembly in the primary structures that are shown in cartoon format in Fig. 1. Once the primary fibers have formed, these can interact in different ways, but this will occur on the nanometre scale (for example, lateral association of fibers will mean that the diameters could be on the order of 10–100 nm, Fig. 2a and b), with the distribution of fibres in space being perhaps homogeneous or heterogeneous on the multi-micron length-scale with potentially different underlying microstructures (Fig. 2c and d). As such, it is necessary to use a range of techniques to access information across these dimensions. We discuss the techniques that have been used below. These techniques have also of course been used for single component systems,[15] but we focus on how these can be used to understand multicomponent systems.

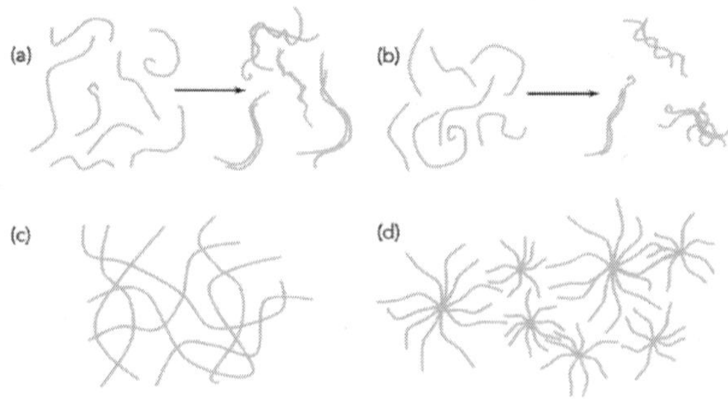

Figure. 2 Assembly of self-sorted fibres could lead to homo-(a) aggregates or (b) hetero-aggregates by (a) lateral association. Different microstructures are

conceptually possible, for example (c) and (d), which would both be consistent with a self-sorted system.

MOLECULAR-LEVEL ASSEMBLY

In terms of our discussion here, we will order the techniques in terms of length-scale. We begin by discussing techniques that can probe the assembly at the molecular level, before moving to the fibre length-scale, and then finally discussing the techniques that can be used on bulk samples.

Spectroscopy

The initial assembly is at the molecular level. In a mixture of gelators, the assembly can be such that the molecules may wish to co-assemble or self-sort (Fig. 1). At this length scale, tools such as infra-red spectroscopy, UV-vis absorption spectroscopy, fluorescence spectroscopy and circular dichroism (CD) spectroscopy can be powerful tools to probe molecular packing, hydrogen bonding etc.

As with most of the techniques discussed here these techniques also rely on the individual components being significantly different to one another or the resulting network being different than the two starting materials. UV-vis absorption spectroscopy in particular can be used to look at aggregation states of molecules and is often used to determine whether aggregates are H-aggregated or J-aggregated, forming extended stacks etc. Conceptually, these spectra can also be used to investigate whether the packing in a multicomponent system is co-assembly or self-sorting. If a system were self-sorted (or at least the majority is self-sorted), then we would expect no change in the absorption of the individual components and the combined spectrum should look like an overlay of the two spectra for the individual components. For example, Sugiyasu et al. have shown that the absorption spectrum for the mixture was simply an addition of the spectra for the two components.[16]. The absence of any new peaks attributable to charge transfer implied that self-sorting had occurred. Monitoring the UV-vis absorption spectra on heating and cooling was used to show that changes occurred at the same temperatures as expected from the data for the single components. A similar overlay of the expected UV-vis absorption spectra was found for a mixture of a gelators.[17]

Alternatively, if there were some co-assembly (either social or random) then we would expect that there could be a change the absorption of the two molecules as the energy levels would be changed by the aggregation. It can be difficult to distinguish whether these systems are socially self-sorted or randomly assembled, or anywhere in between with this technique, or indeed quantify how much of each there is. It may be possible to model the aggregation and resulting spectrum with computational methods, but the number of molecules involved in the stack makes these calculations difficult. However, in some cases, distinct changes in the spectrum can occur on mixing due to charge transfer and this can be used to determine co-assembly. For example, when applied to multicomponent systems, specific co-assembly has been shown for mixtures of electron-rich and electron-poor gelators. For example, Das and Ghosh reported that a mixture of a naphthalene-based gelator could be mixed with a naphthalene diimide based gelator (Fig. 3).[18] The gels formed from the single components were transparent and pale yellow respectively. On mixing, specific co-assembly could be shown by the formation of a charge transfer complex, resulting in a deep red gel. A different mixture was designed such that self-sorting was expected. In this case, the gel was pale yellow, showing the absence of charge transfer.

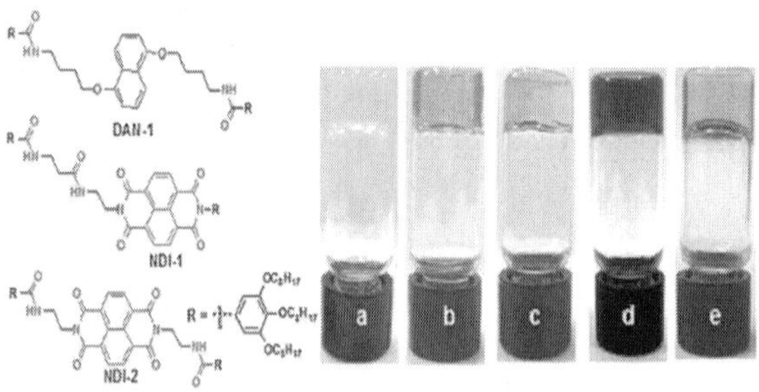

Figure. 3 (left) Structure of the gelators used by Das and Ghosh.[18] DAN-1 and NDI-1 form a self-sorted mixture, whilst DAN-1 and NDI-2 specifically co-assemble. (Right) Photos of gels in methylcyclohexane formed by DAN-1 (a), NDI-1 (b), DAN-1 + NDI-1 (c), DAN-1 + NDI-2 (d) and NDI-2 (e). Reproduced from ref. 18 with permission from The Royal Society of Chemistry.

Dramatic changes in the UV-vis absorption spectra will only be expected if suitably functionalised gelators are used.

Fluorescence spectroscopy can be used in other cases, where energy transfer effects can be used to show that the molecules must be close to one another. This can be used to imply co-assembly. For example, Chen et al. showed that energy transfer could occur in a mixture of a naphthalene-functionalised dipeptide and an anthracene-functionalised dipeptide.[19] Later work by Felip-León et al. however assigned the observation of energy transfer to a self-sorted system.[20] George and co-workers showed both a self-sorted and a co-assembled system which were assigned on the basis on Förster resonance energy transfer and using donor and acceptor type molecules. They showed a chirality driven self-assembly process and used the observation of energy transfer and the CD data to show that the system was co-assembled. They could also visualize their fibers using fluorescence microscopy as a result. Effective energy transfer requires that the donor and acceptor are sufficiently close to one another and this can be achieved either by co-assembly or intimate mixing of fibres. Hence, distinguishing between these two possibilities on the basis of spectroscopy alone is difficult.

CD can provide information on chiral assemblies. For self-sorted systems, the CD spectrum has been shown to be a simple overlap of the sum of those obtained from the individual components.[16] In a case where co-assembly occurred, the CD spectrum was not the same as the expected sum of those of the components.[11]

Fourier-Transform Infrared (FTIR) spectroscopy FTIR works similarly to the above techniques where a difference in the spectrum for the mixed system compared to those for the components indicates co-assembly, whilst an overlay of the two separate components can suggest that co-assembly is occurring.[21] FTIR can also give us an insight as to how interactions are occurring, for example between COOH groups or H-bonding. This technique could therefore be used to understand how the molecules are stacking and even be used to quantify how much co-assembly is occurring. For example, Hao and co-workers used a host–guest system of two gelling components where they could trigger gelation in two different ways giving different types of interactions.[22] This could be seen directly in the IR spectra, with new peaks appearing and other peaks shifting. The spectra also revealed that incomplete self-sorting had occurred in the self-sorted system as a small amount of interaction with the second gelator could be detected. This a rare example of where this has been shown.

Nuclear Magnetic Resonance (NMR) spectroscopy can be used to follow the gelation process. Most commonly, before assembly the molecules can be detected by NMR. As they assemble into fibres, they become NMR-invisible. Hence, this technique can be used to follow the process if the kinetics of gelation are slow, and be used to show that sequential disappearance (indicative of self-sorting)[23] or simultaneous disappearance of the peaks[11] from the different gelators occurs.

For all of these techniques, caution should be used, as they are very concentration dependent. LMWGs can have concentration dependent assembly and so it is important to compare data at the same concentration across all the different techniques. With CD spectroscopy, the HT data needs to below a threshold value in order to be reliable and so should always be shown. Since it is important to use the same concentration for different techniques, dilution is not optimum as a means of (for example) reduce scattering artefacts in CD or the intensity of absorbance in UV-vis absorption spectroscopy. Instead, it is preferable to lower the path-length of cuvettes used.

These techniques can be extremely powerful, but it is important to realize that these can often only be used to show assembly at the molecular level; at best, they imply that the primary structures are formed by co-assembled or self-sorted gelators. To return to Fig. 1, this means that these techniques show whether step (a), (b), or (c) has occurred, but provide little information as to the next length-scale.

FIBRE-LEVEL ASSEMBLY

To probe the next length-scale, most commonly microscopy is used to image the self-assembled fibers, and small angle scattering can also be used to probe these structures.

Microscopy

For most LMWGs, optical microscopy does not have sufficient resolution to allow the network to be imaged. As such, it is most common to probe the fibrous structures using scanning electron microscopy (SEM) or S. Here, the samples are typically dried, so the images are collected for xerogels, not the gels themselves. For hydrogels, cryo-TEM can be used, which has the advantage that the structures are still in a hydrated environment, as opposed to dried under high vacuum, which is the case for TEM. However, the film thickness that is used for cryo-TEM means that it is difficult to truly capture the gel structure; rather it is the structural components that are assumed to be imaged as opposed to the 3D gel network.

For LMWGs, it is not always clear whether the structures that are imaged using SEM or TEM represent those in the gel phase, as drying artefacts can occur, leading to aggregation of structures, and potentially even structural changes. For example, we have shown that drying a dipeptide-based LMWG leads to the fibres becoming much thicker than the primary structures which are observed using cryo-TEM.[24] Further issues can arise from the use of a stain for TEM, which can potentially lead to structural changes. It should also be stated that both SEM and TEM have such high magnifications that only a tiny fraction of a sample can be imaged in a realistic timeframe, meaning that there is always the question as to whether the images are representative. It is also worth highlighting that some specialised gels can be difficult to image, for example those where the solvent is an ionic liquid and therefore difficult to dry.

Saying all of this, microscopy can be an invaluable tool for characterising gels, and in some cases highly informative for multicomponent systems. A wide range of structural types have been imaged for LWMGs, including fibres, tubes, helical structures, and plates, as well as hierarchical structures where primary structures can be imaged forming larger objects. Unfortunately, from the perspective of differentiating between gelators in a multicomponent system, when most LMWGs are imaged, fibrous structures with similar dimensions are found. In our cartoons, it is easy to distinguish between red and blue fibres (for example), but, in reality, often no simple means of differentiation exists. In multicomponent systems, this of course means that it is very difficult to tell if the structures are a mixture of fibres from the individual components or whether new structures are formed. This is often exacerbated by the fact that many gels show the presence of a distribution of fibre widths when imaged. For example, for a mixture of dipeptide-based LMWGs that we were able to show formed a self-sorted system by other techniques, it was extremely difficult to ascertain this by microscopy.[23] Whilst the fibres formed from the two LMWGs formed apparently different fibres when alone, the microscopy for the multicomponent system showed a range of fibres which could not be categorically assigned. Elsewhere, Chen et al. reported simply that the distribution of fibre widths was greater in a mixed system of two peptide-based LMWG than in either of the individual components.[25] From this alone, it is clearly impossible to determine whether there are two networks or whether the distribution of fibres of one (or both) component has been modified by growing in the presence of the other LMWG.

However, in some cases, the microscopy is very clear. Moffat and Smith showed both that images could be collected for a self-sorted system where two clearly different fibre types were visible (Fig. 4a).[26] The two fibre types had very different diameters, and could be seen to be co-existing. In a second case, where co-assembly was expected from gel melting data, SEM showed that a more homogeneous network was formed where the fibre diameters were greater than for either of the two single component systems. Similarly, Escuder's group showed TEM images again of different diameter self-sorted fibres, but with interaction between the two with the thinner fibres being wrapped around the larger one.[27] For two amino acid-based LMWGs, we were able to show by image analysis that the distribution of fibre diameters in a self-sorted mixture was bimodal, mirroring the expected two distributions from the individual components.[28] Similarly, atomic force microscopy (AFM) has been used to image co-existing fibres with different widths.[29]

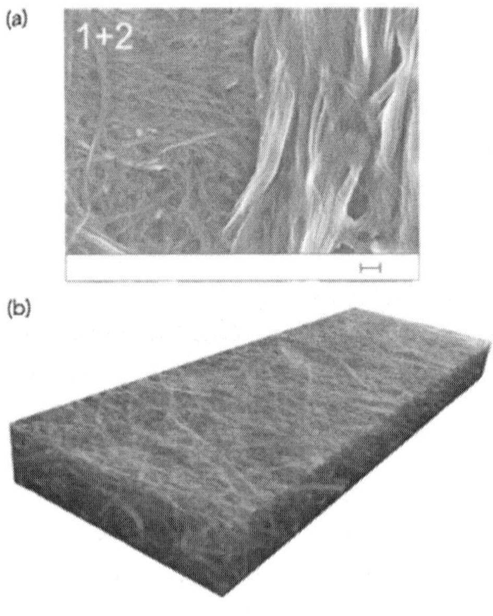

Figure. 4 (a) SEM image showing two different fiber networks in a two-component xerogel. The scale bar represents 200 nm. Adapted from Moffat and Smith.[26] Adapted from ref. 26 with permission from The Royal Society of Chemsitry. (b) A three-dimensional confocal laser scanning microscope image showing self-sorted network in the gel state. The fibres are stained specifically such that one type of fibre has a different colour from the other.[31] Adapted with permission from Onogi et al.ref. 31, Copyright 2016, Nature Publishing Group.

In other cases, SEM was used to show that helical structures were formed in a mixture of two gelators, whilst flat ribbons were imaged in the gels formed from either of the single components.[30] This strongly implies co-assembly has led to the formation of a new type of structure.

The identity of the components forming the fibres that are imaged by TEM, SEM, and AFM has to be inferred on the basis of structure or diameter as it is not trivial to differentiate between the fibres chemically. Conceptually, if one of the constituent molecules contained an element that was not present in the second component, it might be possible to show the location of this component using energy-dispersive X-ray spectroscopy, although to the best of our knowledge this has never been carried out. An alternative approach to localising different components is to label the gelators with a fluorescent dye and use confocal microscopy. Selectively staining only one gelator is difficult as the dye is simply incorporated into all structures in a multicomponent system.

A more effective approach here is to covalently attach the dye to the gelator itself. Here, it is of course very likely that the covalent modification of a gelator will to some degree affect the gelation ability. Nonetheless, if a fluorescently-labelled gelator can be co-assembled with the non-labelled analogue, conceptually this effect should be minimised. Using this approach, Onogi et al. were able to selectively label two different gelators in a two-component mixture. Using confocal laser scanning microscopy (CLSM) and stimulated emission depletion (STED) microscopy, the authors showed the first example where directly imaged self-sorted fibres could be proved in the gel state, and a three-dimensional image constructed (Fig. 4b). This could impressively be extended to imaging the real-time evolution of the self-sorted gel networks.

A similar approach has recently been used to differentiate between co-assembled and self-sorted systems formed from low-complexity domains of a protein.[32] Block structures could also be formed and imaged, and this approach was also used to show that the fibres were dynamic, with mixing of the building blocks occurring even when pre-formed fibres were mixed. A very recent report used stochastic reconstruction microscopy (STORM) to likewise show that mixing two separately labelled samples (albeit of the same peptide) could be used to provide evidence of fibre bundling (which as noted above is very hard to resolve using TEM or AFM) to show little monomer exchange.[33] These high-resolution microscopy techniques are becoming more available, and we anticipate that there will soon be a number of papers showing their power for multicomponent systems.

Small angle scattering

Small angle scattering can be used to probe both primary fibres and some features of the network.[34] The advantages of scattering over most microscopy techniques is that it is non-destructive and can be carried out on the solvated, bulk samples. The disadvantage is that access to large-scale facilities is often needed. Small angle neutron scattering (SANS) also requires contrast, which is generally achieved by using either a deuterated solvent, or a deuterated LMWG. This normally therefore requires that the system is not absolutely the same as for the other techniques used, and also adds cost. SANS and small angle X-ray scattering (SAXS) can be used to look at primary fibres structures[34] and ultra-small angle neutron scattering (USANS) can be used to look at longer length scales, and hence probe some elements of the network[35] (although there seem to currently be no examples of using USANS for multicomponent systems). Generally, the scattering data is fitted to a model, with the choice of model used being based on the best fit to the data. For LMWGs, the best fits usually are best to a cylinder, a flexible cylinder, a hollow cylinder, or some other long, anisotropic structure. The fit provides information as to the radius of the structure, the length, persistence length (if appropriate) etc. It has also recently been shown that SANS can be used to follow the drying process if suitable contrast is achieved using deuterated dipeptide-based gelators.[24] The data showed that the scattering primarily probed the primary fibres, not aggregated fibres that were imaged using SEM.

Such scattering techniques can be used to determine whether self-sorting or co-assembly has occurred if the two networks scatter significantly differently from each other,[36] or the resulting new network is sufficiently different from than those of the two individual components. As an example, a mixture of two dipeptide-based LMWGs was prepared (Fig. 5a). The SANS data from the gels formed from the individual components were significantly different (Fig. 5b).[23]

The data for LMWG **1** could be fitted to a hollow cylinder model, whilst that for LMWG **2** could be fitted to a flexible cylinder. Again, as for the discussion above, these diameters were lower than those imaged by microscopy, implying that SANS probes the primary assemblies. For the gel formed by mixing the two LMWGs, the scattering had key features from the scattering data for both individual gelators, where the core–shell structure could still be seen for LMWG **1** for example by the peak at a Q of 0.1 $Å^{-1}$. A partially-deuterated analogue of LMWG **2** was prepared, which significantly reduced the contribution of this LMWG to the overall scattering. In this mixture, the scattering was very similar to that of LMWG **1** alone, strongly implying that self-sorted fibres had been formed.

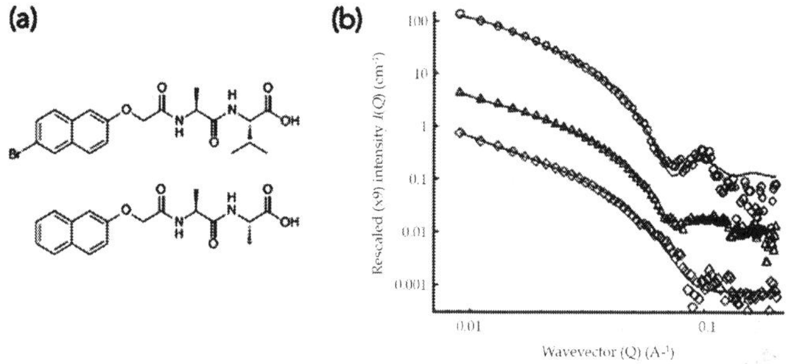

Figure. 5 (a) Structures of functionalised dipeptide LMWGs **1** (top) and **2** (bottom). (b) Small angle neutron scattering (SANS) data. SANS data for (○) **1** alone, (◊) **2** alone and (Δ) **1** and **2**. Adapted with permission from Morris et al.ref. 23, Copyright 2013, Nature Publishing Group.[23]

This latter aspect shows the power of contrast matching in SANS and conceptually it should be possible to modify the ratio of H_2O to D_2O in the solvent mixture to specifically contrast match a single component of a multicomponent system, although this has yet to be shown. The ability to selectively deuterate (and so contrast match) part of a molecule also should allow for a greater understanding of the molecular packing of a LMWG within a fibre, although again this has yet to be fully exploited.

Elsewhere, a co-assembled system was shown to scatter very similarly to one of the individual components, implying that the structures formed by the co-assembly are driven by this gelator.[11] Elsewhere, the SAXS data for a co-assembled system was shown to be different from that expected from either of the individual components, showing that a new type of structure was formed in the co-assembled system.[37]

Computational modelling

Many groups have used computational methods to investigate the molecular packing of these self-assembled or co-assembled materials.[38] Some have even been able to look at stacks of molecules, simulating what a fibre could look like.[9,39] The problems facing computational methods is that the more molecules are involved in the calculations, the longer and more difficult the calculations are to do. It is also difficult to take into account the solvent environment and changing conditions that occur as the molecules are assembling (gels are triggered from a dissolved or dispersed state to a gelled state, so the self-assembly occurs generally in a changing environment). Presently, most people using computational methods are basing their models on data already collected on the samples. In essence, this is post-rationalising the results, rather than predicting whether the sample is self-sorted or co-assembled. There are various calculations that can be used depending on what data has been collected, the number of molecules included and environment. Caneschi and co-workers used Merck Molecular Force Field (MMFF) in vacuo combined with X-Ray Diffraction (XRD) data to model the anti and gauche rotation of the fibres.[40] They could then use the CD spectra within the coupled oscillator framework (DeVoe method) to computationally predict the spectra and compare with the experimental data. The data matched up and they could show the self-sorted fibers had a twisted conformation, although could not predict the handedness of the twist.

For all of these examples, the data are generally useful to describe whether the fibers that have been formed are a result of co-assembly or self-sorting. However, this still does not allow us to fully describe the system. The network forms as a result of the entanglement of the fibers, and the primary fiber assembly does not necessarily allow us to discuss this.

NETWORK-LEVEL ASSEMBLY

It is extremely difficult to access direct information on the gel networks. Generally, electron microscopy and scattering can only inform on the primary fibers, and so understanding the type and relative degrees of crosslinking and or lateral association of fibers is difficult. Generally, therefore the networks have to be differentiated by inference.

Physical properties

First, it can be possible to infer a network type by eye, or by a change in behavior. The change in properties is not necessary able to explain the exact assembled nature of the samples nor the degree to which assembly has occurred, but it can provide a quick suggestion of the assembly type. For example, when mixing two samples which independently give stable transparent gels, if the mixture then gives a turbid gel this could be an indication that some co-assembly could be happening. However, this change in appearance could however be also due to a change in concentration or the samples not mixing effectively and so therefore other techniques would need to be used. The kinetics at which turbidity changes occur on gelation can be indicative of different processes occurring (Fig. 6).[41,42] Likewise, a change in color of the gel could also be an indication that different LMWGs are interacting with each, perhaps suggesting co-assembly, as described above for gels which undergo charge transfer.[18] Self-sorting of a system can be seen by eye in rare cases. For example, if a multicomponent gel network is formed where one component undergoes a gel-to-crystal transition in the single component gel, then the observation that crystallization is occurring in the mixture, with a gel being maintained overall is highly indicative that a self-sorted interpenetrated network has been formed. If the system were co-assembled, crystallization from the gel phase would be expected to destroy the overall gel network as a result. We have observed this crystallization from a multicomponent system,[11] but do stress that this is a highly unusual situation.

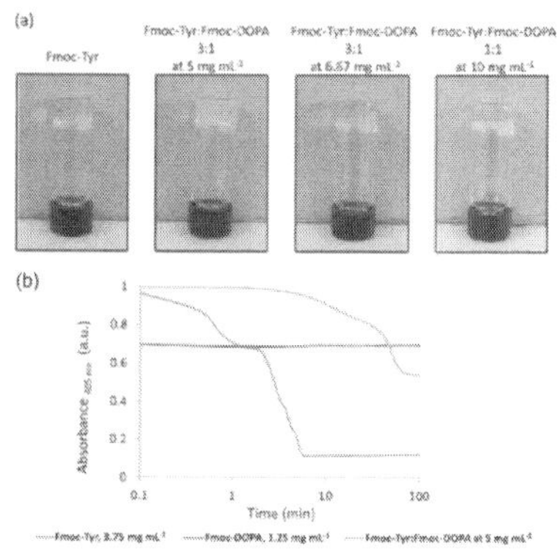

Figure. 6 (a) Photographs showing the change in turbidity upon mixing of two gelators at different concentrations, compared to that of the single component. (b) Graph showing the change in turbidity during gelation of the single components (blue and red) and then the mixed gel (green), showing very different gelation kinetics are occurring.[42] Adapted from ref. 42 with permission from The Royal Society of Chemsitry.

Another example of a change in properties that can arise in multicomponent systems is when using LMWG for electronic applications. The use of a n-type system mixed with p-type gelling system could result in a p–n heterojunction being formed in the gel.[16] This would result in a change in resistance in the material formed. The extent of the change in resistance could indicate whether the sample was most likely self-sorted or co-assembled or even whether the sample was hetero- or homo-aggregated.

For example, Draper et al. used a perylene bisimide based n-type LMWG mixed with a stilbene-based p-type LMWG to make a p–n heterojunction xerogel.[28] The xerogel formed from the n-type component alone only showed a photocurrent when UV light was used, and the single component p-type xerogel did not produce any current. When mixed together, the xerogel was now responsive to blue light. This shows that there was donation of an electron from the p-type LMWG into the n-type LMWG, reducing the amount of energy of light needed to give a response. This indicates that the samples must be self-sorted and not significantly hetero-aggregated. If co-assembly or too intimate a mixing of the two fibres had occurred, one would expect effective recombination of charges, leading to a reduction in the current from the xerogel.

Melting points

The melting temperature of a gel is a common parameter to be measured and quoted. Here, the assumption is that the gel's melting temperature is indicative of the underlying network. Typically, the melting temperature is determined by vial inversion; when the network has sufficiently melted, the sample flows when the vial is turned over. This might happen uniformly throughout the sample, or perhaps sufficient quantities of the network melts to lubricate the material.

Moffat and Smith showed that gel melting temperatures could be used to suggest whether self-sorted or mixed systems are formed in multicomponent systems.[26] In one example, two amino-acid based bolaamphiphile based gelators were used which had a significant difference in melting temperatures. In the mixture, the melting temperature was found to be close to that of the higher melting single component. At the very least, this implies that the network formed by this gelator is similar to that in the single component and hence implies self-sorting has occurred. For another mixture, the melting temperature was higher than the lowest melting component, but lower than that of the highest melting component. This implies that a different network type has formed, and hence that co-assembly has occurred. These data agreed with the microscopy (see above). These data were obtained from bulk melting of the gel. If there were two networks with one melting at a significantly lower temperature, but the second network were able to maintain a self-supporting network, it would not be possible to show this using this method.

Elsewhere, a mixture of two bis(amino acid) gelators was shown to be synergistic, with the mixtures being able to immobilise significantly more solvent than the individual components.[43] The melting point of the gel was around 20–30 °C higher for the mixture as compared to that for the gels formed from the single components.

As alternative methods, differential scanning calorimetry (DSC) or NMR spectroscopy can be used. DSC measures the endo- or exotherms that arise from sol–gel and gel–sol transitions. Using this method, Smith and Smith were able to show that there were two distinct transitions in a two-component mixture at significantly different temperatures, providing clear evidence of two co-existing networks (Fig. 7).[44] The endotherm for the lower melting gelator was around 6 °C below that for the pure component and the authors speculated that this showed that the network was compromised in the mixture by the presence of the other component. For the same mixture, NMR was also used. On heating, the integrals for the peaks from the gelator increase as the network melts and the gelator becomes more mobile in solution. Using variable temperature NMR, it was shown that the integrals increased on heating with a very similar profile for the lower melting gelator in either the single or two-component network. This implies that the networks are independent. An interesting caveat here is that the DSC, the NMR, and the bulk melting data are not perfectly in agreement; the NMR data implies that the lower melting component melts between 30 and 60 °C, the DSC implies that this component melts at 48 °C, and the tube inversion implies that the gel melts at 32 °C. The authors rationalise this on the basis of the size of sample environment and geometry. It could also be the case that different absolute properties are being measured; tube inversion is only possible when a sample tube spanning network is complete, so a slight change in the network might mean that tube inversion is not possible even though the network has not formally melted.

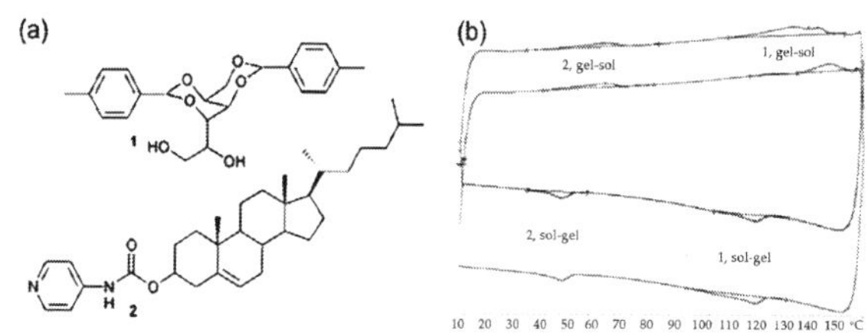

Figure. 7 (a) Structure of gelator **1** (top) and gelator **2** (bottom) (b) DSC plots showing distinct melting points and gelling points in the mixed system for the separate components.[44] Adapted from ref. 44, with permission from The Royal Society of Chemsitry.

Rheology

Rheology can be used to analyses the mechanical properties of bulk gel samples.[45]. The rheological properties are determined by a number of factors, including the morphology of the fibers, the fiber persistence lengths and strengths, the number and type of crosslinks and the distribution of the fibers in space. This is simple to state, but it is often extremely difficult to determine why a specific gel is stiffer than another as any or all of these parameters might have changed.

When two gelators are mixed, a difficult decision is what data should be compared with what. Most examples compare the rheological properties of each single component gel with that of the multi-component gel, with the concentrations of the single component gels being carried over to the multi-component system. Hence, if gelators 1 and 2 are at a concentration of 5 mg mL^{-1} in the single component gel, then this is the concentration also used in the multicomponent systems. However, this means that the multicomponent system is now at a total gelator concentration of 10 mg mL^{-1}; the rheological properties of such gels are generally very concentration dependent (most basically, the more gelator, the higher the number of fibres), so we would expect that it would be very likely that the multicomponent system would have higher values of the storage modulus (G') and loss modulus (G'') no matter what. However, this is not always observed, although it is rare for this to be commented upon.

As such, the absolute values of the G' and G'' alone cannot be used to determine whether a system is self-sorted or co-assembled. There are examples where upon mixing, the moduli increase for the system as compared to the single components. For example, Li et al. showed that a mixture of two gelators formed a gel with a storage modulus an order of magnitude above that of either of the individual components.[46] As mentioned above, this could simply be because there is more gelator in the system. It could also be due to new fibres being formed upon co-assembly making the new system stronger than the individual ones, for example by the incorporation of molecule that provides cross-links making the system stronger.[47] Alternatively, it could be the result of a self-sorted system having fibres with favorable interactions with one another (hence leading to entangling, increased hydrogen bonding etc.) which increases the gel strength. This was shown by Schneider and co-workers where they used different enantiomers to enhance mechanical properties of a self-sorted gel.[48] Kinetics are also important; in some cases for low molecular weight gels, the quicker the network forms, the higher the number of crosslinks and the lower the homogeneity; both of these factors can affect the rheological properties. Hence, sometimes the data are very difficult to deconvolute. For example, Halperin-Sternfield et al. mixed a Fmoc-dipeptide and a Fmoc-amino acid.[41] The rheological data was shown to vary with the ratio of the two LMWG, with the 1: 1 mixture showing significantly higher moduli than either of the two components. However, the kinetics of the assembly process was also shown to vary as the ratio was varied. Hence, it is difficult to link an increase in modulus to one specific effect on the basis of the rheology alone.

There are also examples were the mixing of the two gelators creates a gel that is of a similar stiffness or perhaps weaker than that formed from either of the two individual components. For example, Chen et al. have reported a gel where the values of G′ and G″ are similar to those of the strongest single component gel.[25] Similar data was reported by Horgan et al., where the value of G′ in the mixed gel was very similar to that of the stiffest single component network.[37] However, here the value of G″ was different to that of the stiffest single component gel, meaning that the value of tan δ (G″/G′) was more similar to that of the weaker single component.

The explanation for the mixed gels being no stronger than the individual gels could be simply that the second network is essentially so weak that the first dominates. Alternatively, it could be that co-assembled new fibres are formed that are weaker or less entangled than the network in the two single component gels. In a self-sorted system, the two individual networks could have unfavourable interactions with other or not want to interact or mix with each other, resulting in a weaker overall network. An alternative explanation could be that there are steric crowding effects in the system. As an example of the complexity, Liyanage and Nilsson produced a range of Fmoc-amino acid based LMWG and found the increase or decrease in gel strength did not correlate to whether systems containing two of these LMWGs were co-assembled or self-sorted.[49]

Rheological frequency tests, nanoindentation, and cavitational rheology that give us only G′ and G″ values are not therefore always that useful in providing information on the assembly type. However, if one of the networks can be changed with an external trigger, it can be possible to infer information about the system. For example, Smith and co-workers showed a mixed system where a UV light can be used to trigger the gelation of one of the components in the presence of another gel formed by a change in pH.[13] Hence, here one network is grown first, followed by the second.

This shows that the networks are self-sorted, but does not necessarily show whether the two networks are interpenetrating or whether the second network grows on the first for example.

Draper et al. showed a self-sorted system where the assembly was triggered sequentially by a slow pH change.[12] It was possible to selectively remove one of the networks post-gelation using UV light; one of the gelators underwent a isomerisation under UV light, rendering it ineffective as a LMWG. As a result, the network formed by this gelator fell apart on irradiation. The network remaining after irradiation was found to have the same rheological property as the single component gel. This is a rare example of where is has been shown that one networks is not affected by the other network that already formed and there are apparently two completely independent networks, consistent with Fig. 2a. From other data, they concluded that their system was self-sorted and the two fibres were strongly interacting with one another causing the change in rheological properties. Since this method used UV-light to erase one network, it was possible to pattern a gel, such that one region contained a self-sorted network, and another the single component (Fig. 8).

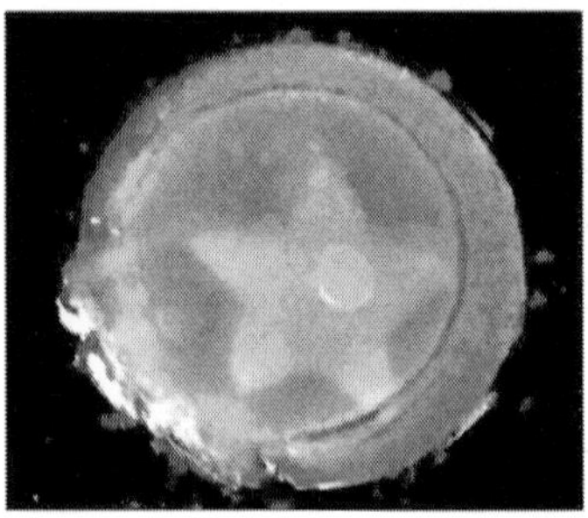

Figure. 8 Photograph of a photopatterned gel under 365 nm UV light. The star has been irradiated with to photo-erase one network, whilst outside the star contains a two-component self-sorted network. Adapted with permission from Draper et al., ref. 12, Copyright 2015, Nature Publishing Group.

Rheological techniques that can be used to show a change in network include using a strain sweep and looking at how the network breaks down. This relies on the two individual components having a different breakdown behavior (maybe in multiple steps, having a sharp break at a specific strain, creaming behavior, or perhaps a large difference in the strain required to break the network). If co-assembly has occurred, a different breakdown behavior may be seen as the fibres in the network will have changed. If self-sorting has occurred, two different breakdown profiles may still be seen even if the G′ and G″ are different compared to the single component systems. At the moment, it is difficult to determine the type of network from these data. Instead, we can only infer that there is a difference. As an example, Che et al. showed that their mixed gel had significantly longer linear viscoelastic region than the single components.[50] As a result, it is clear that the gel network is not simply a combination of the two networks formed from the single components. Self-sorting was proved by other means, and the rheological data were interpreted on the basis of interactions between the different types of fibre. Hence, this correlates with the situation in Fig. 2b. Differences in the strain sweeps were also observed by Xing et al.[22]

Finally, if the gelation method is suitable, it is possible to show the assembly type using a rheological time sweep. We have described a number of systems where a slow pH trigger is used to induce gelation. In many cases, the two LMWG are chosen to have different pH at which assembly and gelation occurs.[12,23] Hence, the slow pH change allows sequential assembly of the networks and this can be seen in a multi-step increase in G′ and G″ during gelation. This can be combined with techniques such as NMR to show that one component becomes NMR-invisible as it gels, and there is a concomitant increase in the rheology at this point. In cases where co-assembly occurs, both LMWG simultaneously become NMR-invisible and there is a single increase in the rheological data.[11]

CONCLUSIONS

In many cases, preparing gels from mixtures of two LMWG is relatively straightforward if the triggering method is the same for both. However, the complexity that can be achieved from even a two-component system is high. This is a fascinating, intellectually stimulating problem, which requires significant input of time and effort to pull apart. For the primary assembled structures, co-assembly, self-sorting, or a combination of these can occur. At the next length scale, self-sorted primary structures can conceptually also co-assemble, self-sort or form a combination. Hence, the simple statement that two LMWG form a specific type of assembly is often extremely difficult to prove across all length scales. This however does not stop such statements being routinely made.Here, we hope that we have shown that a wide range of techniques can be used to probe mixed systems across the length scales. Not all will be appropriate in all cases, and it may be that only a small number of techniques will be able to suggest the assembly type. It is also important to consider the hierarchy level at which the technique is able to provide information. It is also not clear that techniques that show co-assembly has occurred can always pull apart whether the co-assembly is random or specific. Nonetheless, by applying many of these techniques in concert, we believe that it is generally possible to explain the assembly type. This requires using a number of techniques in unison, and often using several pieces of information.

As a single example, we showed that sequential assembly could be proved by the rate at which the peaks from two different gelators disappeared out of the NMR spectrum when a slow gelation trigger was used.[23] This shows only sequential assembly, but does not prove self-sorted fibres are formed. Microscopy was unclear, as this was not a case where two distinct fibre populations could be imaged. SANS was informative, with the scattering being consistent with the system containing a mixture of self-sorted fibres from the two individual components, but was not conclusive. Using fibre X-ray diffraction however, we were able to absolutely demonstrate that the primary fibres were self-sorted. Whilst this was a clear demonstration that sequential assembly could be used to form self-sorted systems, this still does not inform as to how the fibres are distributed in space and whether the self-sorted fibres form homo- or hetero-aggregates. There are limited methods which are able to do this. Hence, there is still a way to go in understanding how to characterise these systems.In some cases, it could be argued that it is unimportant to known definitively. However, we would respond by stating that generally mixed systems are used to add information or signals to a system. As such, design of these is predicated on the understanding of the assembly type. As a single example, if one component has a signalling unit on it, the density of these groups will be very different if the components form co-assembled structures (where the signal will be 'diluted' in a fibre by the other component) or self-sorted fibres.

There are ever increasing numbers of papers discussing mixed systems. In many cases, it seems to us that the assignment of the structure as co-assembled or self-sorted can be at the suggestion of the authors, with little data to categorically demonstrate this. We hope that this review will provide the impetus for the field to move away from the cartoon-level discussion of assembly to the general requirement that proof is provided for each assignment.

CONFLICTS OF INTEREST

There are no conflicts to declare.

ACKNOWLEDGEMENTS

E. R. D. thanks the Leverhulme Trust for funding (ECF-2017-223) and the University of Glasgow for an LKAS Leadership Fellowship. D. J. A. thanks the EPSRC for a Fellowship (EP/L021978/1).

REFERENCES

1. R. G. Weiss J. Am. Chem. Soc., 2014, **136** , 7519 —7530
2. Functional Molecular Gels , B. Escuder and J. F. Miravet, Royal Society of Chemistry, Cambridge, 2014, Search PubMed .
3. M. D. Segarra-Maset , V. J. Nebot , J. F. Miravet and B. Escuder , Chem. Soc. Rev., 2013, **42** , 7086 —7098
4. E. R. Draper and D. J. Adams , Chem, 2017, **3** , 390 —410
5. D. K. Kumar and J. W. Steed , Chem. Soc. Rev., 2014, **43** , 2080 —2088
6. J. B. Matson and S. I. Stupp , Chem. Commun., 2012, **48** , 26 —33
7. L. E. Buerkle and S. J. Rowan , Chem. Soc. Rev., 2012, **41** , 6089 —6102
8. J. Raeburn and D. J. Adams , Chem. Commun., 2015, **51** , 5170 —5180
9. M. Zhou , A. M. Smith , A. K. Das , N. W. Hodson , R. F. Collins , R. V. Ulijn and J. E. Gough , Biomaterials, 2009, **30** , 2523 —2530
10. E. R. Draper , J. R. Lee , M. Wallace , F. Jäckel , A. J. Cowan and D. J. Adams , Chem. Sci., 2016, **7** , 6499 —6505
11. C. Colquhoun , E. R. Draper , E. G. Eden , B. N. Cattoz , K. L. Morris , L. Chen , T. O. McDonald , A. E. Terry , P. C. Griffiths , L. C. Serpell and D. J. Adams , Nanoscale, 2014, **6** , 13719 —13725
12. E. R. Draper , E. G. Eden , T. O. McDonald and D. J. Adams , Nat. Chem., 2015, **7** , 848
13. D. J. Cornwell , O. J. Daubney and D. K. Smith , J. Am. Chem. Soc., 2015, **137** , 15486 —15492
14. Y. J. Adhia , T. H. Schloemer , M. T. Perez and A. J. McNeil , Soft Matter, 2012, **8** , 430 —434
15. G. Yu , X. Yan , C. Han and F. Huang , Chem. Soc. Rev., 2013, **42** , 6697 —6722
16. K. Sugiyasu , S.-I. Kawano , N. Fujita and S. Shinkai , Chem. Mater., 2008, **20** , 2863 —2865
17. S. Prasanthkumar , S. Ghosh , V. C. Nair , A. Saeki , S. Seki and A. Ajayaghosh , Angew. Chem., Int. Ed., 2015, **54** , 946 —950
18. A. Das and S. Ghosh , Chem. Commun., 2011, **47** , 8922 —8924
19. L. Chen , S. Revel , K. Morris and D. J. Adams , Chem. Commun., 2010, **46** , 4267 —4269
20. C. Felip-León , S. Díaz-Oltra , F. Galindo and J. F. Miravet , Chem.

Mater., 2016, **28** , 7964 —7972

21. Y. M. Abul-Haija , S. Roy , P. W. J. M. Frederix , N. Javid , V. Jayawarna and R. V. Ulijn , Small, 2014, **10** , 973 —979
22. P. Xing , X. Chu , S. Li , F. Xin , M. Ma and A. Hao , New J. Chem., 2013, **37** , 3949 —3955
23. K. L. Morris , L. Chen , J. Raeburn , O. R. Sellick , P. Cotanda , A. Paul , P. C. Griffiths , S. M. King , R. K. O'Reilly , L. C. Serpell and D. J. Adams , Nat. Commun., 2013, **4** , 1480 CrossRef PubMed .
24. L. L. E. Mears , E. R. Draper , A. M. Castilla , H. Su , Zhuola , B. Dietrich , M. C. Nolan , G. N. Smith , J. Doutch , S. Rogers , R. Akhtar , H. Cui and D. J. Adams , Biomacromolecules, 2017, **18** , 3531 —3540
25. G. Chen , J. Li , M. Song , Z. Wu , W. Zhang , Z. Wang , J. Gao , Z. Yang and C. Ou , Adv. Funct. Mater., 2017, **27** , 1701798
26. J. R. Moffat and D. K. Smith , Chem. Commun., 2009, 316 —318
27. N. Singh , K. Zhang , C. A. Angulo-Pachon , E. Mendes , J. H. van Esch and B. Escuder , Chem. Sci., 2016, **7** , 5568 —5572
28. E. R. Draper , J. R. Lee , M. Wallace , F. Jackel , A. J. Cowan and D. J. Adams , Chem. Sci., 2016, **7** , 6499 —6505
29. N. Singh , C. Maity , K. Zhang , C. A. Angulo-Pachón , J. H. van Esch , R. Eelkema and B. Escuder , Chem. – Eur. J., 2017, **23** , 2018 —2021
30. J. A. Foster , R. M. Edkins , G. J. Cameron , N. Colgin , K. Fucke , S. Ridgeway , A. G. Crawford , T. B. Marder , A. Beeby , S. L. Cobb and J. W. Steed , Chem. – Eur. J., 2014, **20** , 279 —291
31. S. Onogi , H. Shigemitsu , T. Yoshii , T. Tanida , M. Ikeda , R. Kubota and I. Hamachi , Nat. Chem., 2016, **8** , 743 —752
32. B. An , X. Wang , M. Cui , X. Gui , X. Mao , Y. Liu , K. Li , C. Chu , J. Pu , S. Ren , Y. Wang , G. Zhong , T. K. Lu , C. Liu and C. Zhong , ACS Nano, 2017, **11** , 6985 —6995
33. H. Cox , P. Georgiades , H. Xu , T. A. Waigh and J. R. Lu , Biomacromolecules, 2017, **18** , 3481 —3491
34. J.-B. Guilbaud and A. Saiani , Chem. Soc. Rev., 2011, **40** , 1200 —1210
35. R. A. Hule , R. P. Nagarkar , A. Altunbas , H. R. Ramay , M. C. Branco , J. P. Schneider and D. J. Pochan , Faraday Discuss., 2008, **139** , 251 —264

36. A. M. Castilla , E. R. Draper , M. C. Nolan , C. Brasnett , A. Seddon , L. L. Mears , N. Cowieson and D. J. Adams , Sci. Rep., 2017, **7** , 8380 CrossRef PubMed .
37. C. C. Horgan , A. L. Rodriguez , R. Li , K. F. Bruggeman , N. Stupka , J. K. Raynes , L. Day , J. W. White , R. J. Williams and D. R. Nisbet , Acta Biomater., 2016, **38** , 11 —22
38. R. K. Das , R. Kandanelli , J. Linnanto , K. Bose and U. Maitra , Langmuir, 2010, **26** , 16141 —16149
39. Y. M. Abul-Haija , G. G. Scott , J. K. Sahoo , T. Tuttle and R. V. Ulijn , Chem. Commun., 2017, **53** , 9562 —9565
40. S. Cicchi , G. Ghini , L. Lascialfari , A. Brandi , F. Betti , D. Berti , P. Baglioni , L. Di Bari , G. Pescitelli , M. Mannini and A. Caneschi , Soft Matter, 2010, **6** , 1655 —1661
41. M. Halperin-Sternfeld , M. Ghosh , R. Sevostianov , I. Grigoriants and L. Adler-Abramovich , Chem. Commun., 2017, **53** , 9586 —9589
42. G. Fichman , T. Guterman , L. Adler-Abramovich and E. Gazit , CrystEngComm, 2015, **17** , 8105 —8112
43. Z. Dzolic , K. Wolsperger and M. Zinic , New J. Chem., 2006, **30** , 1411 —1419
44. M. M. Smith and D. K. Smith , Soft Matter, 2011, **7** , 4856 —4860
45. C. Yan and D. J. Pochan , Chem. Soc. Rev., 2010, **39** , 3528 —3540
46. D. Li , Y. Shi and L. Wang , Chin. J. Chem., 2014, **32** , 123 —127
47. S. Boothroyd , A. Saiani and A. F. Miller , Biopolymers, 2014, **101** , 669 —680
48. K. J. Nagy , M. C. Giano , A. Jin , D. J. Pochan and J. P. Schneider , J. Am. Chem. Soc., 2011, **133** , 14975 —14977
49. W. Liyanage and B. L. Nilsson , Langmuir, 2016, **32** , 787 —799
50. X. Che , B. Bai , T. Zhang , C. Zhang , C. Zhang , P. Zhang , H. Wang and M. Li , New J. Chem., 2017, **41** , 8614 —8619

CHAPTER

2

Beyond Covalent Crosslinks: Applications of Supramolecular Gels

Ty Christoff-Tempesta [1], Andrew J. Lew [2], and Julia H. Ortony [1]

[1]Department of Materials Science and Engineering, Massachusetts Institute of Technology, Cambridge, MA 02139, USA

[2]Department of Chemistry, Massachusetts Institute of Technology, Cambridge, MA 02139, USA

ABSTRACT

Traditionally, gels have been defined by their covalently cross-linked polymer networks. Supramolecular gels challenge this framework by relying on non-covalent interactions for self-organization into hierarchical structures. This class of materials offers a variety of novel and exciting potential applications.

Citation: Christoff-Tempesta, T.; Lew, A.J.; Ortony, J.H. "Beyond Covalent Cross-links: Applications of Supramolecular Gels". Gels 2018, 4, 40. https://doi.org/10.3390/gels4020040

This review draws together recent advances in supramolecular gels with an emphasis on their proposed uses as optoelectronic, energy, biomedical, and biological materials. Additional special topics reviewed include environmental remediation, participation in synthesis procedures, and other industrial uses. The examples presented here demonstrate unique benefits of supramolecular gels, including tunability, processability, and self-healing capability, enabling a new approach to solve engineering challenges.

Keywords: supramolecular gel; self-assembly; gels; applied soft matter

INTRODUCTION

Exciting material architectures with unconventional blends of properties exist between the domains of traditional solids, liquids, and gases. In particular, gels are a dispersion of fluids within a solid, which are able to retain a viscoelastic solid-like form despite a majority portion of their mass usually being comprised of some liquid [**1**]. Paul Flory established four canonical categories of gels in 1974 [**2**]: (1) high-order lamellar gels (e.g., phospholipids), (2) disordered covalent networks (e.g., vulcanized rubber), (3) semi-ordered physical networks (e.g., gelatin), and (4) disordered particulate gels (e.g., reticular fiber networks) [**1**]. Gels can be further distinguished by the type of solvent used to yield categories such as hydrogels, organogels, and aerogels.

Supramolecular gels are assembled from distinct constituent molecules, and have been a focus of increased research since the turn of the century [**3**]. In the strictest sense, they are a subcategory of Flory gel category 4, and are comprised of low molecular weight gelators [**4**] self-assembled into a disordered network structure through secondary noncovalent bonds, such as solvophobic interactions, π–π stacking, and hydrogen bonding, among others. The distinction of secondary bonding among molecular units differentiates supramolecular gels and traditional covalently bonded network gels, creating unique properties and functionalities that utilize its ability to more easily and reversibly break and reform bonds between subunits.

However, the moniker of supramolecular gels has also been attributed less strictly by many authors to mean any gel utilizing noncovalent bonding between molecular species [**5**]. Thus, supramolecular gels can be taken to include even polymeric gels with covalent bonding so long as they include other species that are noncovalently bonded to the gel, ionic crosslinks, or physical bonding between polymeric chains (as opposed to exclusively covalent crosslinks). In this view, certain gels within Flory gel categories

1 and 3 as well as assemblies like gel nanocomposites have also been categorized as supramolecular. For the purposes of reporting the wide variety of possibilities in this developing field, here we take a wide approach to supramolecularity and include all of the above into consideration (**Figure 1**).

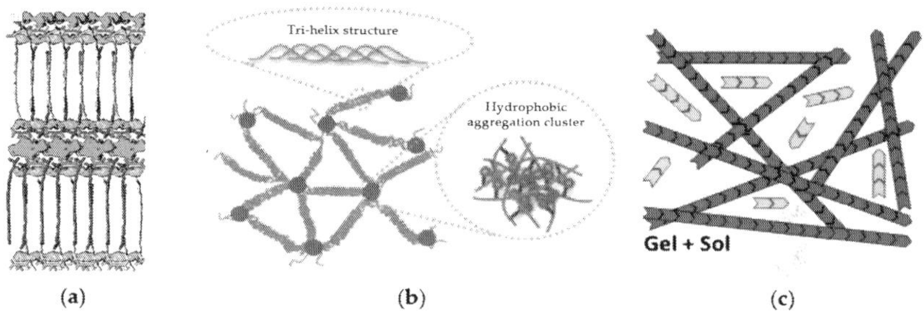

Figure 1. Examples of supramolecular gels are found in three of the four categories of gels defined by Flory [**2**]: (**a**) category 1: high-order lamellar gels [**6**]. Adapted with permission from ref [**6**]. Copyright 2016 Boekhoven et al.; (**b**) category 3: semi-ordered physical networks [**7**]. Adapted with permission from ref [**7**]. Copyright 2018 Springer Nature; and (**c**) category 4: disordered particulate gels [**5**]. Adapted with permission from ref [**5**]. Copyright 2012 Royal Society of Chemistry; The second category identified by Flory, disordered covalent networks, is not considered to be supramolecular.

Supramolecular gels generally have more parameters that can be adjusted to finely tune their properties than their non-supramolecular counterparts. Differences in the eventual mechanical, electrical, and biological properties of the resultant gel are derived from not only the conventional considerations of solvent identity and gelator concentrations, but the utilization of differently functionalized molecular units, the incorporation of additives into the supramolecular matrix, and modifications into self-assembly behavior.

While the boundaries of this material class have previously been vaguely defined, the promise of novel structure-property interactions has spurred exploration into the extent of their use. In this review, we first discuss how supramolecular gels can improve upon traditional polymeric conductors, allow tunable access to unique light-matter interactions, and open new paradigms of optoelectronic devices. The impact of supramolecular gels on advancements in alternative energy is explored next from both the energy generation and energy storage angles. Then, improvements over existing biomedical technologies are discussed through the lens of novel supramolecular gel capabilities in drug delivery, tissue engineering, and paradigms for wound treatment and non-viral vector gene delivery. The capacity to reshape key biological techniques is subsequently discussed, including cell storage and study, electrophoresis, and indication of biological compounds. A further collection of emerging applications is also treated (e.g., environmental remediation and surface coatings, among others), as we look toward the future of this developing field.

Within each section, selected gelators from the presented literature are provided to highlight the diversity of structures used in forming supramolecular gels; these structures are denoted throughout the text. For further information about the synthesis or processing of a specific gelator, readers are directed to the original article.

SUPRAMOLECULAR GELS FOR OPTOELECTRONIC APPLICATIONS

Electrical Conduction

In the realm of electronics, efforts have been raised to utilize the unique properties of supramolecular gels for next-generation electrical conductors. Traditional conductive polymers suffer from reduced flexibility [**8**] and processability [**9**] compared to non-conductive polymers, limiting their potential use in desired applications. The use of supramolecular gels for electronic applications has been demonstrated as a pathway toward alleviating these problems and adding extra functionality. Here we touch on three beneficial aspects of electrically conductive supramolecular gels: their flexibility, processability, and self-healing properties.

Flexibility

Conductive gels in chloroform solvent were synthesized by Pépin-Donat et al. in a reported series of conjugated covalent poly-octylthiophene (POT) gels. Formed by the copolymerization of octylthiophene with trithienylbenzenes, followed by swelling with chloroform and doping with iodine, the resulting gels acted as electrically conductive rubbers due to the external ionic dopant [**8**]. While these still possessed polymeric crosslinks, other approaches have led to flexible supramolecular conductors like the category of nanocomposite bucky gels in imidazolium-ion-based ionic liquids, which are formed by the interactions between imidazolium ions and π-electronic surfaces of ground carbon nanotubes [**10**,**11**]. Ionic liquid-based triblock copolymer gels, for example in butylmethylimidazolium phosphorus hexafluoride as in the work of Lodge et al., have also been identified as flexible conductors by Kang et al. The introduction of ABA triblock copolymers in an ionic liquid with miscible central B blocks and immiscible end A blocks results in a self-assembled gel network of connected micellar structures [**12**]. These flexible and conductive supramolecular gel structures are suitable for applications such as electrical actuators for artificial muscles [**13**,**14**], electromechanical sensors [**15**], and electrolytes for flexible supercapacitors [**16**].

Processability

Gel processing techniques such as the one-pot melt-blending gelation used by Kim et al. provide great advantages in the scalability and cost of electrically conducting materials. The utilization of a eutectic liquid of diphenylamine and benzophenone in the synthesis of bucky gels allows for the dissolution of polymers and gelation of carbon nanotubes, bypassing long and often toxic alternative processes [**17**]. Other supramolecular gels, like the aforementioned triblock copolymer gels, have the added benefit of reversible formation, with temperature [**12**,**18**] or light (demonstrated with UV [**12**] or X-ray [**19**]) acting as reversible and controllable triggers for gelation. This capability allows for easier downstream processing (e.g., alterations for specific applications) with these gels than with traditional conductive polymers like polypyrrole [**9**] or polyaniline [**20**], which are difficult to further process due to their mechanical rigidity and low solubility. Processing by three-dimensional (3D) printing has been achieved with poly(*N*-acryloyl glycinamide) [**21**] and H2010h1 [**22**] hydrogels, adding further versatility to a supramolecular gel approach.

Self-Healing

The capacity for self-healing behavior, in conjunction with flexibility, conductivity, and processability, further expands the scope of applications of conductive supramolecular gels. Work by Wu et al. quantified the healing response in a PNAGA-PAMPS/PEDOT/PSS-3-49 hydrogel (structure shown in **Figure 2**), noting that immersion in a 90 °C water bath for several hours mended a sample of gel that had previously been cut in half and restored both flexible physical and conductive electrical properties. Hydrogen bond-based crosslinking allowed for healing after both multiple cuts and prolonged times between cuts [**21**]. Darabi et al. described self-healing in the aforementioned H2010h1 hydrogel two minutes after cut pieces were brought into contact (**Figure 3**). This was hypothesized to occur because of ionic interactions (between ferric ions and carboxylic groups of poly(acrylic acid)) and hydrogen bonding (between chitosan and poly(acrylic acid)) within the hydrogel [**22**].

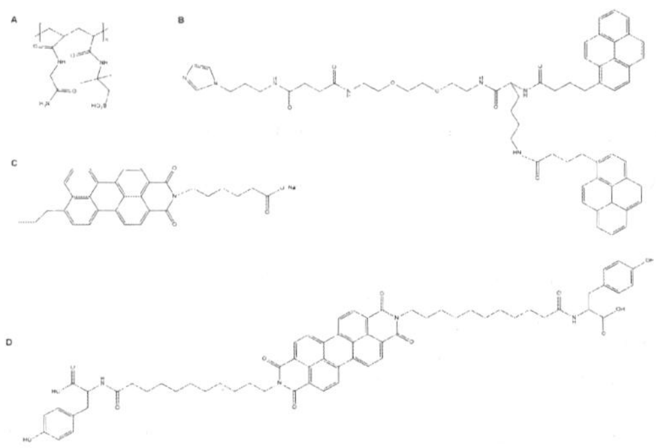

Figure 2. Selected gelators developed for optoelectronic applications: (**A**) a conductive, self-healing PNAGA-PAMPS/PEDOT/PSS-3-49 gelator [**21**]; (**B**) a fluorescent gelator with hydrophobic moieties of pyrenebutyric acid and a hydrophilic moiety of an ethyleneoxy unit coupled with 1-(3-aminopropyl)imidazole [**23**]; (**C**) a photocatalytic chromophore amphiphile gelator [**24**]; and (**D**) a dipeptide-based perylene bisimide derivative gelator whose xerogel exhibits current photo-switching properties [**25**].

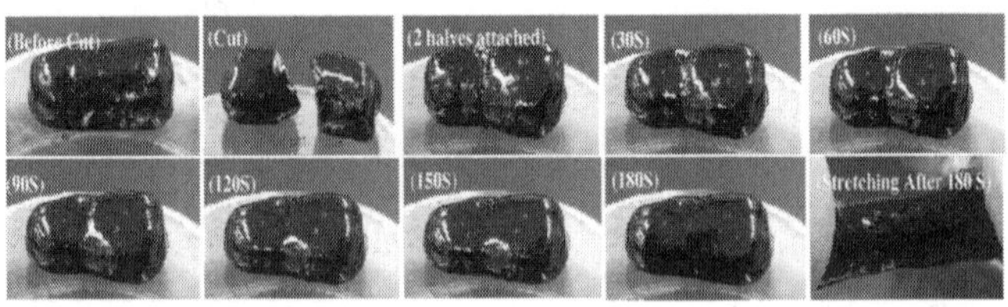

Figure 3. The conductive H2010h1 hydrogel by Darabi et al. exhibits self-healing over minutes (seconds since cutting shown in inset) with, importantly, the restoration of mechanical and electrical properties [**22**]. Adapted with permission from ref [**22**]. Copyright 2017 John Wiley and Sons.

Such materials can be used not only for flexible electrodes or sensors as previously stated, but also for self-repairing circuits and electroactive scaffolding for soft tissue engineering [**21**]. Artificial skin and biomimetic prostheses are a favored biological application where the conduction of electronic signals is critical for functioning [**22**,**26**]. Further research efforts are directed towards balancing the benefits of the aforementioned soft, self-healing type properties with even more robust mechanical properties for applications in fields like soft robotics [**27**] and smart wearables [**28**]. Biomedical applications in particular form an exciting frontier for supramolecular gels, and will be further detailed in a later section.

Photonic Responses

Beyond electrical conduction, supramolecular gels have been materials of interest in photonic applications because of their ability to control their interactions with light. Light-matter interactions are of primary importance in informational, optical, and photocatalytic applications, among others. The deliberate engineering of specific categories of light responses provides a suite of materials able to perform a wide breadth of functions usually associated with other material classes. Specific engineered light-matter interactions covered here include the luminescence, scattering, refraction, and absorbance of light.

Luminescence

A variety of gels have been shown to emit different spectra of light via various modes. For instance, in investigations of low molecular weight supramolecular gelators, Mandal et al. formed a fluorescent gelator with hydrophobic moieties of pyrenebutyric acid and a hydrophilic moiety of an ethyleneoxy unit coupled with 1-(3-aminopropyl)imidazole (structure shown in **Figure 2**). The mixture of these compounds formed both hydrogels and nitrobenzene organogels and, due to the pyrene moieties, emitted blue light from its fiber network under ultraviolet (UV) radiation [**23**]. Strong blue fluorescence upon UV exposure has also been observed from an acylhydrazone-functionalized dual benzimidazole derivative supramolecular organogel with dimethyl sulfoxide (DMSO)-ethylene glycol (1:3) or dimethylformamide (DMF)-water (1:5) solvent, enhanced with one equivalent of rare earth Y^{3+} [**29**].

More general luminescence has been observed with a similar acylhydrazone-functionalized benzimidazole derivative assembled in ethylene glycol and (for a Job plot experiment) DMF-water. When cooled to room temperature, this gel luminesced green, which could be tuned to blue with the addition of one equivalent of Cd^{2+} [**30**]. Luminescence has also been observed with 8-hydroxyquinoline lithium (LiQ) in non-protic solvents like benzene or toluene, with a 3-fold increase in luminescence when modified with cholesterol and a red shift in absorption when gelled into a J-aggregate [**31**]. Kishimura et al. demonstrated red luminescence in an organogel formed from a trinuclear Au(I) pyrazolate complex in hexane, which could be tuned to blue upon doping with Ag^{+}, and green upon subsequent heating [**32**].

Such materials have been explored for sensors, due to the fact that the wavelengths of light emitted can change in the presence of various ions. Lin et al. reported one such system, most stable in ethanol that fluoresces when the gel is coordinated with Pb^{2+} ions. Upon testing 14 anions, it was discovered the system's fluorescence could be selectively quenched only with I^{-}. Further addition of Pb^{2+} to the system allowed for the effect to be reversed [**33**]. Two further supramolecular gels investigated by Lin et al., which operate in a similar fashion and are also most stable in ethanol, when coordinated (1) with Mo^{2+} and Co^{2+} and (2) with Ca^{2+} and Fe^{3+} could reversibly and visually sense Cl^{-} and $H_2PO_4^{-}$, respectively [**34,35**]. In a complementary line of work, Cu^{2+} was detected by a green to blue color transition in a supramolecular thiazolothiazole derivative gel in cyclohexanone by Zhang et al. [**36**] or a yellow color to colorless transition in a salen-linked gelator in DMSO-water (1:1) by Fan et al. [**37**]. Catalysis monitoring has been explored in the work of Harris et al. with a sensor-functionalized gelator that gels in various solvents like DMF, i-PrOH, and water, increasing the local fluorescence in the presence of a desired reaction product [**38**].

The range of possible emission colors has piqued interest in applications toward information storage and display. One frequently cited application is for rewritable, inkless security paper, achievable through the toggling of fluorescence properties via triggers like mechanical grinding, exposure to solvent vapors, electric fields [**39**], heat [**40**], and UV light [**41**].

Scattering

Beyond the emission of light, supramolecular gels have been synthesized to scatter light in deliberate and useful ways. Structural color was achieved by Zhou et al. with supramolecular organogels co-assembled with oleophilic silica nanoparticles and carbon black in a hot chloroform-silicone oil solution. This usage of a self-healing gel architecture for a structural color material resulted in the first reported non-iridescent photonic material that can restore both its modulus and color properties after mechanical damage, i.e., cuts and shear failure. Proposed applications include not only the types of chemical sensing mentioned previously, but also flexible color displays [**42**].

Light scattering supramolecular gels have also been formed by polyfluorene-based π-conjugated polymers. Chen et al. synthesized liquid crystal physical gels using poly(9,9-dioctylfluorene-alt-benzothiadiazole) (F8BT) gelators in nematic liquid crystal E7 (from Fusol material) via a cooling process from isotropic melt.

Such a structure possesses the ability for anisotropic scattering behavior due to the fibrillar aggregation of F8BT, with more scattering occurring for light polarized parallel to the fibers. Notably, this structure is able to switch modes from being transparent to exhibiting anisotropic scattering via the application of a low (2.7 V) driving voltage—much easier than for other liquid crystal gels, which require around 20–60 V [**43,44**].

Upon the removal of the driving field, the gel returns to being transparent. Such a triggerable, reversible scattering response has potential applications as a smart polarizer or in transflective display technology [**45**].

Refraction and Absorption

Other self-healing supramolecular gels have been developed to transmit and refract light. Using mannitol-based diols flanked by ketal moieties, Vidyasagar et al. produced transparent supramolecular organogels with a variety of alkane and aromatic organic solvents (**Figure 4**).

These gels possess the favorable physical properties of flexibility, moldability, and self-healing combined with optical properties of visible light transmittance, UV light filtration, and glass-like indices of refraction.

This wide variety of useful properties has allowed for demonstrated applications in self-healing, shatter-proof, and scratch-free lenses and prisms in place of glass, as well for proposed applications in UV-protective goggles or potential replacements of the oils used in various measurement techniques such as oil immersion microscopy and differential interference contrast microscopy [**46**].

Figure 4. Vidyasagar et al. organogels combine glass-like optical properties with desirable physical properties, including moldability (shown by various shapes in (**a–d**)), robustness (shown in (**e,f**)), and self-healing (shown by UV doping one tube of a self-assembling gel before attaching to another in (**g**)) [**46**]. Reprinted with permission from ref [**46**]. Copyright 2011 John Wiley and Sons.

There also has been research in the direction of gels that absorb light. Following up on the 2014 work by Weingarten et al. [**47**], supramolecular chromophore amphiphile hydrogels (structure shown in **Figure 2**) in the 2017 work of Kazantsev et al. demonstrated a photocatalytic effect. The crystalline phase of this material allows for the formation of excitons under visible light, aiding hydrogen production by the increased photosensitization of a proton reduction catalyst such as $[Mo_3S_{13}]^{2-}$ [**48**]. Another supramolecular gel shown to absorb light was formed from disubstituted acenes. These organogels were shown to self-assemble with a 2,3-di-n-decyloxy derivative gelator into fibrillar structures in a wide range of organic solvents, aligned by magnetic fields or mechanical force, and act as a light-harvesting matrix and optical confining media [**24**]. As might be expected with such a host of electronic and optically relevant properties, supramolecular gels have the potential to impact a variety of optoelectronic fields, with more work to be done in gel stability and device manufacture.

Optoelectronic Devices

Semiconductors

A critical foundation of modern electronic devices is the semiconductor. Traditionally inorganic materials (historically germanium and silicon [**49**], with more recent advancements in III-V compounds like GaAs and InN [**50**]) have been used in the semiconductor industry, but organic semiconductors have been on the rise in the search for flexible [**51**], cheaper [**52**], and potentially more environmentally friendly materials. In particular, semiconductive supramolecular gels have been explored over the past decade.

As an example, Muthusamy et al. produced semiconductive fluorescent organogels from the direct condensation of vinyl esters with functionalized sugar alcohol, using hydrophobic solvents and the lipase Novozyme 435. Doping with graphene resulted in an increase in the current response for a given applied voltage, up to a 1:1 ratio of gel to graphene, allowing the system's semiconductive behavior to be tuned. Their particular process used bio-based plant oils, which have been identified as a particularly economical and renewable category of industrial chemicals [**53**]. Flexible sheets of this material were made by the auto-oxidation of linseed oil, making it useful as a charge transport layer in organic field effect transistors (**Figure 5**) [**54**]. Earlier work by Diring et al. utilizing hydrogen bonding from amide groups to form luminescent ethynyl-pyrene gels in DMF, toluene, and cyclohexane resulted in a field effect transistor prototype with favorable source-to-drain electron/hole transport. While applied gate voltage only negligibly affected the source-to-drain current, making for an insufficient transistor [**55**], the properties of the formed gel were still favorable for other applications, such as fiber-based organic light emitting diodes [**56**]. Further work in this field is directed toward increasing the performance of these devices to match those of traditional semiconductors.

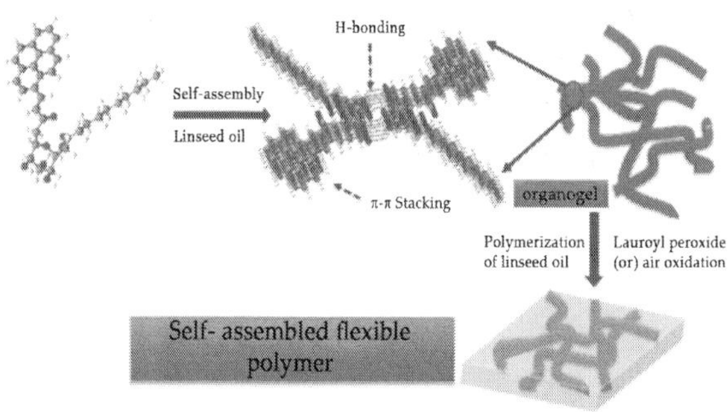

Figure 5. Hierarchical assembly of a semiconductive organogel by Muthusamy et al. from bio-based plant oils, which could be formed into flexible sheets for use as a charge transport layer in organic field effect transistors [**54**]. Adapted with permission from ref [**54**]. Copyright 2016 Royal Society of Chemistry.

Logic Gates and Photo-switching

A computation-focused application of semiconducting supramolecular gels is in the realm of gel-based logic gates. The work of Roy et al. in 2014 demonstrated a dipeptide-based perylene bisimide derivative (PBI-C_{11}-Y) (structure shown in **Figure 2**) that self-assembled into a hydrogel between pH 7.46 to pH 9 and exhibited both semiconductivity and, notably, current photo-switching behavior in its xerogel state. Like the previous semiconducting gel example from Muthusamy et al., the optoelectronic response of this gel can be tuned with the inclusion of graphene. Specifically, incorporation of graphene oxides into the gel allows for a more distinct "on" current state to be accessed upon white light exposure, opening the possibility of a logic gate-esque system based on these self-assembled gels [**25**]. However, such a material was only shown to cycle four times before a roughly 33% degradation in distinction between "on" and "off" states. Further instances of potential logic gate behavior via photo-switching occurred, for example, with a low molecular weight organogelator mixed with trioctylphosphine oxide ligand-capped CdSe/ZnS core-shell nanoparticles in *n*-hexane. This luminescent, transparent organogel was synthesized in 2016 by Schmidt et al. and, with the incorporation of diarylethene, exhibited UV-vis light triggered photo-switching. Notably, diarylethene's intrinsic photochromic properties were retained within the organogel structure with over 20 cycles of reversible switching demonstrated [**57**], making it more robust than the previous example.

While the prior examples have exhibited triggered photo-switching behavior, they alone have not exhibited the ability to simulate specific logic gates such as AND or OR. That said, work by Jung et al. showed a Co^{2+}-imidazole-based supramolecular gel in polar solvents such as DMF, DMF-water (1:1), and DMA with multifactor dependent photo-switching effects. Specifically, the gel possesses an ionochromic effect, turning from red to blue with the addition of Cl^- or Br^- ions, and a piezochromic effect, turning from red to blue when ground with KBr powder [**58**]. The ionochromic effect acts as an OR gate because it requires either or both of the ions' presence to change colors, while the piezochromic effect emulates an AND gate as it demands both the presence of KBr powder and pressure to change color. However, the necessity of mechanically grinding the gel with KBr powder for the piezochromic effect appears to be materially costly and impractical for any large calculation, compared to using conventional transistor-based devices. Thus, while supramolecular gel systems have been explored in proof-of-concept designs for novel optoelectronic switches, progress appears to still be in early stages of development.

Another large class of optoelectronic device applications for semiconducting supramolecular gels, hinted at with the variety of gels previously shown to absorb light, is the photovoltaic cell. Supramolecular gel properties of tunable electrical conductivity and photon absorbance synergize to form an enticing material platform for solar capture and conversion to current. Further perspectives on photovoltaic cells will be detailed in the following section dedicated to energy applications.

SUPRAMOLECULAR GELS FOR ENERGY APPLICATIONS

Energy Generation

A large section of supramolecular gel applications is devoted to the energy sector. Within this realm, solar cells are a particularly enticing application due to the tunability of supramolecular gels' electrical and optical properties. As an example, in the work of Bairi et al., a xerogel was formed by adding anilinium chloride solution and ammonium persulfate solid to a 5,5′-(1,3,5,7-Tetraoxopyrrolo[3,4-f]isoindole-2,6-diyl)-diisophthalic acid (PMDIG) hydrogel. Due to the *p*-type nature of polyaniline (PANI) and the *n*-type PMDIG, the resultant PMDIG-PANI xerogel possessed a *p-n* junction with an I–V curve that showed photocurrent rectification under exposure to white light. When incorporated into an ITO/PMDIG-PANI/graphite device, a power conversion efficiency of 0.1% was obtained [**59**], demonstrating proof of concept but also signifying the gap between gel and traditional solar cell performances.

Supramolecular gels have also found uses in advancing dye-sensitized solar cell technology, serving as an enticing replacement material for liquid electrolytes due to their conductivity and flexibility without the risks of liquid leakage [**60**]. In 2015, Huo et al. demonstrated these benefits with their two-component quasi-solid-state electrolyte gel made from *N,N′*-1, 5-pentanediylbis-dodecanamide and 4-(Boc-aminomethyl)pyridine gelators (structures shown in **Figure 6**) in liquid electrolyte 0.1 M iodine (I_2), 0.1 M anhydrous lithium iodide (LiI), 0.5 M *N*-methylbenzimidazole (NMBI), and 1 M DMPII in 3-methoxypropionitrile (MePN). This two-component electrolyte resulted in a solar cell with a power conversion efficiency of 7.04%, comparable to the baseline of 7.11% for liquid electrolytes and greater than the baseline of 6.59% for a single-component *N,N′*-1, 5-pentanediylbis-dodecanamide electrolyte [**61**]. Huo et al. further reinforced the field in 2017, this time with a slightly different suite of materials: *N*,*N'*-1,8-octanediylbis-dodecanamide and iodoacetamide in a liquid electrolyte 0.1 M LiI, 0.6 M I_2, 0.45 M NMBI, and 0.9 M MePN. Again comparing a two-component versus a single-component system, the two-component electrolyte gave a conversion efficiency of 7.32% versus the single-component electrolyte's 6.24% [**62**]. Such results are comparable with the previous set of materials, but illustrate a power conversion efficiency surpassing that of the baseline liquid electrolyte.

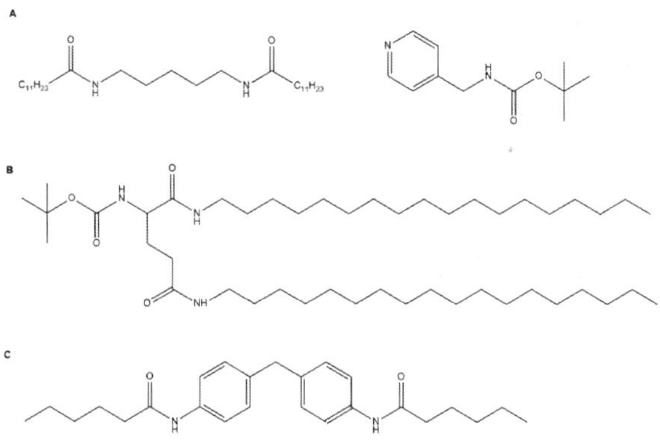

Figure 6. Selected gelators developed for energy applications: (**A**) Co-assembled *N'*-1, 5-pentanediylbis-dodecanamide and 4-(Boc-aminomethyl) pyridine gelators for solar energy generation [**61**]; (**B**) a thylakoid membrane-inspired organogel for photon up-conversion [**63**]; and (**C**) a supercapacitor bis(4-acylaminophenyl)methane gelator [**64**].

Tackling the energy issue from another angle, supramolecular gels have also been used for photon up-conversion. Up-conversion has the potential to increase the efficiency of solar cells by allowing photons with energies less than the semiconductor band gap to contribute to exciton generation [**65,66,67**]. Duan et al. used supramolecular organogels in DMF to perform triplet-triplet annihilation (TTA)-based up-conversion (UC) (**Figure 7**). The usual problems with TTA-UC of triplet deactivation via oxygen exposure or the necessity for high power illumination [**68**] were sidestepped with the supramolecular gel architecture, which shielded donor/acceptor molecules from oxygen deactivation while also presenting efficiently transferred triplet energy. Previous to this work, conventional ways of ameliorating oxygen deactivation would often exacerbate the need for high-power light [**63**]. This utilization of a supramolecular organogel, inspired by thylakoid membranes in natural light harvesting systems (structure shown in **Figure 6**), allowed for the formation of a photon up-conversion system incorporating benefits once thought to be exclusive of each other. Equally important to the generation of energy is the issue of energy storage, which will be discussed in detail in the following section.

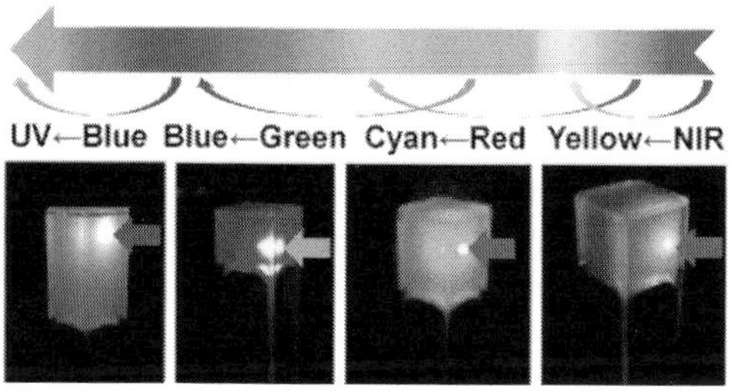

Figure 7. A series of organogels from Duan et al., who performed photon up-conversion for wavelengths across the visual spectrum (color of incident wavelength shown in arrow) [**63**].

Energy Storage

Different technologies of energy storage span a range of specific powers and specific energies, suited to a variety of different applications. The three main energy storage applications detailed here that supramolecular gels have found significant uses in are Li-ion batteries, supercapacitors, and fuel cells.

Li-Ion Batteries

Supramolecular gels can impact the battery industry, expanding the range of effective battery applications. Work by Frischmann et al. detailed a system that uses a perylene bisimide-polysulfide (PBI-PS) gel from catholyte solutions containing 2.5 M S as Li_2S_8 and 0.048 M PBI (5.0% *w*/*w*) in tetraethylene glycol dimethyl ether (TEGDME) with 0.50 M lithium bis(trifluoromethanesulfonyl)imide (LiTFSI) and 0.15 M $LiNO_3$ to improve upon lithium-sulfur hybrid redox flow batteries. π-conjugated organic redox mediators, inspired by nanocarbon current collectors in other systems [**69**], were investigated to assist with charge-transport bottlenecks. Computational simulation identified the aforementioned PBI-PS gel as favorable because it exhibited a similar charge/discharge potential to Li-S as well as the ability to form durable nanowire structures. As a self-assembled, nanostructured, and flowable gel, PBI-PS has been shown to improve sulfur utilization and have potential for opening usage to both longer and larger-scale battery applications [**70**].

Supramolecular gels also have potential in advancing battery electrode technology. Metal oxide nanoparticles can be enticing electrodes for Li-ion batteries [**71**], but periodic volume changes and structural stress due to repeated Li insertion can lead to nanoparticle aggregation and limits on their functional lifetime. Synthetic procedures for hollow sphere structures in supramolecular gels have been investigated for their ability to accommodate space for Li and mitigate these expansive stresses [**72**]. In the work of Lyu et al., a supramolecular hydrogel was used as a template for nanoparticle anode fabrication. Hollow CuO nanoparticles were formed via directed self-assembly in a mixture of chitosan, ascorbic acid, and Cu^{2+} ions, with subsequent calcining at 500 °C. The resultant composite exhibited over 90% retention of discharge capacity after 500 cycles, considered an excellent cycling performance [**73**]. Additionally, Sui et al. reported using supramolecular gels as guiding matrices for manganese dioxide deposition. This resulted in MnO_2/nitrogen-doped graphene hybrid aerogels with increased Li cycling resistance that could similarly be used as an anode material for a Li-ion battery [**74**].

Another method to combat the problem of device degradation after repeated Li-insertion-driven volume changes has been to prevent nanoparticle aggregation directly. Shi et al. synthesized a supramolecular hydrogel binder as a coating for anodic SnO_2 nanoparticles. This gel was formed via reversible ionic bonding between poly(allylamine hydrochloride) (PAH) chains and a phytic acid gelator, and thus exhibited self-healing properties. As an added benefit, the addition of carbon nanotubes into the gel was shown to improve the specific capacity of the battery. The presence of this supramolecular binder coating acted to suppress the aggregation of SnO_2 particles, and its ability to self-heal allowed for a robust structure capable of withstanding volume changes after repeated Li insertion cycles. Specifically, the specific capacity of an uncoated SnO_2 sample was 169 mAh/g after 70 cycles, compared to a coated 60 wt % PAH sample, which retained a specific capacity around 500 mAh/g after 70 cycles [**75**].

Though there have been a variety of studies primarily on Li-ion battery technology, as further work is conducted in this field other types of batteries have the potential to also benefit from supramolecular gels.

Supercapacitors

Aside from traditional batteries, supercapacitors have been a popular device for supramolecular gel research. This is due to supercapacitors' ability to bridge the gap in properties between the high specific power but low specific energy of conventional capacitors and the low specific power but high specific energy of batteries [**76**]. As briefly mentioned previously with the work of Kang et al., supramolecular gels have been studied as electrolytes for flexible supercapacitors. Focused work into the factors affecting the properties of supramolecular gel electrolytes was conducted by Dong et al., who used bis(4-acylaminophenyl)methane (structure shown in **Figure 6**) and bis(4-acylaminophenyl)ether in imidazole-based ionic liquids. It was determined that the length of acyl chains was a strong factor in the resultant gel properties, with longer chains increasing gelation strength but decreasing conductivity [**64**].

Electrolytes, however, are not the full extent of gel involvement in supercapacitor technology. Supramolecular gels have also found use as supercapacitor electrode materials, as in the work of Liang et al. Their aerogel, comprised of polyaniline with manganese oxide, exhibited qualitative flexibility and 96% retention of capacitance after 20,000 charge/discharge cycles (**Figure 8**) [**77**]. Noncovalent interactions, a hallmark of supramolecular gels, were identified to be the source of the gel's physical strength. Continuing in this vein is the work of Li et al., who worked to create similarly durable and robust conductive hydrogels, and further investigated flexibility. By combining polyaniline and polyvinyl alcohol into a supramolecular hydrogel via dynamic boronic acid group bonding, a capacitance of 306 mF/cm^2 was achieved in a flexible supercapacitor. While a slightly lower 90% retention of capacitance was shown after 100 charge/discharge cycles, these gels demonstrated complete retention of capacitance after 1000 mechanical folding cycles [**78**]. Though polyaniline itself is a relatively stiff conductive polymer, its incorporation into a supramolecular gel structure can allow for the tuning of conductive and physical properties to better fit applications. In the realm of graphene-based supercapacitors, similar to the aforementioned idea of a using a supramolecular gel binder coating to prevent the aggregation of anodic nanoparticles in batteries, a gelled structure can improve the quality of devices by preventing π-stacking between graphene layers, which would otherwise decrease ionic diffusion and active surface area [**60**].

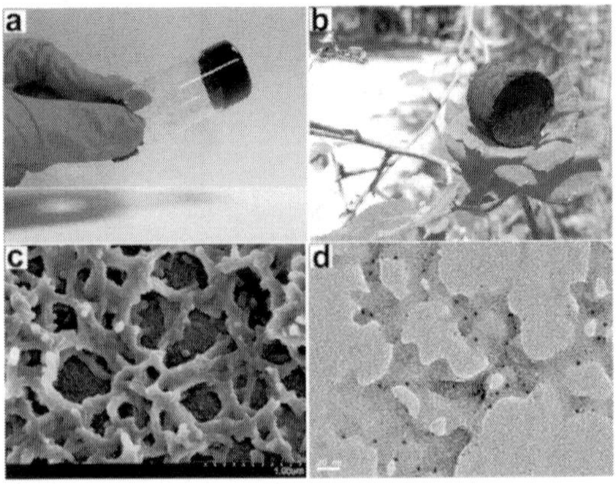

Figure 8. A nanocomposite hydrogel by Liang et al. shown in (**a**) had an extremely low density upon solvent removal, as seen by its aerogel balancing on a flower in (**b**), due to high porosity, as shown in micrographs (**c**) and (**d**). The porosity promotes ion transport and high capacitance [77]. Adapted with permission from ref [77]. Copyright 2015 John Wiley and Sons.

Fuel Cells

Fuel cells present even higher energy density than batteries [**76**] and greater electrical efficiencies than heat engines [**79**]. If utilizing water condensation, fuel cells also have the additional benefit of producing no local waste. As such, they are an attractive form of energy storage and have been investigated in the context of supramolecular gels. In the work of Ye et al., supramolecular assemblies of uracil-terminated telechelic sulfonated polyimides (SPI-U) gelled with 20 wt % of an adenine-based hydrogen bonder, dubbed "SMA-A". The resultant material was used as a proton exchange membrane, allowing for hydrogen ions to pass while excluding methanol. The membranes gelled with SMA-A performed better in terms of stability and selective conductivity than membranes formed of pure SPI, illustrating the benefits of gelation [**80**]. Outside of electrolytes, Kumar et al. created hydrogels for anode applications in microbial fuel cells. Supramolecular assemblies of graphene oxide (GO), carbon nanotubes (CNT), and poly *N*-isopropylacryl-amide (PNIPAM) hydrogel were created with a range of electronic properties dependent on constituent material ratios. Specifically, it was found that the addition of up to 1% CNT did little to improve the low conductivity of PNIPAM, while the addition of only 0.2% GO increased conductivity by a factor of 10^8. Interestingly, using the two in concert with 0.1% of each resulted in a 10 times higher conductivity than the previous 0.2% GO sample. This response was attributed to the CNTs inhibiting stacking between GO and increasing contact with the electrolyte. Fuel cells created with the 0.1% GO, 0.1% CNT mixture exhibited a power density around 400 mW/m^2 over a 300-h operation time [**81**].

Supramolecular gels have also been of ancillary use to fuel cells in the field of oxygen production. Traditionally, noble metal catalysts (e.g., gold and platinum) have been used for oxygen reduction (ORR) and oxygen evolution reactions (OER), respectively [**82**], but less costly and more stable materials are desired for more commercially viable fuel cells. In the work of Wu et al., nitric acid, water, and melamine were mixed and heated before the addition of Vulcan XC-72 and cobalt(II) nitrate hexahydrate, which resulted in the gelation of the solution. This supramolecular hydrogel was then dried and heated under a nitrogen atmosphere to yield an assembly of double-shelled Co@CoO@N–C/C nanoparticles. A range of catalysts with differing doping schemes were created using this type of self-assembly process, but Co@CoO@N–C/C was the best alternative catalyst out of their samples due to its close approximation of the performance of traditional platinum/carbon (Pt/C) catalysts while also providing greater stability [**83**]. A year later, in 2017, Jiang et al. investigated a Co_2P@CoNPG assembly also synthesized through supramolecular hydrogel methods. This catalyst also exhibited greater stability than Pt/C, retaining 91.6% of its current density after 200 min compared to Pt/C, which presented only 39.3% baseline current density after 200 min in alkaline solution. Intriguingly, this system had the added benefit of effectively catalyzing not just the ORR as the previous example but also the OER [**84**], opening the door for more expansive utilization of alternative fuel cell catalysts. As research progresses, the ability for supramolecular gels to self-assemble may lead to novel catalytic materials that can outperform and eventually replace noble metals in fuel cell technology.

Here we conclude our discussion on the more physical science-type supramolecular gel applications of optoelectronics and energy devices. We will now delve into the large group of life science applications, starting briefly with the biomedical field mentioned previously.

SUPRAMOLECULAR GELS FOR BIOMEDICAL APPLICATIONS

Drug Delivery

The large volume of supramolecular gels investigated for drug delivery applications starkly contrasts that of many other specific applications. Due to their high degree of mechanical, physical, biological, and chemical tunability, supramolecular gels are increasingly found at the cross-section of the next generation of drug delivery materials. Consequently, it should be re-emphasized that this review does not include all examples of supramolecular gel applications, but an overview of these applications with examples of such uses.

Oral Drug Delivery

Supramolecular gels have been recognized in oral drug delivery applications for their ability to maintain long-term stability and incorporate environmental sensitivities. Though not explicitly examined in animal models to date, in vitro testing of supramolecular gels promises appropriate properties for ingested drug materials. Hybrid supramolecular gels by Rizzo et al., formed by the self-assembly of Fmoc-l-phenylalanine in the presence of halloysite nanotubes and loaded with the anti-tumor drug camptothecin in aqueous solutions of phosphate buffer saline and phosphate buffer solution, demonstrated notable antiproliferative activity on cervical cancer cells. Camptothecin's coupling to the supramolecular gel overcame common complications, including the maintenance of the drug in its active form and the prevention of its hydrolysis [**85**]. Demonstrating the addition of responsiveness for drug delivery applications, a supramolecular gel formed by the host-guest complexation of cucurbit[8]uril and *N*-(4-diethylaminobenzyl)chitosan in 1 M hydrochloric acid by Lin et al. exhibited tunable and pH-sensitive drug delivery rates [**86**]. Similarly, Ashwanikumar et al. developed a pH-sensitive supramolecular gel with self-assembling RADA-F6 peptides [**87**] which offered sustained delivery of the anti-tumor drug 5-fluorouracil (5-FU) at basic pH values, making it appropriate for the treatment of colon cancers, among others [**88**]. Finally, Felip-León et al. produced a supramolecular hydrogel using a low molecular weight gelator based on a 1,8-naphthalimide unit linked to a dipeptidic glycine-valine unit (structure shown in **Figure 9**) loaded with model dye molecules as particles with dimensions on the nanoscale. These materials showed facile transport into human carcinoma cells, demonstrating the potential for supramolecular gels to be used as an alternative to polymeric nanogels in nanomedicine [**89**].

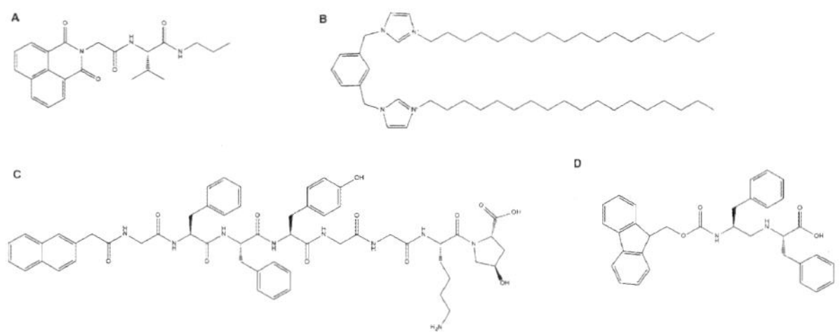

Figure 9. Selected gelators developed for biomedical applications: (**A**) a gelator derived on a 1,8-naphthalimide unit linked to a dipeptidic glycine-valine unit with demonstrated facile transport into human carcinoma cells for oral drug delivery [**89**]; (**B**) a gemini imidazolium amphiphile gelator for topical drug delivery [**90**]; (**C**) a collagen-inspired hydrogelator for tissue engineering [**91**]; and (**D**) an aromatic di-phenylalanine-based gelator silver-based antimicrobial wound treatment [**92**].

With the desirable properties of supramolecular gels for oral drug delivery well established in in vitro studies, further attention should be directed towards elucidating the in vivo behavior of these materials. Considering, e.g., the high salt concentrations which can screen intramolecular interactions and the presence of material-degrading enzymes in animals, future studies should seek to analyze behavior under conditions similar to the desired end use in order to eventually realize clinical applications.

Logic Gate-Based Drug Delivery

Incorporating several sensitivities into a supramolecular gel enables its unique use for logic gate-based drug delivery, where quantitative amounts of drugs are released based on input signals. An OR and XOR gate functionality was achieved in co-assembled phenylalanine-based and bis(pyridinyl)-derived hydrogelators by Liu et al. (**Figure 10**), which use pH, thermal, and light inputs for controlled drug release [**93**]. Supramolecular peptide-based hydrogel-enzyme hybrids by Ikeda et al. achieved OR and AND functionality to sense specific biochemicals and perform drug release based on the activated logic operation [**94**]. Accessing several common logic operators, Komatsu et al. realized AND, OR, NAND, and NOR functionality in a supramolecular gel based on a phosphate-type hydrogelator through sensitivities to temperature, pH, the presence of calcium ions, and light (**Figure 10**) [**95**].

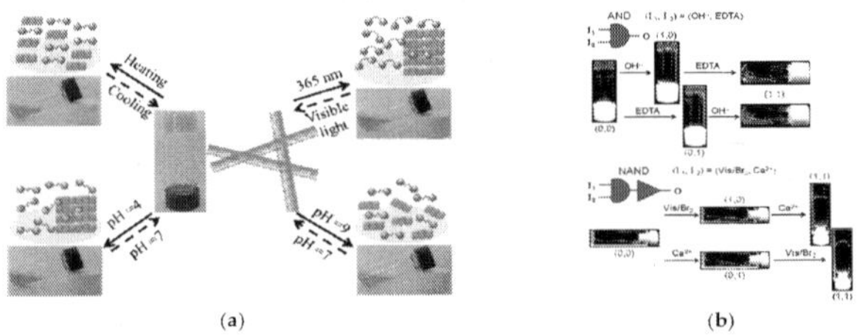

Figure 10. The ability to incorporate several sensitivities into supramolecular gels enables their use in logic-gate based drug delivery. (**a**) OR and XOR functionality in a supramolecular gel by Liu et al. [**93**]. Adapted with permission from ref [**93**]. Copyright 2015 American Chemical Society. (**b**) AND, OR, NAND, and NOR functionality in a supramolecular gel by Komatsu et al. [**95**]. Adapted with permission from ref [**95**]. Copyright 2009 American Chemical Society.

Topical Drug Delivery

Supramolecular gels have been investigated for several applications in the topical application of anti-inflammatory drugs due to their ease of loading and spreadability, improved drug retention times, and topical cooling effect upon evaporation of the solvent. A bis-imidazolium positively charged amphiphile in ethanol-water mixtures produced by Limón et al. is an example of one such system, with the ability to load and topically deliver anionic drugs, such as non-steroidal anti-inflammatory drugs (NSAIDs). The gel demonstrated rapid drug release profile upon contact with the skin, permitted the permeation of the drug into the skin, and promoted the retention of the drug inside the skin, thus exhibiting properties appropriate for the proposed delivery of anti-inflammatory drugs [**96**]. Supramolecular gels reported by Klaewklod et al. based on β-cyclodextrin in a low molecular weight polyethylene glycol (400 g/mol) medium with facile spreadability and high flux of the NSAID diclofenac [**97**], and by Rodrigues et al. using a gemini imidazolium amphiphile (structure shown in **Figure 9**) in ethanol-water mixtures of several anti-inflammatory anionic drugs [**90**] fit a similar bill.

Supramolecular gels have been tested for several other topical drug delivery applications. The anti-allergy drug cetirizine was translated into a supramolecular gel in methyl salicylate (menthol) by Majumder et al. with demonstrated efficacy in relieving allergy symptoms in a mouse trial [**98**]. Marcos et al. formulated a poloxamer-hydroxyethyl cellulose-α-cyclodextrin supramolecular gel in water, which sustained a release of the anti-fungal drug griseofulvin for at least three weeks and demonstrated biocompatibility [**99**]. To deliver a local anesthetic, prilocaine hydrochloride was loaded into a poly(aryl ether) dendron-based gel in DMSO-water solutions by Kannan et al. for localized, sustained release [**100**].

Relative to other drug delivery applications, topical drug delivery likely offers the lowest barrier to clinical translation and may offer the entry point for supramolecular gels to reach applied biomedical use. Further development of these materials should include analyses of material toxicity and permeation through skin, and seek to understand the useable lifetime of a topically applied supramolecular gel.

Injectable Drug Delivery

The abilities of supramolecular gels to include temperature-sensitive functionality, rheological tunability, and stability when used as a platform for drug release have garnered interest in their use for injectable drug delivery applications. A thermally-sensitive folic acid-agar supramolecular hybrid gel in a 1:1 DMSO-water mixture by Song et al. demonstrated mechanical tunability and a drug release profile with a low initial burst and long release time with a model drug [**101**]. With a specific application in mind, Dai et al. developed a supramolecular gel in aqueous media based on host-guest inclusion complexation between α-cyclodextrin and anti-tumor drug-loaded micelles that exhibited several notable properties, including facile preparation, high stability, an attractive drug release profile, low toxicity, and effective in vitro and in vivo anti-tumor activity [**102**]. Recently, Simões et al. successfully incorporated vancomycin into a Pluronic F127-*co*-α-cyclodextrin supramolecular hydrogel for localized antibiotic treatment of infections. The authors highlighted its long-term stability, ease of syringeability while maintaining viscoelastic properties, and controlled and sustained drug release [**103**].

Tissue Engineering

Primarily centered around injection-based designs, supramolecular gels have the opportunity to transform tissue engineering by offering minimally invasive materials which can assemble into solid-like structures in the body that are functionalized for relevant high bioactivity. Peptide amphiphiles, which self-assemble into supramolecular gels, are particularly studied due to their inherent bioactivity. Three such examples are highlighted here. Self-assembling peptide amphiphiles in aqueous media by Shah et al. resulted in extensive cartilage regeneration due to surface functionalization with a transforming growth factor β1 binding sequence [**104**]. Acting as a scaffold for new growth, a guanosine-based hydrogelator by Buerkle et al. offered a self-assembling material with highly tunable mechanical properties with facile injectability and little to no toxicity [**105**]. Demonstrating the value of supramolecular gel tunability, several collagen-inspired hydrogelators (structure shown in **Figure 9**) synthesized by Hu et al. controlled peptide sequence and conformation and mechanical properties to optimize cell growth [**91**]

.

Bone Defect Repair

As a subset of tissue engineering, supramolecular gels offer similar unique advantages for the repair of critical-sized defects in bone (i.e., defects of a size which cannot be remodeled without intervention). Tested with simvastatin and bone morphogenetic protein, a poloxamine and α-cyclodextrin supramolecular gel in aqueous phosphate buffer by Rosario et al. was successfully able to regenerate bone in rat calvarial defects after injection (and thus notably without surgery) [**106**]. Self-assembled protein-based microgels by Sang et al. incubated in calcium and phosphate solutions resulted in a hydroxyapatite-mineralized interconnected porous architecture appropriate for bone regeneration [**107**]. In an effort to closely mimic the properties of bone, O'Leary et al. developed a collagen-mimetic peptide that self-assembled over several length scales into a hydrogel with an organization similar to natural collagen (**Figure 11**). The properties of the resultant material closely matched that of its natural counterpart, including its breakdown by collagenase, furthering its potential as a bioresorbable replacement material [**108**].

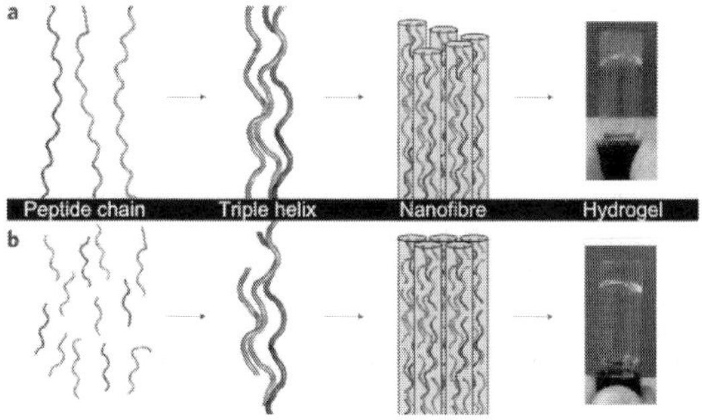

Figure 11. Comparison of self-assembling (**a**) natural rat tail tendon to (**b**) the synthetic collagen-mimetic peptide materials shows notable similarity over several length scales, as demonstrated by O'Leary et al. [**108**]. Adapted with permission from ref [**108**]. Copyright 2011 Springer.

Silver-Based Antimicrobial Wound Treatment

Facile synthesis routes incorporating silver, a widely accepted antimicrobial agent, combined with unique physicochemical properties have sparked interest in using supramolecular gels for wound treatment. Self-assembling aromatic di-phenylalanine-based materials (structure shown in **Figure 9**) loaded with silver by Paladini et al. were developed for use as a wound dressing by maintaining a moist aqueous environment that promotes wound healing while preventing infections with low silver concentrations [**109**]. A silver-infused l-lysine-based amphiphilic hydrogelator by Mandal et al. combined mechanical rigidity with a shear-thinning ability for application as an injectable hydrogel. Notably, the hydrogel was toxic towards bacteria without interrupting the growth of mammalian cells [**110**]. Similarly, an l-cysteine-based hydrogel by Pakhomov et al. demonstrated toxicity at low silver concentrations across numerous bacteria cultures, all while maintaining appropriate mechanical properties for wound and burn treatment [**92**].

Non-Viral Vector Delivery

A fundamental challenge underlying genetic therapies is the lack of safe yet effective vectors for delivering foreign DNA into cells. While viral vectors are highly effective for this purpose, concerns over their safety have instigated great caution in their use. Conversely, while synthetic counterparts are preferred for their lower immunogenicity and higher loading ability, their significantly lower efficiency has prevented successful clinical translation [**111**]. Supramolecular gels offer the opportunity to overcome these challenges through facile formulations of highly tunable materials that remain in the application site (i.e., gel in vivo) for sustained transduction. As this area of research is in its infancy, only one recently reported system is currently published, though it is expected that many more will arrive in the literature as these novel characteristics are exploited. The existing example of this application belongs to Rey-Rico et al., who developed a polypseudorotaxane supramolecular gel in phosphate-buffered saline with demonstrated ability to transfect recombinant adeno-associated virus (rAAV) vectors into human mesenchymal stem cells while maintaining suitable mechanical robustness over time periods of relevance. The addition of α-cyclodextrin into the gels resulted in elevated rAAV concentrations and maintained transgene expression [**112**]. Further expansion of the field into this application can unlock new pathways for gene delivery, a historically difficult but important challenge.

SUPRAMOLECULAR GELS FOR BIOLOGICAL APPLICATIONS

Cell Storage and Study

Fibrous supramolecular gels offer tunable biochemical, mechanical, and physical properties with dimensions on the order of the extracellular matrix (ECM), making them ideal candidates for use in cell storage and study. Self-assembling peptide-based hydrogels are perhaps the most common derivative within this class, offering biomimetic and bioactive building blocks for assembly. With a controllable number of exposed arginylglycylaspartic acid (RGD) tripeptide sequences on the surface, a hydrogel scaffold developed by Zhou et al. is one such classic example of this application, offering a stiff and highly hydrated ECM-like scaffold for anchorage-dependent cells [**113**].

Though the cryopreservation of cells is established as an effective long-term cell storage method for cells, in which cells are brought to subzero temperatures to arrest and thawed to regain biological activity, the process offers several opportunities for damaging cells. These include the formation of intracellular and extracellular ice crystals, cytotoxicity from cryoprotective agents, and toxicity from highly concentrated electrolytes and solutes due to ice formation [**114**]. A novel Boc-*O*-dodecyl-l-tyrosine (BDT) supramolecular gel system (structure shown in **Figure 12**) in a culture medium developed by Zeng et al. overcame many of these issues by confining ice crystal growth into the gel pores, lowering cell volume change rates, depressing osmotic shock, and decreasing the volume of crystalline ice produced (**Figure 13**). Furthermore, the thermo reversibility of the system (unique to supramolecular gels) allows for the facile removal of the gel, as the material is in its liquid state at room and biological temperatures [**115**].

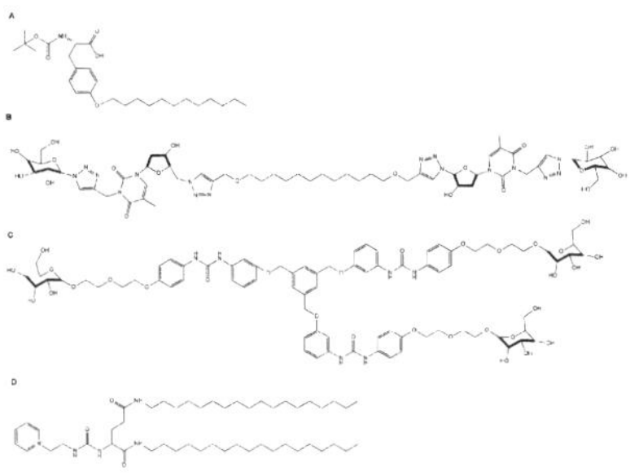

Figure 12. Selected gelators developed for biological applications: (**A**) a Boc-*O*-dodecyl-l-tyrosine gelator for the cryopreservation of cells [**115**]; (**B**) a bole-amphiphile gelator for cell culturing [**116**]; (**C**) a tris-urea hydrogelator for electrophoresis [**117**]; and (**D**) a cationic pyridinium-glutamide amphiphilic gelator for selective visual indication of the presence of adenosine triphosphate (ATP) [**118**].

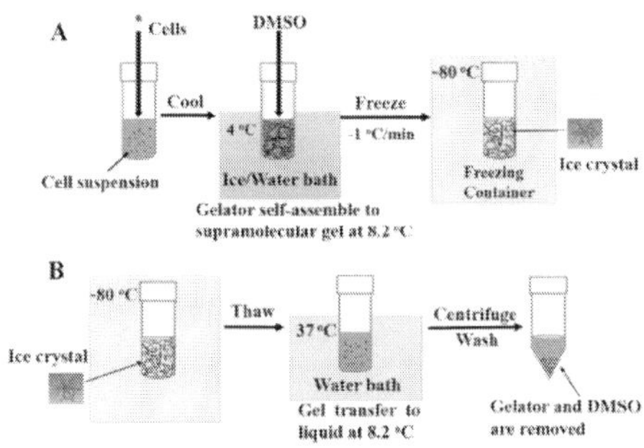

Figure 13. The (**a**) cooling and (**b**) warming processes of the supramolecular gel developed for minimizing cell damage during storage via cryopreservation by Zeng et al. [**115**]. Adapted with permission from ref [**115**]. Copyright 2015 John Wiley and Sons.

Material stiffness is well recognized as a key influencer in many dynamic biological processes, such as cell differentiation, tissue development, and tumor progression [**119**]. A supramolecular hydrogel developed by Zheng et al. allowed clinical studies of such dynamic processes through light modulated stiffness [**120**]. This hydrogel, which stiffens upon the application of near infrared light and expands upon the application of ultraviolet light, offers spatiotemporal resolution of a critical matrix parameter.

A low molecular weight hydrogelator system developed by Poolman et al. demonstrated the versatility of supramolecular gel systems. Through a self-described "toolbox" of modifications to a base hydrozone gelator, their hydrogel could be functionalized with proteins for increased bioactivity or modified for multicolor imaging of mammalian cells and their substructures nested in the gel network [**121**].

Stem Cell Study and Differentiation

Because of the highly tunable nature of supramolecular gels, much interest has been garnered in developing these gels to direct and study stem cell differentiation. A low molecular weight bole-amphiphile gelator (structure shown in **Figure 12**) published by Latxague et al. self-assembled in water into cell culture scaffolds, with transmission electron micrograph and rheological results indicating that the 3D matrix shape affects the behavior of stem cells [**116**]. Jacob and Ghosh et al. furthered this work with the development of an amyloid nanofibril hydrogel which supported cell attachment and spreading over a range of cell types for cell culturing and offered tunable stiffness that notably directed differentiation of mesenchymal stem cells [**122**]. Lastly, Alakpa et al. showed perivascular stem cells cultured on supramolecular peptide gels of 1, 13, and 32 kPa stiffnesses differentiated into neuronal, chondrogenic, and osteogenic cells, respectively. Then, analyzing deviations in the concentration of over 600 metabolites during differentiation on the latter two scaffolds, the authors discovered that a specific lipid was significantly depleted in each process, demonstrating the utility of supramolecular gels for identifying cell-directing metabolites [**123**].

The tunability of supramolecular gels offers the unique ability of cell testing in a manner similar to that acheived by Alakpa et al. Through thoughtful design of the constituents included in the gel structure, supramolecular gels may be used for detailed and insightful analysis of cellular processes ranging from proliferation to endocytosis to movement.

Electrophoresis

Gel electrophoresis is extensively used in biological fields to separate macromolecules such as DNA, RNA, and proteins based on their size and charge [**124**]. Today, sodium dodecyl sulfate polyacrylamide gel electrophoresis (SDS-PAGE) enjoys widespread use for this characterization method. However, supramolecular gels offer notable enticing advantages over their polyacrylamide gel counterparts; namely, significantly easier separation of proteins from the matrix and the ability to tune separation based on the structure of the matrix [**125**]. Dubbed SDS-SUGE (SDS-supramolecular gel electrophoresis), the low molecular weight tris-urea hydrogelator (structure shown in **Figure 12**) matrix developed by Yamamichi et al. provides a prime example for the separation of proteins and their facile recovery through centrifugation [**117**]. A study by Munenobu et al. further demonstrated that the separation of acidic proteins on the same matrix was based on isoelectric point rather than molecular weight and that the proteins retained their native forms during and activity after separation [**126**]. Highlighting the efficiency of this system relative to SDS-PAGE, Tazawa et al. demonstrated that SDS-SUGE could be used in a continuous field manner for the separation of DNA fragments up to 166 kbp, whereas a pulsed field, requiring specialized equipment and a long analysis time, typically must be used for a conventional polyacrylamide matrix [**127**]. The notable advantages of SDS-SUGE over its conventional counterpart may disrupt a well-established bioanalytical technique in favor of tunability and protein recoverability.

Biological Compound Indicator

The collapse of supramolecular gel structures in the presence of specific biological compounds offers enticing applications for the facile detection of these complex molecules without instrumentation. The fabrication of a supramolecular gel from a cationic gelator (a pyridinium tethered covalently to a glutamide amphiphile) (structure shown in **Figure 12**) and the visual indicator methyl orange in a 1:1 mixture by mole of ethanol and water by Yang et al. selectively collapsed in the presence of the key cellular energy molecule adenosine triphosphate (ATP) but not the other related adenosine-phosphates adenosine diphosphate (ADP) and adenosine monophosphate (AMP), allowing visual discriminations between these key metabolic molecules [**118**]. Similarly, a cholesterol-based gel in dimethylformamide (DMF)-water mixtures prepared by He et al. for amino acid detection selectively collapsed in the presence of aspartic acid and glutamic acid [**128**]. A precursor which self-assembled to a hydrogel upon enzyme catalysis was reported by Yang and Xu. When the enzyme was mixed with possible inhibitors before contacting the precursor, actual inhibitors prevented gel formation, and their corresponding minimum inhibition concentration could be determined in a way amenable to running a parallel assay to simultaneously test many inhibitors [**129**]. Supramolecular gels may thus be developed into rapid, useful tests for commonly performed biological assays.

SPECIAL TOPIC APPLICATIONS OF SUPRAMOLECULAR GELS

Though they do not explicitly fit into the other broad categories previously discussed in this review article, supramolecular gels are found at the forefront of research for several additional specific industrial applications. These special topics are discussed here and demonstrate the versatility of supramolecular gel materials to span the breadth of scientific fields.

Environmental Remediation

Water Treatment

The functionalization, high aspect ratio, and high loading capacities of supramolecular gels make them ideal candidates for developing the next generation of water treatment materials. For example, an ambidextrous bifunctional hydrogelator (structure shown in **Figure 14**) synthesized by Basu et al. showed the ability to absorb toxic cationic and anionic organic dyes from contaminated water, and demonstrated continued viability over several cycles [**130**]. Considering the widespread use and presence of mercury despite its toxicity, Lin et al. developed a naphthalimide-functionalized supramolecular polymer gel whose corresponding xerogel showed a 98% removal rate of Hg^{2+} and recyclability [**131**]. A comprehensive purification system was proposed by Yan et al., in which a two component tetrazoyl derivative-octadecylamine supramolecular gel assembled in cyclohexane rapidly co-adsorbed cationic dyes, anionic dyes, and heavy metal ions with demonstrated recyclability and reusability and without significant decreases in efficiency [**132**].

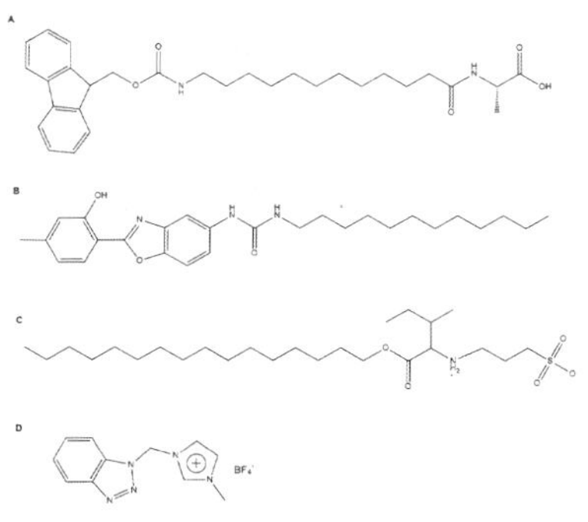

Figure 14. Selected gelators developed for special topic applications: (**A**) an ambidextrous bifunctional hydrogelator for the removal of toxic cationic and anionic organic dyes from water [**130**]; (**B**) an organogelator containing 2-(2′-hydroxyphenyl)benzoxazole for the visual detection of nerve agents [**133**]; (**C**) a zwitterionic gelator for tribology [**134**]; and (**D**) an imidazolium-based gelator containing a benzotriazole group for anticorrosion and lubrication [**135**].

Additional functionalities have been investigated for developing a supramolecular gel which not only cleans water, but can easily be added to and removed from the solution. Zeng et al. incorporated Fe_3O_4 nanoparticles into a benzyl sorbitol-derived supramolecular gel reinforced with halloysite nanotubes and assembled in a 4:1 by weight ethanol-water mixture with dye absorptive properties. In addition to increasing the strength of the gel, the nanoparticles allow for the facile removal of the gels from the solutions they purify by applying a magnetic field [**136**]. A ferrocene-containing supramolecular gel assembled in heptane by Yan et al. efficiently removed iodine from water and was able to undergo a shear-induced gel-sol transition, offering a new method for water purification. Shaking the gel into its sol phase allows it to disperse throughout the matrix, and, within minutes, the gel is reformed containing high quantities of extracted iodine. The gel can then be directly removed, leaving behind purified water [**137**].

The scalability and maintenance of the gelled structure under mechanical stress are of the highest consideration as the field moves towards developing and testing further supramolecular gels for water remediation. Whether a supramolecular gel can withstand significant shear flow, what the maximum rate of aqueous flow through the gel is, and to what degree these gels can be made cost effective will guide their implementation into applied use.

Oil Recovery

The annual transportation of millions of tons of petroleum across waterways worldwide puts the ecosystems they cross at dire risk, particularly as spills are unpredictable and the resources to mitigate their impact are often not readily available [**138**]. Supramolecular gels offer a unique opportunity to combat this issue through the selective gelation of oil products on orders of magnitude larger than their own weight. A dipeptide-based gelator by Raju et al. highlighted this application well, with the phase-selective gelation solely of crude oils with varying gravities and refinery products in mixtures with water. Notably, this system was able to absorb up to 166 times its own weight in hydrocarbons [**139**]. Similarly, an alkyl bicarbamate supramolecular gel by Wang et al. (**Figure 15**) showed rapid (within 15 min) selective removal and the retention of oils from oil-water mixtures [**140**]. For a complete cycle of recovery, a xylitol-based organogelator by Raju et al. (**Figure 15**) demonstrated not only the phase-selective gelation of oil products, but the recovery of oil from the gels as well [**141**].

(a) (b)

Figure 15. Supramolecular gels have been designed for the highly selective separation of crude oil products from water. (**a**) The separation of several oils from water with a xylitol-based organogelator by Raju et al. [**141**] Adapted with permission from ref [**141**]. Copyright Year Copyright Owner's Name. (**b**) The segregation and removal of Russian crude oil from water using an alkyl bicarbamate supramolecular gel by Wang et al. [**140**]. Adapted with permission from ref [**140**]. Copyright 2017 Elsevier.

As with water treatment, scalability looks to be the most pressing concern for these gelator materials. Though the efficacy of these materials is remarkable for oil recovery, vast quantities would be required for real-world oil spill remediation. Supramolecular gels for this application should be further tested in turbulent conditions, simulating the open ocean, to determine when gelation readily occurs. Additionally, the toxicity of these materials should be determined, as quantities which do not participate in the gelation of oil would likely remain in the environment.

Chemical Warfare Agent Sensing and Decontamination

Perhaps one of the most sobering developments of supramolecular gels is their tuning for the detection and collection of chemical warfare agents. Kim et al. noted this application in 2010 with the synthesis of an organogel containing 2-(2′-hydroxyphenyl)benzoxazole (structure shown in **Figure 14**) in chloroform, which enables the detection of nerve agents through demonstrated sensing of the organophosphate diethyl chlorophosphate (DCP) through a visual colorless to yellow color shift of the gel [**133**]. Hiscock et al. developed a system sensitive to dimethyl methylphosphonate (DMMP) and highly sensitive to soman through the interruption of the sol-gel transition in their tren-based tris-urea supramolecular gel system assembled in toluene [**142**]. It should be noted that while DCP and DMMP are not themselves nerve agents, they are commonly used in laboratories as a safe analog to agents such as sarin, soman, and tabun because of their parallel reactivity [**143**].

Hiscock et al. went on to develop oximate-containing organogels in dimethyl sulfoxide, which could decontaminate organophosphorus liquid agents and sense the presence of liquid and vapor agents through color changes. The collapse of the gels in the presence of large volumes of warfare agents could subsequently be employed to release localized bursts of a decontaminant [**144**]. The incorporation of metal oxides into supramolecular gels in the presence of peroxides, notably using off-the-shelf reagents, can alternatively be used for the catalytic degradation of warfare agents, such as the reported degradation of the gaseous sulfur mustard stimulant 2-chloroethyl ethyl sulfide [**145**].

Participation in Synthesis

Catalysis

The high surface area-to-volume ratio of nanostructures, which in turn imparts high reactivity, and the controllable fabrication of supramolecular gels have made sparked interest in their use as scaffolds for catalytic applications, improving upon existing industrial synthesis procedures. Self-assembling glycyrrhizic acid gelators incorporated with graphene oxide and gold nanoparticles in water by Saha et al. illustrated the catalysis of several electron transfer processes, such as the reduction of *p*-nitrophenol to *p*-aminophenol [**146**]. l-glutamic acid-based hydrogelators developed by Jin et al. offered a similar promise when complexed with Cu^{2+} for the catalysis of Diels-Alder cycloaddition reactions, with an accelerated reaction rate and improved control of product chirality relative to existing Cu^{2+}-containing nanostructures [**147**]. Finally, a xerogel formed by the complexation of Cd^{2+} with a cyclohexane-based gelator by Lee et al. exhibited activity as a base catalyst for the Knoevenagel condensation reaction [**148**]. Though all gels presented here utilized a cofactor for catalysis, it should be noted that peptide-based self-assembling gelators are being increasingly studied for catalytic applications due to their inherent specific reactivity [**149**].

Template for Crystal Growth

Gels have historically been used for templating crystal growth because they impart improved optical properties, larger sizes, and fewer defects than crystals grown in the absence of a gel [**150**]. Supramolecular gels improve upon conventional gel templates because they offer readily reversible and stimuli-tunable sol-gel transitions and can gel in a wider variety of solvents. Foster et al. realized four such bis(urea) gelators which offered controlled growth of a diverse range of pharmaceuticals in a variety of organic solvents, including carbamazepine, acetaminophen, and ibuprofen. In many cases, an anion-triggered gel dissolution allowed for the facile recovery of the drug [**151**]. Further work by Foster et al. demonstrated that a bis(urea) supramolecular gel in toluene could be used for the polymorph control of grown pharmaceutical crystals [**152**]. In an unrelated line of research, Daly et al. demonstrated the growth of single-crystal halide nanowires on a supramolecular gel scaffold through a diffusion-driven base growth mechanism; the growth of such structures could not be facilitated by conventional gel scaffolds [**153**].

Surface Coatings

Tribology

Friction and wear frequently stress and damage mechanical equipment while causing unnecessary energy consumption. Lubricants are commonly employed to address these concerns; however, traditional lubricating oils suffer from leakage, volatilization, and climb shift. Semi-solid greases created as an alternative to oils face oil separation and poor lubrication from the use of a thickener [**154**]. Consequently, supramolecular gels have been investigated as an intermediate material between oils and greases to overcome the systemic problems of each category. A zwitterionic gelator (structure shown in **Figure 14**) in base oil by Yu et al. is one such example of this application, which achieved evident friction reduction while maintaining stable mechanical properties and rapid creep recovery [**134**]. Similarly, a thiol-functionalized polyhydric gelator in base oil by Huang et al. offered superior lubrication properties at steel-steel, steel-copper, steel-aluminum, and steel-alumina interfaces [**155**]. Further supramolecular gels by Yu et al. based on amino acid-derivative gelators in base oils showed superior tribological properties to common greases and oils while forming a protective film that prevented the direct contact and consequent severe wear of sliding pairs [**156**].

Anticorrosion

Ionic liquid gels formed through supramolecular self-assembly offer intriguing use for anticorrosion applications, often while maintaining favorable tribological properties. Cai et al. synthesized an imidazolium-based gelator containing a benzotriazole group (structure shown in **Figure 14**), which in turn endowed anticorrosion and antioxidant properties to the material. The thixotropic properties of the gel further offered potential use of the material in high performance lubrication applications, such as for electrical contacts [**135**]. A similar imidazolium-based gelator with a urea group by Yu et al. offered desirable anticorrosion and tribological properties with facile attachment to metal surfaces resulting from its urea-based and imidazole ring functional groups [**157**].

As many tribologically-minded supramolecular gels use similar structures and functionalities to achieve friction resistance to those seen for anticorrosion applications, it seems relevant that many of the former gels be tested for the latter properties. Reporting the additional anticorrosive functionality of these gels can accelerate their pace towards commercial acceptance.

Fat Replacement in Consumer Food Products

Societal, governmental, and industrial pressures to reduce or eliminate saturated and trans fats from consumer food products have sparked interest in using supramolecular gels to create structured liquid oils as an alternative fat source to hard fats or liquid oils structured with saturated fatty acids

[158]. Edible self-assembling small molecule gels are specifically of interest for this application, where small molecules act as the gelator and oils act as the solvent. Co and Marangoni developed one such system with 12-hydroxystearic acid to gel vegetable oils, with shear and thermal fields used to tune the material's mechanical properties and consequent mouth-feel [159]. Similarly, Alvarez-Mitre et al. formulated self-assembling 12-hydroxystearic acid-based ionic gelators for the gelation of safflower oil [160]. Short chain ceramides analyzed by Rogers et al. not only demonstrated the gelation of edible oils like canola oil, but surprisingly decreased the viability of colon, prostate, ovarian, and leukemia cancer cell lines due to the bioactivity of the gelator, demonstrating the translational ability of their material [161]. Such reports highlight the potential of supramolecular gels to disrupt well-established fat sources in the name of consumer health.

CONCLUSIONS

Supramolecular gels have found use in a wide breadth of applications over the past two decades due to their tunable mechanical, electrical, biocompatible, and chemical properties and self-healing abilities. While supramolecular gels have exhibited many specific breakthroughs, further improvements in increased device lifetimes, lower cost and higher throughput processing, and knowledge of long-term biological and environmental effects are needed before these applications can become staple products of a new generation of materials. Additionally, other areas such as logic gate applications, biomimetic prostheses, and non-viral vector gene delivery are still in early development stages. Further research in the field promises the use of supramolecular gels as functional soft materials, forging avenues toward widespread flexible electronics, smarter sensors, revolutions in alternative energy, novel medical treatment methods, greater biological understanding, and expanded industrial capabilities.

ACKNOWLEDGMENTS

This material is based upon work supported by the National Science Foundation Graduate Research Fellowship under Grant No. 1122374. The authors would like to thank Cynthia Lo and Audrey Lai for their input.

CONFLICTS OF INTEREST

The authors declare no conflicts of interest.

REFERENCES

1. Steed, J.W. Supramolecular gel chemistry: Developments over the last decade. *Chem. Commun.* **2011**, *47*, 1379–1383.
2. Flory, P.J. Introductory lecture. *Faraday Discuss. Chem. Soc.* **1974**, *57*, 7.
3. Banerjee, S.; Das, R.K.; Maitra, U. Supramolecular gels "in action". *J. Mater. Chem.* **2009**, *19*, 6649.
4. Sangeetha, N.M.; Maitra, U. Supramolecular gels: Functions and uses. *Chem. Soc. Rev.* **2005**, *34*, 821.
5. Buerkle, L.E.; Rowan, S.J. Supramolecular gels formed from multi-component low molecular weight species. *Chem. Soc. Rev.* **2012**, *41*, 6089.
6. Boekhoven, J.; Brizard, A.M.; Stuart, M.C.A.; Florusse, L.; Raffy, G.; Del Guerzo, A.; van Esch, J.H. Bio-inspired supramolecular materials by orthogonal self-assembly of hydrogelators and phospholipids. *Chem. Sci.* **2016**, *7*, 6021–6031.
7. Feng, Q.; Wei, K.; Zhang, K.; Yang, B.; Tian, F.; Wang, G.; Bian, L. One-pot solvent exchange preparation of non-swellable, thermoplastic, stretchable and adhesive supramolecular hydrogels based on dual synergistic physical crosslinking. *NPG Asia Mater.* **2018**, *10*, e455.
8. Pépin-Donat, B.; Viallat, A.; Blachot, J.F.; Lombard, C. Electromechanical polymer gels combining rubber elasticity with electronic conduction. *Adv. Mater.* **2006**, *18*, 1401–1405.
9. Meng, S.; Rouabhia, M.; Shi, G.; Zhang, Z. Heparin dopant increases the electrical stability, cell adhesion, and growth of conducting polypyrrole/poly(L,L-lactide) composites. *J. Biomed. Mater. Res.* **2008**, *87*, 332–344.
10. Fukushima, T.; Aida, T. Ionic liquids for soft functional materials with carbon nanotubes. *Chem. Eur. J.* **2007**, *13*, 5048–5058.
11. Fukushima, T.; Kosaka, A.; Ishimura, Y.; Yamamoto, T.; Takigawa, T.; Ishii, N.; Aida, T. Molecular Ordering of Organic Molten Salts Trigged by Single-Walled Carbon Nanotubes. *Science* **2003**, *300*, 2072–2074.
12. Lodge, T.P.; Ueki, T. Mechanically Tunable, Readily Processable Ion Gels by Self-Assembly of Block Copolymers in Ionic Liquids. *Acc. Chem. Res.* **2016**, *49*, 2107–2114.
13. Lee, J.; Aida, T. "Bucky gels" for tailoring electroactive materials and

devices: The composites of carbon materials with ionic liquids. *Chem. Commun.* **2011**, *47*, 6757.

14. Kruusamäe, K.; Sugino, T.; Asaka, K. Ionic and viscoelastic mechanisms of a bucky-gel actuator. *J. Appl. Phys.* **2015**, *118*, 14502.
15. Boland, C.S.; Khan, U.; Ryan, G.; Barwich, S.; Charifou, R.; Harvey, A.; Backes, C.; Li, Z.; Ferreira, M.S.; Möbius, M.E.; et al. Sensitive electromechanical sensors using viscoelastic graphene-polymer nanocomposites. *Science* **2016**, *354*, 1257–1260.
16. Kang, Y.J.; Chun, S.J.; Lee, S.S.; Kim, B.Y.; Kim, J.H.; Chung, H.; Lee, S.Y.; Kim, W. All-solid-state flexible supercapacitors fabricated with bacterial nanocellulose papers, carbon nanotubes, and triblock-copolymer ion gels. *ACS Nano* **2012**, *6*, 6400–6406.
17. Kim, C.B.; Jeong, K.B.; Yang, B.J.; Song, J.W.; Ku, B.C.; Lee, S.; Lee, S.K.; Park, C. Facile Supramolecular Processing of Carbon Nanotubes and Polymers for Electromechanical Sensors. *Angew. Chem.* **2017**, *56*, 16180–16185.
18. Huang, Z.; Lee, H.; Lee, E.; Kang, S.K.; Nam, J.M.; Lee, M. Responsive nematic gels from the self-assembly of aqueous nanofibres. *Nat. Commun.* **2011**, *2*.
19. Cui, H.; Pashuck, E.T.; Velichko, Y.S.; Weigand, S.J.; Cheetham, A.G.; Newcomb, C.J.; Stupp, S.I. Spontaneous and X-ray-triggered crystallization at long range in self-assembling filament networks. *Science* **2010**, *327*, 555–559.
20. Huang, L.; Zhuang, X.; Hu, J.; Lang, L.; Zhang, P.; Wang, Y.; Chen, X.; Wei, Y.; Jing, X. Synthesis of biodegradable and electroactive multiblock polylactide and aniline pentamer copolymer for tissue engineering applications. *Biomacromolecules* **2008**, *9*, 850–858.
21. Wu, Q.; Wei, J.; Xu, B.; Liu, X.; Wang, H.; Wang, W.; Wang, Q.; Liu, W. A robust, highly stretchable supramolecular polymer conductive hydrogel with self-healability and thermo-processability. *Sci. Rep.* **2017**, *7*.
22. Darabi, M.A.; Khosrozadeh, A.; Mbeleck, R.; Liu, Y.; Chang, Q.; Jiang, J.; Cai, J.; Wang, Q.; Luo, G.; Xing, M. Skin-Inspired Multifunctional Autonomic-Intrinsic Conductive Self-Healing Hydrogels with Pressure Sensitivity, Stretchability, and 3D Printability. *Adv. Mater.* **2017**, *29*.
23. Mandal, S.K.; Andal, D.; Das, P.K. Synthesis of aLow-Molecular-Weight Fluorescent Ambidextrous Gelator:

Development of Graphene-and Graphene-Oxide-Included Gel Nanocomposites. *ChemPlusChem* **2016**, *81*, 213–221.

24. Olive, A.G.L.; Guerzo, A.; Belin, C.; Reichwagen, J.; Hopf, H.; Desvergne, J.-P. Self-assembly of soluble anthracene, tetracene and pentacene derivatives. *Res. Chem. Intermed.* **2008**, *34*, 137–145.
25. Roy, S.; Maiti, D.K.; Panigrahi, S.; Basak, D.; Banerjee, A. A bolaamphiphilic amino acid appended photo-switching supramolecular gel and tuning of photo-switching behaviour. *Phys. Chem. Chem. Phys.* **2014**, *16*, 6041.
26. Shi, Y.; Wang, M.; Ma, C.; Wang, Y.; Li, X.; Yu, G. A Conductive Self-Healing Hybrid Gel Enabled by Metal-Ligand Supramolecule and Nanostructured Conductive Polymer. *Nano Lett.* **2015**, *15*, 6276–6281.
27. Li, Y.; Hashimoto, M. PVC gel soft actuator-based wearable assist wear for hip joint support during walking. *Smart Mater. Struct.* **2017**, *26*, 125003.
28. Liu, Y.J.; Cao, W.T.; Ma, M.G.; Wan, P. Ultrasensitive Wearable Soft Strain Sensors of Conductive, Self-healing, and Elastic Hydrogels with Synergistic "Soft and Hard" Hybrid Networks. *ACS Appl. Mater. Interfaces* **2017**, *9*, 25559–25570.
29. Ma, X.; Cui, Y.; Liu, S.; Wu, J. A thermo-responsive supramolecular gel and its luminescence enhancement induced by rare earth Y^{3+}. *Soft Matter* **2017**, *13*, 8027–8030.
30. Ma, X.; Zhang, J.; Tang, N.; Wu, J. A thermo-responsive supramolecular organogel: Dual luminescence properties and luminescence conversion induced by Cd^{2+}. *Dalton Trans.* **2014**, *43*, 17236–17239.
31. Kong, S.; Xiao, L.; Chen, Z.; Yan, X.; Qu, B.; Wang, S.; Gong, Q. Enhanced photoluminescence and the self-assembled fibrillar nanostructure of 5-(cholesteryloxy)methyl-8-hydroxyquinoline lithium in a gel state. *New J. Chem.* **2010**, *34*, 325–330.
32. Kishimura, A.; Yamashita, T.; Aida, T. Phosphorescent organogels via "metallophilic" interactions for reversible RGB-color switching. *J. Am. Chem. Soc.* **2005**, *127*, 179–183.
33. Lin, Q.; Liu, L.; Liu, J.; Zheng, F.; Zhang, Y.-M.; Yao, H.; Wei, T.-B. An efficient iodide ion chemosensor and a rewritable dual-channel security display material based on an ion responsive supramolecular gel. *RSC Adv.* **2017**, *7*, 38210–38215.
34. Lin, Q.; Yang, Q.P.; Sun, B.; Fu, Y.P.; Zhu, X.; Wei, T.B.; Zhang, Y.M.

Rewritable security display material and Cl^- sensor: Based on a bimetal competitive coordination controlled supramolecular gel. *Mater. Lett.* **2014**, *137*, 444–446.

35. Lin, Q.; Sun, B.; Yang, Q.-P.; Fu, Y.-P.; Zhu, X.; Zhang, Y.-M.; Wei, T.-B. A novel strategy for the design of smart supramolecular gels: Controlling stimuli-response properties through competitive coordination of two different metal ions. *Chem. Commun.* **2014**, *50*, 10669–10671.
36. Zhang, X.; Choi, H.A.; Lee, S.; Yin, J.; Kim, S.H.; Lee, C.; Yoon, J. A cholesteryl thiazolothiazole derivative for colorimetric sensing of Cu^{2+} and its sol-gel transition. *Dyes Pigments* **2015**, *122*, 109–115.
37. Fan, K.; Song, J.; Li, J.; Guan, X.; Tao, N.; Tong, C.; Shen, H.; Niu, L. Copper(II)-responsive gel-sol phase transition in supramolecular gel systems of salen-appended sorbitol. *J. Mater. Chem. C* **2013**, *1*, 7479.
38. Harris, R.F.; Nation, A.J.; Copeland, G.T.; Miller, S.J. A polymeric and fluorescent gel for combinatorial screening of catalysts. *J. Am. Chem. Soc.* **2000**, *122*, 11270–11271.
39. Sun, H.; Liu, S.; Lin, W.; Zhang, K.Y.; Lv, W.; Huang, X.; Huo, F.; Yang, H.; Jenkins, G.; Zhao, Q.; et al. Smart responsive phosphorescent materials for data recording and security protection. *Nat. Commun.* **2014**, *5*.
40. Kishimura, A.; Yamashita, T.; Yamaguchi, K.; Aida, T. Rewritable phosphorescent paper by the control of competing kinetic and thermodynamic self-assembling events. *Nat. Mater.* **2005**, *4*, 546–549.
41. Wang, W.; Xie, N.; He, L.; Yin, Y. Photocatalytic colour switching of redox dyes for ink-free light-printable rewritable paper. *Nat. Commun.* **2014**, *5*.
42. Zhou, J.; Han, P.; Liu, M.; Zhou, H.; Zhang, Y.; Jiang, J.; Liu, P.; Wei, Y.; Song, Y.; Yao, X. Self-Healable Organogel Nanocomposite with Angle-Independent Structural Colors. *Angew. Chem.* **2017**, *56*, 10462–10466.
43. Suzuki, Y.; Mizoshita, N.; Hanabusa, K.; Kato, T. Homeotropically oriented nematic physical gels for electrooptical materials. *J. Mater. Chem.* **2003**, *13*, 2870.
44. Tong, X.; Zhao, Y.; An, B.K.; Park, S.Y. Fluorescent liquid-crystal gels with electrically switchable photoluminescence. *Adv. Funct. Mater.* **2006**, *16*, 1799–1804.

45. Chen, J.W.; Huang, C.C.; Chao, C.Y. Supramolecular liquid-crystal gels formed by polyfluorene-based π-conjugated polymer for switchable anisotropic scattering device. *ACS Appl. Mater. Interfaces* **2014**, *6*, 6757–6764.

46. Vidyasagar, A.; Handore, K.; Sureshan, K.M. Soft optical devices from self-healing gels formed by oil and sugar-based organogelators. *Angew. Chem.* **2011**, *50*, 8021–8024.

47. Weingarten, A.S.; Kazantsev, R.V.; Palmer, L.C.; McClendon, M.; Koltonow, A.R.; Samuel, A.P.S.; Kiebala, D.J.; Wasielewski, M.R.; Stupp, S.I. Self-assembling hydrogel scaffolds for photocatalytic hydrogen production. *Nat. Chem.* **2014**, *6*, 964–970.

48. Kazantsev, R.V.; Dannenhoffer, A.J.; Weingarten, A.S.; Phelan, B.T.; Harutyunyan, B.; Aytun, T.; Narayanan, A.; Fairfield, D.J.; Boekhoven, J.; Sai, H.; et al. Crystal-Phase Transitions and Photocatalysis in Supramolecular Scaffolds. *J. Am. Chem. Soc.* **2017**, *139*, 6120–6127.

49. Pillarisetty, R. Academic and industry research progress in germanium nanodevices. *Nature* **2011**, *479*, 324–328.

50. Vurgaftman, I.; Meyer, J.R.; Ram-Mohan, L.R. Band parameters for III-V compound semiconductors and their alloys. *J. Appl. Phys.* **2001**, *89*, 5815–5875.

51. Klauk, H. Organic thin-film transistors. *Chem. Soc. Rev.* **2010**, *39*, 2643.

52. Murphy, A.R.; Fréchet, J.M.J. Organic semiconducting oligomers for use in thin film transistors. *Chem. Rev.* **2007**, *107*, 1066–1096.

53. Miao, S.; Zhu, W.; Castro, N.J.; Nowicki, M.; Zhou, X.; Cui, H.; Fisher, J.P.; Zhang, L.G. 4D printing smart biomedical scaffolds with novel soybean oil epoxidized acrylate. *Sci. Rep.* **2016**, *6*.

54. Muthusamy, K.; Sridharan, V.; Maheswari, C.U.; Nagarajan, S. Lipase catalyzed synthesis of fluorescent glycolipids: Gelation studies and graphene incorporated self-assembled sheet formation for semiconductor applications. *Green Chem.* **2016**, *18*, 3722–3731.

55. Horowitz, G. Organic Field-Effect Transistors. *Adv. Mater.* **1998**, *10*, 365–377.

56. Diring, S.; Camerel, F.; Donnio, B.; Dintzer, T.; Toffanin, S.; Capelli, R.; Muccini, M.; Ziessel, R. Luminescent ethynyl-pyrene liquid crystals and gels for optoelectronic devices. *J. Am. Chem. Soc.* **2009**, *131*, 18177–18185.

57. Schmidt, L.C.; Edelsztein, V.C.; Spagnuolo, C.C.; Di Chenna, P.H.; Galian, R.E. Light-responsive hybrid material based on luminescent core-shell quantum dots and steroidal organogel. *J. Mater. Chem. C* **2016**, *4*, 7035–7042.
58. Jung, S.H.; Moon, J.H.; Lee, J.Y.; Jung, J.H. Thermochromic and piezochromic effects of CoII-imidazole-based supramolecular gels as logic gates. *Eur. J. Inorg. Chem.* **2014**, 2350–2355.
59. Bairi, P.; Chakraborty, P.; Shit, A.; Mondal, S.; Roy, B.; Nandi, A.K. A co-assembled gel of a pyromellitic dianhydride derivative and polyaniline with optoelectronic and photovoltaic properties. *Langmuir* **2014**, *30*, 7547–7555.
60. Shi, Y.; Zhang, J.; Pan, L.; Shi, Y.; Yu, G. Energy gels: A bio-inspired material platform for advanced energy applications. *Nano Today* **2016**, *11*, 738–762.
61. Huo, Z.; Tao, L.; Dai, S.; Zhu, J.; Zhang, C.; Chen, S.; Zhang, B. Quasi-solid-state dye sensitized solar cells using supramolecular gel electrolyte formed from two-component low molecular mass organogelators. *Sci. China Mater.* **2015**, *58*, 447–454.
62. Huo, Z.; Wang, L.; Tao, L.; Ding, Y.; Yi, J.; Alsaedi, A.; Hayat, T.; Dai, S. A supramolecular gel electrolyte formed from amide based co-gelator for quasi-solid-state dye-sensitized solar cell with boosted electron kinetic processes. *J. Power Sources* **2017**, *359*, 80–87.
63. Duan, P.; Yanai, N.; Nagatomi, H.; Kimizuka, N. Photon upconversion in supramolecular gel matrixes: Spontaneous accumulation of light-harvesting donor-acceptor arrays in nanofibers and acquired air stability. *J. Am. Chem. Soc.* **2015**, *137*, 1887–1894.
64. Dong, X.; Wang, H.; Fang, F.; Li, X.; Yang, Y. Effect of gelator structures on electrochemical properties of ionic-liquid supramolecular gel electrolytes. *Electrochim. Acta* **2010**, *55*, 2275–2279.
65. Singh-Rachford, T.N.; Castellano, F.N. Photon upconversion based on sensitized triplet-triplet annihilation. *Coord. Chem. Rev.* **2010**, *254*, 2560–2573.
66. Monguzzi, A.; Bianchi, F.; Bianchi, A.; Mauri, M.; Simonutti, R.; Ruffo, R.; Tubino, R.; Meinardi, F. High efficiency up-converting single phase elastomers for photon managing applications. *Adv. Energy Mater.* **2013**, *3*, 680–686.
67. Börjesson, K.; Dzebo, D.; Albinsson, B.; Moth-Poulsen, K. Photon

upconversion facilitated molecular solar energy storage. *J. Mater. Chem. A* **2013**, *1*, 8521.

68. Simon, Y.C.; Weder, C. Low-power photon upconversion through triplet-triplet annihilation in polymers. *J. Mater. Chem.* **2012**, *22*, 20817.
69. Duduta, M.; Ho, B.; Wood, V.C.; Limthongkul, P.; Brunini, V.E.; Carter, W.C.; Chiang, Y.M. Semi-solid lithium rechargeable flow battery. *Adv. Energy Mater.* **2011**, *1*, 511–516.
70. Frischmann, P.D.; Gerber, L.C.H.; Doris, S.E.; Tsai, E.Y.; Fan, F.Y.; Qu, X.; Jain, A.; Persson, K.A.; Chiang, Y.M.; Helms, B.A. Supramolecular Perylene Bisimide-Polysulfide Gel Networks as Nanostructured Redox Mediators in Dissolved Polysulfide Lithium-Sulfur Batteries. *Chem. Mater.* **2015**, *27*, 6765–6770.
71. Reddy, M.V.; Subba Rao, G.V.; Chowdari, B.V.R. Metal Oxides and Oxysalts as Anode Materials for Li Ion Batteries. *Chem. Rev.* **2013**, *113*, 5364–5457.
72. Guo, W.; Sun, W.; Wang, Y. Multilayer CuO@NiO Hollow Spheres: Microwave-Assisted Metal-Organic-Framework Derivation and Highly Reversible Structure-Matched Stepwise Lithium Storage. *ACS Nano* **2015**, *9*, 11462–11471.
73. Lyu, F.; Yu, S.; Li, M.; Wang, Z.; Nan, B.; Wu, S.; Cao, L.; Sun, Z.; Yang, M.; Wang, W.; et al. Supramolecular hydrogel directed self-assembly of *C*- and *N*-doped hollow CuO as high-performance anode materials for Li-ion batteries. *Chem. Commun.* **2017**, *53*, 2138–2141.
74. Sui, Z.-Y.; Wang, C.; Shu, K.; Yang, Q.-S.; Ge, Y.; Wallace, G.G.; Han, B.-H. Manganese dioxide-anchored three-dimensional nitrogen-doped graphene hybrid aerogels as excellent anode materials for lithium ion batteries. *J. Mater. Chem. A* **2015**, *3*, 10403–10412.
75. Shi, Y.; Ma, D.; Wang, W.; Zhang, L.; Xu, X. A supramolecular self-assembly hydrogel binder enables enhanced cycling of SnO_2-based anode for high-performance lithium-ion batteries. *J. Mater. Sci.* **2017**, *52*, 3545–3555.
76. Kötz, R.; Carlen, M. Principles and applications of electrochemical capacitors. *Electrochim. Acta* **2000**, *45*, 2483–2498.
77. Liang, M.; Liu, X.; Li, W.; Wang, Q. A tough nanocomposite aerogel of manganese oxide and polyaniline as an electrode for a supercapacitor. *Chempluschem* **2016**, *81*, 40–43.

78. Li, W.; Gao, F.; Wang, X.; Zhang, N.; Ma, M. Strong and Robust Polyaniline-Based Supramolecular Hydrogels for Flexible Supercapacitors. *Angew. Chem.* **2016**, *55*, 9196–9201.

79. Carrette, L.; Friedrich, K.A.; Stimming, U. Fuel Cells—Fundamentals and Applications. *Fuel Cells* **2001**, *1*, 5–39.

80. Ye, Y.-S.; Huang, Y.-J.; Cheng, C.-C.; Chang, F.-C. A new supramolecular sulfonated polyimide for use in proton exchange membranes for fuel cells. *Chem. Commun.* **2010**, *46*, 7554.

81. Kumar, G.G.; Hashmi, S.; Karthikeyan, C.; GhavamiNejad, A.; Vatankhah-Varnoosfaderani, M.; Stadler, F.J. Graphene oxide/carbon nanotube composite hydrogels—Versatile materials for microbial fuel cell applications. *Macromol. Rapid Commun.* **2014**, *35*, 1861–1865.

82. Lu, Y.C.; Xu, Z.; Gasteiger, H.A.; Chen, S.; Hamad-Schifferli, K.; Shao-Horn, Y. Platinum-gold nanoparticles: A highly active bifunctional electrocatalyst for rechargeable lithium-air batteries. *J. Am. Chem. Soc.* **2010**, *132*, 12170–12171.

83. Wu, Z.; Wang, J.; Han, L.; Lin, R.; Liu, H.; Xin, H.L.; Wang, D. Supramolecular gel-assisted synthesis of double shelled Co@CoO@N-C/C nanoparticles with synergistic electrocatalytic activity for the oxygen reduction reaction. *Nanoscale* **2016**, *8*, 4681–4687.

84. Jiang, H.; Li, C.; Shen, H.; Liu, Y.; Li, W.; Li, J. Supramolecular gel-assisted synthesis Co_2P particles anchored in multielement co-doped graphene as efficient bifunctional electrocatalysts for oxygen reduction and evolution. *Electrochim. Acta* **2017**, *231*, 344–353.

85. Rizzo, C.; Arrigo, R.; D'Anna, F.; Di Blasi, F.; Dintcheva, N.T.; Lazzara, G.; Parisi, F.; Riela, S.; Spinelli, G.; Massaro, M. Hybrid supramolecular gels of Fmoc-F/halloysite nanotubes: Systems for sustained release of camptothecin. *J. Mater. Chem. B* **2017**, *5*, 3217–3229.

86. Lin, Y.; Li, L.; Li, G. A new supramolecular gel via host-guest complexation with cucurbit[8]uril and *N*-(4-diethylaminobenzyl) chitosan. *Carbohydr. Polym.* **2013**, *92*, 429–434.

87. Horii, A.; Wang, X.; Gelain, F.; Zhang, S. Biological designer self-assembling peptide nanofiber scaffolds significantly enhance osteoblast proliferation, differentiation and 3-D migration. *PLoS ONE* **2007**, *2*, e190.

88. Ashwanikumar, N.; Kumar, N.A.; Saneesh Babu, P.S.; Sivakumar,

K.C.; Vadakkan, M.V.; Nair, P.; Saranya, I.H.; Nair, S.A.; Vinod Kumar, G.S. Self-assembling peptide nanofibers containing phenylalanine for the controlled release of 5-fluorouracil. *Int. J. Nanomed.* **2016**, *11*, 5583–5594.

89. Felip-León, C.; Cejudo-Marín, R.; Peris, M.; Galindo, F.; Miravet, J.F. Sizing Down a Supramolecular Gel into Micro- and Nanoparticles. *Langmuir* **2017**, *33*, 10322–10328.
90. Rodrigues, M.; Calpena, A.C.; Amabilino, D.B.; Garduño-Ramírez, M.L.; Pérez-García, L. Supramolecular gels based on a gemini imidazolium amphiphile as molecular material for drug delivery. *J. Mater. Chem. B* **2014**, *2*, 5419.
91. Hu, Y.; Wang, H.; Wang, J.; Wang, S.; Liao, W.; Yang, Y.; Zhang, Y.; Kong, D.; Yang, Z. Supramolecular hydrogels inspired by collagen for tissue engineering. *Org. Biomol. Chem.* **2010**, *8*, 3267.
92. Pakhomov, P.M.; Ovchinnikov, M.M.; Khizhnyak, S.D.; Roshchina, O.A.; Komarov, P.V. A supramolecular medical hydrogel based on l-cysteine and silver ions. *Polym. Sci. Ser. A* **2011**, *53*, 820–826.
93. Liu, G.F.; Ji, W.; Feng, C.L. Installing Logic Gates to Multiresponsive Supramolecular Hydrogel Co-assembled from Phenylalanine Amphiphile and Bis(pyridinyl) Derivative. *Langmuir* **2015**, *31*, 7122–7128.
94. Ikeda, M.; Tanida, T.; Yoshii, T.; Kurotani, K.; Onogi, S.; Urayama, K.; Hamachi, I. Installing logic-gate responses to a variety of biological substances in supramolecular hydrogel-enzyme hybrids. *Nat. Chem.* **2014**, *6*, 511–518.
95. Komatsu, H.; Matsumoto, S.; Tamaru, S.I.; Kaneko, K.; Ikeda, M.; Hamachi, I. Supramolecular hydrogel exhibiting four basic logic gate functions to fine-tune substance release. *J. Am. Chem. Soc.* **2009**, *131*, 5580–5585.
96. Limón, D.; Amirthalingam, E.; Rodrigues, M.; Halbaut, L.; Andrade, B.; Garduño-Ramírez, M.L.; Amabilino, D.B.; Pérez-García, L.; Calpena, A.C. Novel nanostructured supramolecular hydrogels for the topical delivery of anionic drugs. *Eur. J. Pharm. Biopharm.* **2015**, *96*, 421–436.
97. Klaewklod, A.; Tantishaiyakul, V.; Hirun, N.; Sangfai, T.; Li, L. Characterization of supramolecular gels based on β-cyclodextrin and polyethyleneglycol and their potential use for topical drug

delivery. *Mater. Sci. Eng. C* **2015**, *50*, 242–250.

98. Majumder, J.; Deb, J.; Husain, A.; Jana, S.S.; Dastidar, P. Cetirizine derived supramolecular topical gel in action: Rational design, characterization and in vivo self-delivery application in treating skin allergy in mice. *J. Mater. Chem. B* **2015**, *3*, 6634–6644.
99. Marcos, X.; Pérez-Casas, S.; Llovo, J.; Concheiro, A.; Alvarez-Lorenzo, C. Poloxamer-hydroxyethyl cellulose-α-cyclodextrin supramolecular gels for sustained release of griseofulvin. *Int. J. Pharm.* **2016**, *500*, 11–19.
100. Kannan, R.; Muthuvijayan, V.; Prasad, E. In vitro study of a glucose attached poly(aryl ether) dendron based gel as a drug carrier for a local anaesthetic. *New J. Chem.* **2017**, *41*, 7453–7462.
101. Song, Y.; Gao, J.; Xu, X.; Zhao, H.; Xue, R.; Zhou, J.; Hong, W.; Qiu, H. Fabrication of thermal sensitive folic acid based supramolecular hybrid gels for injectable drug release gels. *Mater. Sci. Eng. C* **2017**, *75*, 706–713.
102. Dai, L.; Liu, K.; Wang, L.; Liu, J.; He, J.; Liu, X.; Lei, J. Injectable and thermosensitive supramolecular hydrogels by inclusion complexation between binary-drug loaded micelles and α-cyclodextrin. *Mater. Sci. Eng. C* **2017**, *76*, 966–974.
103. Simões, S.M.N.; Veiga, F.; Torres-Labandeira, J.J.; Ribeiro, A.C.F.; Sandez-Macho, M.I.; Concheiro, A.; Alvarez-Lorenzo, C. Syringeable Pluronic-α-cyclodextrin supramolecular gels for sustained delivery of vancomycin. *Eur. J. Pharm. Biopharm.* **2012**, *80*, 103–112.
104. Shah, R.N.; Shah, N.A.; Del Rosario Lim, M.M.; Hsieh, C.; Nuber, G.; Stupp, S.I. Supramolecular design of self-assembling nanofibers for cartilage regeneration. *Proc. Natl. Acad. Sci. USA* **2010**, *107*, 3293–3298.
105. Buerkle, L.E.; von Recum, H.A.; Rowan, S.J. Toward potential supramolecular tissue engineering scaffolds based on guanosine derivatives. *Chem. Sci.* **2012**, *3*, 564–572.
106. Del Rosario, C.; Rodríguez-Évora, M.; Reyes, R.; Simões, S.; Concheiro, A.; Évora, C.; Alvarez-Lorenzo, C.; Delgado, A. Bone critical defect repair with poloxamine-cyclodextrin supramolecular gels. *Int. J. Pharm.* **2015**, *495*, 463–473.
107. Sang, L.; Huang, J.; Luo, D.; Chen, Z.; Li, X. Bone-like nanocomposites based on self-assembled protein-based matrices with Ca^{2+} capturing

capability. *J. Mater. Sci. Mater. Med.* **2010**, *21*, 2561–2568.

108. O'Leary, L.E.R.; Fallas, J.A.; Bakota, E.L.; Kang, M.K.; Hartgerink, J.D. Multi-hierarchical self-assembly of a collagen mimetic peptide from triple helix to nanofibre and hydrogel. *Nat. Chem.* **2011**, *3*, 821–828.
109. Paladini, F.; Meikle, S.T.; Cooper, I.R.; Lacey, J.; Perugini, V.; Santin, M. Silver-doped self-assembling di-phenylalanine hydrogels as wound dressing biomaterials. *J. Mater. Sci. Mater. Med.* **2013**, *24*, 2461–2472.
110. Mandal, S.K.; Brahmachari, S.; Das, P.K. In situ synthesised silver nanoparticle-infused l-lysine-based injectable hydrogel: Development of a biocompatible, antibacterial, soft nanocomposite. *Chempluschem* **2014**, *79*, 1733–1746.
111. Yin, H.; Kanasty, R.L.; Eltoukhy, A.A.; Vegas, A.J.; Dorkin, J.R.; Anderson, D.G. Non-viral vectors for gene-based therapy. *Nat. Rev. Genet.* **2014**, *15*, 541–555.
112. Rey-Rico, A.; Babicz, H.; Madry, H.; Concheiro, A.; Alvarez-Lorenzo, C.; Cucchiarini, M. Supramolecular polypseudorotaxane gels for controlled delivery of rAAV vectors in human mesenchymal stem cells for regenerative medicine. *Int. J. Pharm.* **2017**, *531*, 492–503.
113. Zhou, M.; Smith, A.M.; Das, A.K.; Hodson, N.W.; Collins, R.F.; Ulijn, R.V.; Gough, J.E. Self-assembled peptide-based hydrogels as scaffolds for anchorage-dependent cells. *Biomaterials* **2009**, *30*, 2523–2530.
114. Karlsson, J.O.M.; Toner, M. Long-term storage of tissues by cryopreservation: Critical issues. *Biomaterials* **1996**, *17*, 243–256.
115. Zeng, J.; Yin, Y.; Zhang, L.; Hu, W.; Zhang, C.; Chen, W. A Supramolecular Gel Approach to Minimize the Neural Cell Damage during Cryopreservation Process. *Macromol. Biosci.* **2016**, *16*, 363–370.
116. Latxague, L.; Ramin, M.A.; Appavoo, A.; Berto, P.; Maisani, M.; Ehret, C.; Chassande, O.; Barthélémy, P. Control of Stem-Cell Behavior by Fine Tuning the Supramolecular Assemblies of Low-Molecular-Weight Gelators. *Angew. Chem.* **2015**, *54*, 4517–4521.
117. Yamamichi, S.; Jinno, Y.; Haraya, N.; Oyoshi, T.; Tomitori, H.; Kashiwagi, K.; Yamanaka, M. Separation of proteins using supramolecular gel electrophoresis. *Chem. Commun.* **2011**, *47*, 10344–10346.
118. Yang, D.; Liu, C.; Zhang, L.; Liu, M. Visualized discrimination of ATP

from ADP and AMP through collapse of supramolecular gels. *Chem. Commun.* **2014**, *50*, 12688–12690.

119. Stowers, R.S.; Allen, S.C.; Suggs, L.J. Dynamic phototuning of 3D hydrogel stiffness. *Proc. Natl. Acad. Sci. USA* **2015**, *112*, 1953–1958.
120. Zheng, Z.; Hu, J.; Wang, H.; Huang, J.; Yu, Y.; Zhang, Q.; Cheng, Y. Dynamic Softening or Stiffening a Supramolecular Hydrogel by Ultraviolet or Near-Infrared Light. *ACS Appl. Mater. Interfaces* **2017**, *9*, 24511–24517.
121. Poolman, J.M.; Maity, C.; Boekhoven, J.; van der Mee, L.; le Sage, V.A.A.; Groenewold, G.J.M.; van Kasteren, S.I.; Versluis, F.; van Esch, J.H.; Eelkema, R. A toolbox for controlling the properties and functionalisation of hydrazone-based supramolecular hydrogels. *J. Mater. Chem. B* **2016**, *4*, 852–858.
122. Jacob, R.S.; Ghosh, D.; Singh, P.K.; Basu, S.K.; Jha, N.N.; Das, S.; Sukul, P.K.; Patil, S.; Sathaye, S.; Kumar, A.; et al. Self healing hydrogels composed of amyloid nano fibrils for cell culture and stem cell differentiation. *Biomaterials* **2015**, *54*, 97–105.
123. Alakpa, E.V.; Jayawarna, V.; Lampel, A.; Burgess, K.V.; West, C.C.; Bakker, S.C.J.; Roy, S.; Javid, N.; Fleming, S.; Lamprou, D.A.; et al. Tunable Supramolecular Hydrogels for Selection of Lineage-Guiding Metabolites in Stem Cell Cultures. *Chem* **2016**, *1*, 512.
124. Righetti, P.G. Electrophoresis: The march of pennies, the march of dimes. *J. Chromatogr. A* **2005**, *1079*, 24–40.
125. Yamanaka, M. Urea derivatives as low-molecular-weight gelators. *J. Incl. Phenom. Macrocycl. Chem.* **2013**, *77*, 33–48.
126. Munenobu, K.; Hase, T.; Oyoshi, T.; Yamanaka, M. Supramolecular gel electrophoresis of acidic native proteins. *Anal. Chem.* **2014**, *86*, 9924–9929.
127. Tazawa, S.; Kobayashi, K.; Oyoshi, T.; Yamanaka, M. Supramolecular gel electrophoresis of large DNA fragments. *Electrophoresis* **2017**, *38*, 2662–2665.
128. He, T.; Li, K.; Wu, M.-Y.; Liao, Y.-X.; Yu, X.-Q. Visual detection of amino acids by supramolecular gel collapse. *RSC Adv.* **2014**, *4*, 2119–2123.
129. Yang, Z.; Xu, B. A simple visual assay based on small molecule hydrogels for detecting inhibitors of enzymes. *Chem. Commun.* **2004**, *1*, 2424–2425.

130. Basu, K.; Nandi, N.; Mondal, B.; Dehsorkhi, A.; Hamley, I.W.; Banerjee, A. Peptide-based ambidextrous bifunctional gelator: Applications in oil spill recovery and removal of toxic organic dyes for waste water management. *Interface Focus* **2017**, *7*, 20160128.
131. Lin, Q.; Mao, P.-P.; Fan, Y.-Q.; Liu, L.; Liu, J.; Zhang, Y.-M.; Yao, H.; Wei, T.-B.; Li, C.; Jia, D.Z. A novel supramolecular polymer gel based on naphthalimide functionalized-pillar[5]arene for the fluorescence detection of Hg^{2+} and I^- and recyclable removal of Hg^{2+} via cation–π interactions. *Soft Matter* **2017**, *112*, 199–204.
132. Yan, L.; Lv, M.; Su, C.; Zheng, L.; Li, J.; Ye, Z. An efficient supramolecular adsorbent for co-adsorption of dyes and metal ions from wastewater and its application in self-healing materials. *Soft Matter* **2017**.
133. Kim, T.H.; Kim, D.G.; Lee, M.; Lee, T.S. Synthesis of reversible fluorescent organogel containing 2-(2′-hydroxyphenyl)benzoxazole: Fluorescence enhancement upon gelation and detecting property for nerve gas simulant. *Tetrahedron* **2010**, *66*, 1667–1672.
134. Yu, Q.; Huang, G.; Cai, M.; Zhou, F.; Liu, W. In situ zwitterionic supramolecular gel lubricants for significantly improved tribological properties. *Tribol. Int.* **2016**, *95*, 55–65.
135. Cai, M.; Liang, Y.; Zhou, F.; Liu, W. Functional ionic gels formed by supramolecular assembly of a novel low molecular weight anticorrosive/antioxidative gelator. *J. Mater. Chem.* **2011**, *21*, 13399.
136. Zeng, X.; Sun, Z.; Wang, H.; Wang, Q.; Yang, Y. Supramolecular gel composites reinforced by using halloysite nanotubes loading with in-situ formed Fe_3O_4 nanoparticles and used for dye adsorption. *Compos. Sci. Technol.* **2016**, *122*, 149–154.
137. Yan, J.; Liu, J.; Lei, H.; Kang, Y.; Zhao, C.; Fang, Y. Ferrocene-containing thixotropic molecular gels: Creation and a novel strategy for water purification. *J. Colloid Interface Sci.* **2015**, *448*, 374–379.
138. Silliman, B.R.; van de Koppel, J.; McCoy, M.W.; Diller, J.; Kasozi, G.N.; Earl, K.; Adams, P.N.; Zimmerman, A.R. Degradation and resilience in Louisiana salt marshes after the BP-Deepwater Horizon oil spill. *Proc. Natl. Acad. Sci. USA* **2012**, *109*, 11234–11239.
139. Kesava Raju, C.S.; Pramanik, B.; Kar, T.; Rao, P.V.C.; Choudary, N.V.; Ravishankar, R. Low molecular weight gels: Potential in remediation of crude oil spillage and recovery. *RSC Adv.* **2016**, *6*, 53415–53420.

140. Wang, Y.; Wu, S.; Yan, X.; Ma, T.; Shao, L.; Liu, Y.; Guo, Z. Alkyl bicarbamates supramolecular organogelators with effective selective gelation and high oil recovery from oil/water mixtures. *Chemosphere* **2017**, *167*, 178–187.

141. Kesava Raju, C.S.; Pramanik, B.; Ravishankar, R.; Chalapathi Rao, P.V.; Sriganesh, G. Xylitol based phase selective organogelators for potential oil spillage recovery. *RSC Adv.* **2017**, *7*, 37175–37180.

142. Hiscock, J.R.; Piana, F.; Sambrook, M.R.; Wells, N.J.; Clark, A.J.; Vincent, J.C.; Busschaert, N.; Brown, R.C.D.; Gale, P.A. Detection of nerve agent via perturbation of supramolecular gel formation. *Chem. Commun.* **2013**, *49*, 9119.

143. Goswami, S.; Manna, A.; Paul, S. Rapid "naked eye" response of DCP, a nerve agent simulant: From molecules to low-cost devices for both liquid and vapour phase detection. *RSC Adv.* **2014**, *4*, 21984–21988.

144. Hiscock, J.R.; Sambrook, M.R.; Wells, N.J.; Gale, P.A. Detection and remediation of organophosphorus compounds by oximate containing organogels. *Chem. Sci.* **2015**, *6*, 5680–5684.

145. Hiscock, J.R.; Bustone, G.P.; Clark, E.R. Decontamination and Remediation of the Sulfur Mustard Simulant CEES with "Off-the-Shelf" Reagents in Solution and Gel States: A Proof-of-Concept Study. *ChemistryOpen* **2017**, *6*, 497–500.

146. Saha, A.; Adamcik, J.; Bolisetty, S.; Handschin, S.; Mezzenga, R. Fibrillar networks of glycyrrhizic acid for hybrid nanomaterials with catalytic features. *Angew. Chem.* **2015**, *54*, 5408–5412.

147. Jin, Q.; Zhang, L.; Cao, H.; Wang, T.; Zhu, X.; Jiang, J.; Liu, M. Self-assembly of copper(II) ion-mediated nanotube and its supramolecular chiral catalytic behavior. *Langmuir* **2011**, *27*, 13847–13853.

148. Lee, H.H.; Jung, S.H.; Park, S.; Park, K.-M.; Jung, J.H. A metal–organic framework gel with Cd^{2+} derived from only coordination bonds without intermolecular interactions and its catalytic ability. *New J. Chem.* **2013**, *37*, 2330.

149. Escuder, B.; Rodríguez-Llansola, F.; Miravet, J.F. Supramolecular gels as active media for organic reactions and catalysis. *New J. Chem.* **2010**, *34*, 1044.

150. Henisch, H.K.; García-Ruiz, J.M. Crystal growth in gels and Liesegang ring formation. II. Crystallization criteria and successive precipitation. *J. Cryst. Growth* **1986**, *75*, 203–211.

151. Foster, J.A.; Piepenbrock, M.O.M.; Lloyd, G.O.; Clarke, N.; Howard, J.A.K.; Steed, J.W. Anion-switchable supramolecular gels for controlling pharmaceutical crystal growth. *Nat. Chem.* **2010**, *2*, 1037–1043.

152. Foster, J.A.; Damodaran, K.K.; Maurin, A.; Day, G.M.; Thompson, H.P.G.; Cameron, G.J.; Bernal, J.C.; Steed, J.W. Pharmaceutical polymorph control in a drug-mimetic supramolecular gel. *Chem. Sci.* **2017**, *8*, 78–84.

153. Daly, R.; Kotova, O.; Boese, M.; Gunnlaugsson, T.; Boland, J.J. Chemical nano-gardens: Growth of salt nanowires from supramolecular self-assembly gels. *ACS Nano* **2013**, *7*, 4838–4845.

154. Cai, M.; Guo, R.; Zhou, F.; Liu, W. Lubricating a bright future: Lubrication contribution to energy saving and low carbon emission. *Sci. China Technol. Sci.* **2013**, *56*, 2888–2913.

155. Huang, G.; Yu, Q.; Cai, M.; Zhou, F.; Liu, W. Highlighting the Effect of Interfacial Interaction on Tribological Properties of Supramolecular Gel Lubricants. *Adv. Mater. Interfaces* **2016**, *3*, 1500489.

156. Yu, Q.; Li, D.; Cai, M.; Zhou, F.; Liu, W. Supramolecular Gel Lubricants Based on Amino Acid Derivative Gelators. *Tribol. Lett.* **2016**, *61*, 16.

157. Yu, Q.; Wu, Y.; Li, D.M.; Cai, M.; Zhou, F.; Liu, W. Supramolecular ionogel lubricants with imidazolium-based ionic liquids bearing the urea group as gelator. *J. Colloid Interface Sci.* **2017**, *487*, 130–140.

158. Rogers, M.A. Novel structuring strategies for unsaturated fats—Meeting the zero-trans, zero-saturated fat challenge: A review. *Food Res. Int.* **2009**, *42*, 747–753.

159. Co, E.; Marangoni, A.G. The formation of a 12-hydroxystearic acid/vegetable oil organogel under shear and thermal fields. *J. Am. Oil Chem. Soc.* **2013**, *90*, 529–544.

160. Alvarez-Mitre, F.M.; Mallia, V.A.; Weiss, R.G.; Charó-Alonso, M.A.; Toro-Vazquez, J.F. Self-assembly in vegetable oils of ionic gelators derived from (R)-12-hydroxystearic acid. *Food Struct.* **2017**, *13*, 56–69.

161. Rogers, M.A.; Spagnuolo, P.A.; Wang, T.M.; Angka, L. A potential bioactive hard-stock fat replacer comprised of a molecular gel. *Food Sci. Nutr.* **2017**, *5*, 579–587.

CHAPTER 3

Low-Molecular-Weight Gels: The State of the Art

Emily R.Draper[1] **and Dave J.Adams**[1]

[1] School of Chemistry, University of Glasgow, Glasgow G12 8QQ, UK

THE BIGGER PICTURE

Gels are ubiquitous soft solids that are widely used in everyday life. Most gels are based on polymeric materials, as either entangled biopolymers or covalently cross-linked networks. Low-molecular-weight gels, or supramolecular gels, are another fascinating and useful class of gel. Here, the assembly of small molecules into a network immobilizes the solvent. These materials are becoming increasingly common and offer an opportunity to prepare interesting, designed, and responsive systems. Specific areas

Citation: Emily R. Draper and Dave J. Adams. "Low-Molecular-Weight Gels: The State of the Art" Chem 3, 390–410, September 14, 2017 https://doi.org/10.1016/j.chempr.2017.07.012

where we believe that there will be advances include exploiting the process-dependent nature of gelation and systems that utilize multiple gelators to access new properties.

SUMMARY

Low-molecular-weight gels are currently a hugely important class of materials that are attracting significant interest. These gels are formed when small molecules self-assemble into one-dimensional structures that entangle and cross-link to form a network that is capable of immobilizing the solvent. Here, we critically discuss the current state of the art and highlight two key areas where we believe there is significant untapped potential. The first is the observation that the properties of the gels are highly process dependent, which means that it is possible to access materials with very different properties from a single gelator. Second, using multiple gelators offers the opportunity to prepare materials with a high degree of information content and with a wider range of properties. We aim to spark thought and discussion on these aspects.

Keywords: gel, supramolecular, Hydrogel, Self-assembly, Multicomponent

INTRODUCTION

Low-molecular-weight gels (LMWGs), or supramolecular gels, are a fascinating and useful class of material. The gels arise from the self-assembly of small molecules into long, anisotropic structures, most commonly fibers.1, 2, 3, 4, 5 At a sufficiently high concentration, these fibers entangle or otherwise form cross-links, leading to the network that is able to immobilize the solvent through surface tension and capillary forces.1, 2 These gels differ from permanently covalently cross-linked polymer gels because the cross-linking can be reversed by the input of energy, for example, by heating.[6] LMWGs have been around for many years but are receiving considerable current interest.[4] They are also used in industrial products,[7] although this seems rarely discussed in the academic literature.

In addition to the industrial applications, many recent advances and uses are being described.8, 9, 10, 11, 12 The specific self-assembly leading to gel formation can be exploited. For example, the fiber formation is a result of molecular stacking, meaning that the self-assembly leads to aggregates that can be suitable for optoelectronic applications.13, 14 The ready reversibility of gelation can be exploited, for example, to release cells from gels on demand in a manner that does not lead to cell death.[15] Ready gel formation by a simple trigger can also be used to allow easy and efficient gel loading.16, 17, 18 Unsurprisingly, therefore, there is significant interest in these materials (Figure 1).

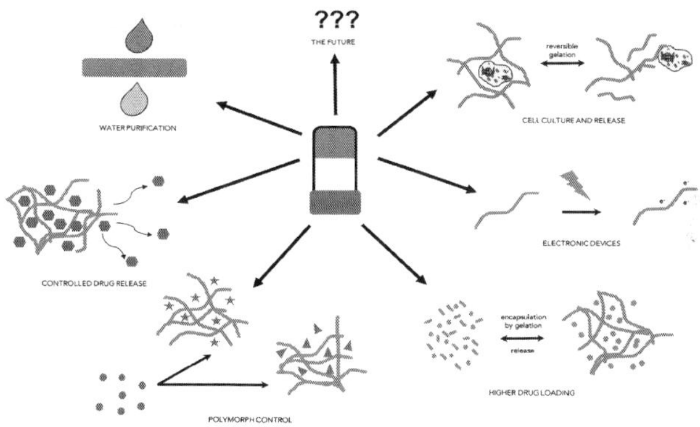

Figure 1. Uses of Low-Molecular-Weight Gels

Cartoon showing some the potential uses of low-molecular-weight gelators, including cell culture and differentiation with non-harmful release from the gelled material, photo responsive semiconducting gel fibers, high loading of a drug particle with a targeted release, slow and controlled drug release upon addition of a stimulus, water purification by the removal of heavy metals, and the control of polymorph by changing the gel network.

This is a fascinating area and in many ways holds attention because of the difficulties in probing and understanding the gels. The gels arise from assembly across many length scales, and understanding all of these is difficult. At the molecular level, the molecules must interact in a manner that leads to the formation of suitable aggregates that can eventually entangle. Thus, one-dimensional growth must be favored. From this perspective, it is very frustrating that it is often extremely difficult to predict whether a molecule will form a gel or not; indeed, gelation has been described as an empirical science.[4] Structurally similar small molecules can exhibit extremely different propensity to form gels. A number of methods have been attempted to overcome this.19, 20 A number of people have attempted to link gelation to crystal structures.21, 22, 23 In many ways, this seems surprising. It is not at all clear that there is (or even that there has to be) a link between the interactions that lead to crystallization in some situations and gelation in others. Indeed, although it has been widely assumed that there is a link, there are a number of examples that now specifically show that packing is different in the gel phase and crystals grown even from the same solvent.24, 25, 26 As a specific example, we showed that fiber X-ray diffraction data from the gel phase did not match the crystal structure of a crystal grown directly from the gel phase.[24] In many cases, the crystals formed in one solvent are compared with a gel formed in another. Here, it is especially difficult to see why there has to necessarily be a link, considering how many properties change on replacing one with another. Even where some crystallinity has been implied in the gel phase,[27] it is not clear how these data can be shown to not arise from simply some (possibly a very small amount of) crystallization within an amorphous gel phase. There is debate here, but to our minds there is no reason to necessarily assume that there is a link between crystallization and gelation.

Other methods that have been used to tackle this inability to predict gelation include using libraries and computational approaches. The library approach simply involves generating a large number of similar molecules and then determining which form gels.28, 29, 30 This brute force approach is often successful in finding gelators, but it is less effective in explaining why some molecules can form gels and others cannot. Presumably, this is because it is difficult to vary only one parameter at a time. For example, changing one functional group to another will result in a change in the steric bulk of the molecule, the ability to pack, possibly the number of hydrogen bond donors or acceptors, the absolute solubility in a particular solvent, etc. Hence, simple lists of gelators and their efficiency often do not capture the complexity of the system. Recent advances in computational approaches have been successful in predicting gelators. Tuttle's group has effectively predicted tripeptide-based gelators,[31] and we have recently been successful in generating a descriptor-based approach that can be used to predict new gelators.[32] However, in both cases the methods again do not explain why some molecules form gels and others do not.

At this point, the simple statement that a specific molecule can form gels is simplistic at best. This statement holds for the molecule in certain solvents, at certain concentrations and temperatures, potentially in the presence of certain additives; most critically, if the same process has been used as the authors describe in the work. Taking these in order, each molecule will form gels in a specific range of solvents (which might be only one!). Clearly this is simply a result of the interactions between molecule and solvent balancing in certain solvents to favor self-assembly into the required one-dimensional aggregates. Recent key work here has shown that this can often be predicted from the properties of the solvent, for example the Hansen solubility parameters (Figure 2A).20, 33, 34 **Second, each** gelator forms gels only above the so-called minimum gelation concentration (mgc). For LMWGs, this tends to be low (typically <1 wt %), but it is therefore possible that when a molecule is stated to not form a gel in a specific solvent, it could simply be that the mgc has not been reached. Also, although it is often assumed that the molecular packing and supramolecular aggregates are the same above and below the mgc, this is rarely proven. Temperature is a key parameter; most LMWGs melt and the range of temperatures over which the gel is stable is variable and again hard to predict and understand. Additives are known to both hinder and promote gelation,35, 36 **often again without a detailed understanding. In some cases, it has been shown that salts can affect aggregation on the basis of the** Hofmeister series,[37] **and certain polymer additives have been shown to affect gelation by specific interactions (Figure 2**B).[38]

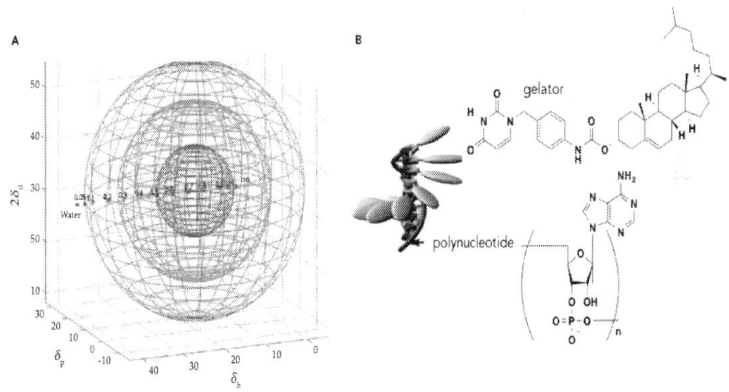

Figure 2. Understanding Gelation

(A) Plot of solubility data for a specific molecule in Hansen space; blue shows where the molecule is soluble, green shows where a gel is formed, and red shows where the molecule is insoluble. Adapted with permission from Yan et al.[39] Copyright 2013 American Chemical Society.

(B) Schematic illustration of a polynucleotide interacting with a gel fiber. Adapted from Numata et al.[40] with permission of the Royal Society of Chemistry.

An interesting issue is the process of gelation (or the formulation used to prepare the gel). In our opinion, a simple statement that X gels in solvent Y only holds for the caveat "using the methods described" unless shown otherwise. How one carries out the gelation can lead to very different materials, as well as examples where a molecule can go from being an effective gelator to being ineffective. We return to this below.

Returning to the issue of understanding gels across length scales, assuming the molecules assemble into fibers, the mechanical properties of individual fibers have been shown to be controlled by intermolecular interactions, and so are related to the chemical structure of the gelator.[41] Once fibers have been formed, these can then conceptually interact in a number of different ways. Lateral association is possible, leading to bundling and fiber thickening. Entanglement is also possible. There are also examples where fiber branching has been shown to occur. At the next length scale, the network that is formed from the fibers can be homogeneous or heterogeneous. Finally, the final (often overlooked) parameter is aging. A number of reports of how gels can change in different ways have been reported,42, 43, 44 but it is extremely common for the time at which the gels were analyzed not to be mentioned. Some interesting observations include gel-to-gel transitions,44, 45 **syneresis,**46, 47 **and transient network formation.**[48]

Hence, to attempt to understand gelation, a wide range of skills and techniques need to be used (Figure 3). To understand the molecular packing, techniques such as nuclear magnetic resonance (NMR) spectroscopy, infrared spectroscopy, circular dichroism, fluorescence, and X-ray diffraction are used.49, 50 **Although all of these can be informative, there are always caveats. Circular** dichroism, for example, is very sensitive to concentration, and so good-quality data can often only be collected at concentrations lower than the mgc. There are therefore questions as to whether the packing in aggregates below the mgc is the same as at and above the mgc. Similarly, fluorescence cannot be easily collected on turbid samples, and again higher concentrations makes this difficult to collect because of quenching, and X-ray diffraction makes the assumption that the diffraction (if any is observed) is from the gel phase as opposed to crystalline impurities. Nonetheless, such techniques can be hugely informative. For example, NMR spectroscopy can be used to both probe the rate of assembly and to infer information about the molecular interactions leading to assembly. Likewise, infrared spectroscopy can be used to show that specific hydrogen bonding is occurring on assembly.

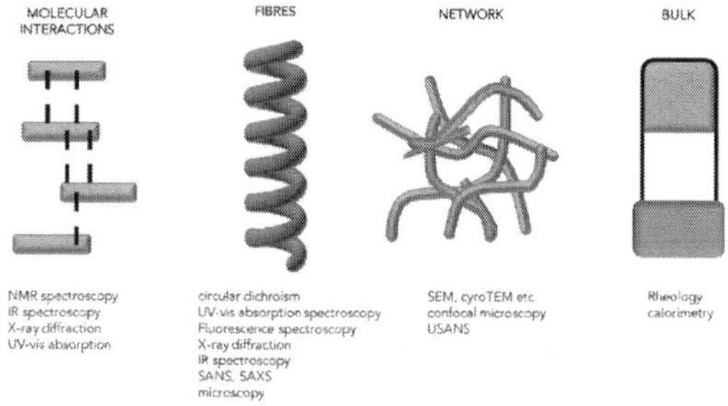

Figure 3. Hierarchical Assembly across Length Scales

Low-molecular-weight gels form as a result of assembly across a range of length scales; different techniques are appropriate for analyzing the structures formed at each length scale.

At the next level of hierarchy, the nature of the fibers is usually probed by microscopy or by small-angle scattering. Any electron microscopy image can only represent a tiny fraction of the sample. Also, unless cryo-transmission electron microscopy (TEM) is used, there could be drying artifacts in the sample preparation. However, in general, microscopy most often shows the presence of long fibers, commonly entangled or branched. In the gel phase, this network is of course be in three dimensions, but collapses to two dimensions when examined by microscopy. Three-dimensional imaging can be captured by confocal microscopy. Although there are significant advances in this area, most examples have relatively low resolution. Nonetheless, confocal imaging can provide useful information on the microstructure and distribution of the fibers in space. Small-angle X-ray or neutron scattering are hugely powerful techniques for imaging gels; the imaging can be done *in situ*.[51] **Although access to a** beam line at a facility is usually required, the quality of more accessible lab-based X-ray equipment is improving constantly to allow access to good-quality data. The scattering data of course have to be fitted to a model, but the advantages are that these are bulk measurements, representing the sample as a whole.

Finally, the mechanical properties of the gels can be measured by techniques such as rheology.[52] A range of different measurements are possible and are able to inform on the gel networks. A typical LMWG will have a storage and a loss modulus (G′ and G″, respectively), which are frequency independent, and will break at relatively low strain. It is surprising how often simple vial inversion is used as a means of proving gelation, despite the concern that this does not necessarily demonstrate that a gel has been formed; viscous liquids can also be stable to inversion, albeit for a relatively short time.[53] It is also apparent to the careful reader that some of the inverted samples seem to be in the process of flowing. From the rheological data, different behavior is often shown, which implies that there are different underlying networks, although significantly more work needs to be done here to understand this.

Hence, at the moment, we believe that these soft materials are hugely interesting, but there are still a significant number of areas where there is a lack of understanding. One reason for this is that there are few reports that truly examine a system across all length scales. A highly readable account of a single gelator system has been prepared by Menger and Caran;[54] the length of this single report shows why perhaps such in-depth studies are not common in today›s publishing regime.

Here, we now wish to focus on what we believe are two real opportunities for the future: controlling the process and mixing LMWGs. To some degree, these have been covered in previous reviews,55, 56, 57, 58, 59, 60 but as the field is developing rapidly we believe that a further critical analysis is warranted. Both of these areas also allow us to further discuss the assembly process. A typical cartoon of gelation is shown in Figure 4; it is often assumed that the molecules are initially freely dissolved. On applying a trigger, the molecules begin to assemble and then form fibers, which entangle to form the network. To our current thinking, the veracity of such a cartoon is unclear, and it is worth considering when probing these soft materials.

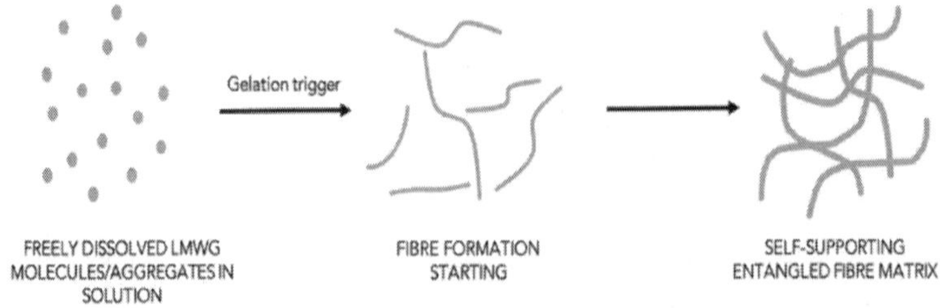

Figure 4. How Gelation Is Thought to Occur

"Standard" cartoon of gelation.

Process of Assembly

Taking these in turn, we touched on the process of gelation above. Essentially here, we are stating that how the gel is formed matters. To some degree, this perhaps seems obvious. A gel is formed by solubilizing or dispersing a molecule, and then triggering the gel formation. It is worth also considering that each solid form of the gelator might not be the same, with different polymorphs potentially having different solubility, etc.[61] It is necessary to initially have the molecule in a more solvated state than at the end. This can be achieved by heating to improve solubility, initially dissolving in a good solvent before adding an anti-solvent, choosing a pH at which the molecule is soluble and then adjusting the pH in one of a number of different ways to render the molecule less soluble,[62] adding a salt to charge screen,[63] using a pro-gelator with a cleavable, solubilizing group,[64] or using a chemical or enzymatic reaction to generate the gelator in situ from suitable precursors.65, 66 To the best of our knowledge, there are no examples where a molecule is simply added to a solvent and gelation results (although there are many examples where "instant" gelation can be induced by mixing two components that react or interact to form a gelator on contact).[67] Because of this, how the gelation is carried out matters. Essentially, in many cases, the rate of gelation becomes competitive with the rate of mixing, and so the gel›s properties are affected by how the gelation occurs. To some degree, this is caught up in the kinetically driven assembly; because at the point of gelation one is passing across a regime of super-saturation, the system is not at equilibrium, and hence error correction is not possible. We stress, however, that we do not necessarily mean that the local assembly of the molecule is affected (although it could be), but rather that the network is affected, resulting in different gel properties.

Taking what might initially seem to be the simplest method—heating and cooling—heating may or may not fully solubilize the molecule; it might be possible for dispersed aggregates to be formed instead. The absolute temperature and absolute duration of heating will both be critical. Finally, the rate of cooling, the final temperature, and the time after cooling could all be critical to whether a gel is formed and the properties of the gel itself. Surprisingly, these parameters are often not varied or not reported. However, to our minds, it is the properties of the gels that are often most important. Pragmatically, for many applications, it perhaps does not actually matter how or why the molecules assemble into fibers; instead, it is most important what properties the entangled networks provide to the system. It is perhaps reductionist to suggest, but if the gels reproducibly have the "right" properties, how they are formed is perhaps less important.With this in mind, rather than approach these gels from the perspective of trying to understand how the molecules interact, or designing a library where some molecules gel and some do not, a more effective approach might be to find extremely robust gelators. Such robust gelators would ideally form gels in anyone's hands and preferably in a wide range of conditions. These could then be used to provide gels with a range of different properties by varying how the gels are formed.

Hence, here, it is not appropriate to simply list gelators and their properties. There are many reviews that already provide this information.1, 2, 3, 5, 68 Rather here we mainly focus on one example where the process of gelation has been used to lead to significantly different outcomes. This example is hydrogelators; this could be one reason why sufficient data exist on the formation of gels under different conditions because, compared with organic solvents, water as a solvent opens up many additional parameters, including variations in ionic strength, pH, and the addition of background salts.

We focus mainly on a N-protected dipeptide. Such protected dipeptides are widely used hydrogelators, such that a significant number of groups have focused on this class of LMWG.69, 70 Naphthalene-protected dipeptides were first reported by Yang et al.[71] We have used this class of material and found that they can be very robust gelators. For example, 2NapFF can form hydrogels by using a range of triggers and is an excellent example of a complicated system that exemplifies how the process of assembly matters (Figure 5).[72]

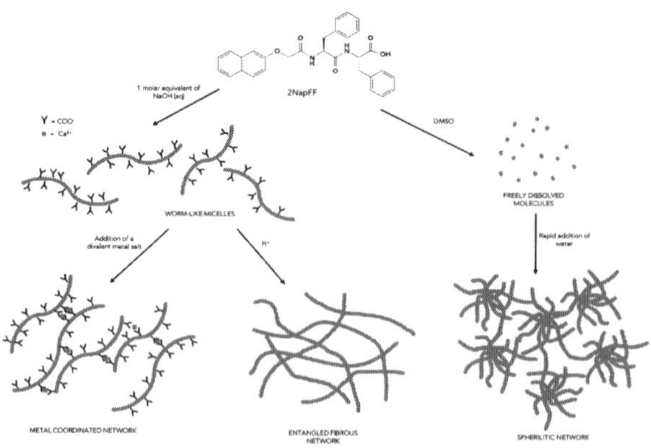

Figure 5. The Process of Assembly Affects the Outcome

The low-molecular-weight gel 2NapFF (at high pH, above the pK_a, the carboxylate will be formed instead) can form a gel in different ways, leading to different types of networks.

First, this LMWG can be dissolved in an organic solvent such as DMSO. Addition of water results in a gel. This seems very simple, but there are many parameters that can be varied. First, the concentration of 2NapFF can be changed. Gels form down to a concentration as low as 1 mg/mL, with more turbid gels forming at 10 mg/mL (the highest concentration tested). As for other related gelators,[73] this method of self-assembly proceeds by the addition of water immediately resulting in a highly turbid solution. This clarifies over a period of minutes as the gel forms. Elsewhere, for a related gelator, we have linked this process to the initial formation of micrometer-sized spherical droplets,[74] which then decrease in concentration as the self-assembled fibers are formed. We have no evidence that the spherical objects are directly transformed to the fibers, although it has been suggested that fibers nucleate at the surface of the spheres for Fmoc-diphenylalanine.[75] We also note that Orbach et al.[76] have another interpretation for this change in turbidity for a related gelator. This phase-separation event implies that how the process is carried out will likely determine the outcome of the self-assembly and the gel properties. Again, for related gelators, we have shown that the choice of organic solvent in which the gelator is initially dissolved affects the outcome,[77] which is unsurprising as the solubility of the gelator will be different in each solvent, and different mixing rates of the solvents with water among other variables. Elsewhere, it has been stated that fresh solutions of the gelator in an organic solvent are always prepared to ensure complete dissolution of the LMWG,[73] but this is never actually proven as far as we can see. Mixing of DMSO and water is an exothermic process, which leads to the mixture heating slightly. Changing the ratio of DMSO to water used results in differences in the temperature increase.

Gels can also be formed from 2NapFF in other ways. Solutions of 2NapFF can be prepared at high pH (typically pH > 9) by addition of a base to deprotonate the carboxylic acid. Gels can then be prepared by the addition of an acid to re-protonate the carboxylic acid. Alternatively, a divalent cation can be added to form a gel at high pH; this is conceptually due to the cross-linking of biopolymers.

In both of these cases, a really important aspect is that the 2NapFF in water at high pH is effectively a surfactant.[78] The preparation of such solutions at high pH as the means of effectively solubilizing the gelator is common, but it is less common to address this surfactancy. It is more common for the schematic to imply that the LMWG is molecularly dissolved. This is not the case. Instead, a range of colloidal aggregates can be formed by this class of LMWG at high pH. These include spherical structures and worm-like micelles. The aggregate formed will depend on the absolute concentration of the gelator, as well as the pH, addition of salts, etc. Thus, there is a phase diagram at high pH of structures. For 2NapFF, we have shown that there is a critical micelle concentration (cmc) at very low concentration, where spherical structures are formed, and then a second cmc at a higher concentration where there is a transition from spherical micelles to worm-like micelles.[78] As the concentration is increased further, the worm-like micelles aggregate to form liquid crystalline phases.

Because of this phase diagram, there is clearly an effect of concentration of the gels of 2NapFF. When a solution of calcium chloride is added to cross-link the 2NapFF at high pH, gels are only formed essentially at concentrations where worm-like micelles are present.[78] Because this cross-linking requires the presence of deprotonated carboxylates on the gelator, the pH is critical. Similarly, the nature of the salt added is important. Weak materials are formed when monovalent cations are added (for example, sodium chloride);[79] in analogy with polymer systems, this is probably simply a charge screening effect, meaning the worm-like micelles do not repel each other as strongly. Changing the divalent cation affects the gel›s mechanical properties, as does changing the counterion for a fixed cation. These data can be linked to the Hofmeister series.

Finally, gels can be formed by lowering the pH of solutions of 2NapFF initially formed at high pH. This again sounds an easy process, but the mechanical properties of the gels and the homogeneity of the gels depend on how this pH drop is achieved. There are many ways that the pH can be decreased described in the literature for a range of low-molecular-weight gelators.62, 80, 81, 82, 83 For example, mineral acids such as HCl can be added most simply as an aqueous solution, although for hydrophobic gelators, this often means that gelation and mixing compete strongly so that it is difficult to prepare reproducible gels.80, 84 It is also possible to rely on diffusion of gaseous acid into a sample, although in our hands we have found this difficult to control such that a specific absolute pH can be targeted. We instead tend to use glucono-δ-lactone (GdL), which dissolves quickly in water and hydrolyzes to gluconic acid to allow a slow, reproducible pH decrease.[80] Likewise, addition of glucose, hydrogen peroxide, and an enzyme can be used to produce gluconic acid in situ.[85] It is also possible to add other precursors that hydrolyze to give acids.[81] In all of this, reproducibility is really important; to our minds, there is little point in being able to form a gel if that gel does not have the same properties each time it is made.

For 2NapFF, with the use of GdL to lower the pH, the gels are again concentration dependent, and their properties will also depend on the final pH.[72] The apparent pK_a of the terminal carboxylic acid of 2NapFF is around 6.0.[86] This is perhaps surprisingly high for the terminal carboxylic acid of a dipeptide, but this is a common observation for this kind of gelator.86, 87, 88 In fact, the apparent pK_a depends on the hydrophobicity of the gelator, the concentration in solution, and the solution temperature, and should probably be interpreted as the apparent pK_a of the self-assembled aggregate. In analogy with other systems, close packing of carboxylates and carboxylic acids results in an increase in the pK_a. For clarity, we always try to refer to this as the "apparent pK_a," because the data are accessed by a slow titration with acid; this always leads to gel formation to some degree so is not a true titration because there is no equilibrium between the species in solution.

At pH values below the apparent pK_a, gels are formed from 2NapFF solutions. The gel strength increases as the pH is decreased further. This can be interpreted as entanglement and fiber formation as sufficient charge is removed from the self-assembled structures. The gel strength increases as the charge is removed further until the rheological properties stabilize. For related systems, we have shown that it is possible for the gel network to start to contract when sufficient charge is removed, leading to syneresis of the gel; this is not observed for 2NapFF.[89]

It is unclear whether the worm-like micelles at high pH simply protonate as the pH is decreased to form the gel or whether there is a structural re-organization. Certainly, the small-angle neutron scattering (SANS) data show that the worm-like micelles at high pH have a hollow core. On decreasing the pH, the core is lost, and the scattering data can best be fit to an elliptical cylinder.[72]

This could be interpreted as the structures re-organizing, or it is also possible that the initial structures fall apart and then new structures form. One interesting observation is that the gels formed at high pH by addition of a calcium salt decrease in gel strength dramatically when the pH is decreased, and then become stronger again as the pH drops below the pK_a.[79] This implies at the very least that there is a rare gel-to-gel transition as opposed to a gradual change from one type of gel to another.

In all of this, we have simply been referring mostly to a gel being formed. We initially stated that the process was important in controlling the gel properties. So, what do we mean by the properties?

There are a number of key aspects. First, we argue strongly that any described property must be reproducible, and we are constantly surprised by how infrequently error bars are included for many reported gel properties. Second, the properties that will be needed will of course be application specific. As such, a range of tests might be needed.

Most simply, all of the mechanical properties of a gel will be the result of the self-assembled network. Hence, it is expected that the type of aggregate formed by the self-assembly (fiber, hollow tube, or helical stack), the propensity to laterally associate, the number and type of cross-links, and the homogeneity of the system will all play a role, and it is often difficult to deconvolute the relative importance of any one of these parameters.

In the example of 2NapFF here, the absolute storage and loss moduli (G' and G'', respectively) can be determined by rheological measurements. There are always potential issues here in terms of how the samples are loaded on to the rheometer; we tend to prepare our gels in plastic cups in which the measurements can be carried out directly to avoid loading issues. Looking at the gels formed by the three methods, at a concentration of 2NapFF of 10 mg/mL, the absolute moduli are similar.

Hence, one might be tempted to say that the process does not matter. However, the strain sweeps clearly show that the gels break down at different absolute strain, with a different shape to the changes in G' and G'' (Figures 6A–6C) and the recovery after a high strain cycle is very different.

Thus, it is likely that were one to want to pass the gel through a syringe for example, the gels formed by one process would be much more effective than the others. Similarly, as mentioned above, the increase in the moduli on increasing the concentration of 2NapFF is not always as expected, and so it would be easier to tune the absolute moduli for gels formed by one method than for another.

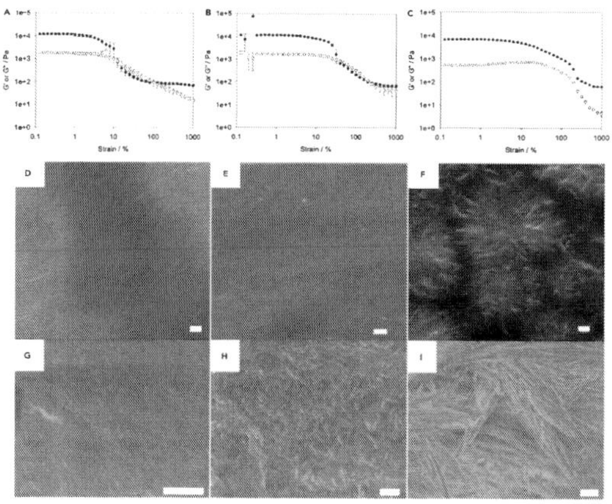

Figure 6. Effect of the Assembly Process on the Final Gel Properties

(A–C) Rheological strain sweeps of gels formed by (A) metal salt, (B) acid-triggered, and (C) solvent switch methods. G′ is the filled shapes and G″ is open shapes.

(D–F) Confocal microscopy images of gels formed by (D) metal salts, (E) acid-triggered, and (F) solvent switch methods. Scale bar represents 5 μm.

(G–I) SEM images of (G) metal salts, (H) acid-triggered, and (I) solvent switch methods.

Scale bar represents 1 μm. Adapted from Colquhoun et al.[72] with permission of the Royal Society of Chemistry.

The effect of pH and concentration can be seen directly in the rheological data. For this kind of gelator, it is generally expected that increasing the concentration will lead to the formation of more self-assembled structures and hence most likely more cross-linking. Hence, increasing the concentration of LMWG should lead to an increase in the rheological properties. At pH 10.5, this is what we have observed on gelation by adding a calcium salt. In fact, the storage modulus (G′) scales with concentration (c) as $G' \propto c^{2.4}$, which is perhaps indicative of a semi-flexible polymer network (if we assume that specific theories hold for this type of gelator).[90] However, at higher pH, G′ likewise scales with concentration in the same manner, but critically only until a certain concentration is reached (4 mg/mL at pH 12). At this point, the gels actually become weaker as the concentration of the 2NapFF is increased.[72] This counter-intuitive observation can be explained in line with the observation of liquid crystalline phases, which presumably result from lateral aggregation of the worm-like micelles at these higher concentrations. As more 2NapFF is added, the lateral association increases, rather than there simply being more and more worm-like micelles present that can entangle. This means that the expected model of gelation is not followed. These data correlate with our work elsewhere on determining the pore sizes of gel networks.[91] The pore size at pH 12 was found to be higher at a concentration of 2NapFF of 10 mg/mL as opposed to at 5 mg/mL.

Because these properties are different, it is interesting to consider why this is the case. The SANS data were different for the gels formed by the three processes, indicating different networks. The fibers imaged by TEM and scanning electron microscopy (SEM) were not significantly different, and so on this length scale, the one-dimensional structures are similar. However, the microstructure, imaged by confocal microscopy, was different in each case. Spherulitic fiber domains were found in the sample formed by dilution of DMSO solution with water. This agrees with our work on other systems gelled by this approach and seems to arise from a nucleation and growth process.[74] The gels formed by addition of calcium salts to a solution of 2NapFF at high pH show significant aligned domains, whereas a more typical cross-linked network is found for the acid-triggered sample. Hence, we can link the rheological properties to the network.

We have focused here on one particular LMWG. However, it is clear that similar observations can be found with other examples. Greenfield et al.[92] have shown that gels with different properties can be formed from the same peptide amphiphile. As with the 2NapFF above, these can be gelled by the addition of a calcium salt to a solution of the LMWG in water or by the addition of acid. For a specific peptide amphiphile, the rheological data for gels formed by the two methods are different, most obviously in the breakage under strain, as well as the relationship between the G′ and concentration. Here, the exponent x in the relationship $G' \propto c^{x}$ is very different, with x being 1.51 for the acid-triggered gels and 2.14 for the calcium-triggered gels, implying a very different underlying network for the gel.

Further interesting results that show how important the process is come from data on heating and cooling. Heating and cooling a solution of a peptide amphiphile results in very different properties, even though the final solution is at the same temperature as the initial solution. For example, the solution viscosity increased significantly.[93] The gel that was formed on addition of a calcium salt was also significantly stiffer than when the salt was added to a solution that had not been heated and cooled before the addition of the salt. The increased viscosity also allowed "noodles" to be drawn out. This was explained by the heating and cooling leading to the formation of aligned bundles of fibers, and so the fibers after heating have a higher diameter than before (Figure 7). These observations show how critical the process is; in many self-assembled, supramolecular systems (especially micellar aggregates as these peptide amphiphiles are), it is often assumed that the state is governed by the energy of the system. Hence, if the temperature is the same (which it is after the cooling cycle), then the structures formed should be the same. However, here, the authors speculate that water molecules associated with the supramolecular structures become bulk water during the heating cycle, leading to dehydration and fusion of the fibers. This is irreversible. Interestingly, we have very recently shown that heating can have a significant effect on the viscosity and extensional viscosity of solutions of 2NapFF.[94] Comparing these systems with the peptide amphiphiles, reduced diameter fibers were formed rather than those with increased diameters, showing that similar macroscopic observations can arise for different reasons.

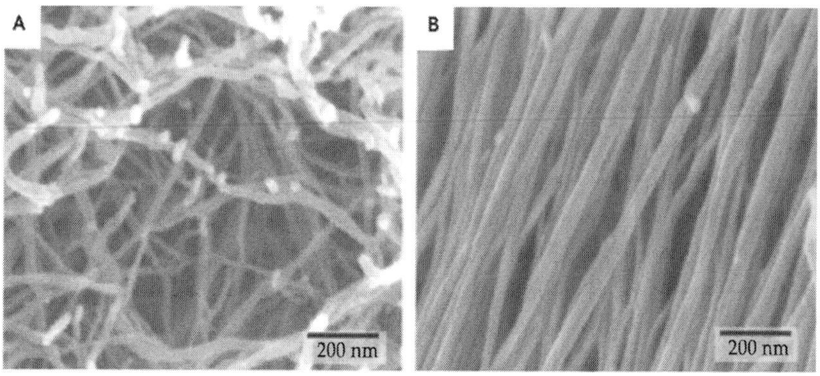

Figure 7. A Heat-Cool Cycle Affects the Solution Properties

(A) Gels formed by adding a calcium salt to a solution of a peptide amphiphile **are formed from an isotropic network of fibers.**

(B) Gels formed by adding the salt to a solution after a heat-cool cycle contain thicker structures that can be aligned by shear.

Adapted with permission from Zhang et al.[93] Copyright 2010 Nature Publishing Group.

This is all important as some applications require specific mechanical properties in addition to (or perhaps rather than) a specific chemistry of the fibers. For example, the ability to pass a gel through a needle and have quick recoverability has been linked to the gel's microstructure.[95] Further changes in the properties of a peptide amphiphile system have been reported recently,[57] where different lengths of self-assembled fibers were prepared by varying the assembly process via dilution and annealing steps. Pre-myoblasts were grown in media containing either short or long fibers, with significantly higher cell death observed in the presence of the shorter fibers. This work clearly shows how an energy landscape can be navigated for a self-assembled system and how sensitive these peptide amphiphiles can be to the process of assembly.

Multicomponent Systems

The other area where we believe that there are significant opportunities for LMWGs is in multicomponent systems. Buerkle and Rowan[59] have defined three classes of multicomponent systems: (1) a two-component gel phase, where both components are needed to form a gel; (2) a two-gelator system, where both components can independently gel; (3) a system comprising a gelator and a non-gelling additive. Here, we specifically focus on system (2), which contain two LMWGs, each of which can individually form a gel. Hence, on mixing there are multiple permutations (Figure 8). First, the system could be designed such that there are strong interactions between the two gelators, leading to self-assembled structures where there are alternating gelators. Alternatively, mixing of LMWGs could still occur, but non-specifically, leading to randomly mixed self-assembled structures. Finally, self-sorting could occur, where the two LWMGs prefer to self-assemble with themselves, leading to self-assembled structures that contain only one of the LMWGs.

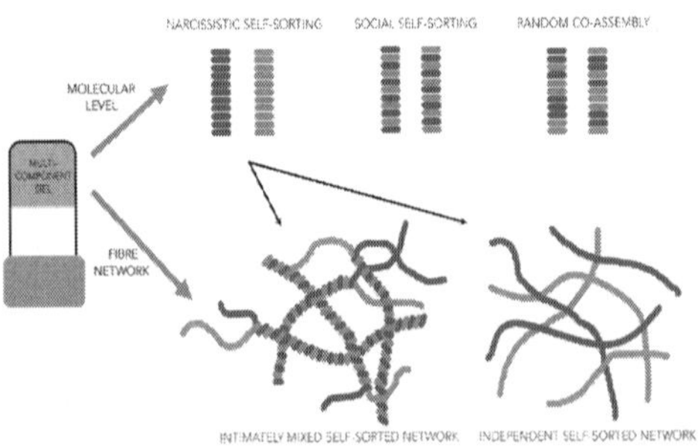

Figure 8. Multicomponent Systems

When both molecules can independently form fibers, mixed systems can lead to different possibilities on a molecular level (top). The properties of the gel will be controlled by this level of assembly and the next hierarchical level of assembly, where (for example) self-sorted fibers could heavily entangle or form an interpenetrating network (bottom).

There are now many examples of mixed LMWGs.56, 59 To our minds, there are two reasons why one might be interested in multicomponent systems. First, intellectually it is an interesting challenge to be able to finely control the assembly of a system to this degree. Second, multicomponent systems ought to be able to provide gels with a higher degree of information content than single-component systems; there are examples where multicomponent systems have been used to adjust the photoresponse of a gel for example, or to provide a gel that can be positively or negatively patterned by adjusting the assembly of only one of the components post-gelation.

We and others have previously reviewed the area of multicomponent gelators.56, 58, 59 We do not simply wish to re-iterate what has been previously said. Rather, we wish to highlight specific aspects. The first key aspect is the idea of design. Because LMWGs are inherently difficult to design, how does one design a multicomponent system? Generating a system with alternating molecules within the self-assembled fiber has been achieved by building in very specific interactions between the molecules. Assuming that there is still a tendency to grow fibrous structures overall, this dimerization leads to fibers with specifically associated gelators.

Generating randomly mixed fibers requires the formation of a system where the two gelators do not mind interacting and then controlling the gelation process in such a manner that assembly occurs in a manner where this mixing is possible. On the other hand, preparing a self-sorted system requires that the interactions between specific gelators are such that they would prefer to assemble only with themselves. Hence, in a mixed system, each fiber only contains a single gelator. This is of course easier to state than to achieve, especially in advance. As stated above, designing gelators from scratch is difficult; it is even more difficult to predict in advance what will occur in a mixed system.

As above, the outcome will be very process dependent. There are a number of examples of self-sorted systems where the gelation is driven by temperature. The self-sorting is brought about by the different gelation temperatures, where generally the two different gelators assemble into fibers at sufficiently different temperatures that the self-sorting can occur. Conceptually, if the molecular design is sufficient, it should be possible to drive the formation of two networks at the same temperature. Of course, this still requires that there is a strong tendency for the two molecules to only want to interact with themselves. This is usually built in by ensuring that the molecular packing is different. For example, Sugiyasu et **al.**[96] **mixed a** perylene bisimide-based gelator and an oligothiophene-based gelator to ensure that temperature-driven self-sorting occurred. This example shows the general rules: that the molecules need to be structurally different, and they need to have a sufficiently different solubility in the solvent to ensure assembly occurs at different temperatures. There could be significant opportunities here in terms of driving varying degrees of self-sorting by controlling the rate of cooling for example.

Controlling the gelation temperature is not always easy to predict. We have recently shown that another approach using the apparent pK_a is possible for hydrogels. As mentioned above, for gelators such as 2NapFF, where the gelation can be controlled by the protonation of the terminal carboxylic acid, we and others have observed that the apparent pK_a of this carboxylic acid is affected by the hydrophobicity of the gelator; the higher the hydrophobicity, the higher the apparent pK_a.[87] Hence, if two gelators with sufficiently different pK_a are mixed and the pH can be decreased in a controlled manner, self-sorting should be favored. Using the hydrolysis of GdL to slowly lower the pH allowed us to produce self-sorted systems. We have shown that this approach can be applied to a number of mixtures provided that the apparent pK_a values are around a unit apart.97, 98 We have proven that self-sorting occurs by a number of techniques. Sequential assembly can be shown by solution-phase NMR. In many cases, at high pH the gelator is detectable by solution-phase NMR; although a self-assembled aggregate is formed, the equilibrium between the freely dissolved molecule and the aggregate is such that the gelator can be observed. On gelation, the molecule becomes undetectable by NMR. Hence, self-sorting can be shown by the sequential disappearance of the signals of each gelator from a series of NMR spectra over time as the pH is slowly decreasing. In the final gel, we used fiber X-ray diffraction to show that the data for the self-sorted gel are essentially an overlay of the expected data from the two individual gelators.[97] Here, we note that in many cases, self-sorting is difficult to prove by microscopy. Although there are some systems where the two individual gelators happen to form self-assembled structures with sufficiently different diameters that microscopy can be used to show that there are two distinct populations,99, 100 in other cases, the structures formed are too similar to distinguish.

We and others have reported a number of systems where self-sorting can be driven in this manner.97, 98, 100, 101, 102 It seems that the self-sorting is essentially pre-programmed into the system on the basis of the differences in the chemical structures. In our first study, we showed that gelation using a mineral acid resulted in a significantly less homogeneous gel, but the diffraction still showed that there was a propensity for self-sorting to occur. Hence, again we highlight how the process of assembly needs to be considered; the molecular assembly is heavily driven by the chemical structure, but the gel properties are controlled by how the gelation is triggered.

We have a small number of examples where self-sorting does not appear to occur. First, two N-protected dipeptides were found to assemble at the same time by NMR despite the differences in the apparent pK_a of the two gelators.[98] Circular dichroism data implied that the assembly was not simple self-sorting. Here, we believe that the two gelators are co-assembling, and our explanation is the relative similarity in molecular structure (only the terminal amino acid was different in this case). This again highlights the design rules above; for effective self-sorting, the molecules should be significantly different in structure.

However, we have also found an unusual case where the two gelators are molecularly distinct.[103] In this example, 2NapFF was mixed with a second N-protected dipeptide. In this (so far) unusual case, it appears that there is a degree of mixing at high pH, and this leads to a degree of mixing as the pH is decreased. Hence, instead of a well-behaved self-sorting system, the first fibrous network seems to incorporate both of the gelators: all of the 2NapFF and some of the second component. After this has completely assembled, the remaining 2NapVG assembles as a single component. The degree of mixing and co-assembly is apparently determined by the relative concentration of the two components.

Our previous data would imply that this mixture should form a self-sorted system. We believe that it does not for two reasons. First, 2NapFF forms worm-like micelles at high pH as mentioned above. The other examples that we have investigated tend not to do this, and we hypothesize that this persistence of structure at high pH results in hydrophobic incorporation of the second component within the micellar structures at high pH that persists as the pH is decreased. Hence, this can be thought of as being to do with the encapsulation of one gelator within the self-assembled aggregate of the second. Once the first has completely assembled, the second then acts as if it is independent and assembles alone. This highlights that simple assignment as either self-sorted or co-assembled might be too simplistic in some cases.

As in the discussion dealing with the process of gelation above, the assignment of the assembly type is only the first level of assembly hierarchy. The mechanical properties of the gels, as well as potentially properties such as optoelectronics in the case of suitably mixed gelators, will be determined not only by the primary fibrous structures but also by how these fibers are located in space. For example, self-sorting might occur such that the primary fibers each consist of only a single gelator. However, in terms of the cross-linked network, the fibers might laterally assemble or otherwise cross-link such that each fiber interacts with only fibers of the same gelator, with only fibers of the other gelator, or randomly. Hence, self-sorting at the molecular level does not necessarily mean that self-sorting will occur at the fiber level. Our data where we proved self-sorting by using fiber diffraction show that this sorting occurs on the molecular level. However, we do not see how this can show any information on the next level up; to our minds, the data are entirely consistent with either random co-assembly or the formation of an interpenetrating two-component network on the fiber level. We were, however, able to show that two independent networks exist by self-sorting two gelators where one gelator was photoresponsive.[104] On irradiation with UV light, this gelator underwent trans-to-cis isomerization. The cis isomer is incapable of forming gel fibers, so this network fell apart. From rheological data, we were able to show that the other network remained intact and had similar properties as if this network had been formed in a single-component system. In addition, the area exposed to irradiation was optically different (Figure 9A). This irradiation represents a negative writing; one component is removed from a two-component system. In related work, Cornwell et al.[101] showed how positive patterning could be carried out in a multicomponent system (Figure 9B).

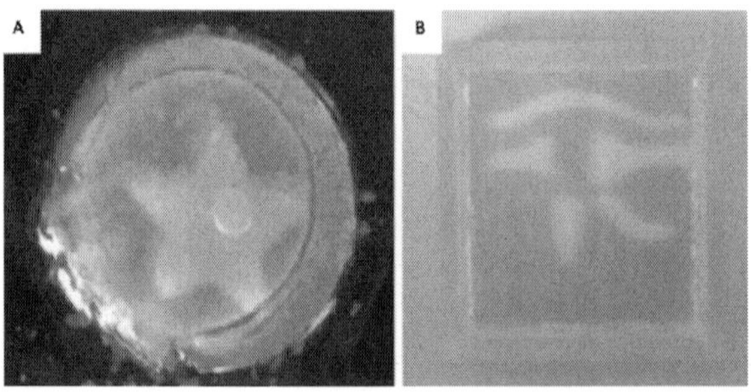

Figure 9. Writing in Multicomponent Systems

(A) Irradiation of a two-component system can be used to negatively "write" into a gel. Adapted with permission from Draper et al.[104] Copyright 2015 Nature Publishing Group.

(B) Positive "writing" is also possible. Adapted with permission from Cornwell et al.[101] Copyright 2015 American Chemical Society.

The importance of the assembly hierarchy for optoelectronic applications is critical. As mentioned above, Sugiyasu et al.[96] have shown that a self-sorted gelling system can be used to prepare the equivalent of p-n heterojunctions where the two types of fiber cross. Hence, to optimize such a system, one would like to know how many p-n heterojunctions there are (i.e., how many points at which the two fibers interact), how these are distributed in space, and how effective different types of interaction are. For example, one could easily imagine that a point where two fibers touch would be very different in efficiency to where two fibers wrap around each other (Figure 8).

Understanding the assembly on the fiber level is however a major challenge. There are almost no methods that are suitable for distinguishing between self-assembled fibers. As we noted above, in some rare cases, it is possible to differentiate between fibers by electron microscopy. However, this requires drying of the sample, and hence it is extremely difficult, to say the least, to interpret how these fibers were arranged in space before drying. Small-angle scattering experiments can potentially be used, but generally the length scales that can be probed mean that one can again show that (for example) there are two different fiber types present, but the information as to fiber interactions is usually interpreted in terms of power law scattering and hence again different to interpret. Here, it might be possible to use selective deuteration as a means of rendering the scattering from each network only detectable by SANS under different ratios of H_2O and D_2O.

A recent impressive insight into self-sorted gel networks was achieved by Onogi et al.[105] who used super-resolution microscopy to independently visualize two-gelator networks in a self-sorted system. Here, the gelators were mixed with fluorescent analogs that were designed to strongly associate with only one of the gelators. On mixing and gelation, impressively two independent fiber networks could be seen (Figure 10). This methodology could really open up the area, allowing an understanding of how the two sets of fibers interact.

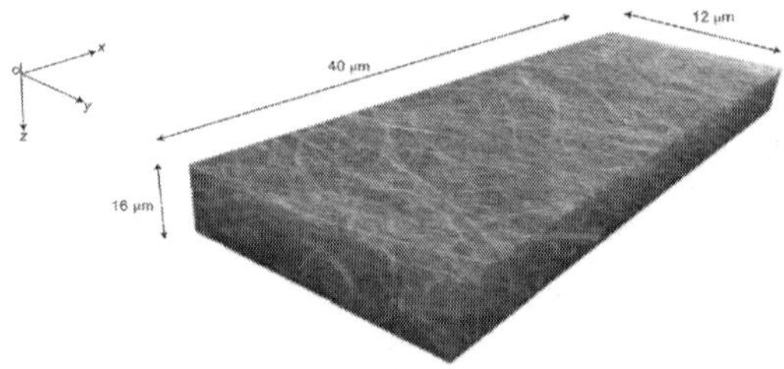

Figure 10. Imaging a Self-Sorted Gel

Three-dimensional confocal laser scanning microscopy image of a self-sorted two-component hydrogel network showing that the fibers are orthogonally assembled. Adapted with permission from Onogi et al.[105] Copyright 2016 Nature Publishing Group.

A final comment is that it is very unclear what the outcome will be on the rheological properties when two gelators are mixed in a multicomponent system. We have examples of self-sorted systems where the absolute rheological moduli are higher than one might expect when comparing with the individual gels; we also have examples where the moduli are around an average of the values of the two individual systems.[98] **When the relative ratios of two** gelators are mixed, we have also found that there can be a non-linear effect on the absolute moduli.[103] **Interestingly, Nagy-Smith et al.[106] have recently shown that mixing two** gelators that are simply the enantiomeric **isomers of each other results in the formation of a co-assembled system. The rheological data show that the co-assembled gel is much stiffer than those formed by either of the** enantiomerically pure gelators. This is due to different packing of the molecules, leading to local stiffening of the fibers. From this short discussion, it can be seen that multicomponent systems can be used to access a range of rheological properties that may not always be that expected from simple addition of the individual components, and further highlights how understanding the assembly across multiple length scales is important.

CONCLUSIONS AND OUTLOOK

We have set out here to critically analyze the current state of the art in low-molecular-weight gelators. There are increasingly more and more reports of such gelators, and the number of applications being described is ever increasing. This is a fascinating area of science, and we believe that there are many more opportunities here than have been currently explored. The idea of adjusting the process of assembly is essentially stating that formulation is key. This is widely applied in industry, where formulation science is a critical aspect. However, the importance of formulation is perhaps less commonly applied in academia labs. Although it is clearly interesting to discover new gelators, a pragmatic approach could very well be to instead wonder whether current gelators might be better able to address a specific need by simple variation of the assembly process. To our minds, the time for papers describing a specific new gelator and an in-depth study into its self-assembly leading to gelation has perhaps past without new insight into why gelation has occurred. Multicomponent systems also offer many potential opportunities. Here, as for the discussion regarding the gelation process, it is important to consider the assembly across multiple length scales, and it might be the case that there are occasions where simple assignment as one type of assembly or another might not be appropriate. Overall, we believe that there are significant opportunities here, but we highlight that some questions that need answers to enable a true understanding of the assembly across multiple length scales are extremely difficult to answer. This in itself provides new opportunities for new techniques to be developed and applied. We expect lots of further exciting developments in the near future and hope that this review stimulates debate and discussion.

ACKNOWLEDGMENTS

D.J.A. thanks the EPSRC for a fellowship (EP/L021978/1), which also funded E.R.D. We gratefully acknowledge the many conversations with other group members (past and present) who have helped crystallize the ideas and opinions expressed in this review.

REFERENCES

1. L.A. Estroff, A.D. HamiltonWater gelation by small organic molecules Chem. Rev., 104 (2004), pp. 1201-1218
2. P. Terech, R.G. WeissLow molecular mass gelators of organic liquids and the properties of their gels Chem. Rev., 97 (1997), pp. 3133-3160
3. X. Du, J. Zhou, J. Shi, B. XuSupramolecular hydrogelators and hydrogels: from soft matter to molecular biomaterials Chem. Rev., 115 (2015), pp. 13165-13307
4. R.G. WeissThe past, present, and future of molecular gels. what is the status of the field, and where is it going? J. Am. Chem. Soc., 136 (2014), pp. 7519-7530
5. N.M. Sangeetha, U. MaitraSupramolecular gels: functions and uses Chem. Soc. Rev., 34 (2005), pp. 821-836
6. C.D. Jones, J.W. SteedGels with sense: supramolecular materials that respond to heat, light and sound Chem. Soc. Rev., 45 (2016), pp. 6546-6596
7. B.O. Okesola, V.M.P. Vieira, D.J.Cornwell,N.K.Whitelaw,D.K.Smith1,3:2,4-Dibenzylidene d sorbitol (DBS) and its derivatives - efficient, versatile and industrially relevant low molecular weight gelators with over 100 years of history and a bright future Soft Matter, 11 (2015), pp. 4768-4787
8. A. Dawn, T. Shiraki, S. Haraguchi, S.-i. Tamaru, S. ShinkaiWhat kind of "soft materials" can we design from molecular gels? Chem. Asian J., 6 (2011), pp. 266-282
9. A.R. Hirst, B. Escuder, J.F. Miravet, D.K. SmithHigh-tech applications of self-assembling supramolecular nanostructured gel-phase materials: from regenerative medicine to electronic devices Angew. Chem. Int. Ed., 47 (2008), pp. 8002-8018
10. W.J. FrithSelf-assembly of small peptide amphiphiles, the structures formed and their applications. (A foods and home and personal care perspective) Philos. Trans. A Math. Phys. Eng. Sci., 374 (2016), p. 20150138
11. B.O. Okesola, D.K. SmithApplying low-molecular weight supramolecular gelators in an environmental setting - self-assembled gels as smart materials for pollutant removal Chem. Soc. Rev., 45 (2016), pp. 4226-4251
12. E.V.Alakpa,V.Jayawarna,A.Lampel, K.V. Burgess, C.C.West,S.C.J.

Bakker,S.Roy,N.Javid,S.Fleming, D.A.Lamprou,*et al.* Tunable supramolecular hydrogels for selection of lineage guiding metabolites in stem cell cultures Chem, 1 (2016), pp. 298-319

13. S.S. Babu, V.K. Praveen, A. Ajayaghosh Functional π-gelators and their applications Chem. Rev., 114 (2014), pp. 1973-2129
14. S. Ghosh, V.K. Praveen, A. Ajayaghosh The chemistry and applications of π-gels Annu. Rev. Mater. Res., 46 (2016), pp. 235-262
15. K.J. Skilling, F. Citossi, T.D. Bradshaw, M. Ashford, B. Kellam, M. MarlowInsights into low molecular mass organic gelators: a focus on drug delivery and tissue engineering applications Soft Matter, 10 (2014), pp. 237-256S. Eskandari, T. Guerin, I. Toth, R.J. Stephenson Recent advances in self-assembled peptides: implications for targeted drug delivery and vaccine engineering Adv. Drug Del. Rev., 110 (2017), pp. 169-187
16. S. Uzan, D. Barış, M. Çolak, H. Aydın, H. Hoşgören Organogels as novel carriers for dermal and topical drug delivery vehicles Tetrahedron, 72 (2016), pp. 7517-7525
17. M.D. Segarra-Maset, V.J. Nebot, J.F. Miravet, B. EscuderControl of molecular gelation by chemical stimuli Chem. Soc. Rev., 42 (2013), pp. 7086-7098
18. D.M. Zurcher, A.J. McNeilTools for identifying gelator scaffolds and solvents J. Org. Chem., 80 (2015), pp. 2473-2478
19. Y. Lan, M.G. Corradini, R.G. Weiss, S.R. Raghavan, M.A. RogersTo gel or not to gel: correlating molecular gelation with solvent parameters Chem. Soc. Rev., 44 (2015), pp. 6035-6058
20. P. DastidarSupramolecular gelling agents: can they be designed? Chem. Soc. Rev., 37 (2008), pp. 2699-2715
21. G.K. Veits, K.K. Carter, S.J. Cox, A.J. McNeilDeveloping a gel-based sensor using crystal morphology prediction J. Am. Chem. Soc., 138 (2016), pp. 12228-12233
22. A.D. Martin, J.P. Wojciechowski, M.M. Bhadbhade, P. ThordarsonA capped dipeptide which simultaneously exhibits gelation and crystallization behavior Langmuir, 32 (2016), pp. 2245-2250
23. K.A. Houton, K.L. Morris, L. Chen, M. Schmidtmann, J.T.A. Jones, L.C. Serpell, G.O. Lloyd, D.J. AdamsOn crystal versus fiber formation in dipeptide hydrogelator systems Langmuir, 28 (2012), pp. 9797-9806
24. E.R. Draper, K.L. Morris, M.A. Little, J. Raeburn, C. Colquhoun, E.R.

Cross, T.O. McDonald, L.C. Serpell, D.J. AdamsHydrogels formed from Fmoc amino acids CrystEngComm, 17 (2015), pp. 8047-8057

25. E.C. Barker, A.D. Martin, C.J. Garvey, C.Y. Goh, F. Jones, M. Mocerino, B.W. Skelton, M.I. Ogden, T. BeckerThermal annealing behaviour and gel to crystal transition of a low molecular weight hydrogelator Soft Matter, 13 (2017), pp. 1006-1011
26. E. Ostuni, P. Kamaras, R.G. WeissNovel X-ray method for in situ determination of gelator strand structure: polymorphism of cholesteryl anthraquinone-2-carboxylate Angew. Chem. Int. Ed., 35 (1996), pp. 1324-1326
27. S. Awhida, E.R. Draper, T.O. McDonald, D.J. AdamsProbing gelation ability for a library of dipeptide gelators J. Colloid Interface Sci., 455 (2015), pp. 24-31
28. D.R. Trivedi, P. DastidarInstant gelation of various organic fluids including petrol at room temperature by a new class of supramolecular gelators Chem. Mater., 18 (2006), pp. 1470-1478
29. K.K. Carter, H.B. Rycenga, A.J. McNeillImproving Hg-triggered gelation via structural modifications Langmuir, 30 (2014), pp. 3522-3527
30. W.J.M. FrederixPim, G.G. Scott, Y.M. Abul-Haija, D. Kalafatovic, C.G. Pappas, N. Javid, N.T. Hunt, R.V. Ulijn, T. TuttleExploring the sequence space for (tri-)peptide self-assembly to design and discover new hydrogels Nat. Chem., 7 (2015), pp. 30-37
31. J.K. Gupta, D.J. Adams, N.G. BerryWill it gel? Successful computational prediction of peptide gelators using physicochemical properties and molecular fingerprints Chem. Sci., 7 (2016), pp. 4713-4719
32. S. Wu, J. Gao, T.J. Emge, M.A. RogersInfluence of solvent on the supramolecular architectures in molecular gels Soft Matter, 9 (2013), pp. 5942-5950
33. M. Raynal, L. BouteillerOrganogel formation rationalized by Hansen solubility parameters Chem. Commun., 47 (2011), pp. 8271-8273
34. J.L. Li, X.Y. Liu, C.S. Strom, J.Y. XiongEngineering of small molecule organogels by design of the nanometer structure of fiber networks Adv. Mater., 18 (2006), pp. 2574-2578
35. J.-L. Li, X.-Y. LiuArchitecture of supramolecular soft functional materials: from understanding to micro-/nanoscale engineering Adv. Funct. Mater., 20 (2010), pp. 3196-3216

36. G.O. Lloyd, J.W. SteedAnion-tuning of supramolecular gel properties Nat. Chem., 1 (2009), pp. 437-442
37. Y.J. Adhia, T.H. Schloemer, M.T. Perez, A.J. McNeilUsing polymeric additives to enhance molecular gelation: impact of poly(acrylic acid) on pyridine-based gelators Soft Matter, 8 (2012), pp. 430-434
38. N. Yan, Z. Xu, K.K. Diehn, S.R. Raghavan, Y. Fang, R.G. WeissHow do liquid mixtures solubilize insoluble gelators? Self-assembly properties of pyrenyl-linker-glucono gelators in tetrahydrofuran–water mixtures J. Am. Chem. Soc., 135 (2013), pp. 8989-8999
39. M. Numata, K. Sugiyasu, T. Kishida, S. Haraguchi, N. Fujita, S.M. Park, Y.J. Yun, B.H. Kim, S. ShinkaiCreation of polynucleotide-assisted molecular assemblies in organic solvents: general strategy toward the creation of artificial DNA-like nanoarchitectures Org. Biomol. Chem., 6 (2008), pp. 712-718
40. Y. Li, Y. Sun, M. Qin, Y. Cao, W. WangMechanics of single peptide hydrogelator fibrils Nanoscale, 7 (2015), pp. 5638-5642
41. M.P. Conte, N. Singh, I.R. Sasselli, B. Escuder, R.V. UlijnMetastable hydrogels from aromatic dipeptides Chem. Commun., 52 (2016), pp. 13889-13892
42. M.M. Smith, D.K. SmithSelf-sorting multi-gelator gels-mixing and ageing effects in thermally addressable supramolecular soft nanomaterials Soft Matter, 7 (2011), pp. 4856-4860
43. E.R. Draper, T.O. McDonald, D.J. AdamsA low molecular weight hydrogel with unusual gel aging Chem. Commun., 51 (2015), pp. 6595-6597
44. V.A. Mallia, P.D. Butler, B. Sarkar, K.T. Holman, R.G. WeissReversible phase transitions within self-assembled fibrillar networks of (R)-18-(n-alkylamino)octadecan-7-ols in their carbon tetrachloride gels J. Am. Chem. Soc., 133 (2011), pp. 15045-15054
45. A.M. Castilla, M. Wallace, L.L.E. Mears, E.R. Draper, J. Doutch, S. Rogers, D.J. AdamsOn the syneresis of an OPV functionalised dipeptide hydrogel Soft Matter, 12 (2016), pp. 7848-7854
46. S.-L. Zhou, S. Matsumoto, H.-D. Tian, H. Yamane, A. Ojida, S. Kiyonaka, I. HamachipH-responsive shrinkage/swelling of a supramolecular hydrogel composed of two small amphiphilic molecules Chem. Eur. J., 11 (2005), pp. 1130-1136
47. T. Heuser, E. Weyandt, A. WaltherBiocatalytic feedback-driven

temporal programming of self-regulating peptide hydrogels Angew. Chem. Int. Ed., 54 (2015), pp. 13258-13262

48. V.J. Nebot, D.K. SmithTechniques for the characterisation of molecular gels B. Escuder, J.F. Miravet (Eds.), Functional Molecular Gels, The Royal Society of Chemistry (2014), pp. 30-66
49. G. Yu, X. Yan, C. Han, F. HuangCharacterization of supramolecular gels Chem. Soc. Rev., 42 (2013), pp. 6697-6722
50. J.-B. Guilbaud, A. SaianiUsing small angle scattering (SAS) to structurally characterise peptide and protein self-assembled materials Chem. Soc. Rev., 40 (2011), pp. 1200-1210
51. C. Yan, D.J. PochanRheological properties of peptide-based hydrogels for biomedical and other applications Chem. Soc. Rev., 39 (2010), pp. 3528-3540
52. S.R. Raghavan, B.H. CiprianoGel formation: phase diagrams using tabletop rheology and calorimetry R.G. Weiss, P. Terech (Eds.), Molecular Gels, Springer (2005), p. 233
53. F.M. Menger, K.L. CaranAnatomy of a gel. Amino acid derivatives that rigidify water at submillimolar concentrations J. Am. Chem. Soc., 122 (2000), pp. 11679-11691
54. J. Raeburn, A. Zamith Cardoso, D.J. AdamsThe importance of the self-assembly process to control mechanical properties of low molecular weight hydrogels Chem. Soc. Rev., 42 (2013), pp. 5143-5156
55. J. Raeburn, D.J. AdamsMulticomponent low molecular weight gelators Chem. Commun., 51 (2015), pp. 5170-5180
56. F. Tantakitti, J. Boekhoven, X. Wang, R.V. Kazantsev, T. Yu, J. Li, E. Zhuang, R. Zandi, J.H. Ortony, C.J. Newcomb, et al.Energy landscapes and functions of supramolecular systems Nat. Mater., 15 (2016), pp. 469-476
57. M.M. Safont-Sempere, G. Fernández, F. WürthnerSelf-sorting phenomena in complex supramolecular systems Chem. Rev., 111 (2011), pp. 5784-5814
58. L.E. Buerkle, S.J. RowanSupramolecular gels formed from multi-component low molecular weight species Chem. Soc. Rev., 41 (2012), pp. 6089-6102
59. D.B. Amabilino, D.K. Smith, J.W. SteedSupramolecular materials Chem. Soc. Rev., 46 (2017), pp. 2404-2420
60. S. Diaz-Oltra, C. Berdugo, J.F. Miravet, B. EscuderStudy of the effect of polymorphism on the self-assembly and catalytic performance of an l-proline

based molecular hydrogelator New J. Chem., 39 (2015), pp. 3785-3791
61. B. Ding, Y. Li, M. Qin, Y. Ding, Y. Cao, W. WangTwo approaches for the engineering of homogeneous small-molecule hydrogels Soft Matter, 9 (2013), pp. 4672-4680
62. B. Ozbas, J. Kretsinger, K. Rajagopal, J.P. Schneider, D.J. PochanSalt-triggered peptide folding and consequent self-assembly into hydrogels with tunable modulus Macromolecules, 37 (2004), pp. 7331-7337
63. Z. Yang, G. Liang, B. XuEnzymatic hydrogelation of small molecules Acc. Chem. Res., 41 (2008), pp. 315-326
64. R. Huang, Y. Wang, W. Qi, R. Su, Z. HeChemical catalysis triggered self-assembly for the bottom-up fabrication of peptide nanofibers and hydrogels Mater. Lett., 128 (2014), pp. 216-219
65. F. Trausel, F. Versluis, C. Maity, J.M. Poolman, M. Lovrak, J.H. van Esch, R. EelkemaCatalysis of supramolecular hydrogelation Acc. Chem. Res., 49 (2016), pp. 1440-1447
66. A.R. Hirst, D.K. SmithTwo-component gel-phase materials—highly tunable self-assembling systems Chem. Eur. J., 11 (2005), pp. 5496-5508
67. M. de Loos, B.L. Feringa, J.H. van EschDesign and application of self-assembled low molecular weight hydrogels Eur. J. Org. Chem., 2005 (2005), pp. 3615-3631
68. G. Fichman, E. GazitSelf-assembly of short peptides to form hydrogels: design of building blocks, physical properties and technological applications Acta Biomater., 10 (2014), pp. 1671-1682
69. S. Fleming, R.V. UlijnDesign of nanostructures based on aromatic peptide amphiphiles Chem. Soc. Rev., 43 (2014), pp. 8150-8177
70. Z. Yang, G. Liang, M. Ma, Y. Gao, B. XuConjugates of naphthalene and dipeptides produce molecular hydrogelators with high efficiency of hydrogelation and superhelical nanofibers J. Mater. Chem., 17 (2007), pp. 850-854
71. C. Colquhoun, E.R. Draper, R. Schweins, M. Marcello, D. Vadukul, L.C. Serpell, D.J. AdamsControlling the network type in self-assembled dipeptide hydrogels Soft Matter, 13 (2017), pp. 1914-1919
72. A. Mahler, M. Reches, M. Rechter, S. Cohen, E. GazitRigid, self-assembled hydrogel composed of a modified aromatic dipeptide Adv. Mater., 18 (2006), pp. 1365-1370
73. L. Chen, J. Raeburn, S. Sutton, D.G. Spiller, J. Williams, J.S. Sharp, P.C. Griffiths, R.K. Heenan, S.M. King, A. Paul, et al.Tuneable

mechanical properties in low molecular weight gels Soft Matter, 7 (2011), pp. 9721-9727

74. N.A. Dudukovic, C.F. ZukoskiEvidence for equilibrium gels of valence-limited particles Soft Matter, 10 (2014), pp. 7849-7856
75. R. Orbach, I. Mironi-Harpaz, L. Adler-Abramovich, E. Mossou, E.P. Mitchell, V.T. Forsyth, E. Gazit, D. SeliktarThe rheological and structural properties of fmoc-peptide-based hydrogels: the effect of aromatic molecular architecture on self-assembly and physical characteristics Langmuir, 28 (2012), pp. 2015-2022
76. J. Raeburn, C. Mendoza-Cuenca, B.N. Cattoz, M.A. Little, A.E. Terry, A. Zamith Cardoso, P.C. Griffiths, D.J. AdamsThe effect of solvent choice on the gelation and final hydrogel properties of Fmoc-diphenylalanine Soft Matter, 11 (2015), pp. 927-935
77. A.Z. Cardoso, L.L.E. Mears, B.N. Cattoz, P.C. Griffiths, R. Schweins, D.J. AdamsLinking micellar structures to hydrogelation for salt-triggered dipeptide gelators Soft Matter, 12 (2016), pp. 3612-3621
78. L. Chen, G. Pont, K. Morris, G. Lotze, A. Squires, L.C. Serpell, D.J. AdamsSalt-induced hydrogelation of functionalised-dipeptides at high pH Chem. Commun., 47 (2011), pp. 12071-12073
79. D.J. Adams, M.F. Butler, W.J. Frith, M. Kirkland, L. Mullen, P. SandersonA new method for maintaining homogeneity during liquid-hydrogel transitions using low molecular weight hydrogelators Soft Matter, 5 (2009), pp. 1856-1862
80. E.R. Draper, L.L.E. Mears, A.M. Castilla, S.M. King, T.O. McDonald, R. Akhtar, D.J. AdamsUsing the hydrolysis of anhydrides to control gel properties and homogeneity in pH-triggered gelation RSC Adv., 5 (2015), pp. 95369-95378
81. E.K. Johnson, L. Chen, P.S. Kubiak, S.F. McDonald, D.J. Adams, P.J. CameronSurface nucleated growth of dipeptide fibres Chem. Commun., 49 (2013), pp. 8698-8700
82. J. Raeburn, T.O. McDonald, D.J. AdamsDipeptide hydrogelation triggered via ultraviolet light Chem. Commun., 48 (2012), pp. 9355-9357
83. W. Helen, P. de Leonardis, R.V. Ulijn, J. Gough, N. TirelliMechanosensitive peptide gelation: mode of agitation controls mechanical properties and nano-scale morphology Soft Matter, 7 (2011), pp. 1732-1740
84. X.-D. Xu, B.-B. Lin, J. Feng, Y. Wang, S.-X. Cheng, X.-Z. Zhang,

R.-X. ZhuoBiological glucose metabolism regulated peptide self-assembly as a simple visual biosensor for glucose detection Macromol. Rapid Commun., 33 (2012), pp. 426-431

85. L. Chen, S. Revel, K. Morris, L. C Serpell, D.J. AdamsEffect of molecular structure on the properties of naphthalene–dipeptide hydrogelators
86. Langmuir, 26 (2010), pp. 13466-13471
87. C. Tang, R.V. Ulijn, A. SaianiEffect of glycine substitution on fmoc–diphenylalanine self-assembly and gelation properties Langmuir, 27 (2011), pp. 14438-14449
88. C. Tang, A.M. Smith, R.F. Collins, R.V. Ulijn, A. SaianiFmoc-diphenylalanine self-assembly mechanism induces apparent pKa shifts Langmuir, 25 (2009), pp. 9447-9453
89. M. Wallace, J.A. Iggo, D.J. AdamsUsing solution state NMR spectroscopy to probe NMR invisible gelators Soft Matter, 11 (2015), pp. 7739-7747
90. F.C. MacKintosh, J. Käs, P.A. JanmeyElasticity of semiflexible biopolymer networks Phys. Rev. Lett., 75 (1995), pp. 4425-4428
91. M. Wallace, D.J. Adams, J.A. IggoAnalysis of the mesh size in a supramolecular hydrogel by PFG-NMR spectroscopy Soft Matter, 9 (2013), pp. 5483-5491
92. M.A. Greenfield, J.R. Hoffman, M. Olvera de la Cruz, S.I. StuppTunable mechanics of peptide nanofiber gels Langmuir, 26 (2010), pp. 3641-3647
93. S. Zhang, M.A. Greenfield, A. Mata, L.C. Palmer, R. Bitton, J.R. Mantei, C. Aparicio, M.O. de la Cruz, S.I. StuppA self-assembly pathway to aligned monodomain gels Nat. Mater., 9 (2010), pp. 594-601
94. E.R. Draper, H. Su, C. Brasnett, R. Poole, S. Rogers, H. Cui, A. Seddon, D.J. AdamsOpening a can of worm(-like micelle)s: the effect of temperature of solutions of functionalised dipeptides Angew. Chem. Int. Ed. (2017), 10.1002/anie.201705604
95. C. Yan, A. Altunbas, T. Yucel, R.P. Nagarkar, J.P. Schneider, D.J. PochanInjectable solid hydrogel: mechanism of shear-thinning and immediate recovery of injectable [small beta]-hairpin peptide hydrogels
96. Soft Matter, 6 (2010), pp. 5143-5156
97. K. Sugiyasu, S.-i. Kawano, N. Fujita, S. ShinkaiSelf-sorting organogels with p–n heterojunction points Chem. Mater., 20 (2008), pp. 2863-2865

98. K.L. Morris, L. Chen, J. Raeburn, O.R. Sellick, P. Cotanda, A. Paul, P.C. Griffiths, S.M. King, R.K. O'Reilly, L.C. SerpellChemically programmed self-sorting of gelator networks Nat. Commun., 4 (2013), p. 1480
99. C. Colquhoun, E.R. Draper, E.G. Eden, B.N. Cattoz, K.L. Morris, L. Chen, T.O. McDonald, A.E. Terry, P.C. Griffiths, L.C. SerpellThe effect of self-sorting and co-assembly on the mechanical properties of low molecular weight hydrogels Nanoscale, 6 (2014), pp. 13719-13725
100. J.R. Moffat, D.K. SmithControlled self-sorting in the assembly of 'multi-gelator' gels Chem. Commun. (2009), pp. 316-318
101. E.R. Draper, J.R. Lee, M. Wallace, F. Jäckel, A.J. Cowan, D.J. AdamsSelf-sorted photoconductive xerogels Chem. Sci., 7 (2016), pp. 6499-6505
102. D.J. Cornwell, O.J. Daubney, D.K. SmithPhotopatterned multidomain gels: multi-component self-assembled hydrogels based on partially self-sorting 1,3:2,4-dibenzylidene-d-sorbitol derivatives J. Am. Chem. Soc., 137 (2015), pp. 15486-15492
103. M. Tena-Solsona, B. Escuder, J.F. Miravet, V. Casttelleto, I.W. Hamley, A. DehsorkhiThermodynamic and kinetic study of the fibrillization of a family of tetrapeptides and its application to self-sorting. What takes so long? Chem. Mater., 27 (2015), pp. 3358-3365
104. E.R. Draper, M. Wallace, R. Schweins, R.J. Poole, D.J. AdamsNonlinear effects in multicomponent supramolecular hydrogels Langmuir, 33 (2017), pp. 2387-2395
105. E.R. Draper, E.G. Eden, T.O. McDonald, D.J. AdamsSpatially resolved multicomponent gels Nat. Chem., 7 (2015), pp. 848-852
106. S. Onogi, H. Shigemitsu, T. Yoshii, T. Tanida, M. Ikeda, R. Kubota, I. HamachiIn situ real-time imaging of self-sorted supramolecular nanofibres Nat. Chem., 8 (2016), pp. 743-752
107. K. Nagy-Smith, P.J. Beltramo, E. Moore, R. Tycko, E.M. Furst, J.P. SchneiderMolecular, local, and network-level basis for the enhanced stiffness of hydrogel networks formed from coassembled racemic peptides: predictions from Pauling and Corey ACS Cent. Sci., 3 (2017), pp. 586-597

CHAPTER

4

Tuning Gel State Properties of Supramolecular Gels by Functional Group Modification

Dipankar Ghosh [1], Matthew T. Mulvee [2] and Krishna K. Damodaran [1]

[1]Department of Chemistry, Science Institute, University of Iceland, Dunhagi 3, 107 Reykjavík, Iceland

[2]Department of Chemistry, Durham University, South Road, Durham DH1 3LE, UK

ABSTRACT

The factors affecting the self-assembly process in low molecular weight gelators (LMWGs) were investigated by tuning the gelation properties of a well-known gelator *N*-(4-pyridyl)isonicotinamide (4PINA). The

Citation: Ghosh, D.; Mulvee, M.T.; Damodaran, K.K. "Tuning Gel State Properties of Supramolecular Gels by Functional Group Modification". Molecules 2019, 24, 3472. https://doi.org/10.3390/molecules24193472

N—H···N interactions responsible for gel formation in 4PINA were disrupted by altering the functional groups of 4PINA, which was achieved by modifying pyridyl moieties of the gelator to pyridyl *N*-oxides. We synthesized two mono-*N*-oxides (INO and PNO) and a di-*N*-oxide (diNO) and the gelation studies revealed selective gelation of diNO in water, but the two mono-*N*-oxides formed crystals. The mechanical strength and thermal stabilities of the gelators were evaluated by rheology and transition temperature (T_{gel}) experiments, respectively, and the analysis of the gel strength indicated that diNO formed weak gels compared to 4PINA. The SEM image of diNO xerogels showed fibrous microcrystalline networks compared to the efficient fibrous morphology in 4PINA. Single-crystal X-ray analysis of diNO gelator revealed that a hydrogen-bonded dimer interacts with adjacent dimers via C—H···O interactions. The non-gelator with similar dimers interacted via C—H···N interaction, which indicates the importance of specific non-bonding interactions in the formation of the gel network. The solvated forms of mono-*N*-oxides support the fact that these compounds prefer crystalline state rather than gelation due to the increased hydrophilic interactions. The reduced gelation ability (minimum gel concentration (MGC)) and thermal strength of diNO may be attributed to the weak intermolecular C—H···O interaction compared to the strong and unidirectional N—H···N interactions in 4PINA.

Keywords: LMWGs; hydrogel; structural modification; pyridyl amides; N-oxides

INTRODUCTION

Supramolecular gels based on low molecular weight gelators (LMWGs) [1,2,3,4,5,6,7,8,9,10] are an excellent class of soft materials with tunable gel state properties and potential applications [9,10,11,12,13,14,15,16], such as dynamic gels, cell culture, drug delivery, and media of crystal growth. LMWGs are formed by the immobilization of the solvent molecules in the three-dimensional (3-D) network of the gelator, which are stabilized by various noncovalent interactions [1,2,3,4,5,6,7,8,9], such as hydrogen bonding, van der Waals interactions, π–π stacking, etc. The gelation properties of LMWGs depend on various factors such as concentration [17,18], sonication [19,20,21], additives [22], and seeding [21,23]. The understanding and prediction of the gel structure and self-assembly process of LMWGs to control or tune the gelation properties are difficult because of the dynamic nature of noncovalent interactions [1,2,3,4,5,6,7,8,9,10]. This is due to the low molecular order of the gel state as a whole because of the variation in the length scales (ranging from nano to micro) of the gel structure. Efforts have been made to investigate the gelation mechanism and unveil the factors influencing the gel using various methods [8,19,24,25, 26,27,28,29,30,31,32], such as ultraviolet–visible (UV–Vis) spectroscopy, nuclear magnetic resonance spectroscopy (NMR), Fourier transform infrared spectroscopy (FT-IR), scanning electron microscopy (SEM), atomic force microscopy (AFM), transmission electron microscopy (TEM), rheology and X-ray diffraction techniques including X-ray diffraction (XRD), small-angle X-ray scattering (SAXS), small-angle neutron scattering (SANS), etc. Although the characterization of supramolecular gels from molecular to mesoscopic scale has been challenging, X-ray diffraction techniques can be used to correlate the crystal structure of the gelator with the powder diffraction pattern of either native gel or the xerogel [4,8,33,34,35]. However, the removal of solvent to prepare a xerogel can result in artifacts due to dissolution, recrystallisation, and changes in morphology or polymorphic phase transition, but this approach still remains as one of the practical methods to gain insight into the gelator structure and aggregation behavior. Furthermore, this technique enables us to correlate the intermolecular interactions observed in the single-crystal structure with the molecular aggregation in the gel state, which may be different from its crystalline network [34]. We were interested in studying the effect of hydrogen-bonding motifs in the self-assembly process of LMWGs, to enable us to tune the gel state properties. However, the design, mechanism, and understanding of gel structure and the self-assembly process of new LMWGs are challenging because the structure and properties rely mostly on the geometry and spatial arrangement of the building blocks and also the nature of intermolecular noncovalent interactions [36,37]. Thus, modifying an existing supramolecular gel can be considered as a good strategy to analyze the self-assembly, enabling us to compare the structural and gelation properties of new LMWGs with the parent gelator [38].

Amide and urea moieties have been used extensively as supramolecular synthons [39] to generate LMWGs with tunable properties [40,41,42]. Amide-based LMWGs contain an amide N−H donor and C=O acceptor resulting in a strong N−H···O=C interaction, which self-assemble to a 3-D network via cooperative and unidirectional hydrogen bonding involving amide units [43]. These 3-D networks are capable of immobilizing solvent molecules to form organo/hydrogels [34,44,45,46,47,48,49]. Recently, we reported the structural modification of trimesic amide-based gelator N^1,N^3,N^5-tri(pyridin-

3-yl)benzene-1,3,5-tricarboxamide [50] to corresponding tris-*N*-oxide compound (L-3Nox) [38]. The gelation properties suggested that L-3Nox is a weaker gelator compared to the parent amide and the structure–property correlation was not performed due to the lack of a single-crystal X-ray structure. This prompted us to select low molecular weight pyridyl amide gelators, which are classified as stimuli-responsive supramolecular gels due to their tunable properties to external stimuli such as pH, coordination, salts/ions, etc. [6,51,52,53,54,55,56,57,58,59,60,61,62,63,64,65]. The advantage of using pyridyl amide is twofold: (a) the opportunity to modify each ring selectively by simple organic reactions (Scheme 1) and (b) the easiness of obtaining the crystal structure. Thus, a pyridyl amide-based hydrogelator, namely *N*-(4-pyridyl)isonicotinamide (4PINA) [34], with remarkably low molecular weight and minimum gel concentration was selected. The derivatives of 4PINA are ideal candidates to study the molecular aggregation enabling us to compare the role of non-bonding interactions in gel formation with 4PINA. The aggregation mode of molecules in the xerogel fibrils was found to be N—H···N hydrogen bonding, which had a prominent effect in the self-assembly of 4PINA. The structural modification of 4PINA was achieved by oxidizing the pyridine ring to the corresponding *N*-oxide, which may restrict the N—H···N hydrogen bonding-based molecular aggregation.

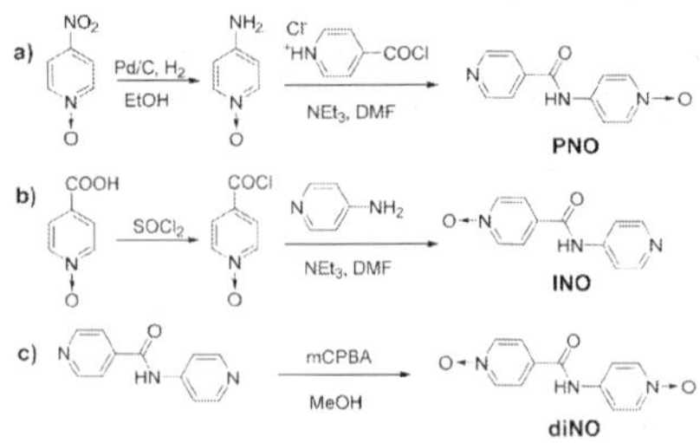

Scheme 1. Synthesis of the *N*-oxide amides: (**a**) 4-(isonicotinamido) pyridine-1-oxide (**PNO**), (**b**) 4-(pyridin-4-ylcarbamoyl) pyridine-1-oxide (**INO**),and(**c**) 4-((1-oxidopyridin-4-yl)carbamoyl) pyridine-1-oxide (**diNO**).

RESULTS

Design and Synthesis

We synthesized three *N*-oxides compounds by modifying the pyridyl group of *N*-(4-pyridyl)isonicotinamide (**4PINA**) [**34**]. This included a di-*N*-oxide (4-((1-oxidopyridin-4-yl)carbamoyl)pyridine-1-oxide, **diNO**) and two mono-*N*-oxides, namely 4-(isonicotinamido)pyridine-1-oxide (PNO) and 4-(pyridin-4-ylcarbamoyl)pyridine-1-oxide (INO) with the *N*-oxide group at the aminopyridine and isonicotinoyl end, respectively. Mono-*N*-oxide amides were synthesized by reacting acid chlorides with corresponding amines in anhydrous DMF in the presence of triethylamine (Scheme 1a,b). The precipitates obtained for INO and PNO were washed with aqueous sodium bicarbonate solution followed by water to ensure the removal of salts. The di-*N*-oxide (diNO) was synthesized by oxidizing 4PINA with *m*-chloroperoxybenzoic acid (Scheme 1c) and the product was washed with water and dried. All compounds were characterized by NMR, HRMS, IR, and X-ray diffraction (single-crystal and powder).

Gelation Experiments

The gelation experiments for diNO, INO, and PNO were tested in various solvents. In a typical experiment, a 10.0 mg portion of the compound was heated in 1.0 mL of solvent (1.0 wt %) to get a clear solution and was cooled to room temperature. The hydrogel formation of these amides tested at 1.0 wt % indicated no gelation, which prompted us to increase the concentration to 4.0 wt %. The mixture was heated (80.0–90.0 °C) and on cooling to room temperature, gelation was observed only for the di-*N*-oxide (diNO) after 1 h, which was confirmed via inversion test. Crystalline materials were obtained for both the mono-*N*-oxides (PNO and INO). The gelation experiment performed with an equimolar mixture of PNO and INO resulted in concomitant crystallization of the individual compounds, which was confirmed by single-crystal X-ray diffraction (SCXRD). The gel strength of diNO was analyzed by sol–gel transition temperature (T_{gel}) and minimum gel concentration (MGC) experiments. The lowest concentration at which the gel was obtained was recorded as the minimum gel concentration (MGC). The MGC experiment of the hydrogel was performed in 1.0 mL of water by weighing different amounts of diNO gelator (30.0 to 50.0 mg) in a 7.0 mL vial. The mixture was heated and then cooled to room temperature, and the vial was left undisturbed for gel formation. After 24 h, the gel formation was confirmed by the inversion test and the MGC of the gelator was found to be 4.0 wt %.

Thermal Stability

The thermal stability of the gel network of **diNO** was evaluated using gel-to-solution transition temperature (T_{gel}) experiments, which is the critical temperature at which the gel converts to a solution. The experiment was performed by placing a small spherical glass ball (92.0 mg) on top of the **diNO** preformed hydrogel in a standard vial. The vial was heated and the temperature at which the ball touched the bottom of the vial was recorded as T_{gel}. The experiments were performed at 4.0 and 6.0 wt % and the T_{gel} did not vary much with the concentration of **diNO**, which was found to be 78.0 °C and 80.0 °C at 4.0 and 6.0 wt %, respectively.

Rheology

Rheological analysis was undertaken to determine the mechanical properties of the supramolecular hydrogels (4.0 wt %) [**66,67**]. The oscillatory amplitude sweeps, at a constant frequency (1.0 Hz), demonstrated a linear viscoelastic region (LVR) in which the storage (elastic) G' modulus was approximately an order of magnitude higher than the loss (viscous) G" modulus for both samples (Figure 1a) [66,68]. Thus, both samples had a linear elastic response to low stress amplitudes. This behavior is indicative of a solid-like network throughout the sample gelling the aqueous solvent. At increased shear stresses, a stress softening behavior was observed, the viscous moduli (G") increased, and at a crossover point, there was a concomitant sharp decrease in the storage moduli (G') indicating a transition from an elastic gel to a viscous fluid [69,70]. At low shear stresses, 4PINA had an elastic modulus of approximately 140,000 Pa and a yield stress of approximately 500 Pa. diNO was a significantly softer gel with an elastic modulus of approximately 5000 Pa and was also weaker, with a yield stress of approximately 60 Pa.

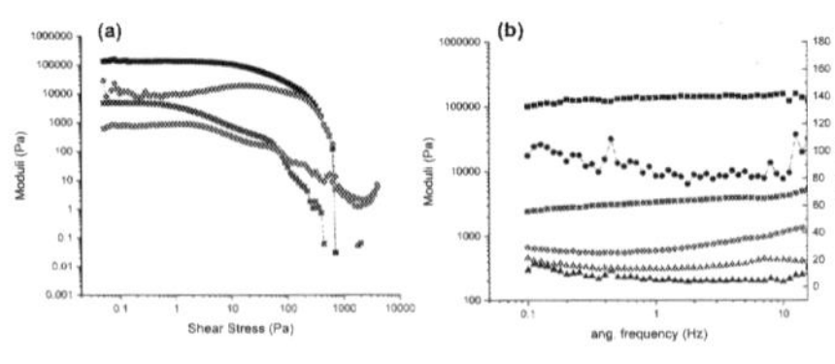

Figure 1. (**a**) Oscillatory amplitude sweeps and (**b**) oscillatory frequency sweeps, for *N*-(4-pyridyl)isonicotinamide (4PINA) (black) and diNO (red). For both, ▪ refers to G' and ● refers to G", the storage modulus and viscous modulus, at 4.0 wt %, respectively. ▲ refers to the phase lag (δ).

Oscillatory frequency sweeps (Figure 1b) at a fixed amplitude (0.05 Pa) demonstrated that G', arising from the elastic network, was higher than G" across all frequencies for 4PINA, which is characteristic of the presence of a solid-like network of filaments that are temporally persistent in the viscoelastic gel. These moduli were effectively indifferent to the frequency and this was true for diNO as well at low frequencies (<10 Hz). However, with increasing frequencies G' increased, the typical behavior of viscoelastic liquids rather than supramolecular gels [66], which further indicates that diNO formed weaker gels than 4PINA. Furthermore, the ratio of G':G" is not at least an order of magnitude, the generally accepted ratio for supramolecular gels. Instead the ratio is roughly 7:1, further indicating that the modifications made to diNO severally affect its gelling ability. However, the phase angle, the lag between the applied shear stress and the measured strain, for diNO of approximately 15.0 δ is comparable to that of the original gelator, 4PINA, with a measured phase lag of approximately 7.0 δ. Moreover, the phase angle is not dependent on the frequency, a key Watson–Chambon criterion for gels. Thus, as the diNO samples do not meet the moduli requirements but demonstrated other viscoelastic behavior typical of a physical gel, it can, with caveats, be considered a very weak gel.

Scanning Electron Microscopy (SEM)

The morphology of gel fibers was studied by analyzing the SEM images of dried **diNO** gel. The gel was prepared at 4.0 wt % in water, filtered after 24 h, and dried under a fume hood. A small portion of the dried gel was placed on a carbon tab and was coated with gold for 2 min. The SEM images revealed that the xerogel displayed a rod-type architecture with needle-shaped microcrystalline materials in the network. The diameter of the small needle-shaped fibers ranged from 1.2 to 2.5 μm and the width of the larger microcrystalline rods ranged from 8.0 to 30.0 μm (**Figure 2**).

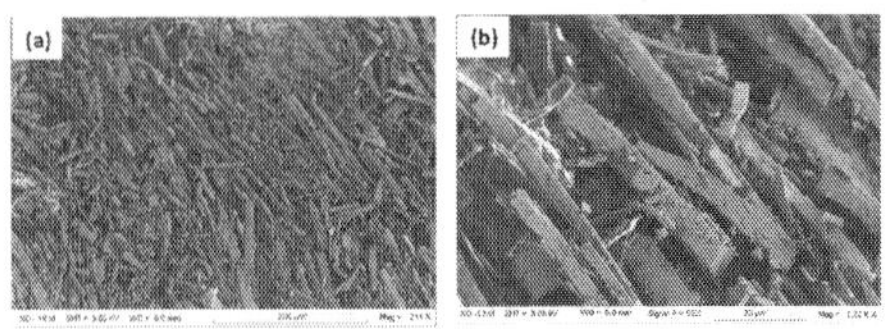

Figure 2. SEM images of **diNO** show the microcrystalline fibrous network: (**a**) overview; (**b**) magnified images.

Crystal Structure

The crystallization experiments of *N*-oxides and an equimolar mixture of two mono-*N*-oxides were performed in water and various organic solvents.

The compounds were proved to be sparingly soluble in most of the organic solvents, hence the experiments were carried out in a water or aqueous solution of methanol, ethanol, acetonitrile, and tetrahydrofuran. About 20.0 mg of the material were dissolved by heating in 1.0 mL of water, and the solution was cooled to room temperature. Crystallization from mixed solvents (1:1, v/v) such as methanol/water, ethanol/water, acetonitrile/water, and tetrahydrofuran/water resulted in precipitates. In some cases, needle-shaped crystals were obtained with low yield and the crystal quality was not good for single-crystal X-ray diffraction. The crystallization experiments performed in water produced X-ray quality single crystals for all *N*-oxides. The crystals of diNO were found to be needle-shaped, whereas block-shaped crystals were observed for both mono-*N*-oxides. Interestingly, crystallization of a 1:1 mixture of PNO and INO also resulted in block-shaped crystals.

The gelator diNO crystallized in a triclinic space group ($P\bar{1}$) with the carbonyl moiety of the amide group slightly deviated from the amide plane (15.74(8)°) and the pyridyl *N*-oxide rings were twisted at an angle of 58.71(3)° (Figure 3a). The N-H moiety was involved in hydrogen-bonding interaction (N—H···O) with the oxygen atom of the adjacent molecule to form a dimer. This dimer was further stabilized by a strong π–π interaction (3.6563(9) Å) between the isonicotinoyl *N*-oxide ring of diNO (Figure 3b) and also N—H···O interactions (2.8178(16) Å) between the amide N-H moiety and isonicotinoyl *N*-oxide moieties. The aminopyridine *N*-oxide moieties of the diNO dimer interact with the isonicotinoyl *N*-oxide moieties of adjacent dimers via offset π–π interactions (3.8745(9) and 3.8219(9) Å) and also display bifurcated C—H···O interactions (3.1789(18) and 3.1621(19) Å) with adjacent dimers (Figure 3c).

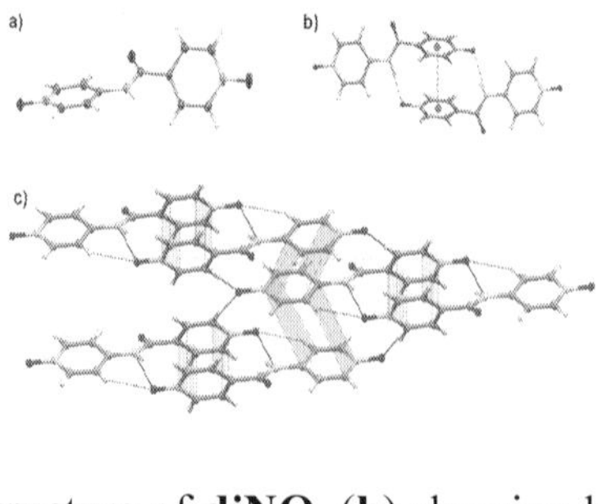

Figure 3. (**a**) Crystal structure of **diNO**, (**b**) showing hydrogen-bonded dimer stabilized with π–π interaction and (**c**) interaction of the individual dimer with adjacent dimer via various non-bonding interactions (dotted lines); black lines indicate hydrogen bonding, red line for π–π interactions of the individual dimer, and blue line for π–π interactions with adjacent dimers.

Compound INO crystalized in a triclinic space group ($P\bar{1}$) and the twisting of the pyridyl *N*-oxide ring (58.86°) was similar to diNO (Figure 4a). The hydrogen-bonding pattern was exactly the same as in diNO resulting in a dimer, which was stabilized by N—H···O and π–π interactions. The only difference between these two structures was the connection between adjacent dimers and the dimer in INO propagates via C—H···N interaction (3.4391(16) and 3.5489(16) Å), whereas C—H···O and π–π interactions are observed in diNO. The solvated form of INO was also isolated (INO.2H_2O) and the crystals belong to a monoclinic $C2_{/c}$ space group. The asymmetric unit contains one INO and two molecules of water, and the isonicotinoyl *N*-oxide and pyridyl moieties are coplanar to the amide groups compared to the de-solvated form. The hydrogen-bonding pattern of INO.2H_2O was different from INO due to the presence of dimeric water clusters. The isonicotinoyl *N*-oxide was hydrogen-bonded to two water molecules via O—H···O interactions (2.793(3) and 2.796(2) Å), and the pyridyl nitrogen atom interacted with a water molecule through O—H···N interaction (2.771(2) Å). This resulted in the formation of a hydrogen-bonded macrocycle involving two INO molecules and four water molecules. The amide moieties of the macrocycle were hydrogen bonded to adjacent macrocycles via N—H···O (2.921(2) Å) interactions to form a hydrogen-bonded 3-D network.

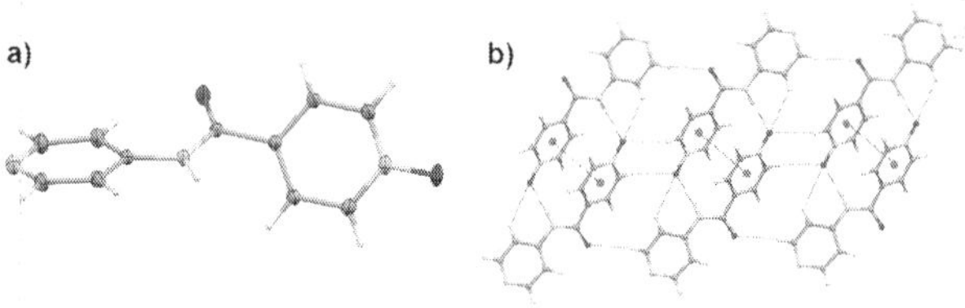

Figure 4. (**a**) Crystal structure of INO and (b) interaction of individual dimers via various non-bonding interactions; black dotted lines indicate hydrogen bonding and red dotted lines represent the π–π interactions.

The structural analysis of the pyridyl *N*-oxide compound (PNO) revealed that the compound crystallized in a similar space group as diNO but with two solvent water molecules (PNO.2H_2O, Figure 5a). The pyridyl *N*-oxide and isonicotinoyl moieties are almost planar to each other but slightly twisted from the amide plane (16.70 and 18.72°). The N-H moiety of the amide group, the pyridyl nitrogen atom, and the N-oxide moieties are hydrogen bonded to the water molecules. The amide acts as donor (N—H···O = 2.8617(15) Å), whereas the *N*-oxide (O—H···O

= 2.7412(15) and 2.7701(16) Å) and isonicotinoyl groups (O—H···N = 2.8598(17) Å) act as acceptor, resulting in four water molecules interacting with one PNO molecule. The hydrogen bonding between water molecules and PNO resulted in a hexagonal architecture of oxygen atoms formed by *N*-oxide oxygen atoms of two PNO and four water molecules, which was stabilized by various interactions with amide and pyridyl nitrogen atoms of adjacent PNO molecules (Figure 5b).

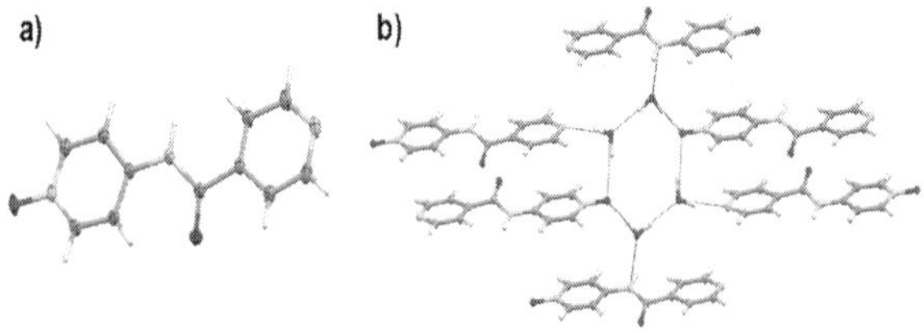

Figure 5. Single-crystal structure of PNO: (a) Asymmetric unit (water molecules not shown); (b) two-dimensional (2-D) sheet constructed by H-bonding between the ligand and solvent (water) molecules.

We also performed the crystallization of a 1:1 mixture of INO and PNO in water (2.0 wt %) and block-shaped crystals appeared on cooling after 1 h. The SCXRD data matched the PNO.2H_2O crystals. The crystals of INO.2H_2O were also obtained from the same mixture after 24 h. The experiments repeated by varying the ratios of INO and PNO (1:3 and 3:1) also resulted in the co-crystallization of the individual components.

X-ray Powder Diffraction (XRPD)

The phase purity of the compounds was further analyzed by comparing the XRPD pattern of the bulk crystals with the simulated pattern obtained from the single-crystal data. XRPD was performed on the bulk solid of recrystallized diNO from water and the xerogel obtained from water at 4.0 wt %. The powder X-ray pattern of the re-crystallized diNO matches perfectly well the simulated graph obtained from the single-crystal data (Figure 6). However, the powder X-ray pattern of the xerogel did not match the calculated pattern from the crystal structure of diNO. The XRPD patterns of mono-*N*-oxides were performed with the recrystallized samples from water, which were filtered and dried in air. The XRPD pattern of PNO revealed that the pattern of bulk solid was virtually super-imposable on the simulated pattern of PNO. The XRPD pattern of the bulk crystalline material of INO was also similar to the simulated pattern of INO. We also performed the XRPD of the hydrated form of INO, but the pattern was different from the calculated pattern of the INO.2H_2O crystal structure. Interestingly, the INO.2H_2O bulk crystal pattern matched the simulated pattern of INO.

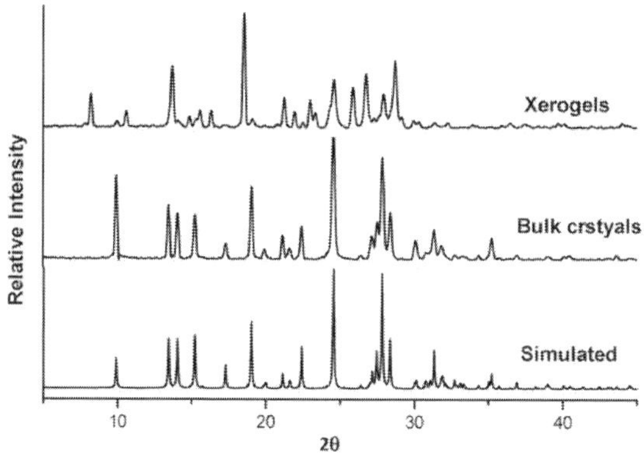

Figure 6. X-ray powder diffraction (XRPD) comparison of **diNO**: simulated pattern from single-crystal X-ray diffraction (SCXRD) data, bulk crystals, and xerogel obtained from water.

DISCUSSION

The ability of pyridyl amides to form a one-dimensional (1-D) hydrogen-bonded network is well established, which makes them ideal candidates in designing LMWGs with tunable properties. The availability of single-crystal X-ray structures of many organo/hydrogelators based on pyridyl amides enabled us to identify the key factors responsible for the formation of gel networks, for example, the formation of 1-D networks. The structural analysis of pyridyl amides indicates that these compounds self-assemble through mainly two types of hydrogen bonding, namely N—H···O synthon of the amide moiety and N—H···N synthon involving the amide and pyridyl nitrogen. These two types of interactions are predominant in the molecular aggregation of *N*-(4-pyridyl)isonicotinamide (4PINA) [34] along with various other non-bonding interactions. However, the impact of these types of interactions was not studied in detail. For example, what is the role of N—H···N synthon in gel network stabilization? What happens if these interactions are replaced by another synthon? We addressed this question by restricting the N—H···N synthon by modifying the pyridyl nitrogen to pyridyl-*N*-oxide moiety. Thus, we modified the pyridyl groups of 4PINA [34] to *N*-oxides;

two mono-*N*-oxides were synthesized by oxidizing the nitrogen atoms at the pyridyl amine end (PNO) and the isonicotinic acid end (INO), and oxidation of both pyridyl nitrogen atoms resulted in diNO. This restricts N—H···N interactions between the gelators and hinders the one-dimensional growth of the gel fibril, which may facilitate the formation of either a two-dimensional (2-D) or 3-D network of gelator. Another advantage of pyridyl-*N*-oxides is the ease of crystallization in an aqueous medium, enabling us to compare the crystal structures of mono-*N*-oxides and di-*N*-oxide for structure–property correlation.

The gelation properties of all the *N*-oxides in various solvents resulted in selective gelation of diNO in the water at higher concentration (4.0 wt %). The formation of needle-shaped crystals at lower concentration revealed that crystallization and gelation properties [71] were concentration dependent. The mono-*N*-oxides (PNO and INO) did not form gel in any solvents but formed crystals indicating the importance of similar groups at both ends in gel formation, which corroborates well with the gelation properties of 4PINA [34]. We also prepared a mixture of PNO and INO in different ratios to test the ability of these compounds to form multi-component gels. Such gels are obtained by mixing two or more compounds (gelator/non-gelator) where individual molecules interact either constructively or destructively to form well-ordered fibers containing individual components (self-sorting), both components (specific co-assembly), or a mixture of both (random co-assembly) [72,73]. We showed that mixing enantiomeric supramolecular gels based on bis(urea) compounds tagged with a phenylalanine methyl ester leads multi-component gels with enhanced thermal and mechanical strength [74]. However, the experiments performed with the mixtures of PNO and INO resulted in concomitant crystallization of individual compounds. INO was proved to be more soluble between the two mono-*N*-oxides leading to slow crystallization. The crystallization of the individual compounds suggests self-sorting of the components in the mixture.

The minimum gel concentration (MGC) of diNO (4.0 wt %) indicates that the gel network formed was weaker compared to 4PINA (MGC 0.37 wt %) [34]. This may be attributed to the fact that converting the pyridyl group of 4PINA to pyridyl-*N*-oxide increased the hydrophilic interactions preferring crystalline form, which hindered the gel formation at low concentration [75]. The gel-to-solution transition temperature (T_{gel}) experiments performed at MGC of diNO to evaluate the thermal stability of the gel revealed that the gel network collapsed at 78.0 °C. This indicates that the gel network of diNO is slightly weaker compared to 4PINA and the T_{gel} of diNO network did not change drastically with concentration. This is in agreement with the rheological measurements that revealed that diNO gels were weaker compared to the 4PINA gelator. The rheology experiments enabled us to elucidate the information regarding factors controlling the gelation, gel strength, and the solid-like properties of pure gels, and the mechanical strength of the 4PINA and diNO gels was evaluated using rheology. The rheological measurements performed for both 4PINA and diNO gels revealed that diNO gels were significantly softer than 4PINA gels. At a low strain of 1.0%, both diNO and 4PINA samples had an effectively infinite shear stress relaxation time, typical of a temporally persistent supramolecular elastic network. On the other hand, at a higher strain of 10.0% the relaxation time for the 4PINA sample, which had sheared, now displayed a finite relaxation time and behaved like a viscous liquid. The morphologies of the gel fibers were analyzed using SEM images of the dried gel, and the results indicated that a microcrystalline network was observed in diNO gels. The xerogels of 4PINA at a similar concentration displayed fibrous morphology with tape-type architecture with widths ranging from 0.57 to 5.7 μm [34], which was different from diNO gels. This is quite interesting because changing the pyridyl groups to *N*-oxide resulted in a drastic change in the morphology of the gel fibers. The stronger gelation properties of 4PINA may be attributed to the ability of 4PINA to immobilize the solvent molecules more efficiently in the fibrous network compared to the *N*-oxide network.

The crystallization experiments of the *N*-oxides performed in aqueous solutions of polar solvents resulted in either needle-shaped crystals or precipitate, which prompted us to explore the crystallization in pure water. The crystallization of INO and PNO in water resulted in blocked-shaped crystals in 3–5 h, but needle-shaped diNO crystals were formed in 1 to 2 days. The solid-state structures of the *N*-oxides were analyzed by single-crystal X-ray diffraction. The structure of *N*-oxides with both ends modified (diNO) was compared to the parent 4PINA structure and these two structures differ mainly in the type of non-bonding interactions. The crystal structure of 4PINA showed that the pyridyl nitrogen atoms and the amide moieties were hydrogen bonded via N—H···N interaction to form a 1-D hydrogen-bonded network. However, discrete dimers were formed in diNO by N—H···O hydrogen bonding, which was stabilized by π–π interactions. These dimers interact with each other via bifurcated C—H···O interactions and π–π interactions to form a 2-D supramolecular structure. The comparison of the intermolecular interactions in both the structures indicated that the N—H···N interaction in 4PINA is replaced by bifurcated C—H···O interactions in diNO. This explains why 4PINA is a good gelator, which indicates the importance of N—H···N interaction in the formation of supergelators. The phase purity of the *N*-oxide compounds was analyzed by comparing the diffraction pattern of the crystal structures with the powder X-ray pattern of the bulk materials. The analysis of the pattern revealed that the recrystallized form of diNO matched the single-crystal structure of diNO, indicating an identical structure. However, the XRPD pattern of the xerogels of diNO did not match the pattern obtained from the crystal structure of diNO. This is presumably due to the presence of hydrogen-bonded solvent molecules in the xerogels. The XRPD data of the bulk materials of INO and PNO.2H_2O matched their corresponding crystal structures. However, the XRPD pattern of INO.2H_2O matched the crystal structure of the INO pattern, confirming the transformation of INO.2H_2O crystals due to the loss of water molecules during the drying process.

The crystal structure analysis of INO obtained from water revealed the presence of two forms, namely INO and INO.2H_2O. These two forms were crystallized from water and the rapid crystallization process produced large block-shaped crystals of INO.2H_2O. The gelation experiments performed at higher concentration (>40.0 mg in 1.0 mL water) also favored the crystallization of INO.2H_2O. The selective crystallization of INO was obtained by the slow crystallization in dilute aqueous solution over a period of five days. This indicates that INO.2H_2O was the kinetically favored form and these crystals were slowly converted to the thermodynamically stable INO form, which was confirmed by powder X-ray analysis. The solid-state structure of these two forms was different and the molecule was planar in INO.2H_2O compared to INO. This was due to the hydrogen-bonding interaction of the water molecules resulting in a hydrogen-bonded macrocycle involving two INO molecules and four water molecules. The structures of INO and diNO were compared to analyze the importance of identical groups on both ends in the gelation process. The solid-state structures of INO and diNO were similar resulting in the dimer formation, but the non-bonding interaction responsible for the propagation of the dimers was different. The dimer in diNO propagated via C—H···O interactions (3.1789(18) and 3.1621(19) Å) and this was replaced by a C—H···N interaction (3.4391(16) and 3.5489(16) Å) in INO. The selective gelation of diNO may be attributed to the strong C—H···O interaction compared to the non-bonding interaction in INO. The structure analysis of PNO revealed the presence of two water molecules in the crystal structure, which helped the molecule to retain the molecular planarity. The water molecules play a crucial role in overall packing of the crystal, and a hexagonal architecture of oxygen atoms was formed by the hydrogen bonding between PNO and the water molecules. The structural analysis of 4PINA and the *N*-oxides confirm the importance of the non-bonding interactions and the presence of identical groups at both ends of the compounds in gel formation.

MATERIALS AND METHODS

Chemicals and Reagents

All starting materials and solvents were purchased from Sigma-Aldrich (MEDOR ehf, Reykjavik, Iceland) and were used as supplied. Deionized water was used for all the experiments and anhydrous methanol was obtained by distilling the solvent over Mg turnings and iodine. 4-aminopyridine-1-oxide [**76**] and *N*-(4-pyridyl)isonicotinamide (**4PINA**) [**34**] were synthesized following the reported procedures. ^{1}H NMR and IR spectra were recorded on a Bruker Advance 400 spectrometer (Rheinstetten, Germany) and a Nicolet iN10 (Thermo Fisher Scientific, Hvidovre, Denmark), respectively. Single-crystal X-ray diffraction (SCXRD) was performed on a Bruker D8 VENTURE (Karlsruhe, Germany), and X-ray powder diffraction (XRPD) was carried out using a Bruker D8 Focus instrument (Karlsruhe, Germany). The morphology of the xerogel was analyzed by scanning electron microscopy (SEM) using a Leo Supra 25 Microscope (Carl Zeiss, Oberkochen, Germany).

Synthesis

General procedure for INO and PNO: The acid chloride was prepared by stirring the carboxylic acid (10.0 mmol) with 5.0 mL of thionyl chloride overnight in a round-bottom flask at 65.0 °C. The pale-yellow solution obtained was cooled and the excess thionyl chloride was removed by distillation. The resulting white solid/crystal was transferred to a 100 mL two-neck RB flask containing 4-aminopyridine/4-aminopyridine-1-oxide (10.0 mmol). A pale yellowish solution was obtained by adding 20.0 mL anhydrous DMF to the flask, the solution was cooled to 0 °C, and triethylamine (1.5 mL, 10.8 mmol) was added dropwise to this solution. The resulting mixture was stirred overnight at room temperature and the solution was added to a beaker containing 200 mL diethyl ether resulting in a white suspension. The mixture was filtered, and the residue was stirred with 4.0% $NaHCO_3$ for 5 h, filtered, and washed with cold water. The resulting white solid was recrystallized from water to obtain the desired product.

4-(pyridin-4-ylcarbamoyl)pyridine-1-oxide (INO): Isonicotinic acid-*N*-oxide (1.39 g, 10.0 mmol) and 4-aminopyridine (0.94 g, 10.0 mmol). Yield: 72.0% (1.55 g, 7.2 mmol). ^{1}H NMR (400 MHz, DMSO-d_6) δ: 10.68 (1H, s), 8.47 (2H d, J = 8.0), 8.36 (2H d, J = 8.0), 7.95 (2H d, J = 8.0), 7.73 (2H d, J = 8.0). MS (ESI) m/z for $C_{11}H_9N_3O_2Na^+$: expected 238.20, found 238.20.

4-(isonicotinamido)pyridine-1-oxide (PNO): Isonicotinic acid (1.23 g, 10.0 mmol) and 4-aminopyridine-1-oxide (1.10 g, 10.0 mmol). Yield: 79.0% (1.70 g, 7.9 mmol). ^{1}H NMR (400 MHz, DMSO-d_6) δ: 10.98 (1H, s), 8.80 (2H d, J = 6.0), 8.19 (2H d, J = 7.6), 7.85 (2H d, J = 4.4, 1.4), 7.82 (2H dd, J = 5.4, 2.2). MS (ESI) m/z for $C_{11}H_9N_3O_2Na^+$: expected 238.20, found 238.19.

4-((1-oxidopyridin-4-yl)carbamoyl)pyridine-1-oxide (diNO): To a solution of *N*-(4-pyridyl)isonicotinamide (1.35 g, 6.8 mmol) in MeOH (25 mL), *m*-chloroperoxybenzoic acid (3.80 g, 17.0 mmol) was added in portions for 15 min. The reaction mixture was refluxed overnight at 70.0 °C and cooled to room temperature. The solid was collected by filtration, washed with hot methanol, and then dried. Yield 0.88 g, 56.0%. ^{1}H NMR (400 MHz, DMSO-d_6) δ: 10.81 (1H, s), 8.39 (2H d, J = 7.2), 8.18 (2H d, J = 7.2), 7.95 (2H d, J = 7.2), 7.79 (2H d, J = 7.2). MS (ESI) m/z for $C_{11}H_9N_3O_3Na^+$: expected 254.05, found 254.05.

Gelation Studies

Gelation experiments were initially carried out by weighing 10.0 mg of the *N*-oxide ligands (INO, PNO, diNO, and an equimolar mixture of INO and PNO) in a 7.0 mL vial and 1.0 mL of distilled water was added to the compounds. The mixtures were heated until a clear solution was obtained; cooling to room temperature resulted in no gelation in any case after 24 h. Since the compounds were soluble only in water, gelation experiments were carried out in aqueous solution of organic solvents. An aqueous solution (1:1, v/v) was prepared by mixing 500 μL of organic solvents (methanol, ethanol, tetrahydrofuran, acetonitrile, or nitrobenzene) with 500 μL of water. These solutions were added to 10.0 mg of the *N*-oxide compounds and the mixtures were heated; cooling to room temperature resulted in a white precipitate after 24 h in all cases. The higher solubility of the compounds in water compared to other solvents prompted us to check the gelation experiments at higher concentration. Crystals were obtained when the gelation experiments were performed in 1.0 mL water with 20.0 mg compounds (2.0 wt %). The gelation ability tested in mixed solvents at 2.0 wt % mostly produced precipitate, whereas some crystalline materials were observed in a few experiments (in methanol/water).

The gelation experiments at higher concentration (4.0 wt %) also resulted in crystals for INO, PNO, and their equimolar mixture but an opaque gel was formed under identical conditions (confirmed by inversion test) for diNO. The gelation experiments at higher concentration in a mixed solvent were avoided due to insolubility of these materials. The solubility of diNO was found to be approximately 60.0 mg/mL in water, and the solution formed gel at concentrations between 4.0 and 6.0 wt %. The two mono-*N*-oxides and their equimolar mixture at high concentration (>4.0 wt %) were dissolved in boiling water, which readily formed crystals on cooling.

Minimum Gel Concentration (MGC)

The MGC was performed by taking various amounts (30.0 to 50.0 mg) of **diNO** in a standard 7.0 mL vial and the vial was sealed after adding 1.0 mL of distilled water. The mixtures were heated to dissolve the compound and left undisturbed to form gel. The minimum concentration at which gel was obtained was noted as MGC. The MGC of **diNO** was found to be 4.0 wt % in pure water.

T_{gel} Experiments

The gel-to-solution transition temperature (T_{gel}) experiment was performed at two different concentrations (4.0 and 6.0 wt %) of **diNO**. The required amount (40.0 or 60.0 mg) of **diNO** was placed in a 7.0 mL vial and 1.0 mL of water was added. The vial was sealed, and the mixture was heated to obtain a clear solution and cooled to room temperature to form the gel. After 24 h, a small spherical glass ball (92.0 mg) was carefully placed over the gel. The vial was gradually heated in an oil bath equipped with a magnetic stirrer and a thermometer. The temperature at which the glass ball touched the bottom of the vial was recorded as T_{gel}. The T_{gel} of **diNO** was found to be 78.0 °C and 80.0 °C at 4.0 and 6.0 wt %, respectively.

Rheology

The rheology experiments were performed at Durham University using a TA Instruments AR 2000 (New Castle, DE, USA) fitted with a rough Peltier plate and a 25 mm rough plate geometry (gap width of 2500 μm) at 25 °C. Supramolecular gels were prepared as discussed previously at a concentration of 4.0 wt %. The top plate was lowered and the normal force was allowed to reach equilibrium. Oscillatory amplitude sweeps were performed at a constant frequency (1.0 Hz) and the oscillatory frequency sweeps were performed at a constant shear stress (0.5 Pa). These experiments were performed in triplicate and the mean moduli and standard deviation were calculated.

Scanning Electron Microscopy (SEM)

The gelator diNO (40.0 mg) was dissolved in 1.0 mL of water by heating and cooled to room temperature to form the gel. After 24 h, the yellowish opaque gel was filtered through a filter paper and the residue was air dried in a fume hood. The xerogels were gold-coated for 2 min and SEM was performed using a Leo Supra 25 Microscope.

Crystallography

X-ray quality single-crystals were isolated from water, immediately immersed in cryogenic oil, and mounted. The diffractions were collected using MoK$_{\alpha}$ radiation (λ = 0.71073 Å) on a Bruker D8 VENTURE (Photon100 CMOS detector) diffractometer equipped with a Cryostream (Oxford Cryosystems, Oxford, UK) open-flow nitrogen cryostat at room temperature. The unit cell determination, data collection, data reduction, structure solution/refinement, and empirical absorption correction (SADABS) were carried out using Apex III (Bruker AXS, Madison, WI, USA). The structure was solved by a direct method and refined by the full-matrix least-squares on F^2 for all data using SHELXTL [**77**] (Version 2017/1, University of Göttingen, Göttingen, Germany) and Olex2 [**78**] (Version 1.2, OlexSys Ltd., Durham, U.K.) software. All non-disordered non-hydrogen atoms were refined anisotropically, and the hydrogen atoms were placed in the calculated positions and refined using a riding model.

X-ray Powder Diffraction (XRPD)

The *N*-oxide compounds (20.0 mg) were dissolved in 1.0 mL of hot water and left undisturbed for crystallization. After crystallization, the mixture was filtered and the residue was dried in air. The crystals were ground to fine powder and XRPD was carried out using a Bruker D8 Focus instrument. The xerogel of **diNO** was prepared by heating 40.0 mg of the compound in 1.0 mL of water (4.0 wt %), and the gel formed was filtered after 24 h followed by drying the residue overnight in a fume hood.

CONCLUSIONS

We synthesized two mono-*N*-oxides (INO and PNO) and a di-*N*-oxide (diNO) by altering the pyridyl moieties of a known hydrogelator (4PINA) to *N*-oxide moieties.

The gelation ability of these compounds tested in various solvent/solvent mixtures indicated that selective gelation of diNO was observed in water. The mechanical strength and the thermal stability of diNO were evaluated using rheology and gel-to-solution transition temperature (T_{gel}) experiments, respectively. The gelation properties of diNO were compared with the parent 4PINA gelator, which indicated that diNO is a weaker gelator. SEM images were used to visualize the changes in the morphology of the gel fibers due to the alteration in the functional groups. The effect of various non-bonding interactions in the crystalline state and dried gel was studied using X-ray diffraction. Single-crystal X-ray analysis of the di-*N*-oxide structure revealed the formation of a hydrogen-bonded dimer, which interacted with adjacent dimers via C—H···O interactions to form the extended network, and similar dimers observed in non-gelator INO were interconnected through C—H···N interaction. The strong interaction of the mono-*N*-oxide with water molecules indicates the increased hydrophilic nature of these compounds to form solvated crystals. The reduced gelation ability (MGC) and thermal strength of diNO may be attributed to the weak intermolecular C—H···O interaction. The absence of strong N—H···N interactions in the *N*-oxide compounds compared to 4PINA clearly indicate the role of these interactions in gel formation. The tuning of gelation ability (MGC) and thermal/mechanical strength of 4PINA by alerting the functional group proves the importance of non-bonding interactions and functional groups in designing LMWGs with tunable properties, which contributes to the ongoing efforts to identify the key structural features of gel network formation.

ACKNOWLEDGMENTS

We thank the University of Iceland Research Fund and the Science Institute for funding. D.G. thanks the University of Iceland for the Doctoral Research and Teaching Assistantship grant. We thankfully acknowledge Jonathan Steed, Durham University, Durham, UK for rheology experiments, Sigurjur Sveinn Jonsson, ISOR-Iceland for X-ray powder diffraction analysis, and Sigridur Jonsdottir, University of Iceland for NMR and mass spectroscopy. We thank Rannís Iceland for granting an infrastructure grant for a single-crystal X-ray diffractometer.

CONFLICTS OF INTEREST

The authors declare no conflict of interest.

REFERENCES

1. Estroff, L.A.; Hamilton, A.D. Water Gelation by Small Organic Molecules. *Chem. Rev.* **2004**, *104*, 1201–1218.
2. De Loos, M.; Feringa, B.L.; van Esch, J.H. Design and Application of Self-Assembled Low Molecular Weight Hydrogels. *Eur. J. Org. Chem.* **2005**, *2005*, 3615–3631.
3. George, M.; Weiss, R.G. Molecular Organogels. Soft Matter Comprised of Low-Molecular-Mass Organic Gelators and Organic Liquids. *Acc. Chem. Res.* **2006**, *39*, 489–497.
4. Dastidar, P. Supramolecular gelling agents: Can they be designed? *Chem. Soc. Rev.* **2008**, *37*, 2699–2715.
5. Hirst, A.R.; Escuder, B.; Miravet, J.F.; Smith, D.K. High-Tech Applications of Self-Assembling Supramolecular Nanostructured Gel-Phase Materials: From Regenerative Medicine to Electronic Devices. *Angew. Chem. Int. Ed.* **2008**, *47*, 8002–8018.
6. Piepenbrock, M.-O.M.; Lloyd, G.O.; Clarke, N.; Steed, J.W. Metal- and Anion-Binding Supramolecular Gels. *Chem. Rev.* **2010**, *110*, 1960–2004.
7. Steed, J.W. Anion-tuned supramolecular gels: A natural evolution from urea supramolecular chemistry. *Chem. Soc. Rev.* **2010**, *39*, 3686–3699.
8. Yu, G.; Yan, X.; Han, C.; Huang, F. Characterization of supramolecular gels. *Chem. Soc. Rev.* **2013**, *42*, 6697–6722.
9. Kumar, D.K.; Steed, J.W. Supramolecular gel phase crystallization: Orthogonal self-assembly under non-equilibrium conditions. *Chem. Soc. Rev.* **2014**, *43*, 2080–2088.
10. Banerjee, S.; Das, R.K.; Maitra, U. Supramolecular gels 'in action'. *J. Mater. Chem.* **2009**, *19*, 6649–6687.
11. Foster, J.A.; Damodaran, K.K.; Maurin, A.; Day, G.M.; Thompson, H.P.G.; Cameron, G.J.; Bernal, J.C.; Steed, J.W. Pharmaceutical polymorph control in a drug-mimetic supramolecular gel. *Chem. Sci.* **2017**, *8*, 78–84.
12. Terech, P.; Weiss, R.G. Low Molecular Mass Gelators of Organic Liquids and the Properties of Their Gels. *Chem. Rev.* **1997**, *97*, 3133–3160.
13. Du, X.; Zhou, J.; Shi, J.; Xu, B. Supramolecular Hydrogelators and Hydrogels: From Soft Matter to Molecular Biomaterials. *Chem.*

Rev. **2015**, *115*, 13165–13307.

14. Babu, S.S.; Praveen, V.K.; Ajayaghosh, A. Functional π-Gelators and Their Applications. *Chem. Rev.* **2014**, *114*, 1973–2129.
15. Worthington, P.; Pochan, D.J.; Langhans, S.A. Peptide Hydrogels—Versatile Matrices for 3D Cell Culture in Cancer Medicine. *Front. Oncol.* **2015**, *5*, 92.
16. Truong, W.T.; Su, Y.; Meijer, J.T.; Thordarson, P.; Braet, F. Self-Assembled Gels for Biomedical Applications. *Chem. Asian J.* **2011**, *6*, 30–42.
17. Li, J.-L.; Wang, R.-Y.; Liu, X.-Y.; Pan, H.-H. Nanoengineering of a Biocompatible Organogel by Thermal Processing. *J. Phys. Chem. B* **2009**, *113*, 5011–5015.
18. Li, J.-L.; Liu, X.-Y.; Wang, R.-Y.; Xiong, J.-Y. Architecture of a Biocompatible Supramolecular Material by Supersaturation-Driven Fabrication of its Fiber Network. *J. Phys. Chem. B* **2005**, *109*, 24231–24235.
19. Afrasiabi, R.; Kraatz, H.-B. Sonication-Induced Coiled Fibrous Architectures of Boc-L-Phe-L-Lys(Z)-OMe. *Chem. Eur. J.* **2013**, *19*, 1769–1777.
20. Dhibar, S.; Dey, A.; Majumdar, S.; Ghosh, D.; Mandal, A.; Ray, P.P.; Dey, B. A supramolecular Cd(ii)-metallogel: An efficient semiconductive electronic device. *Dalton Trans.* **2018**, *47*, 17412–17420.
21. Wang, R.-Y.; Liu, X.-Y.; Li, J.-L. Engineering Molecular Self-Assembled Fibrillar Networks by Ultrasound. *Cryst. Growth Des.* **2009**, *9*, 3286–3291.
22. Alexander, S.L.M.; Korley, L.T.J. Nucleation effects of high molecular weight polymer additives on low molecular weight gels. *Polym. J.* **2018**, *50*, 775–786.
23. Wang, R.-Y.; Liu, X.-Y.; Narayanan, J.; Xiong, J.-Y.; Li, J.-L. Architecture of Fiber Network: From Understanding to Engineering of Molecular Gels. *J. Phys. Chem. B* **2006**, *110*, 25797–25802.
24. Draper, E.R.; Adams, D.J. How should multicomponent supramolecular gels be characterised? *Chem. Soc. Rev.* **2018**, *47*, 3395–3405.
25. Cao, X.; Meng, L.; Li, Z.; Mao, Y.; Lan, H.; Chen, L.; Fan, Y.; Yi, T. Large Red-Shifted Fluorescent Emission via Intermolecular π–π Stacking in 4-Ethynyl-1,8-naphthalimide-Based Supramolecular Assemblies. *Langmuir* **2014**, *30*, 11753–11760.

26. Wang, C.; Zhang, D.; Zhu, D. A Chiral Low-Molecular-Weight Gelator Based on Binaphthalene with Two Urea Moieties: Modulation of the CD Spectrum after Gel Formation. *Langmuir* **2007**, *23*, 1478–1482.
27. Wang, C.; Zhang, D.; Xiang, J.; Zhu, D. New Organogels Based on an Anthracene Derivative with One Urea Group and Its Photodimer: Fluorescence Enhancement after Gelation. *Langmuir* **2007**, *23*, 9195–9200.
28. Wang, G.; Cheuk, S.; Yang, H.; Goyal, N.; Reddy, P.V.N.; Hopkinson, B. Synthesis and Characterization of Monosaccharide-Derived Carbamates as Low-Molecular-Weight Gelators. *Langmuir* **2009**, *25*, 8696–8705.
29. Gao, W.; Liang, Y.; Peng, X.; Hu, Y.; Zhang, L.; Wu, H.; He, B. In situ injection of phenylboronic acid based low molecular weight gels for efficient chemotherapy. *Biomaterials* **2016**, *105*, 1–11.
30. Das, U.K.; Banerjee, S.; Dastidar, P. Remarkable Shape-Sustaining, Load-Bearing, and Self-Healing Properties Displayed by a Supramolecular Gel Derived from a Bis-pyridyl-bis-amide of L-Phenyl Alanine. *Chem. Asian J.* **2014**, *9*, 2475–2482.
31. Tang, Y.-T.; Dou, X.-Q.; Ji, Z.-A.; Li, P.; Zhu, S.-M.; Gu, J.-J.; Feng, C.-L.; Zhang, D. C2-symmetric cyclohexane-based hydrogels: A rational designed LMWG and its application in dye scavenging. *J. Mol. Liq.* **2013**, *177*, 167–171.
32. Li, P.; Dou, X.-Q.; Tang, Y.-T.; Zhu, S.; Gu, J.; Feng, C.-L.; Zhang, D. Gelator-polysaccharide hybrid hydrogel for selective and controllable dye release. *J. Colloid Interface Sci.* **2012**, *387*, 115–122.
33. Ghosh, D.; Lebedytė, I.; Yufit, D.S.; Damodaran, K.K.; Steed, J.W. Selective gelation of *N*-(4-pyridyl)nicotinamide by copper(ii) salts. *CrystEngComm* **2015**, *17*, 8130–8138.
34. Kumar, D.K.; Jose, D.A.; Dastidar, P.; Das, A. Nonpolymeric Hydrogelator Derived from *N*-(4-Pyridyl)isonicotinamide. *Langmuir* **2004**, *20*, 10413–10418.
35. Sahoo, P.; Kumar, D.K.; Raghavan, S.R.; Dastidar, P. Supramolecular Synthons in Designing Low Molecular Mass Gelling Agents: L-Amino Acid Methyl Ester Cinnamate Salts and their Anti-Solvent-Induced Instant Gelation. *Chem. Asian J.* **2011**, *6*, 1038–1047.
36. Bhattacharjee, S.; Bhattacharya, S. Role of synergistic [small pi]-[small pi] stacking and X-H[three dots, centered]Cl (X = C, N, O)

H-bonding interactions in gelation and gel phase crystallization. *Chem. Commun.* **2015**, *51*, 7019–7022.

37. Jones, C.D.; Steed, J.W. Gels with sense: Supramolecular materials that respond to heat, light and sound. *Chem. Soc. Rev.* **2016**, *45*, 6546–6596.
38. Ghosh, D.; Ferfolja, K.; Drabavičius, Ž.; Steed, J.W.; Damodaran, K.K. Crystal habit modification of Cu(ii) isonicotinate–*N*-oxide complexes using gel phase crystallisation. *New J. Chem.* **2018**, *42*, 19963–19970.
39. Desiraju, G.R. Supramolecular Synthons in Crystal Engineering—A New Organic Synthesis. *Angew. Chem. Int. Ed.* **1995**, *34*, 2311–2327.
40. Weiss, R.G.; Terech, P. *Molecular Gels: Materials with Self-Assembled Fibrillar Networks*; Springer: Heidelberg, Germany, 2006; p. 978.
41. Fages, F.; Voegtle, F.; Zinic, M. Systematic design of amide- and urea-type gelators with tailored properties. *Top. Curr. Chem.* **2005**, *256*, 77–131.
42. Isare, B.; Pensec, S.; Raynal, M.; Bouteiller, L. Bisurea-based supramolecular polymers: From structure to properties1. *C. R. Chim.* **2016**, *19*, 148–156.
43. Gale, P.A.; Busschaert, N.; Haynes, C.J.E.; Karagiannidis, L.E.; Kirby, I.L. Anion receptor chemistry: Highlights from 2011 and 2012. *Chem. Soc. Rev.* **2014**, *43*, 205–241.
44. Dzolic, Z.; Cametti, M.; Dalla Cort, A.; Mandolini, L.; Zinic, M. Fluoride-responsive organogelator based on oxalamide-derived anthraquinone. *Chem. Commun.* **2007**, *34*, 3535–3537.
45. Kotova, O.; Daly, R.; dos Santos, C.M.G.; Boese, M.; Kruger, P.E.; Boland, J.J.; Gunnlaugsson, T. Europium-Directed Self-Assembly of a Luminescent Supramolecular Gel from a Tripodal Terpyridine-Based Ligand. *Angew. Chem. Int. Ed.* **2012**, *51*, 7208–7212.
46. Feng, L.; Cavicchi, K.A. Investigation of the relationships between the thermodynamic phase behavior and gelation behavior of a series of tripodal trisamide compounds. *Soft Matter* **2012**, *8*, 6483–6492.
47. Mukhopadhyay, S.; Ira; Krishnamoorthy, G.; Maitra, U. Dynamics of Bound Dyes in a Nonpolymeric Aqueous Gel Derived from a Tripodal Bile Salt. *J. Phys. Chem. B* **2003**, *107*, 2189–2192.
48. Li, Z.; Cao, J.; Li, H.; Liu, H.; Han, F.; Liu, Z.; Tong, C.; Li, S. Self-assembled drug delivery system based on low-molecular-weight bis-amide organogelator: Synthesis, properties and in vivo evaluation. *Drug*

Deliv. **2016**, *23*, 3168–3178.

49. Bradberry, S.J.; Dee, G.; Kotova, O.; McCoy, C.P.; Gunnlaugsson, T. Luminescent lanthanide (Eu(iii)) cross-linked supramolecular metallo co-polymeric hydrogels: The effect of ligand symmetry. *Chem. Commun.* **2019**, *55*, 1754–1757.
50. Kumar, D.K.; Jose, D.A.; Dastidar, P.; Das, A. Nonpolymeric Hydrogelators Derived from Trimesic Amides. *Chem. Mater.* **2004**, *16*, 2332–2335.
51. Meazza, L.; Foster, J.A.; Fucke, K.; Metrangolo, P.; Resnati, G.; Steed, J.W. Halogen-bonding-triggered supramolecular gel formation. *Nat. Chem.* **2013**, *5*, 42–47.
52. Pal, A.; Ghosh, Y.K.; Bhattacharya, S. Molecular mechanism of physical gelation of hydrocarbons by fatty acid amides of natural amino acids. *Tetrahedron* **2007**, *63*, 7334–7348.
53. Kawamoto, K.; Grindy, S.C.; Liu, J.; Holten-Andersen, N.; Johnson, J.A. Dual Role for 1,2,4,5-Tetrazines in Polymer Networks: Combining Diels–Alder Reactions and Metal Coordination to Generate Functional Supramolecular Gels. *ACS Macro Lett.* **2015**, *4*, 458–461.
54. Das, U.K.; Banerjee, S.; Dastidar, P. Primary Ammonium Monocarboxylate Synthon in Designing Supramolecular Gels: A New Series of Chiral Low-Molecular-Weight Gelators Derived from Simple Organic Salts that are Capable of Generating and Stabilizing Gold Nanoparticles. *Chem. Asian J.* **2013**, *8*, 3022–3031.
55. Jin, Q.; Zhang, L.; Liu, M. Solvent-Polarity-Tuned Morphology and Inversion of Supramolecular Chirality in a Self-Assembled Pyridylpyrazole-Linked Glutamide Derivative: Nanofibers, Nanotwists, Nanotubes, and Microtubes. *Chem. Eur. J.* **2013**, *19*, 9234–9241.
56. Karak, S.; Kumar, S.; Bera, S.; Diaz, D.D.; Banerjee, S.; Vanka, K.; Banerjee, R. Interplaying anions in a supramolecular metallohydrogel to form metal organic frameworks. *Chem. Commun.* **2017**, *53*, 3705–3708.
57. Kartha, K.K.; Praveen, V.K.; Babu, S.S.; Cherumukkil, S.; Ajayaghosh, A. Pyridyl-Amides as a Multimode Self-Assembly Driver for the Design of a Stimuli-Responsive π-Gelator. *Chem. Asian J.* **2015**, *10*, 2250–2256.
58. Liu, Q.; Wang, Y.; Li, W.; Wu, L. Structural Characterization and Chemical Response of a Ag-Coordinated Supramolecular

Gel. *Langmuir* **2007**, *23*, 8217–8223.

59. Lee, H.H.; Jung, S.H.; Park, S.; Park, K.-M.; Jung, J.H. A metal-organic framework gel with Cd2+ derived from only coordination bonds without intermolecular interactions and its catalytic ability. *New J. Chem.* **2013**, *37*, 2330–2335.
60. Aoyama, R.; Sako, H.; Amakatsu, M.; Yamanaka, M. Palladium ion-induced supramolecular gel formation of tris-urea molecules. *Polym. J.* **2015**, *47*, 136–140.
61. Ghosh, K.; Panja, S.; Bhattacharya, S. Naphthalene linked pyridyl urea as a supramolecular gelator: A new insight into naked eye detection of I- in the gel state with semiconducting behaviour. *RSC Adv.* **2015**, *5*, 72772–72779.
62. Kumar, D.P. Synthesis of gold nanoparticles and nanoclusters in a supramolecular gel and their applications in catalytic reduction of p-nitrophenol to p-aminophenol and Hg(ii) sensing. *RSC Adv.* **2014**, *4*, 45449–45457.
63. Byrne, P.; Lloyd, G.O.; Applegarth, L.; Anderson, K.M.; Clarke, N.; Steed, J.W. Metal-induced gelation in dipyridyl ureas. *New J. Chem.* **2010**, *34*, 2261–2274.
64. Pandurangan, K.; Kitchen, J.A.; Blasco, S.; Paradisi, F.; Gunnlaugsson, T. Supramolecular pyridyl urea gels as soft matter with antibacterial properties against MRSA and/or *E. coli*. *Chem. Commun.* **2014**, *50*, 10819–10822.
65. Stanley, C.E.; Clarke, N.; Anderson, K.M.; Elder, J.A.; Lenthall, J.T.; Steed, J.W. Anion binding inhibition of the formation of a helical organogel. *Chem. Commun.* **2006**, *30*, 3199–3201.
66. Goodwin, J.W.; Hughes, R.W. *Rheology for Chemists: An Introduction*, 2nd ed.; Royal Society of Chemistry: Cambridge, UK, 2008; p. 264.
67. Guenet, J.-M. Rheological Aspects. In *Organogels: Thermodynamics, Structure, Solvent Role, and Properties*; Guenet, J.-M., Ed.; Springer International Publishing: Cham, Switzerland, 2016; pp. 83–94.
68. Sathaye, S.; Mbi, A.; Sonmez, C.; Chen, Y.; Blair, D.L.; Schneider, J.P.; Pochan, D.J. Rheology of peptide- and protein-based physical hydrogels: Are everyday measurements just scratching the surface? *Wiley Interdiscip. Rev. Nanomed. Nanobiotechnol.* **2015**, *7*, 34–68.

69. Zuidema, J.M.; Rivet, C.J.; Gilbert, R.J.; Morrison, F.A. A protocol for rheological characterization of hydrogels for tissue engineering strategies. *J. Biomed. Mater. Res. B* **2014**, *102*, 1063–1073.
70. Gaharwar, A.K.; Avery, R.K.; Assmann, A.; Paul, A.; McKinley, G.H.; Khademhosseini, A.; Olsen, B.D. Shear-Thinning Nanocomposite Hydrogels for the Treatment of Hemorrhage. *ACS Nano* **2014**, *8*, 9833–9842.
71. Dasgupta, D.; Thierry, A.; Rochas, C.; Ajayaghosh, A.; Guenet, J.M. Key role of solvent type in organogelation. *Soft Matter* **2012**, *8*, 8714–8721.
72. Colquhoun, C.; Draper, E.R.; Eden, E.G.B.; Cattoz, B.N.; Morris, K.L.; Chen, L.; McDonald, T.O.; Terry, A.E.; Griffiths, P.C.; Serpell, L.C.; et al. The effect of self-sorting and co-assembly on the mechanical properties of low molecular weight hydrogels. *Nanoscale* **2014**, *6*, 13719–13725.
73. Raeburn, J.; Adams, D.J. Multicomponent low molecular weight gelators. *Chem. Commun.* **2015**, *51*, 5170–5180.
74. Tómasson, D.A.; Ghosh, D.; Kržišnik, Z.; Fasolin, L.H.; Vicente, A.A.; Martin, A.D.; Thordarson, P.; Damodaran, K.K. Enhanced Mechanical and Thermal Strength in Mixed-Enantiomers-Based Supramolecular Gel. *Langmuir* **2018**, *34*, 12957–12967.
75. Sauvée, C.; Ström, A.; Haukka, M.; Sundén, H. A Multi-Component Reaction towards the Development of Highly Modular Hydrogelators. *Chem. Eur. J.* **2018**, *24*, 8071–8075.
76. Gardner, J.N.; Katritzky, A.R. 875. *N*-oxides and related compounds. Part V. The tautomerism of 2- and 4-amino- and -hydroxy-pyridine 1-oxide. *J. Chem. Soc.* **1957**, 4375–4385.
77. Sheldrick, G.M. Crystal structure refinement with SHELXL. *Acta Crystallogr. Sect. C Struct. Chem.* **2015**, *71 Pt 1*, 3–8.
78. Dolomanov, O.V.; Bourhis, L.J.; Gildea, R.J.; Howard, J.A.K.; Puschmann, H. OLEX2: A complete structure solution, refinement and analysis program. *J. Appl. Crystallogr.* **2009**, *42*, 339–341.

CHAPTER 7

Supramolecular Gels with High Strength by Tuning of Calix[4] Arene-Derived Networks

Ji Ha Lee[1], Jaehyeon Park[1], Jin-Woo Park[2], Hyo-Jun Ahn[2], Justyn Jaworski[3],[4] and Jong Hwa Jung[1]

[1]Department of Chemistry and Research Institute of Natural Sciences Gyeongsang National University, Jinju 660-701, Korea.

[2] School of Materials Science and Engineering, Gyeongsang National University, Jinju 660-701, Korea.

[3]Department of Chemical Engineering, Hanyang University, Seoul 133-791, Korea.

[4] Institute of Nanoscience and Technology, 222 Wangsimni-ro, Seoul 133-791, Korea

Citation: Lee, J. H. et al. "Supramolecular gels with high strength by tuning of calix[4]arene-derived networks". Nat. Commun. 6:6650 (2015). https://doi.org/10.1038/ncomms7650

ABSTRACT

Supramolecular gels comprised of low-molecular-weight gelators are generally regarded as mechanically weak and unable to support formation of free-standing structures, hence, their practical use with applied loads has been limited. Here, we reveal a technique for *in situ* generation of high tensile strength supramolecular hydrogels derived from low-molecular-weight gelators. By controlling the concentration of hydrochloric acid during hydrazone formation between calix-[4]arene-based gelator precursors, we tune the mechanical and ductile properties of the resulting gel. Organogels formed without hydrochloric acid exhibit impressive tensile strengths, higher than 40 MPa, which is the strongest among self-assembled gels. Hydrogels, prepared by solvent exchange of organogels in water, show 7,000- to 10,000-fold enhanced mechanical properties because of further hydrazone formation. This method of molding also allows the gels to retain shape after processing, and furthermore, we find organogels when prepared as gel electrolytes for lithium battery applications to have good ionic conductivity.

INTRODUCTION

Soft material development has seen a wealth of applications for viscoelastic systems such as organogels and hydrogels, particularly for areas including controlled release1 and soft tissue reconstruction2. Aside from biomedical applications, gel systems have been implemented in energy capture and storage3 as well as sensing areas4 by revealing exciting responsive properties to external stimuli such as temperature, pH, light and electric field5. In a typical gel system, a gelator component makes use of extended structures to facilitate the formation of thermodynamically stable networks via physical or chemical interactions. In contrast to most chemically formed gels that have irreversible shapes due to permanent network structures, physical gels offer reversible networks capable of functional responsiveness and interesting self-healing properties6. These extended network structures often utilize polymers or fibrous supramolecular assemblies, which in the case of a hydrogel are closely associated with a large proportion of water-molecules that remain within the networked structure. Similarly, in the case of an organogel, apolar solvent molecules are associated with the network depending on the properties of the component gelators implemented.

Traditionally, hydrogel and organogel systems derived from polymers have been used in application areas where their characteristically weak mechanical properties would not be a limitation. Recent works have looked to forming hydrogels using polymeric composites with clay[7,8,9], polymers reinforced with silica[10] or peroxidized microspheres[11], and mechanophores capable of shear force activated crosslinking[12]. Despite improved compressive strengths, in general the tensile strength remained largely unaffected. By employing a secondary phase of embedded microgel particles, a striking enhancement in tensile strength was observed by the group of Jian Ping Gong in polymeric hydrogels[13]; although, network rupture in the microgel components was found to cause irreversible gel damage. The same group later reported that polymers exhibiting a mixture of cationic and anionic repeats could provide a mixture of strong and weak bonds allowing ductile hydrogels to have tunable tensile strength up to 2 MPa (ref. 14). Recently, polyurethane-urea-derived hydrogels have moved past this boundary by demonstrating controllable tensile strength from 3.3 to 34 MPa depending on the diisocyanate content[15]. Although such advances provide a bright outlook for implementation of polymeric hydrogel systems in high strength application areas, similar advances have yet to be made for conventional supramolecular hydrogel systems.

Compared with conventional polymeric gelator systems, supramolecular gels such as hydrogels derived from cyclodextrin-based gelators[16] have yet to yield mechanically robust materials for applications requiring high strength. In one example, cyclodextrin-derived supramolecular hydrogels revealed tensile strengths of 18 kPa; however, this system instead provided an impressive reversible elasticity up to 180% strain[17]. For many supramolecular gels, the inherently weak nature of non-covalent interactions that comprise the self-assembled network structures may cause such low-mechanical strengths. Attempts to overcome this limitation by strengthening supramolecular hydrogels using multivalent interactions have had some success. In a specific example, researchers developed hydrogels from zwitterionic amphiphiles encompassing an amino-acid head group and glycolipid hydrogelator[18]. Such a molecular gelator that can allow multiple orthogonal binding sites, may potentially yield a ‘cluster effect’ wherein complexes exhibit increased stability[19] that may translate to enhanced mechanical properties. Although overcoming some of the mechanical fragility associated with supramolecular gels, the mechanical properties of the aforementioned hydrogel were nevertheless still only on par with that of agarose-based gels.

Aside from the mechanical limitation put forth, supramolecular hydrogels derived from low-molecular-weight gelators may face additional issues related to solubility in water, as many macrocyclic gelators are predominantly hydrophobic. Reports using calixarene-based hydrogels, although rare, provide evidence that proper functionalization of these macrocycles can yield effective hydrogelators[20,21,22,23]; however, we could not find any reports on their mechanical properties. The scarcity of studies on the mechanical properties of organogels and hydrogels prepared from self-assembled low-molecular-weight gelators may stem from these materials being very weak.

Herein we reveal an alternative strategy for generating calix[4]arene-derived hydrogels with high-mechanical strength for a supramolecular gel. The tensile strengths of the hydrogels approached 25 MPa (and greater than 40 MPa in the case of the organogel), which surpasses the mechanical properties of many polymer-based physical gels. In this system, we circumvent solubility issues by first generating an organogel from a calix[4]arene-derived gelator in dimethyl sulfoxide (DMSO) and subsequently utilizing solvent exchange after gel formation for *in situ* hydrogel formation. The sensitivity of the gelator components to the conditions of the hydrazone reaction allows the effective tuning of not only the conversion pathway of the gelator but also the network structure to provide a controlled means of strengthening the hydrogels mechanical properties. The material is ductile and offers retention in shape, even after solvent exchange; which makes this strategy using a monomeric gelator system effective for molding of freely-standing objects. When implementing the organogels as gel electrolytes, we observe good ionic conductivities. In addition, we find the extracted liquid from the processed hydrogels to be non-toxic under *in vitro* human cell culture conditions. As we will discuss, this new material offers the potential to extend the possible applications of supramolecular gel systems to more mechanically demanding roles.

RESULTS

Supramolecular organogels formed by oligomerized gelators

To provide a gelator that could be formed *in situ*, we first synthesized calix[4]arene-derivative (**1**) and diphenyl terephthalate-derivative (**2**) precursors possessing reactive hydrazide and aldehyde groups, respectively. The hydrazide moieties of precursor **1** upon reaction with the aldehyde groups of precursor **2** result in hydrazone formation, wherein these linkages can offer amenable hydrogen-bonding sites to facilitate non-covalent assembly. The resulting gelator consists of 1,3 alternate-con-type calix[4]arene as the core component bordered by hydrazone groups linking with the hydrophobic diphenyl terephthalate derivative. Subsequent oligomerization can furthermore provide a larger network-structured gelator that is inherently less susceptible to crystallization. Importantly, this can be carried out *in situ* by the formation of hydrazone bonds between any tetrahydrazide-bearing **1** and up to four molecules of the aldehyde functionalized **2**. The linking of gelator precursors through hydrazone bonds yields a short oligomer network (Fig. 1) that can be controlled by addition of different catalytic amounts of acid. The hydrogen-bonding capabilities of the oligomeric gelator components provide a framework for inter-gelator hydrogen bonding into an extended network for supramolecular gel formation.

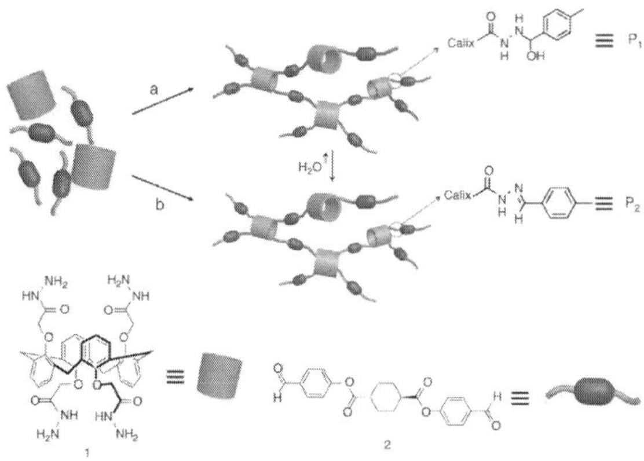

Figure 1: Schematic of gel formation by hydrazone reaction.

The formation of oligomers from calix[4]arene-based gelator precursors by hydrazone reaction (a) without addition of acid and (b) with addition of acid.

Effects of hydrazone formation rate on organogel properties

To observe the effect on gel formation, HCl was systematically added to solutions containing **1** and **2** in DMSO (3wt% with respect to **1**) at a 1:2 molar ratio to provide equivalent proportions of hydrazide and aldehyde functional groups. Transparent organogel-7 (generated in the absence of acid) increased viscosity after 1 day, forming a weak gel capable of handling 1.5 days after initiating the reaction. In contrast, increasing the amount of HCl in the solutions to 10 nmol (organogel-5) and 5 nmol (organogel-6) resulted in the rapid onset of gel formation by 11 and 15 min, respectively. Although the addition of HCl clearly accelerated formation of oligomerization in the organogels, it was also found to elicit a yellow hue, believed to result from carbazone formation[24,25], and also the appearance of turbidity in the resulting gels (Fig. 2). Longer reaction times and higher HCl content resulted in an increased yellow hue for the gels. Upon measuring the sample absorbance at 450 nm, almost no loss in transparency was observed, even after several days, for the organogel prepared in the absence of acid, referred to herein as organogel-7.

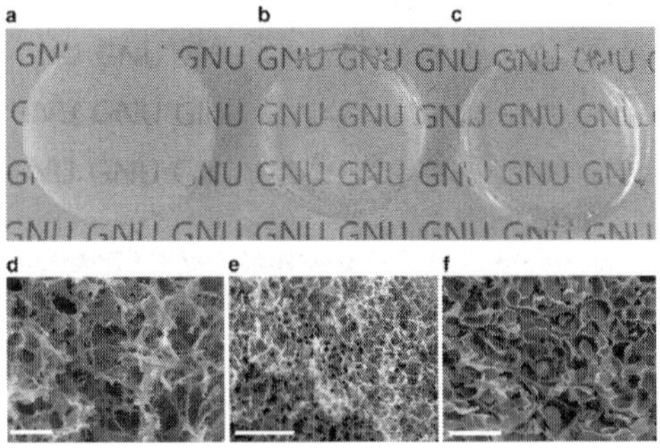

Figure 2: Photograph and SEM images of organogels at different concentration of HCl.

(**a**) Organogel-5 and (**d**) SEM of corresponding xerogel (scale bar, 2 μm) formed using 10 nmol HCl (after 24 h reaction), (**b**) organogel-6 and (**e**) SEM of corresponding xerogel (scale bar, 2 μm) formed using 5 nmol HCl (after 24 h reaction), as well as (**c**) organogel-7 and (**f**) SEM of corresponding xerogel (scale bar, 50 μm) formed without addition of HCl (after 36 h reaction) were all produced using a mixture of **1** (42 mmol) and **2** (84 mmol) in DMSO.

In contrast, significant turbidity was observed shortly after gel formation when gel formation was conducted in the presence of 10 nmol HCl (organogel-5) and 5 nmol HCl (organogel-6). We suspected that the turbidity was strongly related to the structural formation within the gel, hence, we observed the respective organogels network morphology by scanning electron microscope (SEM; Fig. 2). Freeze-dried organogels formed with different amounts of HCl showed very distinct microstructured networks as revealed by SEM. Although organogel-7 revealed a honeycomb-like structure with large open plate-like networks (Fig. 2f), the ogranogel-6 and organogel-5 samples had comparatively much smaller pore structures and their networks appeared more like heterogeneous fibrils (Fig. 2d,e). Despite the heterogeneity for the acidic conditions, the micropores still appear to be highly interconnected.

In an effort to follow changes in the molecular structure of the gelator associated with oligomerization during organogel formation, the evolution of the chemical species over time with respect to different catalytic amounts of HCl was determined by nuclear magnetic resonance spectroscopy (NMR). Measurements of the ^{1}H NMR spectra at various points during gelation allowed the extent of hydrazone bond formation to be evaluated from the ratio of the aldehyde proton of precursor **2** (peak 7) to the –NH proton peak present in the gelator product after coupling. The ratio of the integrals of peak 12 and peak 7 are increasing in the spectra because of the formation of the hydrazone group as the reaction proceeds. The integral ratio of peak 12 to peak 7 may seem to decrease because of broadening of the NMR peak resulting from strong gel formation when the hydrazone reaction reaches equilibrium. From these data, the degrees of hydrazone linkage formation were estimated to reach 81%, 73% and 40% reaction for organogel-5, -6 and -7, respectively. In addition, the reactions were found to reach equilibrium within 30 min for the 10 nmol condition of organogel-5 and within 60 min for the 5 nmol condition of organogel-6 thereby resulting in coupled gelators comprised of 4-mers and 3-mers, respectively, based on an average across samples. The gel formed without addition of acid (organogel-7), in contrast, required at least 60 h to reach equilibrium of the oligomerization reaction, resulting in gelators comprised of predominantly 4-mers.

To confirm that the hydrazone reaction did not continue to polymerize the gelator into a covalently cross-linked network structure, we verified the reversibility of gel formation. Repeated measurements of the sol-gel transition were examined using differential scanning calorimetry (DSC) to reveal endothermic reaction temperatures corresponding to a T_{gel} of 105±3 °C, 100±2 °C and 98±2 °C for organogel-5, -6 and -7, respectively. These sol-gel transition temperatures were reproducible over the three repeated observations of the organogels. The endothermic peaks occur at the transition of the organogel into a solution phase and have a relatively broadened thermogram in comparison to purely monomeric organogels, which is due to the occurrence of a distribution of multi-meric species arising from our hydrazone reaction. Thus, we can be certain the resulting supramolecular gel was formed from discrete gelator species that were multi-meric rather than extended polymer network structures. In addition, a tube inversion method with heating and cooling of organogel-5 provided further visual confirmation of the reversible sol-gel transition indicating these are supramolecular gels and offered supporting assessment of the sol-gel transition temperature.

Mechanical properties of organogels

Having witnessed the organogels as being robust, we examined their mechanical properties in detail by first evaluating the storage (G') and loss (G'') moduli by rheometric measurement. Strain sweep, frequency sweep and continuous step strain tests were conducted for each of the organogels, and the mechanical properties of organogels during evolution of gel formation as well as subsequent ageing were also examined (Fig. 3 and Fig. 11). Among the different treatment conditions, organogel-5 showed the highest storage modulus as impacted by the increased rate of oligomerization under these conditions. Strain amplitude sweeps of the samples demonstrated an elastic response typical of gels (Fig. 3). The loss modulus of organogel-5 was found to rapidly decrease above the critical strain region (γ=20%), indicating a collapse of the gel state to a quasi-liquid state. After large-amplitude oscillatory break-down, the recovery of the rheological properties occurred quickly, and under the application of the large-amplitude oscillatory force (γ=1,000%; frequency, ω=6.0 rad s^{-1} (1.0 Hz)), we observe a decrease in the G' value from 0.45 MPa to 0.03 KPa, resulting in a quasi-liquid state (tan $\delta=G''/G'\approx4.5$). When the amplitude is decreased (γ=0.5%) at the same frequency (1.0 Hz), we see the G' recovers quickly to the initial value and the system returns to a quasi-solid state (tan $\delta=G''/G'\approx0.21$). Upon comparison with organogel-6, the G' and G'' values of organogel-5 were slightly higher than those observed for organogel-6 after reaching equilibrium.

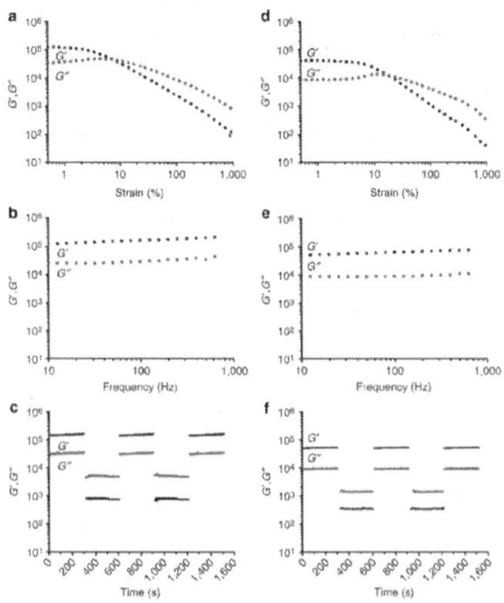

Figure 3: Rheological properties of organogels prepared by hydrazine reaction for 24 h.

Storage (G') and loss (G'') values of organogel-5 (**a–c**), formed using 10 nmol HCl, and (organogel-6 (**d–f**), formed using 5 nmol HCl. Measurements of strain sweep tests at 0.5–1,000% for (**a**) organogel-5 and (**d**) organogel-6. Frequency sweep tests (from 10 to 1,000 Hz) for (**b**) organogel-5 and (**e**) organogel-6. Continuous step strain measurements at 0.5 and 1,000% for (**c**) organogel-5 and (**f**) organogel-6. All experiments conducted at 25 °C.

At equilibrium, each of the organogels exhibited typical strain thinning responses owing to alignment of the microstructure; however, a weak strain overshoot could be observed when measurements were taken before complete equilibrium gel formation as was most pronounced in the case of organogel-7 at 36 h (1.5 day) after initiating the reaction. Such behavior arises during the destruction and formation of network junctions, which may be expected to occur given the reaction does not yet reach equilibrium after 1.5 days26. For each of the organogels, we found the elastic response, G', to dominate across all frequencies, and each of the organogels exhibited immediate recovery of G' and G'' in repeated step strain tests alternating above and below their critical strain amplitudes. The magnitude of the gels elastic response, G', appeared to depend on the extent of acid-induced oligomerization; however, upon ageing, the gels were observed to have different behaviors. In looking more closely at the ageing behaviour of organogel-7, the dynamic modulus values gradually increased after initiating the reaction until the third day when the hydrazone reaction equilibrium had been reached. Even though the reaction had reached equilibrium by this point, further ageing continued to improve the rheological properties through stronger network formation, as we found after 1 week an increase in both the storage and loss moduli by two orders of magnitude. For each of the organogels, we could observe a quick recovery to their original shape once removing the applied load.

To further confirm the interesting progression in the toughness and ductility with ageing, we examined the tensile stress-strain profiles of organogel-7 at additional time points throughout gel formation (Fig. 4a,c). After 1.5 days, the tensile strength of organogel-7 reached ~5 MPa (based on the engineering stress) with an almost 70% maximum strain, but upon continued ageing the gel strength exhibits a continued increase to ~20 MPa by day 3 and over 42 MPa by day 7, albeit with a simultaneous decrease in the maximum strain. This far exceeds the tensile strengths observed for organogel-5 and -6 and is one of the highest among any previously published supramolecular gel14·20·21·22·27, which is expectedly due to the strength of the calix[4]arene-based gelator network. The maximum strain of organogel-7 when examined with respect to reaction time over the course of 1 week reveals a gradual decline over time. Nonetheless, at longer reaction times, organogel-7 exhibited astounding tensile strengths, much higher than even many polymeric hydrogels such as those derived from poly(*N*-isopropylacrylamide) 28 and polyampholytes14. This enhanced tensile strength of organogel-7 is due to the characteristics of the hydrazone reaction under non-acidic conditions, which produces –OH groups at the coupling site of precursors **1** and **2**. As a result, organogel-7 offers enhanced intermolecular hydrogen-bonding interactions in comparison to organogel-5 and -6 that have no auxiliary –OH groups and progressed at faster reaction rates relative to organogel-7. In the case of the stress-strain behaviour of organogel-5 and -6 after 24 h reaction (Fig. 4b), the organogel-6 was found to have a higher maximum strain and tensile strength compared with organogel-5. We find these observed maximum strains of our organogels surpass that of many hydrogels, even those composed of polyampholytes14. We may suspect that the improved ductility may in part arise due to more extensive connectivity in the network structure formed by the gelators when given sufficient time as was observed under slower reaction conditions.

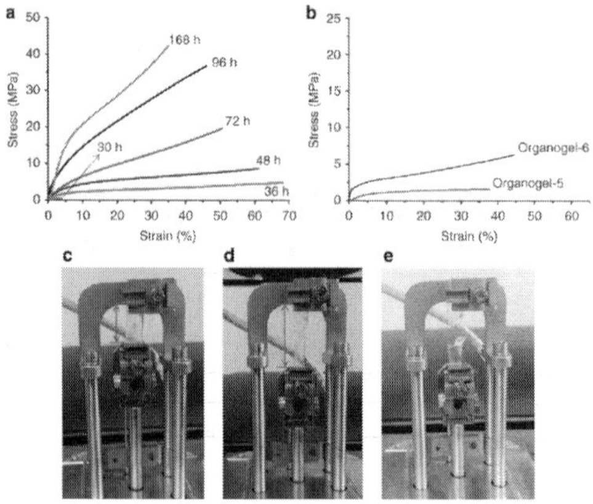

Figure 4: The engineering stress-strain tensile behaviour of organogels.

(**a**) Tensile stress-strain curves of organogel-7 (having no HCl) prepared over (**a**) 30 h (light blue trace, designated by the black arrow), 36 h (1.5 days), 48 h (2 days), 72 h (3 days), 96 h (5 days) or 168 h (7 days) hydrazone reaction. (**b**) Tensile stress-strain curves of organogel-5 (red trace) and organogel-6 (black trace) prepared by hydrazone reaction for 24 h. Example of static tensile stress-strain test of organogel-7 after 36 h reaction illustrating (**c**) the initial state of the gel (scale bar, 2.5 cm), (**d**) the gel during pulling (scale bar, 4 cm), and (**e**) after mechanical failure of the gel.

Hydrogel formation and characterization

Although the surprisingly high strength of organogel-7 is most certainly worthy of merit, because of solvent toxicity the application of this supramolecular gel to biomedical applications would be highly inappropriate. Therefore, we examined the formation of a hydrogel from organogel-7 by exchanging the DMSO solvent in the gel with H_2O (thereafter referred to as hydrogel-7). Exchange of the solvent by 24 h immersion in water provided hydrogel-5, -6 and -7, which maintained their shape, although hydrogel-5 exhibited increased turbidity. After incubation in water, we found the DMSO was almost completely exchanged with H_2O without losing the integrity of the gel network as determined by NMR spectroscopy.

It is important to note that such a hydrogel cannot directly form from precursors **1** and **2**; however, using this distinctive solvent exchange technique we can easily prepare a gelator from these precursors to generate an organogel that can thus undergo replacement with water for forming a functional hydrogel. Moreover, we found an increase in the storage and loss moduli after hydrogel formation. After a hydrazone reaction time of 24 h for organogel-5 and -6 and 36 h for organogel-7, the organogels were then immersed in water for 24 h to allow solvent exchange to realize the corresponding hydrogels.

By weighing the original organogels before and after solvent exchange with water and also after drying, we could determine the solvent content of the gels. Specifically, the water content was determined to be on the order of 93–95% of the total weight of the hydrogel. Significantly larger storage and loss moduli were observed in the resulting hydrogels compared with that of the original organogels after the corresponding hydrazone formation reactions. Hydrogel-7 exhibited the largest increase in elastic and viscous properties compared with the original organogel-7 from which it was produced (7.000- to 10,000-fold). The rheological properties of this hydrogel-7 were almost the same as those found in the samples of organogel-7 aged for 1 week. Hydrogel-6 also had increased elastic and viscous properties and, to a similar extent, so did hydrogel-5. For hydrogels prepared from acidic conditions (hydrogel-5 and -6), we expect the addition of water acted to further catalyze the formation of hydrazone linkages to the enol form (similar to that of oligomers formed under non-acidic conditions as noted in Fig. 1, P_1), thereby increasing the elastic properties via entangled gelator networks formed from extended oligomers. In addition, an increase in intermolecular hydrogen bonding could result from –OH formation as was noted for the reactions conducted in the absence of acid that were described above. This may be justified as the hydrogels exhibited more of a quasi-solid state as compared with that of the organogels, with the exception of organogel-7 and hydrogel-7, which gave similar rheological measurements.

To observe the sol-gel transition temperatures of the hydrogels, we carried out three repeated measurements by DSC, which indicated an increase in T_{gel} compared with that of the organogels. Owing to the possibility of water affecting the heat flow signal, we used a tube inversion method with heating and cooling of the gels to confirm the sol-gel transition temperature. The reversibility of the sol-gel transition of the hydrogels can also be observed. To ensure that the DSC peaks arose from the sol-gel transition, we examined the powder XRD pattern to ensure it was not attributed to melting of any crystalline domains within the gel samples. We found that the various organogels and hydrogels each revealed only a single broad peak at 12–26°. This suggests that the gels are amorphous; hence, we could associate our observed DSC peak with the sol-gel transition. In existing literature[29,30,31,32], the occurrence of X-ray scattering patterns can often be seen at small angles for organogels and hydrogels that possess lamellar structures when the small molecular gelators exist in well-ordered arrangements. In our case, however, the fairly large size of the 3- and 4-mers produced from the hydrazone reaction between **1** and **2** did not yield any sharp peaks in the powder XRD pattern indicating a disordered arrangement. As was the case for the original organogel-7, we also found that enhanced tensile strengths of the hydrogels occurred with an associated decrease in the maximum strain. It may be the case that the increased intermolecular hydrogen bonding or the extent of gelator entanglement could be the basis for the observed decrease in maximum strain.

In addition to rheological examination, we also measured the tensile stress-strain curves of the hydrogels prepared by exchange with water. The tensile strengths and maximum strains of both hydrogel-5 and hydrogel-6 were significantly enhanced compared with that of the original organogel-5 and -6.

The increased tensile strengths of both hydrogel-5 and hydrogel-6 occurred to a greater extent than did the increase in maximum strain, which may be characteristically attributed to the entanglement between gel fibres provided in a chemical cross-linked structure. In contrast, the stress-strain behavior of hydrogel-7 was decreased in terms of the maximum strain relative to the original organogel-7 after 36 h hydrazone reaction. This result indicates that hydrogel-7 became more brittle perhaps resulting from cross-linking within the structure.

With the aim of investigating the molecular level source of the hydrogel's enhanced mechanical properties, we examined the NMR spectra of the organogels before and after the addition of water. As we could expect, the aldehyde proton peak decreased after addition of water. More importantly, the -OH proton ratios of organogel-5 and -6 after addition of H_2O increased by approximately 1.4- to 1.5-fold relative to those of organogel-5 and -6 before addition of H_2O, as attributed to water serving as catalyst. In contrast, organogel-7, which had similar mechanical properties before and after addition of H_2O, showed no significant change in either their NMR spectra or Fourier transform infrared (FT-IR) spectra. The FT-IR spectra of xerogels produced by freeze drying of hydrogel-5 and -6, on the other hand, showed changes in their –OH vibrational peaks around 3,400 cm^{-1}, which may indicate production of –OH during the hydrazone reaction without HCl. This supports the notion that additional intermolecular interactions may be occurring through hydrogen bonding of the extended gelators to provide enhanced tensile strength. As such, there would seem to be two different reaction pathways present depending on the presence of acid in the gel system.

Another fundamental transformation was discovered by SEM after hydrogel formation, which may offer an explanation for the strong mechanical properties. Specifically, we found the network structure of the organogel to be completely transformed into a lamellar structure after hydrogel formation in water as seen in the SEM of the corresponding freeze-dried xerogels. From the perspective of the multi-meric gelator fibres, a favourable realignment of the gel fibre could occurs through increased intermolecular hydrogen bonding as a result of the additional –OH groups formed upon immersion in water (Fig. 1). It remains to be seen if this change in microstructure may alternatively be attributed to unfavourable interaction parameters between the core calix[4]arene domain (**1**), the solvent, and the –OH bearing coupling domain (**2**). Prior works on gels comprised of hydrazone bonds have alluded to their aggregation as imposed by hydrogen bonding27 and the possibility of their dynamic nature allowing reversible covalent recombination33. The realignment of the gelator fibres in the lamellar structure of the hydrogel provides a clear indication for improvements in tensile strength as compared with the network structure seen in the organogels. As outlined schematically, the building blocks derived from the coupling reaction of **1** and **2** during organogel formation provided intermolecular hydrogen-bonding interactions between –C=O and –NH groups on nearby fibres. However, the partial –C=N bond that resulted from the hydrazone reaction for organogel formation was transformed upon submersion in water to produced new –NH and –OH groups on the surface of fibres during hydrogel formation. The additional intermolecular hydrogen bonds between –NH, –C=O and –OH groups may thereby induce alignment of fibres, due to strong fibre–fibre coupling, and this alignment of fibres could result in the formation of lamellar sheet. Thus, the mechanical properties of hydrogels with lamellar structures would be expected to exhibit a significant enhancement.

In getting back to the question of applicability to biomedical devices, we examined the cytotoxicity of liquid extracted from the hydrogels by addition of the liquid to *in vitro* cultures of adherent human cells, specifically using HeLa cells. MTT (3-(4,5-dimethylthiazol-2-yl)-2,5-diphenyltetrazolium bromide) assays reveal a noticeable decrease in cell viability with increasing amount of extracted liquid from samples that still contained trace amounts of DMSO, which is to be expected as long-term exposure to even low concentration of DMSO are known to be cytotoxic. In contrast, after sufficient replacement of DMSO with water by repeated immersion of the hydrogel, the extracted liquid showed no cytotoxicity, wherein the cell viability was similar to that of addition of pure distilled water. Further evaluations will be carried out in the future to determine the potential *in vivo* use of this new supramolecular hydrogel material towards biomedical applications.

Organogels for battery applications

Examining other relevant areas for use of these unique gel systems, we prepared two different composite of organogel-7 as a gel electrolyte towards battery applications. Gel electrolyte discs were produced from organogel-7 with either $LiNO_3$ (referred to as composite-I) or with a mixture of $LiPF_6$ in EC/DEC (composite-II). Photographs of the composite gel electrolytes, which reveal that the gel electrolyte can be produced as a freely standing and easily handled robust material. Although the composite-I gel electrolyte was colorless and transparent, we found the composite-II gel electrolyte to exhibit an opaque white color. Using electrochemical impedance spectroscopy, the ionic conductivities of the gel electrolytes were estimated for each composite placed between a set of steel electrodes, and the corresponding Nyquist plot is displayed in Fig. 5. We find that the gel electrolyte composites showed good ionic conductivities of 8.05×10^{-3} and 4.29×10^{-3} S cm^{-1} for composite-II and composite-I, respectively. Although the conductivities were less than that of a corresponding liquid electrolyte, these conductivities were nonetheless on par with previously developed polymer gel type electrolytes34. The difference in ionic conductivity of composite-II (ca two times higher than that of composite-I) may be attributed to the mobility effects by the solvent35 and subtle variations in gel phase structure, as two different molding techniques were employed for these composites. Accordingly, further optimization of such supramolecular gel electrolytes may prove useful in future lithium battery applications, as bendable and flexible materials remain a desired design component for batteries used in wearable devices36.

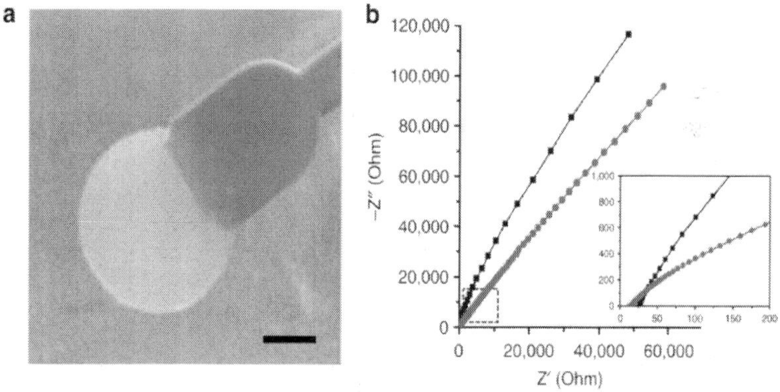

Figure 5: Photograph and electrochemical impedance spectroscopy of the gel electrolytes.

(**a**) Photograph of a gel electrolyte disc (composite-II; scale bar, 2.5 mm) and (**b**) electrochemical impedance spectra of gel electrolytes prepared from two different composites of organogel-7. The inset provides an expanded view of the lower range measurements.

In conclusion, ductile and high tensile strength hydrogels were produced by a robust new approach, which provides an alternative to supramolecular gel formation that have in the past been mechanically weak and unable to be molded into free standing objects7. According to our literature search, the work presented here is the first example of on-site supramolecular hydrogel formation via an organogel intermediate. Given the wealth of responsive and switchable gels that have been developed by incorporation of macrocycles, such as cyclodextrin among others32·37·38·39, the potential may exist for expanding our *in situ* hydrogel preparation method for the development of future smart-materials or dual-use functional/structural hydrogels by implementing calixarene-based gelators or related macrocyclic derivatives.

METHODS

General characterization

Using a Bruker ARX 300, the ^{1}H and ^{13}C NMR spectra of the samples were obtained. A Shimadzu FT-IR 8400S was used to measure the FT-IR spectra in ATR mode over the range of 400–4,000 cm^{-1}. A Hitachi U-2900 was used to determine the optical absorption spectra while a JEOL JMS-700 mass spectrometer was used to obtain the mass spectra. For DSC measurements, gel samples were hermetically sealed in a silver pan and referenced to a pan containing alumina. We employed a high-sensitivity differential scanning calorimeter (Hitachi DSC 7020) utilizing liquid N_2 cooling unit with thermograms obtained at a 1.0 °C min^{-1} heating rate. Powder XRD patterns were measured on a Bruker AXS D8 DISCOVER.

Preparation of supramolecular organogels and hydrogels

Compounds **1** and **2** were synthesized according to the detailed steps outlined. Compound **1** (15 mg, 42 mM) and compound **2** (16 mg, 84 mM) were first dissolved in DMSO (0.5 ml).

Two samples (organogel-5 and -6) were then prepared in the presence of 10 and 5 nmole of HCl, respectively. Organogel-7 was prepared without addition of HCl. The gels were formed after simple mixing and incubation at room temperature.

For preparation of hydrogel-5, -6 and -7, the corresponding organogels (3 wt% for **1**, 0.5 ml) previously prepared from DMSO (organogel-5 and -6 aged for 24 h; organogel-7 aged for 36 h) were immersed in H_2O (50 ml) for 24 h at room temperature. After solvent exchange, the amounts of DMSO removed were evaluated by ^{1}H NMR.

Preparation of xerogels from organogels and hydrogels

Organogels and hydrogels were transferred into liquid N_2 for 10 min and samples were then frozen at −40 °C overnight and freeze dried by vacuum at 0.1 Pa to yield the dry xerogels according to standard protocols30·40·41·43. The resulting xerogels were subsequently used for FT-IR, XRD and SEM measurements.

Preparation of gel samples for rheological measurements

Compounds **1** (60 mg, 42 mmol) and **2** (64 mg, 84 mmol) were dissolved in 2 ml of DMSO with and without HCl in a 4.5-cm diameter round container. The reaction mixtures were maintained at a constant reaction temperature of 25 °C for a set period of time to allow gel formation. The organogels and hydrogels prepared by hydrazone reaction without or with subsequent solvent exchange were loaded onto the rheometer plate and then cut to exactly 4 cm diameter (equal to the 4 cm plate diameter) standard for measuring rheological properties.

SEM observation

An FE-SEM, Philips XL30 S FEG field emission SEM was used to obtain images of the freeze-dried gel samples using an accelerating voltage 5–15 kV and an emission current of 10 μA. The observed gel samples were freeze dried to provide the corresponding xerogels and were observed from the side after cutting.

Rheological properties

Rheological test of gels were carried out by using An AR-2000ex (TA Instruments Ltd) implemented with a 40-mm diameter parallel plate that was attached to a transducer. The gap in the setup for rheological testing of the gels was 1.0 mm and experiments were conducted at 25 °C. Strain sweep tests were performed with increasing amplitude oscillation up to 100% apparent strain on shear. Frequency sweeps were performed from 10–1,000 Hz. The recovery properties of the gels in response to applied shear force were investigated with the following 1,500 s procedure: 0.5% (300 s)→1,000% (300–600 s)→0.5% (600–900 s)→1,000% (900–1,200 s)→0.5% (1,200–1,500 s).

Static tensile stress-strain test of organogels/hydrogels

Using a DMA Q800 (TA Instruments Ltd), we performed static mechanical testing on the gel samples under tensile loading at 25 °C. In our system, the tensile stress was considered to be the engineering stress as the force applied to the gel divided by the initial cross-sectional area of the gel, while the tensile strain was defined as the ratio of gauge length change of the gel relative to the initial gauge length. For static tensile stress-strain testing, gels with dimensions of 25 × 5 × 2 mm^3 were used, and averages of the mechanical property tested were obtained over at least three specimens.

Hydrazone reaction rate between 1 and 2

Compound **1** (42 mmol) was dissolved in DMSO-d_6. Compound **2** (84 mmol) was added to the compound **1** solution. Next, a constant amount of DCl (0, 5 and 10 nmol) was added to a given reaction mixture at room temperature. The rates for the hydrazone reaction between **1** and **2** with or without DCl were evaluated by ^{1}H NMR. After a range of set reaction times, the rates for hydrazine reaction between **1** and **2** were evaluated. In addition, for a set reaction time upon addition of H_2O (25 µl), we examined the ^{1}H NMR spectra to investigate the influence of H_2O on the hydrazone reaction.

MTT cell viability assay

HeLa cells were cultured according to ATCC methods to a confluency of 50% before media replacement and addition of liquid test samples of DMSO, distilled water or liquid extracted from hydrogel-7 prepared samples. After 24 h of incubation, MTT reagent was added for 1 h at 37 °C. The assay/spectra acquisition were then carried out according to the manufacturer's protocol (EZ-Cytox, Dail Lab Service).

Preparation of gel electrolyte

Gel electrolytes were prepared through two different solution casting methods. In the first method, gelator precursors were dissolved in DMSO solvent as described above along with 0.5 M lithium nitrate ($LiNO_3$) and stirring at room temperature for 1 h. The solution was next heated at 80 °C for 10 min. The transparent solutions were then cast into glass beakers and kept at room temperature overnight until the formations of the gel type electrolytes. Finally, the gel electrolyte was cut into circular discs with areas of 1.53 cm^2. In the second method, gelator precursors were also added into DMSO solvent as described; however, they were first heated at 80 °C for 10 min before addition of Li. Once completely dissolved, a specific amount of 1 M $LiPF_6$ in EC/DEC (1:1 (v/v)) was added to the gelator solution and the gel was cast in the same manner as described above to yield the gel electrolyte disc. All processes were conducted in an argon-filled glove box.

Electrochemical measurement

The ionic conductivity of the gel polymer electrolytes were determined from the impedance spectrum using a blocking cell of which the electrolyte was sandwiched between two stainless steel electrodes in a Swagelok cell. Electrochemical impedance spectrum measurements were performed using a VMP3 (BioLogic) over a frequency range of 1 MHz to 100 mHz with a potentiostatic signal amplitude of 10 mV.

ACKNOWLEDGEMENTS

This work was supported by a grant from the NRF (2012R1A2A2A01002547 and 2012R1A4A1027750) supported from the Ministry of Education, Science and Technology, Korea. In addition, this work was partially supported by a grant from the Next-Generation BioGreen 21 Program (SSAC, grant#: PJ011177022015), Rural Development Administration, Korea.

REFERENCES

1. Kim, H.-J., Zhang, K., Moore, L. & Ho, D. Diamond nanogel-embedded contact lenses mediate lysozyme-dependent therapeutic release. *ACS Nano* **8**, 2998–3005 (2014) .
2. Rose, S. et al. Nanoparticle solutions as adhesives for gels and biological tissues. *Nature* **505**, 382–385 (2014) .
3. Germain, J., Hradil, J., Fréchet, J. & Svec, F. High surface area nanoporous polymers for reversible hydrogen storage. *Chem. Mater.* **18**, 4430–4435 (2006) .
4. Lee, J., Kang, S., Lee, J., Jaworski, J. & Jung, J. Instant visual detection of picogram levels of trinitrotoluene by using luminescent metal–organic framework gel□coated filter paper. *Chem. Eur. J.* **19**, 16665–16671 (2013) .
5. Stuart, M. et al. Emerging applications of stimuli-responsive polymer materials. *Nat. Mater.* **9**, 101–113 (2010) .
6. Deng, G., Tang, C., Li, F., Jiang, H. & Chen, Y. Covalent cross-linked polymer gels with reversible sol-gel transition and self-healing properties. *Macrocmolecules* **43**, 1191–1194 (2010) .
7. Tamesue, S. et al. Linear versus dendritic molecular binders for hydrogel network formation with clay nanosheets: studies with ABA triblock copolyethers carrying guanidinium ion pendants. *J. Am. Chem. Soc.* **135**, 15650–15655 (2013) .
8. Wang, Q. et al. High-water-content mouldable hydrogels by mixing clay and a dendritic molecular binder. *Nature* **463**, 339–343 (2010) .
9. Haraguchi, K. & Takehisa, T. Nanocomposite hydrogels: a unique organic-inorganic network structure with extraordinary mechanical, optical, and swelling/de-swelling properties. *Adv. Mater.* **14**, 1120–1124 (2002) .
10. Lin, W.-C., Fan, W., Marcellan, A., Hourdet, D. & Creton, C. Large strain and fracture properties of poly (dimethylacrylamide)/silica hybrid hydrogels. *Macromolecules* **43**, 2554–2563 (2010) .
11. Huang, T. et al. A novel hydrogel with high mechanical strength: a macromolecular microsphere composite hydrogel. *Adv. Mater.* **19**, 1622–1626 (2007) .
12. Ramirez, A. L. B. et al. Mechanochemical strengthening of a synthetic polymer in response to typically destructive shear forces. *Nat. Chem.* **5**, 757–761 (2013) .

13. Hu, J. et al. Microgel-reinforced hydrogel films with high mechanical strength and their visible mesoscale fracture structure. *Macromolecules* **44**, 7775–7781 (2011) .
14. Sun, T. L. et al. Physical hydrogels composed of polyampholytes demonstrate high toughness and viscoelasticity. *Nat. Mater.* **12**, 932–937 (2013) .
15. Deng, C., Cui, Y., Zhao, T., Tan, M., Huang, H. & Guo, M. Mechanically strong and stretchable polyurethane-urea supramolecular hydrogel using water as an additional in situ chain extender. *RSC Adv.* **4**, 24095–24102 (2014) .
16. Suzaki, Y., Taira, T. & Osakada, K. Physical gels based on supramolecular gelators, including host-guest complexes and pseudorotaxanes. *J. Mater. Chem.* **21**, 930–938 (2011) .
17. Kakuta, T., Takashima, Y. & Harada, A. Highly elastic supramolecular hydrogels using host–guest Inclusion complexes with cyclodextrins. *Macromolecules* **46**, 4575–4579 (2013) .
18. Komatsu, H., Ikeda, M. & Hamachi, I. Mechanical reinforcement of supramolecular hydrogel through incorporation of multiple noncovalent interactions. *Chem. Lett.* **40**, 198–200 (2011) .
19. Badjic, J. D., Nelson, A., Cantrill, S. J., Turnbull, W. B. & Stoddart, J. F. Multivalency and cooperativity in supramolecular chemistry. *Acc. Chem. Res.* **38**, 723–732 (2005) .
20. Zhang, J., Guo, D.-S., Wang, L.-H., Wang, Z. & Liu, Y. Supramolecular binary hydrogels from calixarenes and amino acids and their entrapment–release of model dye molecules. *Soft Matter* **7**, 1756–1762 (2011) .
21. Goh, C. Proline-functionalised calix [4] arene: an anion-triggered hydrogelator. *Chem. Commun.* 3900–3902 (2008) .
22. Song, S., Wang, J., Feng, H.-T., Zhu, Z.-H. & Zheng, Y.-S. Supramolecular hydrogel based on amphiphilic calix[4]arene and its application in the synthesis of silica nanotubes. *RSC Adv.* **4**, 24909–24913 (2014) .
23. Verdejo, B., Rodriguez-Llansola, F., Escuder, B., Miravet, J. F. & Ballester, P. Sodium and pH responsive hydrogel formation by the supramolecular system calix[4]pyrrole derivative/tetramethylammonium cation. *Chem. Commun.* **47**, 2017–2019 (2011) .

24. Rostamizaden, S. & Housaini Ghasem, S. A. Microwave assisted syntheses of 2,5-disubstituted 1,3,4-oxadiazoles. *Tetrahedron Lett.* **45**, 8753–8756 (2004) .
25. Kaim, E. I. L., Menestrel, I. L. & Morgentin, R. Trichloroacetic acid hydrazones I: New formation of 1,3,4-oxadiazoles from aldehydes. *Tetrahedron Lett.* **39**, 6885–6888 (1988) .
26. Sim, H. G., Ahn, K. H. & Lee, S. J. Large amplitude oscillatory shear behavior of complex fluids investigated by a network model: a guideline for classification. *J. Nonnewton. Fluid Mech.* **112**, 237–250 (2003) .
27. Boekhoven, J. et al. Catalytic control over supramolecular gel formation. *Nat. Chem.* **5**, 433–437 (2013) .
28. Xia, L.-W. et al. Nano-structured smart hydrogels with rapid response and high elasticity. *Nat. Commun.* **4**, 2226 (2013) .
29. Jung, J. H., Kobayashi, H., Masuda, M., Shimizu, T. & Shinkai, S. Helical ribbon aggregate composed of a crown-appended cholesterol derivative which acts as an amphiphilic gelator of organic solvents and as a template for chiral silica transcription. *J. Am. Chem. Soc.* **123**, 8785–8789 (2001) .
30. Jung, J. H., Shinkai, S. & Shimizu, T. Spectral characterization of self-assemblies of aldopyranoside amphiphilic gelators: what is the essential structural difference between simple amphiphiles and bolaamphiphiles? *Chem. Eur. J.* **8**, 2684–2690 (2002) .
31. John, G., Jung, J. H., Masuda, M. & Shimizu, T. Unsaturation effect on gelation behavior of aryl glycolipids. *Langmuir* **20**, 2060–2065 (2004) .
32. Yamaguchi, H. et al. Photoswitchable gel assembly based on molecular recognition. *Nat. Commun.* **3**, 603 (2012) .
33. Sreenivasachary, N. & Lehn, J.-M. Gelation-driven component selection in the generation of constitutional dynamic hydrogels based on guanine-quartet formation. *Proc. Natl. Acad. Sci. USA* **102**, 5938–5943 (2005) .
34. Tarascon, J. M. & Armand, M. Issues and challenges facing rechargeable lithium batteries. *Nature* **414**, 359–367 (2001) .
35. Hikmet, R. A. M. & Peeters, M. P. J. A study of the effect of the polymer network and the solvent on the ion conductivity in gels. *Solid State Ionics* **126**, 25–39 (1999) .

36. Koo, M. et al. Bendable inorganic thin-film battery for fully flexible electronic systems. *Nano Lett.* **12**, 4810–4816 (2012) .
37. Zheng, Y., Hashidzume, A., Takashima, Y., Yamaguchi, H. & Harada, A. Switching of macroscopic molecular recognition selectivity using a mixed solvent system. *Nat. Commun.* **3**, 831 (2012) .
38. Takashima, Y. et al. Expansion–contraction of photoresponsive artificial muscle regulated by host–guest interactions. *Nat. Commun.* **3**, 1270 (2012) .
39. Nakahata, M., Takashima, Y., Yamaguchi, H. & Harada, A. Redox-responsive self-healing materials formed from host–guest polymers. *Nat. Commun.* **2**, 511 (2012) .
40. Jung, S. H., Jeon, J., Kim, H., Jaworski, J. & Jung, J. H. Chiral arrangement of achiral Au nanoparticles by supramolecular assembly of helical nanofiber templates. *J. Am. Chem. Soc.* **136**, 6446–6452 (2014) .
41. Hsu, S.-M., Lin, Y.-C., Chang, J.-W., Liu, Y.-H. & Lin, H.-C. Intramolecular interactions of a phenyl/perfluorophenyl pair in the formation of supramolecular nanofibers and hydrogel. *Angew. Chem. Int. Ed.* **53**, 1921–1927 (2014) .
42. Yu, G., Yan, X., Hana, C. & Huang, F. Characterization of supramolecular gels. *Chem. Soc. Rev.* **42**, 6697–6722 (2013) .
43. Jung, J. H. et al. Self-assembly of a sugar-based gelator in water: its remarkable diversity in gelation ability and aggregate structure. *Langmuir* **17**, 7229–7232 (2001) .

SECTION 2
METAL–ORGANIC GELS

CHAPTER

8

Metal–Organic Framework Gels and Monoliths

Jingwei Hou,[a] Adam F. Sapnik[a] and Thomas D. Bennett[a]

[a]Department of Materials Science & Metallurgy, University of Cambridge, 27 Charles Babbage Road, Cambridge, UK

ABSTRACT

The synthesis of metal–organic frameworks (MOFs) has, to date, largely been in the form of crystalline powders. However, interest in different physical morphologies of this class of materials is growing. In this perspective, we provide an overview of the structure, properties and applications of MOF monoliths. In particular, we explore the complex synthetic landscapes associated with MOF crystallization and discuss the synthetic factors leading

Citation: Jingwei Hou, Adam F. Sapnik and Thomas D. Bennett. "Metal–organic framework gels and monoliths" Chem. Sci., 2020, 11, 310 https://doi.org/10.1039/C9SC04961D

to the formation of MOF gels, *i.e.* the precursor to sol–gel MOF monoliths. Finally, we provide our thoughts on the future development of this field, and attempt to highlight the importance of the MOF gel state in the discovery of new functional materials.

INTRODUCTION

Metal–organic frameworks (MOFs) are obtained through the self-assembly of metal nodes with organic ligands, and represent a huge emerging class of functional materials.

They are considered potential solutions for numerous environmental and energy challenges, including access to clean water, air pollution reduction and renewable energy supplies.[1–4] Interest in the family is primarily because design, modification and alteration of their chemical structures are possible *via* the use of a host of different chemical building units, regulation of synthetic parameters or utilization of post-synthetic reaction chemistry.

These changes result in different chemical and structural properties, which impact upon potential applications in separation, adsorption, catalysis and gas storage.[5–9]

Attention has slowly been shifting from the chemical structure and properties of microcrystalline MOF powders, to the relationship between physical morphology, structure and applications of these interesting materials (Fig. 1).

The majority of MOFs are synthesized as polydisperse microcrystalline powders. These may suffer from inherent problems such as poor handling properties, mass transfer limitations, and mechanical instability.[10,11]

For example, the "free" powder form of a MOF packed within an adsorption column can lead to a significant pressure drop over time, caused by the gradual compaction of the powder with pressure, resulting in higher mass resistance within the column.[12] In the area of catalysis, the use of MOFs in powder form usually results in difficulty recycling the catalyst.[13,14] Importantly, while the search for high surface area MOFs is given great gravitas, it is gravimetric and volumetric gas adsorption capacities that are of higher importance for solid-state storage systems in industry.[15,16]

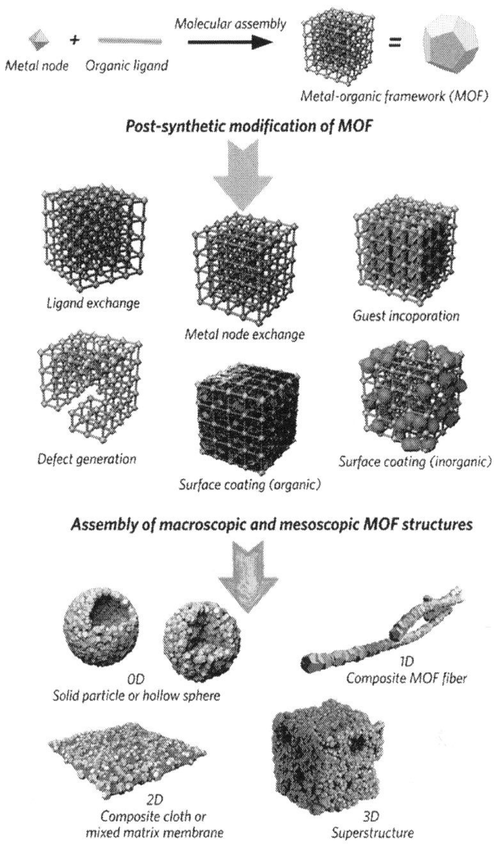

Figure. 1 Schematic overview of post-synthetic modifications and meso/macroscopic morphological assemblies of MOFs.

The United States Department of Energy (DOE) 2015 technical target for an on-board hydrogen storage system is given as 0.09 kg H_2 kg^{-1}, or 0.081 kg H_2 L^{-1}, respectively.[17] Large quantities of void space between individual discrete crystallites in powders thus reduces the packing density and therefore compromises volumetric adsorption capacities.[18,19] A bottleneck for the use of MOFs is therefore the assembly of practical devices with a high powder packing density, or in structurally coherent, robust, continuous morphologies.[20]

The fabrication of bulk architectures has been approached in a number of ways,[21] including the use of binders, mechanical compression, or through the blending of active MOF components in polymeric substrates. These mixed-matrix membranes,[22,23] for example, offer enhanced process abilities and mechanical stabilities compared to purely crystalline powders, though effective loadings are limited due to reduced colloidal stabilities during the fabrication process. Blocking of the internal porosity of the MOF component is also possible, and reduces the accessibility of the microspores to guest molecules.[24] Methodologies for shaping MOFs without the use of a secondary component, and producing "pure" bulk MOF materials are therefore highly sought after.

The formation of sol–gel-derived MOF monoliths, in particular, has come to prominent attention,[12,25,26] due to the ability to structure different pore size regimes within the same material, and the increased capabilities of volumetric uptakes of gas. However, the formation mechanism of these bulk architectures is poorly understood, despite several publications showing that their existence is linked to the prior formation of a MOF gel state.

The gel state is formally defined by IUPAC as a "non-fluid colloidal network or polymer network that is expanded throughout its whole structure by a fluid".[27] The MOF gel state has been shown, by several publications, to be a colloidal network of discrete crystalline nanoparticles that aggregate *via* weak non-covalent interactions throughout a liquid phase. For clarity, the MOF gel state is a subset of coordination polymer gels,[28] *i.e.* a type of metal–organic gel (MOG), that is comprised solely of discrete nanoparticles of crystalline MOF. Hence, with smaller particle sizes favouring the formation of MOF gels, the propensity to form a MOF gel is itself linked to the factors behind the crystallization of MOFs.

An important aim in this field is the ability to generalize monolith formation. To do so, one must first generalize MOF gel formation, where the avoidance of microcrystalline powder products is a pre-requisite. An understanding of the crystallization process is therefore essential in order to obtain particles sizes which favor the formation of MOF gels.This perspective aims to summarize the state-of-the-art work on the fabrication of MOF-based monolithic architectures, and concentrates on linking the formation of sol–gel MOF monoliths to the mechanistic aspects of MOF gel formation. This focus clearly delineates from the excellent summary of the wider area of sol–gel processing of MOFs by Falcaro *et al.*[29] Finally, we present open questions for this area, and hypothesize that many of the 70 000 known MOF structures will be accessible in the monolithic state.[30]

MECHANISTIC ASPECTS OF MOF GEL FORMATION

Formation of the MOF gel state, *i.e.* the precursor to MOF monoliths, has been linked to the supramolecular aggregation of MOF nanoparticles. These may be monodisperse, though this is not necessary. This can be achieved by first understanding, and subsequently controlling, the crystallization process. This not only determines the periodic structure of the MOF, but also the crystallite size and distribution. Hence, controlling crystallization processes facilitates the targeted synthesis of monodisperse MOF nanoparticles.

Crystallization of MOFs

The self-assembly of MOFs can be understood in terms of classical crystallization, which consists of two processes: nucleation and crystal growth.[31,32] Classical crystallization typically occurs from a supersaturated solution. Once super saturation has been reached, nucleation can begin to occur. As nuclei are introduced into the system crystal growth starts to occur. Often, both processes occur in competition with each other, both depleting the concentration of reactants in the solution. Once the concentration has decreased sufficiently, both processes will terminate. The rates of nucleation and crystal growth can be quantified using the Avrami and Gualtieri models.[31,33] In addition to the rates, the Avrami exponent, n, can be extracted from crystallization data.

This gives an indication of the mechanism and dimensionality of the crystal growth.[31,34–36] An $n > 1$ indicates the crystal growth is phase-boundary-limited, *i.e.* it occurs through successive monomer-attachment on the surface of the particles. An $n = 4$ suggests a monomer-attachment process occurring in three dimensions alongside continuous nucleation. On the other hand, $n < 1$ suggests a diffusion-limited mechanism of growth.[37] These models go some way to providing a mechanistic way to rationalise the size and distribution of crystallites in terms of kinetic parameters. In practice, it is a delicate balance between the relative rates of nucleation and crystal growth that dictates the crystallite size and dispersity. For example, a comparatively fast nucleation rate with respect to crystal growth results in the formation of small, monodisperse, crystallites. This is well understood in terms of the LaMer "burst" mechanism.[38] Hence, understanding how the rates of nucleation and crystal growth are affected by synthetic parameters can allow for targeted syntheses of monodisperse crystalline nanoparticles.

Despite the use of relatively simple models to understand crystallization, it is nonetheless a complex phenomenon and even in the simplest cases is not fully understood. For example, Yeung *et al.* recently highlighted disparities in the literature surrounding the crystallization of a prototypical MOF, ZIF-8 ($Zn(mIm)_2$, where mIm = 2-methylimidazolate).[39] Specifically, *in situ* powder X-ray diffraction (PXRD), pH and turbidity measurements were used to show that the rate of crystallization decreased with increasing reactant concentration. A pre-equilibrium model consisting of metastable intermediate clusters was proposed in order to rationalise their somewhat counterintuitive findings. Within this framework they could explain how increasing reaction concentration inhibits both nucleation and crystal growth. Further analysis showed that the kinetics of crystallization also varied as a function of reaction progress. Such a dynamic mechanism for a seemingly simple system indicates the level of complexity associated with the crystallization of MOFs.

MOF gel formation

The formation of MOF gels is based on colloidal chemistry, beginning with the generation of crystalline MOF nanoparticles with a narrow size distribution. Weak non-covalent interactions between Nano crystals may then compete with continued crystal growth, maximizing the potential for gelation (Fig. 2).

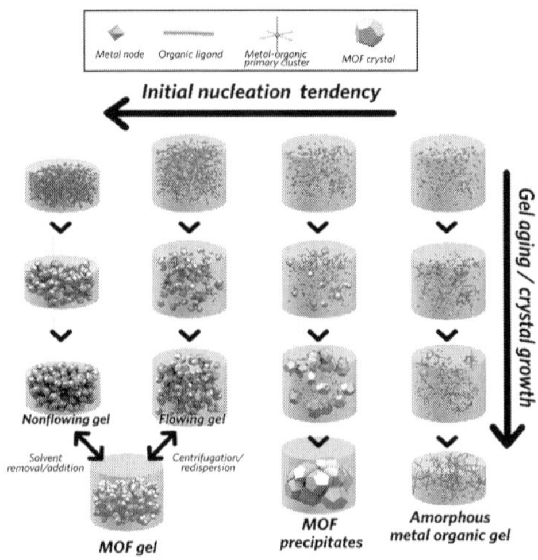

Figure. 2 Mechanism of the formation of a MOF gel, precipitate and metal–organic gel.

Key examples of gel formation of several prototypical MOFs; ZIF-8,[40] MIL-53 (Al(OH)BDC, where BDC = benzene-1,4-dicarboxylate),[41] HKUST-1 ($Cu_3(BTC)_2$, where BTC = benzene-1,3,5-tricarboxylate)[12] and several Zr-based frameworks,[42] have recently been reported. Such cases are often represented by the "test-tube inversion test", *i.e.* inverting a test tube and using the non-flowing nature to identify the gel state. This is not a rigorous test, however, and can lead to the misidentification of viscous liquids as gels.[43]

In work on the gelation properties of several Zr-based MOFs with different ligands, including both pristine and functionalized UiO-66 ($Zr_6O_4(OH)_4(BDC)_6$), UiO-67 (biphenyl dicarboxylate), MOF-801 (fumarate), MOF-808 (benzene-1,3,5-tricarboxylate) and NU-1000 (1,3,6,8-tetrakis(*p*-benzoate)pyrene), Bennett *et al.* proposed gelation to be a consequence of rapid formation of a high concentration of nanocrystals in solution. This promptly consumes the reactants within the precursor solution, impeding further growth to form microcrystalline MOF species and subsequent precipitation. As the viscosity increases, the nanocrystals gradually aggregate to form a gel state, with a weakly, non-covalently bonded colloidal network structure throughout the liquid volume, that adopts the shape of the container. The non-covalent interactions between discrete nanocrystals are mainly van der Waals forces. Possible coordinative crosslinking and intergrowth between the nanocrystals may occur, though no conclusive evidence exists at present.[42] In this case, transmission electron microscopy (TEM), scanning electron microscopy (SEM) and atomic pair distribution function (PDF) analysis indicated the absence of any other interactions, or indeed a separate amorphous or crystalline phase, which holds the crystalline nanoparticles together at the interface.

The MOF gel fabricated by Bennett *et al.* exhibits a viscoelastic property, and requires certain (external) stress to overcome internal interparticle interactions. Both flowing and non-flowing gels were fabricated, by controlling the crystalline volume fraction within the gel (Fig. 2). Interconversion between the two states was also achieved, for example by removing excess solvent *via* centrifugation and reducing the viscosity of a flowing gel sufficiently to form a non-flowing gel. Conversely, by applying enough shear forces, a non-flowing gel can be evenly redispersed in a larger solvent volume to form a flowing gel.[42]

Other researchers have converged upon the same mechanism for MOF gel formation.[12,41] They also state that gelation is a consequence of the high initial nucleation rate of discrete MOF nanoparticles: the strong coordinative bonding between metal ions and organic ligands rapidly facilitates the assembly of MOF clusters, which then aggregate to form nanocrystals. Small crystalline MOF nanoparticles (<100 nm) have been experimentally observed at the early stages of gelation.[41] Here, the dominance of supramolecular assembly over crystal growth leads to aggregation of particles and the formation of a gel. The key is to find the conditions under which the formation of nanocrystals can allow gelation to outcompete crystal growth and subsequent precipitation.

The formation process has been experimentally verified by monitoring the gel formation process of Al–BDC using solid-state NMR which showed identical subunit structure and metal node chemical environments during different stages of the gelation.[41] Furthermore, *ex situ* PXRD of the gel contained Bragg peaks corresponding to the positions expected from MIL-53(Al). This also suggests the presence of crystalline components at the early stages of crystallization, and the broadness of the diffraction peaks suggests that crystalline particles have a size of around 10.4 nm based on Scherrer broadening analysis.[41]

PXRD can be useful in characterizing MOF gels, though data can often be misinterpreted and should be used in conjunction with other techniques. The scattering from crystalline nanoparticles can be very broad, though remains centred around the positions of the Bragg reflections. This must be differentiated from the broad diffuse scattering that is associated with amorphous materials. Sufficiently small nanoparticles can give rise to scattering that can easily be mistaken for an amorphous material. In these instances TEM, PDF analysis and extended X-ray absorption fine structure (EXAFS) analysis are useful.[44]

Synthetic control of MOF gels

MOF gel formation consists of two separate steps: the nucleation of a large amount of discrete MOF nanoparticles, and subsequent gelation to form a colloidal network structure. Both aspects require careful regulation, especially the latter which usually competes against more thermodynamically favorable processes of precipitation. A key objective is thus to find a suitable set of conditions under which initial nucleation and subsequent gelation dominate. Bennett *et al.* highlighted several parameters necessary for the formation of a set of Zr-based MOF gels, including UiO-based series, MOF-801, MOF-808 and NU-1000.[42] Here, the choice of metal source, the presence of water and concentration of the reactant were all found to play key roles.

In the case of using $ZrOCl_2·8H_2O$ as a metal source, gels were favored upon increasing reactant concentrations. Products progressed from microcrystalline powders at low concentration, to "flowing" gels and eventually to "non-flowing" opaque gels. Within a diluted precursor solution, the limited number of nuclei tend to grow and precipitate as large MOF particles rather than forming gels. The beneficial influence of high reactant concentration upon gel formation aligns with classical crystallization theory, where the nucleation rate is exponentially dependent on the reactant supersaturation.[45] Conversely, lower initial concentrations of reactants in solution results in crystallization, linked to decreased nucleation relative to crystal growth.[42]

On the other hand, when $ZrCl_4$ is used as the metal source, increasing reactant concentration alone is not sufficient to induce gelation, and water is required.[42] This is not the case for the $ZrOCl_2·8H_2O$, where the water content of the precursor appears enough to promote initial $[Zr_6O_4(OH)_4]^{12+}$ cluster formation. Higher nucleation propensity and subsequent smaller MOF nanoparticles can therefore be expected with the progressive addition of water in the reactant solution.[46,47]

At the same time, controlling the gelation kinetics, rather than the initial nucleation, can be the dominating factor for some MOF gel systems. Li *et al.* prepared a series of Al(iii), Fe(iii), Ga(iii) and In(iii) MOF gels with carboxylic acid ligands, including H_3BTC and H_2BDC. While the metal–ligand precursor mixture remains stable at room temperature, gelation was observed to occur at an elevated temperature.[41] This was rationalized by the higher temperatures facilitating greater reversibility of the coordination bonding, making it comparable to the weak non-covalent interactions which lead to gelation.

Extensive work has also been performed on the formation of HKUST-1 gels. For example, the deprotonation of the H_3BTC ligand by KOH prior to addition of a Cu^{2+} solution results in a gel product, whilst the use of NaOH, LiOH or NEt_3 does not.[48] Solvent choice is also important, as gelation is observed with DMSO, though not DMF, ethanol or methanol.[49] This was ascribed to stronger interactions between DMSO and Cu^{2+}, which can facilitate the supramolecular association–dissociation process and results in the gelation of metal–organic clusters.[49] The balance between gelation, and crystallization in MOFs is thus multi-faceted and influenced by multiple competing factors. At present, a detailed understanding is only gained on a case-by-case basis.

Routes to MOF monoliths

The word monolith is derived from the ancient Greek word monolithos, meaning a single block of stone. Thus, a monolith of a material is of macroscopic size, with a continuous morphology.[50] Conventionally, monolithic structures are categorized into organic polymeric, and inorganic-based categories.[51,52] Early literature used a series of alternative names to describe these polymeric materials, such as "continuous polymer bed",[53] "continuous polymer rod"[54] and "continuous column support".[55] A functionalized cellulose sponge was, however, termed 'monolithic' in 1993.[56] The application of such sponges has mostly been as the stationary phases in chromatography columns, but they are hampered by their limited mechanical stability, which can compromise the porous structure over time under the high operating pressure of the chromatograph.[57]

Inorganic monoliths, for example the silica gel-based structures reported by Nakanishi and Soga in 1991,[58] together with carbon, ceramic and metallic monoliths,[59,60] have been the focus of much more extensive research. Their higher mechanical stabilities are advantageous in industrial processes, compared with beads or pellets, due to the reduced abrasion, lower flow resistance and pressure drop within the packed bed and flow through reactors.[52] They are extensively used as catalyst supports for the combustion of methane, hydrogen and carbon monoxide, and in nitric monoxide reduction reactions.[61–63]

The assembly of hybrid materials, and specifically MOFs, into monolithic structures is advantageous in terms of (i) easier handling associated with higher structural rigidity, (ii) lower mass transfer resistance (associated with fewer surface barriers), and (iii) higher volumetric adsorption capacities and volumetric BET surface areas. The improved mechanical stability of continuous MOF bodies would also benefit devices in which the sorbent material faces friction against a container and/or operational vibrations.[40] Such considerations have led to an increase in research on the fabrication and application of monolithic MOFs, with the goal of producing shaped microporous materials in the form of mechanically robust macrostructures, which retain their intrinsic microscale porous texture and inherent chemical and physical functionalities. The assembly, shaping and processing of such MOF monoliths is still however challenging, given that conventional solvothermal, microwave and mechanochemical synthetic methods result in fine powders.

Non gel-derived methods of monolithic MOF formation

Monolithic MOF structures have been accessed *via* a variety of methods, including (i) the formation of composite materials through the use of binders, (ii) mechanical densification and (iii) coordination replication. Pioneering approaches for the formation of MOF composites include the coating of MOF-binder mixtures onto substrates, termed wash-coating (Fig. 3a), or growing MOFs on substrates in multiple cycle syntheses. Both inorganic scaffolds such as silica,[64] and organic templates such as porous carbon[65] have been used. PolyHIPE (high internal phase emulsion) foams are the most extensively investigated support to host MOF crystals, due to their commercial availability, good chemical stability and macroporous structures with high void fractions. For example, a multiple cycle impregnation procedure has been applied to fabricate an HKUST-1@HIPE composite monolith, which achieved a maximum MOF loading of 62.3 wt% after three impregnation cycles.[66] Some composite monoliths can be fabricated using industrially available manufacturing processes. Küsgens *et al.*, for example, prepared monoliths of HKUST-1, by first homogenizing the MOF with binding and plasticization agents. This mixture was then extruded in an industrial ram extruder and cut to obtain 200 mm pieces (Fig. 3b).[67]

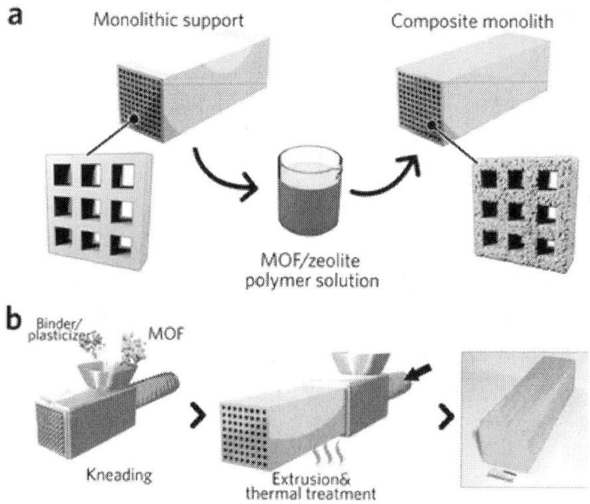

Figure. 3 Fabrication of composite monolithic MOFs. (a) Wash coating to deposit MOFs on monolithic supports. (b) Extrusion approach for the fabrication of monolithic MOFs, with an optical photo of the resulting MOF monolith. Reproduced from ref. 67 with permission from the John Wiley & Sons, copyright 2010.

Pathways to 'pure' MOF monoliths, *i.e.* those which contain only the active MOF component, are highly sought after. Mechanical densification is arguably the most straightforward technique amongst these. Typically, MOF powders are filled into a cylindrical die, which is closed using a dowel. The assembly is then placed between the jaws of an hydraulic press to apply external pressure (Fig. 4a).[68] The compressed samples may have an improved volumetric gas adsorption capacity, together with an increased bulk density *versus* the free powder form (*e.g.* 0.09 and 0.23 cm^3 (STP) cm^{-3} for powder and pellets, respectively, of UiO-66).[69]

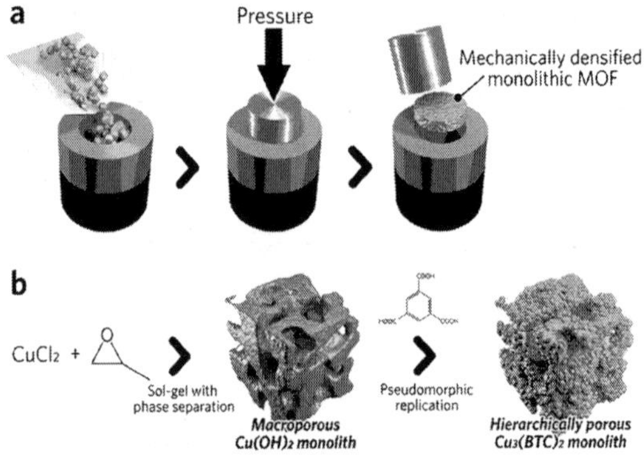

Figure. 4 Schematic of methods available to fabricate single component MOF monoliths. (a) Mechanical densification and (b) coordination replication. (b) is reproduced from ref. 77 with permission from the Royal Society of Chemistry, copyright 2015.

A key objective of any effective compaction process is to reduce the intergranular void volume without destroying the intrinsic micro porous structure. The optimum conditions required for compaction without structural collapse have been shown to be highly dependent upon the exact identity of the MOF. For example, UiO-66 has been considered as one of the most "rigid" MOFs due to its high degree of coordination between Zr–O secondary building units (SBUs) and organic ligands (each $Zr_6O_4(OH)_4$ cluster is coordinated to 12 organic linkers).[70] The minimal shear modulus of the framework is an order of magnitude higher than HKUST-1, where each copper atom in the paddlewheel SBU is coordinated by one water molecule and four oxygens from organic linkers.[71,72] The UiO-66 structure remains unaffected upon pressurization to 700 MPa, whereas HKUST-1 experiences a loss of □50% in its BET surface area after densification at just 70 MPa.[73,74] Slower pressurization rates and the presence of solvent within the MOF micropores have both been reported to be beneficial for the preservation of MOF microporosity under pressure.[69] Recently, a more advanced pressure-assisted sintering technique (field-assisted sintering technique, FAST) has been introduced to fabricate monolithic ZIF-4 ($Zn(Im)_2$, where Im = imidazolate) tablets with adjustable microporosity and macroporosity.[75]

A second approach to the fabrication of single component monoliths is based on the morphological replacement of a shaped, sacrificial, metal oxide mesoscopic template (Fig. 4b). This is also known as a pseudomorphic mineral replacement, in more general terms: the metastable metal oxide product dissolves at the solid/liquid interface, and immediately re-crystallizes as a more thermodynamically stable crystalline phase at the same site in the presence of organic ligands.[76] For example, $Cu(OH)_2$ monoliths have been used as a template for fabricating monolithic HKUST-1.[77] In this case, full conversion of a $Cu(OH)_2$ precursor under mild conditions within 6 min was observed.[77]

Sol–gel MOF monoliths

Another important approach to the formation of pure monolithic MOF species is *via* the MOF gel state. Solvent removal from a MOF gel affords a monolithic MOF state, in which the absence of an inert material renders questions of active component loading obsolete, and avoids the issue of pore-blocking. They also provide advantages compared to monoliths prepared by densification, in that processing *via* the gel state allows greater control over product morphology, and consideration does not need to be paid to conditions which avoid structural collapse.[42]

The formation of sol–gel monoliths of MOFs is not new. For example, Kaskel *et al.* formed monolithic bodies of Fe-BTC in 2009,[78] followed by the work of Furukawa and others on HKUST-1 and Al-multicarboxylate systems.[41,77] Recently, Fairen-Jimenez, Bennett and others have reported sol–gel monoliths derived from Zr-based and HKUST-1 MOF gels.[12,42,79] In the latter case, an increased capacity for volumetric adsorption of methane of 259 cm^3 (STP) cm^{-3} was noted for the monolithic state over the crystalline powder material. The formation and structures of both MOF gels, and their resultant monolithic phases, are however relatively poorly understood, especially when compared with those prepared by conventional inorganic sol–gel processes, such as carbon, titania, silica and alumina.[80]

Monolith formation from gels

In short, removal of solvent from a MOF gel improves the likelihood of nanoparticle aggregation, and the resultant formation of bulk monolithic structures. The non-uniform, non-covalent interactions between nanoparticles result in large, interparticulate void spaces, in addition to the nanopores within each particle.

The presence of this hierarchical porosity facilitates rapid mass transfer, as well as high adsorptive capacities. As discussed, the formation of nanoparticles is an important pre-requisite for the formation of MOF gels and is, therefore, also an influential factor in formation of sol–gel MOF monoliths.

For example, several HKUST-1 gels were created containing initial particle sizes of 51 to 73 nm. Gels composed of larger particles were found to form a monolith only after drying at low temperatures, whereas those gels formed of smaller initial particles were able to be heated at either low or higher temperatures to remove the solvent, and still formed monolithic pieces. During the drying process, there exists mechanical stress at the gas–liquid meniscus interface due to the surface tension.

Therefore a slower solvent removal rate, or smaller primary nanoparticles can better accommodate this stress, leading to the formation of dense MOF monolith.[12]

In other systems, mesopore size, and resultant monolith morphology, may be altered through alternative solvent removal conditions. For example, solvent evaporation from a UiO-66 gel (Fig. 5a) under high temperatures in air is likely to result in the formed monolith (xerogel) experiencing a significant shrinkage in volume due to the capillary force exerted during the drying process (Fig. 5b), which reduces mesopore size.[78] In comparison, slow and controllable removal of the solvent from the gel produces aerogels (Fig. 5c), *e.g.* by sub/supercritical CO_2 extraction,[41,42] results in avoidance of mesopore collapse and formation of high porosity, low density and large internal surface areas.

Different macroscopic structures have also been prepared *via* different approaches of solvent removal for HKUST-1 monoliths.[12] Materials of intermediate porosity have been formed from conventional inorganic sol–gels,[81–84] through freeze-drying. Such cryogels have not yet however been formed from MOF gels.

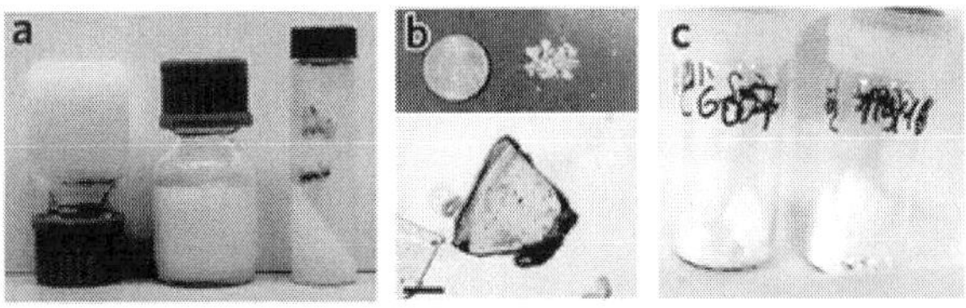

Figure. 5 Transforming MOF gel to MOF monolith. (a–c) Optical photos of (a) "non-flowing" UiO-66 gel, (b) UiO-66 xerogel fabricated from UiO-66 gel by solvent removal at high temperature in air, and (c) UiO-66 aerogel fabricated with UiO-66 gel solvent removal by critical CO_2 drying. Reproduced from ref. 42 with permission from the Royal Society of Chemistry, copyright 2017.

MONOLITH STRUCTURE AND PROPERTIES

TEM experiments on monoliths of Al–BDC [41] and UiO-66 [42] clearly indicate that the structures are comprised of irregularly packed nanoparticles (Fig. 6a and b), *i.e.* there is no change to the structure of the nanoparticles upon drying of the precursor MOF gel. In particular, for a UiO-66 xerogel, annular dark field (ADF) scanning transmission electron microscopy (STEM) shows the monolith consists entirely of aggregated crystalline nanoparticles of around 10 nm in size (Fig. 6c).[42] The presence of mesoporous interparticle spaces (10–30 nm) is confirmed by tomographic reconstruction from electron microscopy data (Fig. 6d).

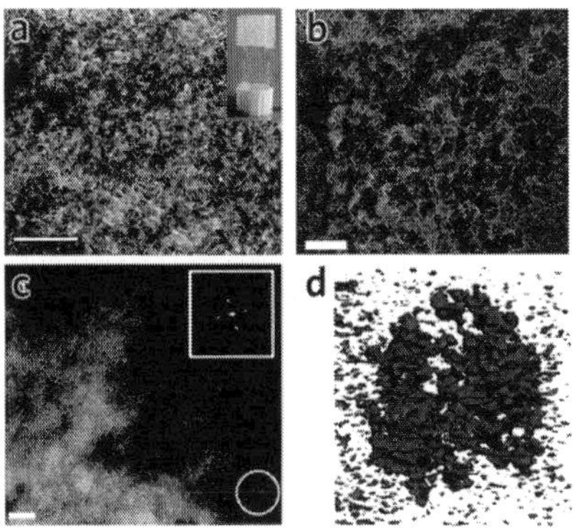

Figure. 6 TEM images and tomography of MOF monoliths. (a) TEM image of an Al–BDC aerogel. The inset is the photo of the MOF gel before solvent extraction. The scale bar is 100 nm. (a) Is reproduced from ref. 41 with permission from the Springer Nature, copyright 2013. (b) TEM image and (c) ADF-STEM image of UiO-66 xerogel, showing the randomly packed nanocrystals

with interparticle pores. The scale bar is (b) 50 nm and (c) 10 nm. (d) Electron tomographic reconstruction of the UiO-66 xerogel particle of around 100 nm in size. Solid matter is represented in red. (b)–(d) are reproduced from ref. 42 with permission from the Royal Society of Chemistry, copyright 2017.

Sol–gel monolithic MOFs typically exhibit Type IV, mesoporous, N_2 sorption behaviour, showing pore condensation with adsorption–desorption hysteresis at high pressures (Fig. 7a and b). This observation confirms the presence of both micro- and mesoporosity.[41,42] Based on N_2 isotherm and mercury intrusion measurements, monolithic UiO-66 possesses 3–4 times higher total porosity compared with the bulk MOF powder, which is mainly ascribed to the presence of considerable mesoporosity.[42] Monoliths formed from smaller nanoparticles may, however, exhibit smaller surface areas, due to a reduction in micropore surface area and volume compared to the external surface area.[85]

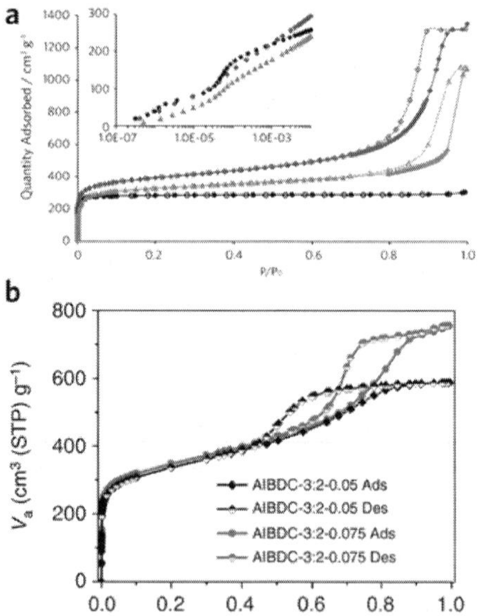

Figure. 7 Nitrogen physisorption isotherms at 77 K for UiO-66 crystalline powder (back), xerogel (red) and aerogel (blue). Full symbols are adsorption branch and open symbols are desorption branch. The inset is a logarithmic representation of the adsorption branch at low pressure region, showing the stepwise nitrogen uptake by smaller tetrahedral (6 Å) and larger octahedral cages (8 Å). (a) is reproduced from ref. 42 with permission from the Royal Society of Chemistry, copyright 2017. (b) Nitrogen physisorption isotherms for Al–BDC monoliths fabricated with different precursor concentrations. (b) is reproduced from ref. 41 with permission from the Springer Nature, copyright 2013.

It has also been proposed in the case of HKUST-1 that, unlike other MOF monoliths, a dense structure is formed upon solvent evacuation. A mechanism was proposed whereby the gradual removal of solvents under mild temperatures allows the unreacted precursors to nucleate at the interstitial space interface, leading to epitaxial growth to fill the space, forming a near-continuous structure. Within this monolithic sample, there exists no boundary or interphase between the primary particles, resulting in a continuous phase without mesopores.[12] Unexpectedly, the densification does not significantly compromise adsorption kinetics. For example, both powdered and monolithic samples show rapid equilibrium with methane, with only slightly slower transport diffusivity for the dense MOF due to the absence of mesoporosity. The monolithic MOF exhibits improved mechanical stability towards irreversible plastic deformation (as suggested by the significantly higher Young's Modulus of the material), which can be attributed to its higher density.[12]

As in microcrystalline MOF powders, the preservation of microporosity is closely related to the nature of the organic ligand. For example, for a monolith fabricated with Cr^{3+} and different bridging carboxylic acids, the incorporation of H_2BDC or H_3BTC leads to BET surface area of over 400 $m^2\ g^{-1}$.

However, the monoliths fabricated with bulkier 9-fluorenone-2,7-dicarboxylic acid (H_2FDC) and bent ligand 9,10- anthracenedicarboxylic acid (H_2ADC) have significantly lower BET surface areas. A more significant loss of the surface area is observed with ligands possessing an aliphatic substituent, indicating the microporosity may also be reduced by pore blockage from aliphatic side groups.[86]

Architectural and macroscale shaping of MOF monoliths

The similarity between MOF and inorganic nanoparticle sol–gel processes has inspired the use of supramolecular templating during the gelation stage, in order to regulate mesostructure.[87] The surfactant cetyltrimethylammonium bromide (CTAB), together with auxiliary agent 1,3,5-trimethylbenzene (TMB), has been applied in this way and resulted in monoliths possessing a narrower mesopore distribution.[41] Other soft (amphiphiles, block copolymers, ionic liquids, biopolymers and biomacromolecules) and hard (colloidal nanoparticles, bacterial filaments or cellulose nanocrystals) templates may also be able to regulate MOF monolith mesostructures,[80] though reports in this area are still rare.

Control over the macroporous structure of MOF monoliths is harder to achieve. One promising avenue may be to employ a controlled freeze, and subsequent freeze-drying, approach. This technique, also known as ice-templating, has been applied for the fabrication of various porous materials.[88] Work on HKUST-1 demonstrated that it is possible to fabricate a monolithic species with aligned pores, *via* a directional freezing process (Fig. 8). The aligned macropores offer enhanced mass transport and low-pressure drop across the monolithic column. Water and organic solvents (*e.g.* DMSO) may also be applied as frozen templates, and further control of macropore size exhibited by varying freezing rate or temperature. Formation of the inter-connected monolith structure in this case was, however, reliant upon the bridging of separated nanocrystals by the remaining reactants, as, unlike slow evaporation, freeze drying of the gel alone did not form a continuous network.[89]

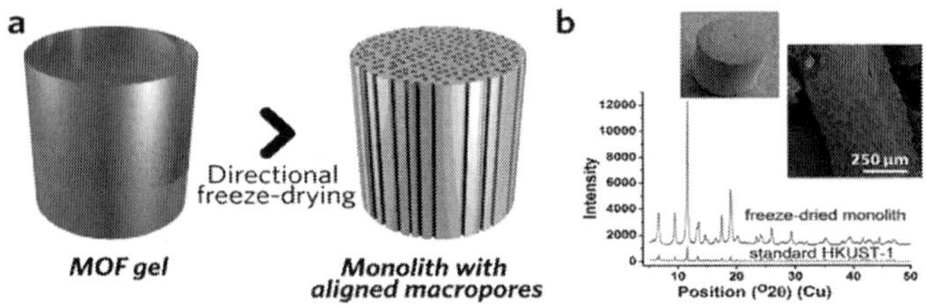

Figure. 8 Freeze drying to introduce aligned macropores to monolithic MOF. (a) Schematic diagram of the freeze-drying process. (b) PXRD, optical photo and SEM image of the HKUST-1 monolith with aligned macropores. Reproduced from ref. 89 with permission from the Royal Society of Chemistry, copyright 2015.

The gel state does however provide a flexible setting for the fabrication of solid monolithic bodies with pre-defined shapes.[25] MOF gels usually exhibits a viscoelastic behavior, and can be shaped using a specific mould, forming monolithic solids for different industrial settings like heterogeneous catalysis, adsorption and membrane/film fabrication.[90] Some gels exhibit stimuli-responsive behavior: reversible transitions between gel and colloidal suspension (slurry-like fluid) have been observed, which can be triggered by thermal treatment or mechanical shaking (thixotropic responsiveness).[90] In addition, the micro scale morphological structures of MOF gels can be shifted by adding metal ion or organic ligand into the formed MOF gel, *e.g.* transforming HKUST-1 from a granular to fibrous shape.[48]

Bennett *et al.* attempted to produce shaped monolithic species of UiO-66 in the form of spherical, monodispersed beads. An oil-drop granulation process was employed, where the flowing form of a MOF gel in DMF solvent was dispensed by a perfusion pump into immiscible hot silicone oil.[42] The droplets underwent syneresis, and formed a xerogel after 10 min at 150 °C. Subsequent annealing in air leads to the formation of the uniform monolithic MOF beads.

The final monolith preserved the original MOF crystalline structure and porosity, and sizes were regulated by using different sized needles to dispense the gel solution. The significant drawback of the method was however the time taken to produce the beads.

APPLICATIONS

Guest adsorption

The critical limitation of using natural, or hydrogen gas as a transportation fuel is the low storage density at ambient temperature, which leads to a significantly lower volumetric energy density compared with traditional fossil fuels.[91] The US DOE has set an ambitious volumetric storage target for methane of 263 cm^3 (STP) cm^{-3} at room temperature and 65 bar.[92] In terms of the methane and hydrogen storage, some MOFs with large surface areas also exhibit good promise for practical application.

For example, DUT-49 (Cu_2(BBCDC), where BBCDC = 9,9′-([1,1′-biphenyl]-4,4′-diyl)bis(9*H*-carbazole-3,6-dicarboxylate)), with a specific surface area of 5476 m^2 g^{-1}, showed an H_2 excess uptake of 80 mg g^{-1} at 77 K under 50 bar, and an exceptionally high methane storage capacity of 308 mg g^{-1} at 298 K under 110 bar. In reality however, the problem of low packing density and low mechanical strength may limit the applicability of MOFs.[93]

The hierarchical porous structure of sol–gel MOF monoliths is crucial for the reversible adsorption of small guest molecules, with minimal mass transport resistance.

The presence of a mesoporous structure also facilitates a higher gas adsorption capacity, especially for condensable gases.[41] Fairen-Jimenenz *et al.* reported a key example, where a pure monolithic sample of HKUST-1 was demonstrated to meet the DOE target for methane uptake (Fig. 9a).[79]

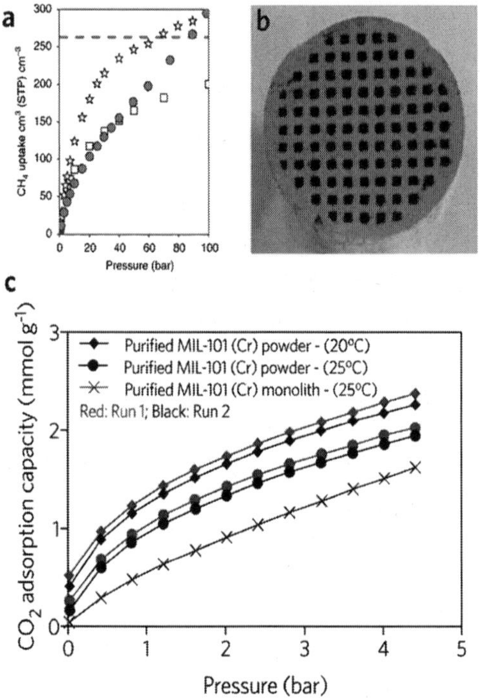

Figure. 9 Monolithic MOF for gas storage and gas adsorption. (a) Methane gas adsorption at 298 K for HKUST-1 (white stars) and UiO-66 (green circles) monolithic materials. Comparison is made against computationally simulated purely microporous/defect-free UiO-66 (white squares) at 298 K. The DOE target of methane uptake is represented by the red dashed line in the graph. Reproduced from ref. 79 with permission from the Springer Nature, copyright 2019. (b and c) CO_2 adsorption for biogas upgrade with MIL-101 (Cr) monolith. (b) Optical image of the cross-sectional image of the MIL-101 (Cr) monolith, and (c) CO_2 adsorption isotherms of MOF powders and MOF monolith. (b and c) are reproduced from ref. 94 with permission from the Elsevier, copyright 2015.

In another example, a monolithic MIL-101 (Cr) ($Cr_3O(OH)(BDC)_3(H_2O)_2$) sample was reported by Hong *et al.*, exhibiting an equilibrium CO_2 adsorption capacity of 1.95 mmol g^{-1} under ambient conditions. The low-pressure drop effect for CO_2 when packed into an adsorption column, combined with stability under humid feed gas conditions, demonstrated promise for removal of CO_2 from industrial gas streams. In comparison, the loss of CO_2 uptake capacity for a column filled with pure MIL-101 (Cr) is observed during cycling (Fig. 9b and c). One drawback of the use of the monolith is that, with increasing effective loadings (from 60 to 75 wt%), longer times are required to reach equilibrium in the dynamic adsorption process, indicating an increase in resistance to mass transfers.[94] A fine balance between adsorptive kinetics and capacity is therefore required.

Wastewater treatment using monolithic MOFs mainly focuses on micropollutants (*e.g.* pharmaceutically active compounds, pesticides, endocrine disrupting agents), which are difficult to remove using conventional water treatment techniques such as filtration, sedimentation, membrane bioreactors and even advanced oxidization.[95,96] It is proposed that monolithic MOFs can be dispersed in wastewater for the adsorption of such pollutants, and then easily recovered *via* mesh filtration due to their mascroscopic size. Then the active monolithic MOF can be regenerated after micropollutant removal, usually by solvent extraction.[97] For example, microcystine is produced by cyanobacteria in water, and can cause liver damage even at low dosage. It is a chemically stable compound that displays resistance to conventional water treatment procedures once produced. Monolithic MIL-100 (Al) ($Al_3O(OH)(BTC)_2(H_2O)_2$), in the form of both aerogels and xerogels, has been applied for the adsorption of microcystine from aqueous environments, and was reported to possess a larger adsorption capacity and faster adsorption rate than other comparable porous materials including mesoporous silica SBA-15 and active carbon.[97]

Separations

HKUST-1 monoliths have, in the past, been employed as the stationary phase within a chromatography column, and used in liquid phase separations.[98] For example, fast separation of ethylbenzene and styrene by an HKUST-1 monolith is achieved within 2 min, with a back pressure of 134 bar. Notably, the monolith significantly outperforms powdered MOF counterparts, where over 2.5 h is required to achieve the separation.[98] This rapid separation originates mainly from specific interactions between the analyte and the stationary phase, combined with the dynamic flow of the mobile phase through the monolithic MOF porous structures. The column also exhibits satisfactory performance over a long-term operation (6 weeks with over 60 injections) and under a high back pressure of 300 bar.[99]

Monolithic MOFs with large surface areas, tuneable wettabilities and good water resistance have been shown to play an important role in oil–water separations.[100] In a recent study, a superhydrophobic (water contact angle of above 170°), water-stable MOF containing Cu^{2+} and hexafluorinated dicarboxylate linkers was prepared. After coating on a porous support, the monolithic MOF film exhibits excellent oil absorption capacities and reusability, during the separation of hexadecane, bio-diesel, toluene and crude oil from water (Fig. 10a and b).[101] Composites may also be applied to tackle the same problem. For example, highly fluorinated graphene oxide (HFGO) has been mixed into a ZIF-8 precursor solution, where the GO allows selective nucleation of ZIF-8 on its surface, and serves as structure directing agent in the production of a mesoporous structure. The resultant material exhibits high absorptive selectivity and rapid kinetics for oil absorption from water. The composite has an excellent oil absorption capacity of 150 to 500% of its initial dry weight, which surpasses pristine MOF gels and all other reported crystalline pure MOF materials (Fig. 10c and d).[102]

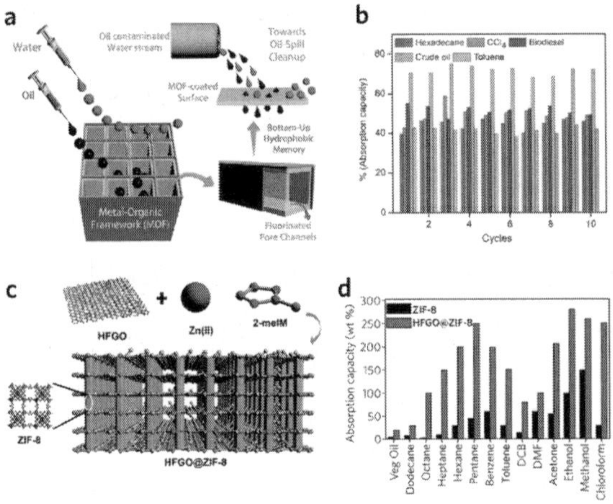

Figure. 10 Oil–water separation using monolithic MOF materials. (a) Schematic diagram of oil-water separation with the MOF containing fluorinated linkers. (b) Oil absorption capacities of the monolithic MOF containing fluorinated linkers. (a and b) are reproduced from ref. 101 with permission from the John Wiley & Son, copyright 2016. (c) Illustration showing the concept of the highly fluorinated graphene oxide and ZIF-8 composite. (d) Absorption of oil and organic solvents with ZIF-8 and composite ZIF-8. (c and d) are reproduced from ref. 102 with permission from the John Wiley & Sons, copyright 2016.

Further applications

Monoliths have also been suggested to possess promise in other areas. For example, post-synthetic modification has been used to produce luminescent sensing xerogels.[103] In the area of catalysis, on top of the controllable hierarchical structure mentioned above, monolithic MOFs have several inherent advantages. Due to the nature of the MOF gel formation process, the formed primary nanocrystals may contain a large proportion of under-coordinated metal sites, which are expected to promote their catalytic and adsorptive efficacy.[90,104] For example, the presence of coordinative unsaturated Cu^{2+} sites endows HKUST-1 with attractive Lewis acid catalytic capabilities, such as Diels–Alder reactions, cyclization of citronellal and the Friedländer reaction.[9,105–107] The high surface area of MOFs can also locally concentrate the reactants to promote the reaction. Therefore, there are driving forces to assemble MOFs into practical catalytic reactors. The continuous flow catalytic reactors enabled by monolithic MOFs offer easy implementation, direct product separation, long-term stable performance and easier catalyst regeneration. In addition, due to the good preservation of the nanosized MOF crystals, the catalytic efficiency can be usually higher than the conventional batch reactors with a slurry of MOF particles.[64]

In addition, the functional groups or the micropores within the MOFs can serve as anchoring sites for catalytically active guest species.[108] Mehta *et al.* reported the fabrication of a chemically and mechanically robust monolithic ZIF-8 material *via* a sol–gel process. Photocatalytically active SnO_2 nanoparticles were encapsulated *in situ* within the monolith during the crystallization process. Compared with the pure MOF powders, the composites with monolithic MOFs demonstrated improved photocatalytic activities and reusabilities in dye degradation: the micropores of ZIF-8 provide size-exclusion for the organics, only allowing the diffusion of water and hydrogen peroxide as carriers for the indirect photo-degradation reaction. The structure and catalytic mechanism prevents the poisoning and leaching of the expensive catalytic nanoparticles.[109]

The application of MOF monoliths and gels in energy-related fields may be an intriguing avenue for future research. For example, redox-active Fe-based heteroatom electrodes have been prepared from Fe-based aerogels. By applying suitable carbonization temperatures, Fe_3O_4 active sites were maintained, and the final resultant electrodes exhibit stable performance over 5000 charge–discharge cycles with a high current density (8 A g^{-1}).[110]

OUTLOOK

The fabrication, optimization and design of MOF monoliths is an intriguing field. It has shown great promise for improving the industrial viability of MOFs. However, there are significant areas for further work, and large knowledge gaps to bridge in order to fully exploit their potential. Our understanding of the formation mechanisms of pure MOF monoliths *via* sol–gel syntheses remains poor, and is limited to a case-by-case basis. In particular, the crystallization process and precise role of interactions between nanoparticles needs further exploration. A better understanding of these processes may enable the fabrication of pure sol–gel monoliths from a whole array of MOF structures.

It is important that possible misconceptions in the analysis of PXRD data are addressed, if the fields of MOF gels and amorphous MOFs are to be correctly delineated. In particular, we highlight that the absence of sharp diffraction peaks in powder diffraction data does not distinguish between extremely small nanoparticle-based structures, and those which are three dimensional amorphous solids. Electron microscopy, amongst other techniques, should be used in such cases to distinguish between the two and confirm, or rule out, the presence of crystalline MOF nanoparticles.[111]

The gel and monolithic states hold great promise for the future of materials discovery in the field. For example, Maspoch *et al.* recently demonstrated how several functional forms of ZIF-8 monoliths may be prepared using only small changes in reaction conditions,[112] whilst Furukawa *et al.* have investigated the linking and self assembly of metal–organic polyhedra together, to form amorphous gel structures.[113] Such reports highlight the potential of the MOF–gel state in the production of new functional materials.

The multi-stage hierarchical structure properties of MOF gels and monoliths may also facilitate tandem multi-stage catalytic reactions. However, bringing monolithic MOFs and gels closer to commercialization requires scalable fabrication. More effort, therefore, must be dedicated to investigating the fabrication techniques for mass production of the complicated monolithic structures, ideally with a solvent-free approach or a benign solvent synthesis process. High mechanical robustness of these materials are also needed. Further investigation is therefore required to fabricate monolithic MOFs which are able to withstand the application of external stress, without breaking up into their constituent nanoparticles.

It is abundantly clear from the examples presented in this perspective that there is still much to be discovered in the field of MOF gels and monoliths. However, with a greater understanding of MOF crystallization and gelation we shall be able to access a host of morphologically diverse monolithic materials, with highly tuneable hierarchical porosities, that are well suited to range of industrial applications.

CONFLICTS OF INTEREST

There are no conflicts to declare.

ACKNOWLEDGEMENTS

TDB would like to acknowledge the Royal Society for a University Research Fellowship (UF150021) and the University of Canterbury|Te Whare Wānanga o Waitaha, New Zealand, for a Visiting Cambridge Canterbury Fellowship. JH and TDB are extremely grateful to the EPSRC (EP/R015481/1). JH would like to thank the Australian Research Council (DE190100803). AFS acknowledges the EPSRC for a PhD studentship award under the industrial CASE scheme, along with Johnson Matthey PLC (JM11106).

NOTES AND REFERENCES

1. H. Wang , Q.-L. Zhu , R. Zou and Q. Xu , Chem, 2017, 2 , 52 —80
2. H. Kim , S. Yang , S. R. Rao , S. Narayanan , E. A. Kapustin , H. Furukawa , A. S. Umans , O. M. Yaghi and E. N. Wang , Science, 2017, 356 , 430 —434
3. J. B. DeCoste and G. W. Peterson , Chem. Rev., 2014, 114 , 5695 —5727 .
4. P. Kumar, K.-H. Kim , E. E. Kwon and J. E. Szulejko , J. Mater. Chem. A, 2016, 4 , 345 —361 .
5. X. Li , Y. Liu , J. Wang , J. Gascon , J. Li and B. Van der Bruggen , Chem. Soc. Rev., 2017, 46 , 7124 —7144 .
6. Z. Kang , L. Fan and D. Sun , J. Mater. Chem. A, 2017, 5 , 10073 —10091 .
7. J.-R. Li , R. J. Kuppler and H.-C. Zhou , Chem. Soc. Rev., 2009, 38 , 1477 .
8. H. Li , K. Wang , Y. Sun , C. T. Lollar , J. Li and H.-C. Zhou , Mater. Today, 2018, 21 , 108 —121 .
9. J. Gascon , A. Corma , F. Kapteijn and F. X. Llabrés i Xamena , ACS Catal., 2014, 4 , 361 —378 .
10. J. Ren , H. W. Langmi , B. C. North and M. Mathe , Int. J. Energy Res., 2015, 39 , 607 —620 .
11. N. Stock and S. Biswas , Chem. Rev., 2012, 112 , 933 —969 .
12. T. Tian , Z. Zeng , D. Vulpe , M. E. Casco , G. Divitini , P. A. Midgley , J. Silvestre-Albero , J.-C. Tan , P. Z. Moghadam and D. Fairen-Jimenez , Nat. Mater., 2017, 17 , 174 —179 .
13. J. Lee , O. K. Farha , J. Roberts , K. A. Scheidt , S. T. Nguyen and J. T. Hupp , Chem. Soc. Rev., 2009, 38 , 1450 .
14. P. García-García , M. Müller and A. Corma , Chem. Sci., 2014, 5 , 2979 .
15. R. E. Morris and P. S. Wheatley , Angew. Chem., Int. Ed., 2008, 47 , 4966 —4981 .
16. O. K. Farha , A. **Özgür** Yazaydın , I. Eryazici , C. D. Malliakas , B. G. Hauser , M. G. Kanatzidis , S. T. Nguyen , R. Q. Snurr and J. T. Hupp , Nat. Chem., 2010, 2 , 944 —948 .
17. S. Niaz , T. Manzoor and A. H. Pandith , Renewable Sustainable Energy Rev., 2015, 50 , 457 —469 .

18. N. L. Rosi Science, 2003, 300 , 1127 —1129 .
19. J. L. C. Rowsell and O. M. Yaghi , Angew. Chem., Int. Ed., 2005, 44 , 4670 —4679 .
20. B. Valizadeh , T. N. Nguyen and K. C. Stylianou , Polyhedron, 2018, 145 , 1 —15 .
21. M. Rubio-Martinez , C. Avci-Camur , A. W. Thornton , I. Imaz , D. Maspoch and M. R. Hill , Chem. Soc. Rev., 2017, 46 , 3453 —3480 .
22. M. S. Denny , J. C. Moreton , L. Benz and S. M. Cohen , Nat. Rev. Mater., 2016, 1 , 16078 .
23. B. Seoane , J. Coronas , I. Gascon , M. E. Benavides , O. Karvan , J. Caro , F. Kapteijn and J. Gascon , Chem. Soc. Rev., 2015, 44 , 2421 —2454 .
24. H. Zhu , X. Yang , E. D. Cranston and S. Zhu , Adv. Mater., 2016, 28 , 7652 —7657 .
25. G. J. H. Lim , Y. Wu , B. B. Shah , J. J. Koh , C. K. Liu , D. Zhao , A. K. Cheetham , J. Wang and J. Ding , ACS Mater. Lett., 2019, 1 , 147 —153 .
26. S. M. F. Vilela , P. Salcedo-Abraira , L. Micheron , E. L. Solla , P. G. Yot and P. Horcajada , Chem. Commun., 2018, 54 , 13088 —13091 .
27. IUPAC Compendium of Chemical Terminology: Gold Book , M. Nič, J. Jirát, B. Košata, A. Jenkins and A. McNaught, IUPAC, Research Triagle Park, NC, 2009, Search PubMed .
28. J. Zhang , Y. Hu and Y. Li , Gel Chemistry , Springer Singapore, Singapore, 2018, vol. vol. 96, pp. 61–118 Search PubMed .
29. K. Sumida , K. Liang , J. Reboul , I. A. Ibarra , S. Furukawa and P. Falcaro , Chem. Mater., 2017, 29 , 2626 —2645 .
30. P. Z. Moghadam , A. Li , S. B. Wiggin , A. Tao , A. G. P. Maloney , P. A. Wood , S. C. Ward and D. Fairen-Jimenez , Chem. Mater., 2017, 29 , 2618 —2625 .
31. M. Avrami J. Chem. Phys., 1939, 7 , 1103 —1112 .
32. R. E. Morris ChemPhysChem, 2009, 10 , 327 —329 .
33. A. F. Gualtieri Phys. Chem. Miner., 2001, 28 , 719 —728 .
34. M. Avrami J. Chem. Phys., 1940, 8 , 212 —224 .
35. M. Avrami J. Chem. Phys., 1941, 9 , 177 —184 .
36. J. D. Hancock and J. H. Sharp , J. Am. Ceram. Soc., 1972, 55 , 74 —77 .

37. T. Pradell , D. Crespo , N. Clavaguera and M. T. Clavaguera-Mora , J. Phys.: Condens. Matter, 1998, 10 , 3833 —3844 .
38. V. K. LaMer and R. H. Dinegar , J. Am. Chem. Soc., 1950, 72 , 4847 —4854 .
39. H. H.-M. Yeung , A. F. Sapnik , F. Massingberd-Mundy , M. W. Gaultois , Y. Wu , D. X. Fraser , S. Henke , R. Pallach , N. Heidenreich , O. Magdysyuk , N. T. Vo and A. L. Goodwin , Angew. Chem., Int. Ed., 2018, 58 , 566 —571
40. T. Tian , J. Velazquez-Garcia , T. D. Bennett and D. Fairen-Jimenez , J. Mater. Chem. A, 2015, 3 , 2999 —3005 .
41. L. Li , S. Xiang , S. Cao , J. Zhang , G. Ouyang , L. Chen and C.-Y. Su , Nat. Commun., 2013, 4 , 1774 .
42. B. Bueken , N. Van Velthoven , T. Willhammar , T. Stassin , I. Stassen , D. A. Keen , G. V. Baron , J. F. M. Denayer , R. Ameloot , S. Bals , D. De Vos and T. D. Bennett , Chem. Sci., 2017, 8 , 3939 —3948 .
43. E. R. Draper and D. J. Adams , Chem, 2017, 3 , 390 —410 .
44. S. J. L. Billinge and I. Levin , Science, 2007, 316 , 561 —565 .
45. S. L. Girshick and C. Chiu , J. Chem. Phys., 1990, 93 , 1273 —1277 .
46. F. Ragon , P. Horcajada , H. Chevreau , Y. K. Hwang , U.-H. Lee , S. R. Miller , T. Devic , J.-S. Chang and C. Serre , Inorg. Chem., 2014, 53 , 2491 —2500 .
47. A. Schaate , P. Roy , A. Godt , J. Lippke , F. Waltz , M. Wiebcke and P. Behrens , Chem. – Eur. J., 2011, 17 , 6643 —6651 .
48. P. Liao , H. Fang , J. Zhang , Y. Hu , L. Chen and C.-Y. Su , Eur. J. Inorg. Chem., 2017, 2017 , 2580 —2584 .
49. A. K. Chaudhari , I. Han and J.-C. Tan , Adv. Mater., 2015, 27 , 4438 —4446 .
50. R. Gvishi The Sol-Gel Handbook , D. Levy and M. Zayat, Wiley-VCH Verlag GmbH & Co. KGaA, Weinheim, Germany, 2015, pp. 317–344 Search PubMed .
51. R. M. Schoenherr , M. Ye , M. Vannatta and N. J. Dovichi , Anal. Chem., 2007, 79 , 2230 —2238 .
52. J. L. Williams Catal. Today, 2001, 69 , 3 —9 .
53. S. Hjertén , Y. Yi-Ming Li , J.-L. Liao , J. Mohammad , K. Nakazato and G. Pettersson , Nature, 1992, 356 , 810 —811 .
54. F. Svec and J. M. J. Frechet , Anal. Chem., 1992, 64 , 820 —822 .

55. S. M. Fields Anal. Chem., 1996, 68 , 2709 —2712 .
56. J. Kennedy and O. Williams, Cellulosics: Materials for Selective Separations and other Technologies, Ellis Horwood, 1993, Search PubMed .
57. H. Zou , X. Huang , M. Ye and Q. Luo , J. Chromatogr. A, 2002, 954, 5 —32 .
58. K. Nakanishi and N. Soga , J. Am. Ceram. Soc., 1991, 74 , 2518 —2530 .
59. Z. Walsh , B. Paull and M. Macka , Anal. Chim. Acta, 2012, 750 , 28 —47 .
60. E. L. Pfaunmiller , M. L. Paulemond , C. M. Dupper and D. S. Hage , Anal. Bioanal. Chem., 2013, 405 , 2133 —2145 .
61. F. Rezaei and P. Webley , Sep. Purif. Technol., 2010, 70, 243 —256.
62. S. Siemund , J. P. Leclerc , D. Schweich , M. Prigent and F. Castagna , Chem. Eng. Sci., 1996, 51 , 3709 —3720 .
63. S. Govender and H. Friedrich, Catalysts, 2017, 7, 62 .
64. A. Sachse , R. Ameloot , B. Coq , F. Fajula , B. Coasne , D. De Vos and A. Galarneau , Chem. Commun., 2012, 48 , 4749 .
65. D. Qian , C. Lei , G.-P. Hao , W.-C. Li and A.-H. Lu , ACS Appl. Mater. Interfaces, 2012, 4 , 6125 —6132 .
66. M. G. Schwab , I. Senkovska , M. Rose , M. Koch , J. Pahnke , G. Jonschker and S. Kaskel , Adv. Eng. Mater., 2008, 10 , 1151 —1155 .
67. P. Küsgens , A. Zgaverdea , H.-G. Fritz , S. Siegle and S. Kaskel , J. Am. Ceram. Soc., 2010, 93 , 2476 —2479 .
68. H. Oh , D. Lupu , G. Blanita and M. Hiher , Adv., 2014, 4 , 2648 —2651 .
69. J. Dhainaut , C. Avci-Camur , J. Troyano , A. Legrand , J. Canivet , I. Imaz , D. Maspoch , H. Reinsch and D. Farrusseng , CrystEngComm, 2017, 19 , 4211 —4218 .
70. Z. Su , Y.-R. Miao , G. Zhang , J. T. Miller and K. S. Suslick , Chem. Sci., 2017, 8 , 8004 —8011 .
71. H. Wu , T. Yildirim and W. Zhou , J. Phys. Chem. Lett., 2013, 4 , 925 —930 .
72. C. Prestipino , L. Regli , J. G. Vitillo , F. Bonino , A. Damin , C. Lamberti , A. Zecchina , P. L. Solari , K. O. Kongshaug and S. Bordiga , Chem. Mater., 2006, 18 , 1337 —1346 .

73. G. W. Peterson , J. B. DeCoste , T. G. Glover , Y. Huang , H. Jasuja and K. S. Walton , Microporous Mesoporous Mater., 2013, 179 , 48 —53 .
74. S. E. Bambalaza , H. W. Langmi , R. Mokaya , N. M. Musyoka , J. Ren and L. E. Khotseng , J. Mater. Chem. A, 2018, 6 , 23569 —23577 .
75. R. N. Widmer , G. I. Lampronti , B. Kunz , C. Battaglia , J. H. Shepherd , S. A. T. Redfern and T. D. Bennett , ACS Appl. Nano Mater., 2018, 1 , 497 —500 .
76. J. Reboul , S. Furukawa , N. Horike , M. Tsotsalas , K. Hirai , H. Uehara , M. Kondo , N. Louvain , O. Sakata and S. Kitagawa , Nat. Mater., 2012, 11 , 717 —723 .
77. N. Moitra , S. Fukumoto , J. Reboul , K. Sumida , Y. Zhu , K. Nakanishi , S. Furukawa , S. Kitagawa and K. Kanamori , Chem. Commun., 2015, 51 , 3511 —3514 .
78. M. R. Lohe , M. Rose and S. Kaskel , Chem. Commun., 2009, 6056 .
79. B. M. Connolly , M. Aragones-Anglada , J. Gandara-Loe , N. A. Danaf , D. C. Lamb , J. P. Mehta , D. Vulpe , S. Wuttke , J. Silvestre-Albero , P. Z. Moghadam , A. E. H. Wheatley and D. Fairen-Jimenez , Nat. Commun., 2019, 10 , 2345 .
80. A. E. Danks , S. R. Hall and Z. Schnepp , Mater. Horiz., 2016, 3 , 91 —112 .
81. N. Job , A. Théry , R. Pirard , J. Marien , L. Kocon , J.-N. Rouzaud , F. Béguin and J.-P. Pirard , Carbon, 2005, 43 , 2481 —2494 .
82. K. M. S. Khalil , R. M. El-Khatib , T. T. Ali , H. A. Mahmoud and A. A. Elsamahy , Powder Technol., 2013, 245 , 156 —162 .
83. J. Livage and C. Sanchez , J. Non-Cryst. Solids, 1992, 145 , 11 —19 .
84. C. J. Brinker and G. W. Scherer , Sol–gel science: the physics and chemistry of sol–gel processing , Academic Press, Boston, 1990, Search PubMed .
85. J. Cravillon , S. Münzer , S.-J. Lohmeier , A. Feldhoff , K. Huber and M. Wiebcke , Chem. Mater., 2009, 21 , 1410 —1412 .
86. S. Xiang , L. Li , J. Zhang , X. Tan , H. Cui , J. Shi , Y. Hu , L. Chen , C.-Y. Su and S. L. James , J. Mater. Chem., 2012, 22 , 1862 —1867 .
87. K. M. Choi , H. J. Jeon , J. K. Kang and O. M. Yaghi , J. Am. Chem. Soc., 2011, 133 , 11920 —11923 .
88. S. Deville J. Mater. Res., 2013, 28 , 2202 —2219 .
89. A. Ahmed , T. Hasell , R. Clowes , P. Myers , A. I. Cooper and H.

Zhang , Chem. Commun., 2015, 51 , 1717 —1720 .
90. J. Zhang and C.-Y. Su , Coord. Chem. Rev., 2013, 257 , 1373 —1408 .
91. Y. Peng , V. Krungleviciute , I. Eryazici , J. T. Hupp , O. K. Farha and T. Yildirim , J. Am. Chem. Soc., 2013, 135 , 11887 —11894 .
92. Y. He , W. Zhou , G. Qian and B. Chen , Chem. Soc. Rev., 2014, 43 , 5657 —5678 .
93. J. Yu , L.-H. Xie , J.-R. Li , Y. Ma , J. M. Seminario and P. B. Balbuena , Chem. Rev., 2017, 117 , 9674 —9754 .
94. W. Y. Hong , S. P. Perera and A. D. Burrows , Microporous Mesoporous Mater., 2015, 214 , 149 —155 .
95. J. Hou , G. Dong , Y. Ye and V. Chen , J. Membr. Sci., 2014, 469 , 19 —30 .
96. M. B. Asif , F. I. Hai , J. Hou , W. E. Price and L. D. Nghiem , J. Environ. Manage., 2017, 201 , 89 —109 .
97. W. Xia , X. Zhang , L. Xu , Y. Wang , J. Lin and R. Zou , Adv., 2013, 3 , 11007 .
98. R. Ahmad , A. G. Wong-Foy and A. J. Matzger , Langmuir, 2009, 25 , 11977 —11979 .
99. A. Ahmed , M. Forster , R. Clowes , P. Myers and H. Zhang , Chem. Commun., 2014, 50 , 14314 —14316 .
100. A. Wang , M. Jin , N. Li , Y. Ma , L. Chen , D. Ma , B.-X. Chen, F. Gao , Y.-Q. Tian and Y.-K. Shi , New J. Chem., 2017, 41 , 8031 —8035 .
101. S. Mukherjee , A. M. Kansara , D. Saha , R. Gonnade , D. Mullangi , B. Manna , A. V. Desai , S. H. Thorat , P. S. Singh , A. Mukherjee and S. K. Ghosh , Chem. – Eur. J., 2016, 22 , 10937 —10943 .
102. K. Jayaramulu , K. K. R. Datta , C. Rösler , M. Petr , M. Otyepka , R. Zboril and R. A. Fischer , Angew. Chem., Int. Ed., 2016, 55 , 1178 —1182 .
103. Z.-S. Qin , W.-W. Dong, J. Zhao , Y.-P. Wu, Q. Zhang and D.-S. Li, Inorg. Chem. Front., 2018, 5, 120 —126 .
104. F. Vermoortele , B. Bueken , G. Le Bars , B. Van de Voorde , M. Vandichel , K. Houthoofd , A. Vimont , M. Daturi , M. Waroquier , V. Van Speybroeck , C. Kihhock and D. E. De Vos , J. Am. Chem. Soc., 2013, 135 , 11465 —11468 .
105. E. V. Ramos-Fernandez , M. Garcia-Domingos , J. Juan-Alcañiz , J. Gascon and F. Kapteijn , Appl. Catal., A, 2011, 391 , 261 —267 .

106. Z. Fang , B. Bueken , D. E. De Vos and R. A. Fischer , Angew. Chem., Int. Ed., 2015, 54 , 7234 —7254 .
107. A. S. Shalygin , A. L. Nuzhdin , G. A. Bukhtiyarova and O. N. Martyanov , J. Sol-Gel Sci. Technol., 2017, 84 , 446 —452
108. Y. Chen , X. Huang , S. Zhang , S. Li , S. Cao , X. Pei , J. Zhou , X. Feng and B. Wang , J. Am. Chem. Soc., 2016, 138 , 10810 —10813 .
109. J. P. Mehta , T. Tian , Z. Zeng , G. Divitini , B. M. Connolly , P. A. Midgley , J.-C. Tan , D. Fairen-Jimenez and A. E. H. Wheatley , Adv. Funct. Mater. 2018, 28 , 1705588
110. A. Mahmood , R. Zou , Q. Wang , W. Xia , H. Tabassum , B. Qiu and R. Zhao , ACS Appl. Mater. Interfaces, 2016, 8 , 2148 —2157 .
111. C. Dazon, O. Witschger , S. Bau , V. Fierro and P. L. Llewellyn , Environ. Sci.: Nano, 2019, 6 , 152 —162
112. C. Avci , I. Imaz , A. Carné-Sánchez , J. A. Pariente , N. Tasios , J. Pérez-Carvajal , M. I. Alonso , A. Blanco , M. Dijkstra , C. López and D. Maspoch , Nat. Chem., 2017, 10 , 78 —84 .
113. A. Carné-Sánchez , G. A. Craig , P. Larpent , T. Hirose , M. Higuchi , S. Kitagawa , K. Matsuda , K. Urayama and S. Furukawa , Nat. Commun., 2018, 9 , 2506

CHAPTER 9

Recent Bio-Advances in Metal-Organic Frameworks

Isobel Tibbetts and George E. Kostakis

Department of Chemistry, School of Life Sciences, University of Sussex, Brighton BN1 9QJ, UK

ABSTRACT

Metal-organic frameworks (MOFs) have found uses in adsorption, catalysis, gas storage and other industrial applications. Metal Biomolecule Frameworks (bioMOFs) represent an overlap between inorganic, material and medicinal sciences, utilising the porous frameworks for biologically relevant purposes. This review details advances in bioMOFs, looking at the synthesis, properties and applications of both bioinspired materials and

Citation: Tibbetts, I.; Kostakis, G.E. "Recent Bio-Advances in Metal-Organic Frameworks". Molecules 2020, 25, 1291. https://doi.org/10.3390/molecules25061291

MOFs used for bioapplications, such as drug delivery, imaging and catalysis, with a focus on examples from the last five years.

Keywords: Bio metal-organic frameworks (bioMOFs), drug delivery, imaging, catalysis, bioinspired

INTRODUCTION

Metal-organic frameworks (MOFs) are a type of porous coordination polymer (CP), consisting of metal ions or clusters coordinated to organic linkers. Alkali and transition metals, lanthanides and actinides are all used in the construction of MOFs, with different metals adopting alternative coordination geometries or having potentially advantageous luminescent or fluorescent properties [1]. The first MOF was reported in 1989 when Robson et al. reported a '*framework consisting of three-dimensionally linked rod-like segments*' [2]. The term MOF was introduced in 1995 [3], and the field began to gain momentum in the late 1990s. In 1999, Williams et al. published on a MOF known as HKUST-1, made from copper-based clusters and benzene tricarboxylate linkers [4]. This was followed by Yaghi et al. who reported a MOF known as MOF-5, made from zinc-based clusters and benzene dicarboxylate linkers, later in the year [5], which broke records with its extremely high porosity.

Unlike other porous materials, the internal surface chemistry of MOFs can be tuned, allowing properties such as hydrophilicity and acidity to be altered as desired [6]. However, the defining feature of MOFs is their ultrahigh porosity (up to 90% free volume) and large internal surface areas (ISA), with some MOFs reported to have over 10,000 $m^2\ g^{-1}$ ISA [7]. MOFs are commonly synthesised using hydro- or solvothermal reactions [7]. Recently, this has expanded into more environmentally friendly procedures, utilising microwave, flow or mechano techniques [8,9,10]. In terms of the tuneable structures of MOFs, avoiding interpenetration—where multiple lattices can become entwined and affect the pores of the MOF—may be required when pore volume is the key factor [11]. Additionally, defects can be engineered in to provide additional adsorption or catalytic sites [12]. However, the functionalisation of the framework can alter the properties of a MOF most drastically. For example, catalysts could feature the active site at the ligand, at the metal node, or as a guest embedded within the pores [13]. Metal leaching has been observed in some cases, which could diminish catalytic activity, disrupt the MOF framework or limit use in vivo due to toxicity concerns [14].

These properties make MOFs attractive options for purposes such as fuel gas storage, the capture of greenhouse gases, and new catalytic procedures, in addition to improving upon the roles that zeolites previously fulfilled [15]. More recently, MOFs have also been used for biomedical purposes, such as exploiting the porosity of the material for drug delivery systems (DDS) [16]. Water stability and biocompatibility of MOFs is a particular issue in this area; degradation of the MOF or metal leaching could lead to unwanted toxicity, but water sensitivity could alternatively be advantageous; moisture was shown to trigger the release of encapsulated allyl isothiocyanate from a zinc-based MOF in 2016 [17]. The commercial viability of MOFs are evident, and a spin-off company, 'MOFgen', which utilises MOFs to release antimicrobial NO gas for medicinal purposes, was set up in 2013 [18,19]. Reportedly the first clinical trial involving MOFs was announced in 2018 [20], aiming to enhance the efficiency of X-ray radiotherapy (RT) as a cancer therapeutic. MOFs have been explored for applications such as drug delivery, biosensing, biocatalysis, imaging and as antimicrobial materials [21,22]. Toxicity issues can be addressed by the synthesis of Metal Biomolecule Frameworks, or bioMOFs, which incorporate biocompatible metals and active pharmaceutical ingredients (APIs) or endogenous molecules to form the porous materials [21].

The popularity of these materials for biomedical applications has been highlighted in recent reviews [22,23]. Chedid and Yassin explored recent examples of the use of MOFs and covalent organic frameworks (COFs) for biomedical applications, focusing on their synthesis and use as DDS [23]. Similar to MOFs, COFs have been used for energy and catalysis purposes [24,25], in addition to applications including biosensors and DDS [26]. More recently, Yang and Yang surveyed the current field of MOF-based biomedical materials, reporting advances in five areas including DDS, biosensing and biocatalysis [22]. Both reviews examined future challenges, such as regulatory difficulties and a need for further in vivo investigations. In addition to discussing these areas, this review explores applications, such as remediation of pharmaceutical waste, catalytic synthesis of API intermediates and the synthesis and applications of MOFs incorporating APIs or endogenous small molecules as part of their framework. As depicted in Figure 1, examples of both MOF bioapplications and bioinspired MOFs will be discussed and the prospects and challenges of the future of this field are explored.

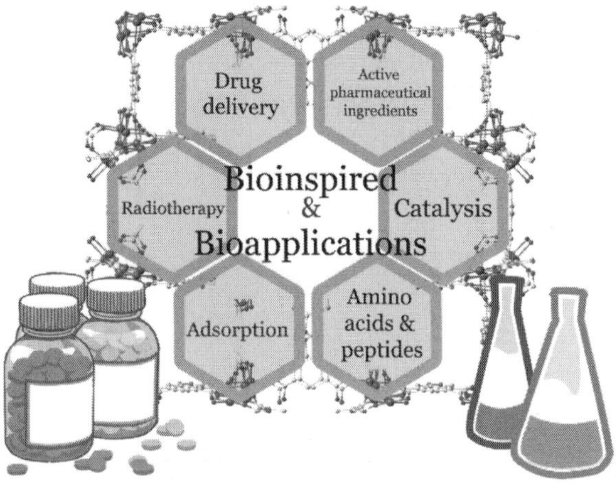

Figure 1. Schematic representation of bioapplications and biomaterials or active species used in the synthesis of metal-organic frameworks (MOFs).

METAL-ORGANIC FRAMEWORKS (MOFS) BIOAPPLICATIONS

MOFs and Drug Delivery

The porous nature of MOFs can be used for a variety of host-guest chemistry, including drug delivery. This was first achieved in 2006 by Férey et al., who encapsulated isobutylphenylpropanoic acid (ibuprofen) in MIL-100 and -101; thus, forging the way for further work using MOFs as DDS [16]. Férey initially proposed MIL-100(Cr) be used for hydrogen storage [27]; more recently, MIL-100(Fe) nanoparticles have been coated in a lipid bilayer and used to store dye molecules [28]. It has also been used in cutaneous patches with biopolymers to deliver molecules including caffeine or ibuprofen [29]. Biocompatibility—the property of a material that is not harmful to living tissue—is an essential factor in the design and synthesis of MOFs that are to be used as a DDS in humans. The search for biocompatible MOFs has resulted in a shift from the use of metals, such as chromium in early MOFs, to non-toxic analogues, such as iron, zinc and zirconium [30]. MOFs have been used to encapsulate both small molecule inhibitors and larger, more complex molecules, such as hormones. In 2018, Farha et al. used a MOF, NU-1000, to develop a potential alternative formulation for insulin, with the framework protecting the hormone from degrading in the presence of stomach acid and pepsin, a digestive enzyme [31]. As DDS, MOFs can provide protective formulations [31], enhance the solubility of an API [32], allow slow release of a drug for a more stable pharmacokinetic profile [33], and act synergistically through the design of a bioactive cation and linker [34].

Structure-Activity Relationship between MOFs and Their Cargo: The Effect of Hydrophilicity/Hydrophobicity and Pore Size

Many MOF-drug delivery studies have focused on model drugs, such as ibuprofen, to achieve proof of concept. However, a more profound understanding is required to allow for a set of principles to be used in MOF DDS research and to allow for the encapsulation of a range of drugs with different properties. A comprehensive study by Horcajada and co-workers aimed to understand the effect of the features of both MOF and drug on the drug encapsulation and release processes [33]. The relationship between three MOFs—MIL-100 (Fe), MIL-127 (Fe) and UiO-66 (Zr)—and two drugs, ibuprofen and acetylsalicylic acid (aspirin), was studied, focusing on parameters such as hydrophilicity/hydrophobicity and pore size (Figure 2). Furthermore, cutaneous administration was studied, rather than intravenous which had been the focus of most MOF studies of this kind [35,36,37].

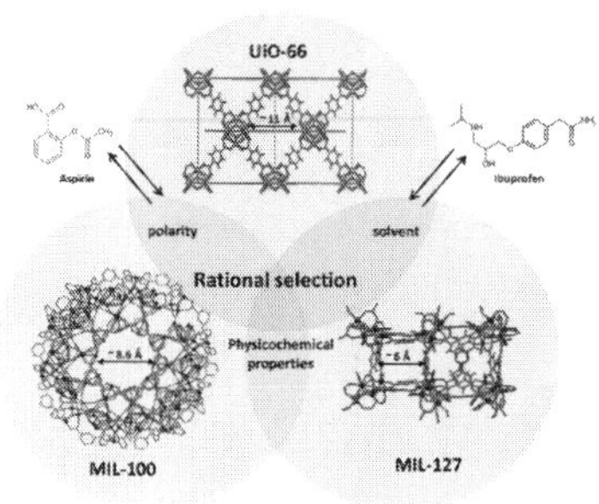

Figure 2. Structures of acetylsalicylic acid (aspirin) (left) and isobutylphenylpropanoic acid (ibuprofen) (right) with structural representation of UiO-66, MIL-100 and MIL-127. Zirconium polyhedra are cyan, iron polyhedra are green, nitrogen is blue, oxygen is red, and carbon is grey. Hydrogen atoms have been removed for clarity [33] Reprinted from S. Rojas, I. Colinet, D. Cunha, T. Hidalgo, F. Salles, C. Serre, N. Guillou and P. Horcajada, ACS Omega, 2018, 3, 2994–3003, available at https://pubs.acs.org/doi/10.1021/acsomega.8b00185. Any further permissions related to the material excerpted should be directed to the ACS.

MOFs consist of polar metal ions and nonpolar organic ligands which typically results in an amphiphilic interior. This, combined with the large pores characteristic of MOFs, makes them well suited to the encapsulation of large molecules such as pharmaceuticals. There is a need to develop structure-activity relationship (SAR) data for MOFs and the drugs they encapsulate and to understand the process of drug incorporation and delivery concerning existing MOFs and their properties. Previous works have studied the effects of linker properties, MOF topology and polarity on the DDS process, but only more recently are researchers focusing on the nature of the drug and its impact on the drug delivery process [35,36,37].

The MOFs were synthesised using solvothermal and reflux methods as detailed in Scheme 1.

To mimic the cutaneous environment, the drug release experiments were performed in distilled water at 37 °C. It was found that the hydrophobicity/hydrophilicity of the drug and the accessibility of the MOF pores were the main parameters affecting the rate of drug delivery. This offers the potential to tailor the MOF to a chosen drug, depending on its properties. The properties of the MOFs are detailed in Table 1; fast delivery was seen with the more hydrophilic, mesoporous MIL-100, slow delivery was a result of the narrow pored, amphiphilic MIL-127, and the rate of delivery depended on the hydrophobicity or hydrophilicity of the drug in the case of the microporous, hydrophobic UiO-66. Aspirin has a logP of 1.19, is hydrophilic and was released quickly, while ibuprofen has a logP of 3.97—hydrophobic by comparison—and was released slowly from this MOF.

Table 1. Properties of the MOFs used in the study.

MOF	Cage Size (Å)	Window Size (Å)	Hydrophobic or hydrophilic?	Notes
MIL-100 (Fe)	25 and 29	4.8-5.8 and 8.6	Hydrophilic	Mesoporous cages accessed through microporous windows
MIL-127 (Fe)	6 (channel system) 10 (cages)	6 3	Hydrophobic Hydrophilic	Two pore types; an accessible one-dimensional channel system, and hydrophilic cages accessed through narrow apertures
UiO-66 (Zr)	11 and 8	5–7	Slightly hydrophobic	Octahedral and tetrahedral cavities accessed through triangular micropores

The suitability of MOFs to a cutaneous approach was also proved due to the high drug loading seen (up to 35.5 *wt* %) and aqueous stability, associated with the ability to release a drug within 1–7 days. It is clear that a focus on the physicochemical properties of the encapsulated drug is required when considering DDS, rather than solely that of the MOF.

Sugar-Based MOFs: The Benefits of a Renewable, Soluble Material

In 2010, Stoddart et al. reported the synthesis of a MOF made from edible materials—namely, cyclodextrins (CDs) (Scheme 2) [38]. The MOF was made of renewable materials (CDs are produced from starch) and was water-soluble, the desired property in oral drugs.

Recently, this MOF has been used as a separation medium (2016) [39] and later, in a medicinal context—in 2019, CD-MOF was used to encapsulate folic acid (FA), used as a model molecule, achieving molar ratios of up to 2:1 FA:CD [32]. FA within CD-MOF demonstrated a 1450-fold increase in solubility compared to free FA, and in rats, the bioavailability of FA was increased by a factor of 1.48.

CD-MOF has also been used to offer the potential for longer-lasting pain relief [40]. Ibuprofen, a commonly used painkiller, as a free acid, is poorly soluble in physiological environments. Therefore, increasing the half-life of the drug and its duration of action would be beneficial for patients. As cyclodextrins are known to improve the bioavailability of drugs by improving their solubility [41], CD-MOF (Figure 3) was perfectly suited to this application.

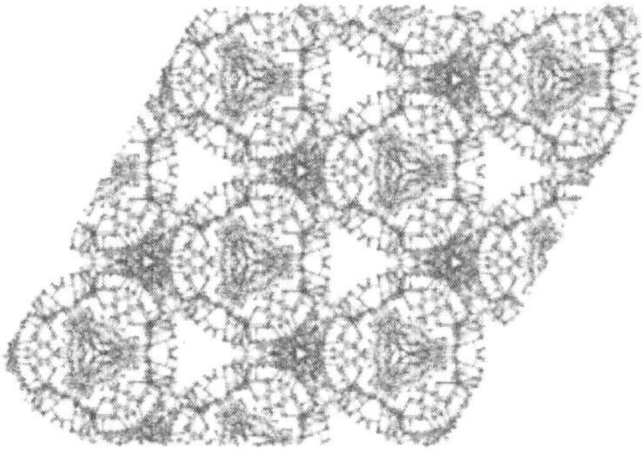

Figure 3. Space-filled representation of the extended solid-state structure of CD-MOF-1 where C is grey, O is red and K is purple [40].

Incorporation of the drug was achieved using two methods; firstly, via crystallisation, using the potassium salt of ibuprofen as the source of the alkali metal cations required to form the MOF. This method utilises the fact that MOFs consist of a metal ion and an organic ligand—here, the salt of a drug, and a sugar. The incorporation of ibuprofen anions achieved charge balance. Secondly, the free acid form of ibuprofen was absorbed in a pre-synthesised MOF. In this case, potassium hydroxide was used as the alkali metal source. However, ibuprofen loading was similar in both cases (23 *wt* % vs. 26 *wt* %, respectively).

The potassium salt of ibuprofen is soluble in water, unlike the free acid form. Formulations of this type have faster uptake rates but are only available as gel capsules due to the hygroscopic nature of the potassium salt. Therefore, the ability of CD-MOF to encapsulate ibuprofen in pharmaceutically relevant quantities is valuable.

The MOF demonstrated similar bioavailability and rapid blood plasma uptake to the pure potassium salt, showing that CD-MOF can act as an active delivery agent for ibuprofen. Moreover, when the drug was encapsulated in the sugar-based MOF, the duration of ibuprofen in blood plasma was doubled compared to ibuprofen salts alone.

CD-MOF-1 is nanoporous and has a Langmuir surface area of $1320 m^2 \cdot g^{-1}$ and an estimated total pore volume of 54% [38]. The cavity of cyclodextrin is hydrophobic, and in solution, they are known to host a variety of substrates [42].

The pore diameters in CD-MOF-1 range from 0.42 to 0.78 nm with a central pore of 1.7 nm in each $(\gamma\text{-CD})_6$ cube. These values are similar to the amphiphilic, microporous MIL-127 (Fe) from which slow delivery of ibuprofen was observed and reinforce the conclusions drawn by Horcajada et al. that MOF hydrophilicity/hydrophobicity and pore size play a large part in the rate of release of a drug [33].

Since Férey et al. first encapsulated ibuprofen in a MOF in 2006, ibuprofen has often been used as a ‘model drug’ in these studies. However, few have focused on oral administration or in vivo activity [40]. By utilising food-grade, soluble γ-cyclodextrins in the synthesis of this MOF, the effects of oral administration could be studied, a desirable property for a MOF DDS. There is clear potential for creating other CD-MOF-drug co-crystals and extending this to other therapeutic agents. The ease of synthesis of this non-toxic, biocompatible MOF make it an attractive option as a DDS.

MOFs as Cancer Treatments: Radiotherapy and Transport of Reactive Oxygen Species

In addition to applications in drug encapsulation and delivery, MOFs have been used as radioenhancers, agents which sensitise tumour cells towards radiation therapy and are used to increase its effectiveness. Roentgen reported the discovery of X-rays in 1895, and radiotherapy has been used as a cancer treatment since soon after that discovery [43,44]. While an important therapeutic tool, side-effects including damage to neighbouring, healthy organs, has resulted in a search for compounds and techniques to enhance a tumour-targeting effect [45,46]. Heavy metal-based nanoparticles (NPs) have been studied for this purpose, as elements of high atomic number have corresponding high X-ray absorption coefficients [47]. Increased uptake by tumours leads to an increased radiosensitivity difference between the tumour and surrounding healthy tissue.

Lin et al. theorised that the electron-dense secondary building units (SBUs)—inorganic polynuclear clusters—in nanoMOFs could preferentially absorb X-rays, thereby increasing radiosensitivity [48]. Additionally, the porous nature of the chosen MOF would allow the transport and diffusion of reactive oxygen species (ROS), used to exert a cytotoxic effect in photodynamic therapy (Figure 4). Furthermore, they hoped to combine nanoMOF radiotherapy with checkpoint blockade immunotherapy, a treatment which enhances antitumour immune system responses by targeting T cell inhibitory pathways.

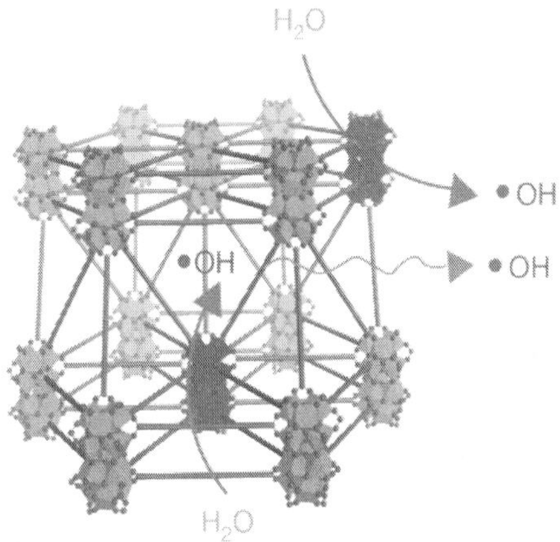

Figure 4. Representation of hydroxyl radical (ROS) generation upon X-ray irradiation and diffusion through Hf_{12}-DBA. Adapted from [48] This article

The MOFs, Hf_6-DBA ($Hf_6(\mu_3\text{-}O)_4(DBA)_6$) and Hf_{12}-DBA ($Hf_{12}(\mu_3\text{-}O)_8(\mu_3\text{-}OH)_8(\mu_2\text{-}O)_6(DBA)_9$) (Figure 5), were synthesised using a solvothermal reaction (Scheme 3). The electron-dense $Hf_6O_4(OH)_4$ and $Hf_{12}O_8(OH)_{14}$ SBUs were chosen as X-ray absorbers that would generate ROS and allow diffusion from the channels.

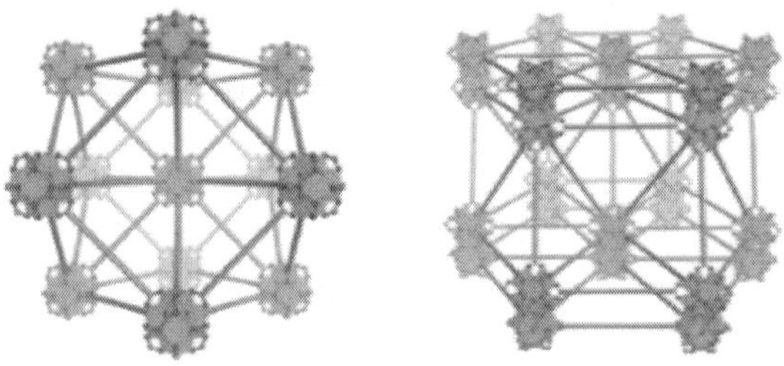

Figure 5. Structural representations of Hf_6-DBA or Hf_6-DBAn (left) and Hf_{12}-DBA or Hf_{12}-DBAn (right). Adapted from [48] This article is licensed under a Creative Commons Attribution 4.0 International License, which permits use, sharing, adaptation, distribution and reproduction in any medium or format, as long as appropriate credit is given to the original author(s) and the source, a link to the Creative Commons license is provided, and changes made are indicated. http://creativecommons.org/licenses/by/4.0/ This image is a cropped version of the original figure in K. Ni, G. Lan, C. Chan, B. Quigley, K. Lu, T. Aung, N. Guo, P. La Riviere, R. R. Weichselbaum and W. Lin, Nat. Comms., 2018, 9, 2351.

Hydroxyl radical generation studies demonstrated that Hf_6-DBA and Hf_{12}-DBA both were more efficient in radioenhancing than HfO_2 NPs. They were able to combine the nanoMOF-mediated RT with the checkpoint blockade therapy and thereby increase the RT effects from local to distant tumours.

As seen in previous examples, the unique properties of the MOF made this a successful experiment. Unlike MOFs designed for gas storage, maximised surface area does not always represent the ideal medicinal MOF. N_2 sorption measurements were used to determine the Brunauer–Emmett–Teller (BET) surface area of the MOFs; Hf_6-DBA was 804 $m^2 g^{-1}$ while Hf_{12}-DBA was 464 $m^2 g^{-1}$. The surface area of the Hf MOFs are low in comparison to MOF-210, known to have an Langmuir surface area of over 10,000 $m^2 g^{-1}$ and a BET surface area of 6240 $m^2 \cdot g^{-1}$ [7], but for use as a radioenhancer, this parameter held less importance. Due to strong coordination between Hf^{4+} ions and carboxylate groups, both nanoMOFs were stable in aqueous suspensions, the cell culture medium and upon X-ray irradiation. The electron-dense SBUs of $Hf_6O_4(OH)_4$ and $Hf_{12}O_8(OH)_{14}$ absorbed X-rays to generate ROS, which cause oxidative stress to cells. The thin nanoplate structure of Hf_{12}-DBA led to higher X-ray absorption and improved ROS diffusion compared to Hf_6-DBA, highlighting the importance of considering structural effects in addition to electronic effects. This team were also behind the development of RiMO-301, a MOF which entered Phase I clinical trials in 2018 and have commented that they hope to advance further MOF-based therapies to clinical trials soon [20,49].

MOFs as Cancer Treatments: Carbon Monoxide Therapy and Magnetic Nuclear Resonance Imaging Through MOF Modification

Further use of MOFs in medicine involves carbon monoxide (CO), a gas which, due to its toxicity, has the potential to be used in combination with chemotherapy as a cancer treatment [50]. This was first achieved in 2013 when Otterbein et al. reported that the use of CO targeted mitochondrial activity and metabolism in cancer cells resulting in the inhibition of tumour growth [51]. However, there is a need for a method of controlled administration and intracellular release of CO to minimise the risk of its toxic effects being felt by the lungs rather than the desired tumour. Previous work on precise intracellular delivery of CO has utilised metal carbonyl-caged graphene oxide (a nanomedicine), or iron carbonyl derived Prussian blue NPs [52, 53]. The porous nature and tuneable structures of MOFs make them well suited to this therapy, as was recently demonstrated by Shen and co-workers [54].

The chosen MOF, MIL-100 (Fe), had multiple purposes. Its pores could encapsulate the chosen chemotherapy agent, Doxorubicin (DOX) and transport it to the site of action. Additionally, the magnetic carbon nanoparticles that were coated in the MOF enabled tumour imaging such as magnetic resonance imaging (MRI) to be conducted. The MOF-coated NPs were then modified with NH_2-PEG-2000 to increase dispersion and with 4,4′-diamino-2,2′-bipyridine (DABPY), designed to chelate to $Mn(CO)_5$ as the source of CO (Figure 6). This was dispersed within $Mn(CO)_5Br$/EtOH solution to obtain the CO and form MCM@PEG-CO-DOX, the therapeutic agent (Scheme 4).

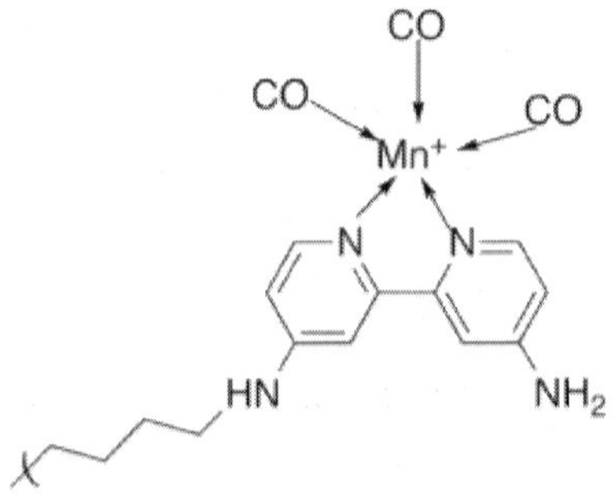

Figure 6. A representation of the DABPY/PEG coating on the MOF surface chelating to Mn carbonyl, the CO source.

Using gas chromatography-mass spectrometry (GC-MS) to determine the release ability of CO from the MCM@PEG-CO-DOX NPs, they found that irradiation with an 808 nm laser for 30 min led to release of 94.8% of the CO. Control experiments found that heat was the driving force in the release, rather than the irradiation, and additionally that this temperature rise did not affect the cell viability. This clarified that the beneficial results were due to the impact of the MOF and its cargo, rather than the conditions required for release.

Clear differentiation between the MCM@PEG-DOX NPs (without CO) and MCM@PEG-CO-DOX NPs was demonstrated, with cell survival rates of 54.9% and 38% respectively, illustrating that MOF-released CO could enhance the chemotherapeutic effect of DOX. Quantitative analysis of the T_2 signal at the tumour site from the MRI was possible, in addition to measurement of the increase in the photoacoustic imaging intensity, demonstrating that the MCM@PEG-CO-DOX NPs were suitable for use as an imaging agent. This illustrates the benefits of using MOFs in medicine; the large pore volume (its Langmuir surface area is estimated to be >2800 $m^2 g^{-1}$) [55] allowed for encapsulation of the chemotherapy drug, while the use of iron (II) (with ferrocene as the source) as a component resulted in the desired magnetic properties. Further modification in the form of a PEG and DABPY coating further demonstrates the versatility of these materials.

MOFs as Detoxifying Agents: Adsorption and Removal of Drugs

MOFs are not solely used to bring drugs into the body—they can remove them as well. A recent paper highlighted the possibility of using MOFs as detoxifying agents [56]. Conventional treatment methods for overdose or intoxication include antidotes, gastric lavage or the use of porous media such as activated charcoal [57,58,59,60]. However, the efficacy of these methods is limited; antidotes are specific to certain toxins, and gastric lavage and charcoal remove unabsorbed drug, therefore requiring fast administration. Nanomaterials have been posited as an alternative option, utilising ligands or micelles to encapsulate or adsorb toxins in the bloodstream [60]. The use of MOFs represents a further advance in this field. High porosity and simple synthesis are valuable in a detoxifying agent, but biocompatibility and high chemical stability in the differing pH zones of the body are additional requirements for use in humans.

MIL-127, or $Fe_3(OH)_{0.66}Cl_{0.33}(C_{16}N_2O_8H_6)_{1.5}(H_2O)_3{\cdot}nH_2O$, was first reported by Eddaoudi et al. in 2007, using indium(III) rather than iron [61]. This MOF is both water stable and biocompatible, and has found other medicinal uses since its synthesis, such as the delivery of NO gas, reported by Serre et al. in 2014 [62]. More recently, Horcajada et al. repurposed MIL-127 as a detoxifying agent, focusing on aspirin (Figure 7) [56]. Aspirin is hydrolysed in the body to form salicylic acid, the active ingredient. Pain medication is frequently implicated in adult poison exposures, and antidotes and treatments are limited [63]. Therefore, there is an apparent demand for additional detoxifying agents and the porous nature of MOFs makes them ideally suited to this purpose.

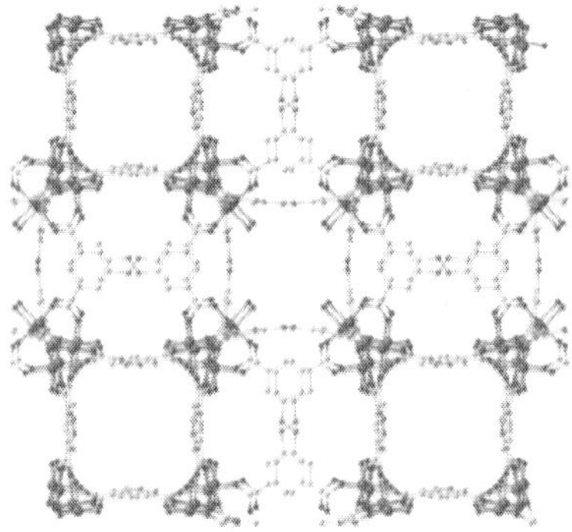

Figure 7. MIL-127 (Fe) crystal structure [56].

The MOF was synthesised solvothermally using a metallic Parr bomb. As discussed in Section 2.1.1. (Structure-activity relationship between MOFs and their cargo: the effect of hydrophilicity/hydrophobicity and pore size), MIL-127 contains both hydrophilic and hydrophobic pores, has a microporous structure, and is based on iron octahedral trimers and 3,3′,5,5′-azobenzenetetracarbozylate anions [64]. MIL-127 is reported to have high porosity, with a BET surface area of 1400 $m^2 g^{-1}$ and pore volume of 0.7 $cm^3 g^{-1}$. This is an essential parameter in the case of adsorption of guest molecules.

Structural stability is required to ensure biocompatibility. This was confirmed after incubation in gastric (2 h) and intestinal (24 h) media containing aspirin. Regardless of the aspirin: MIL-127 ratio, < 9% MOF degradation was seen. Additionally, biodistribution studies demonstrated that, due to the large particle size and high chemical and structural stability, MIL-127 was not adsorbed in the intestines.

Experiments were conducted under simulated gastrointestinal (GI) conditions to determine an appropriate dose of MIL-127 in the case of an aspirin overdose. At a 1.5:1 ratio of aspirin: MOF, 39.4% of the salicylates were adsorbed, compared to 25.2% at a 3:1 ratio. Despite the (expected) trend between higher MOF content and higher amount of removed aspirin, the drug loading (g drug per g MOF) did not depend on the amount of MOF. The maximum aspirin capacity reached was around 0.14 $g g^{-1}$ in both cases. Importantly, after exchange from gastric medium to intestinal medium, there was no detected release of salicylates, which indicated both gastric stability and high affinity of the drug to the MOF. Therefore, while in terms of adsorption, MIL-127 showed lower levels than activated carbon, there was a definite advantage seen in the GI tract, as MIL-127 was able to retain the aspirin.

When administered orally in rats, after 24 h, the plasma and urine concentrations of salicylates were reduced threefold in the presence of MIL-127, reaching similar results as previously published for the adsorption of aspirin by activated carbon [65]. This work illustrates the diverse range of biomedical applications that make use of the porous nature of MOFs.

MOFs as Catalysts for the Degradation of Pharmaceutical Contaminants in Wastewater

MOFs are valuable catalysts, including in medicinal contexts. Sulfonamide antibiotics, such as sulfachloropyradazine (SCP), have toxic effects on marine life, yet overuse has resulted in frequent detection of SCP in wastewater [66]. Parabens (esters of para-hydroxybenzoic acid (*p*-HBA) are found in pharmaceuticals, glues, animal feeds, personal care products and other substances [67]. These degrade into various derivatives, including *p*-HBA, which has been shown to have adverse health effects in men and women. These adverse effects demand the removal of these water contaminants. Activated carbon and zeolites have been used in the removal of parabens and their derivatives, but issues including recycling difficulties and low adsorption levels remain [68]. Therefore, water-stable MOFs are essential materials to control levels of water contaminants via adsorption and catalytic reactions. The development of a highly water stable, biocompatible, cobalt and adenine based MOF, bio-MOF-11-Co, facilitated the degradation of parabens, via activation of the oxidising agent peroxymonosulfate (PMS) (Figure 8) [68].

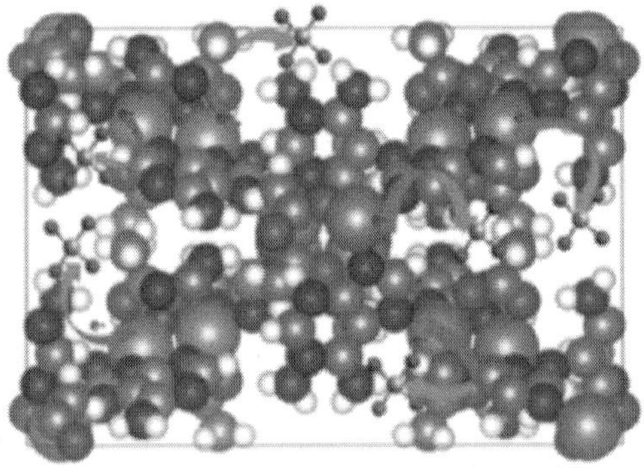

Figure 8. Representation of activation of peroxymonosulfate (PMS), used in catalytic degradation processes, by bio-MOF-11 where cobalt is purple, carbon is grey, oxygen is red, hydrogen is white and nitrogen is blue [68]. Reprinted from Chemosphere, 196, M. R. Azhar, P. Vijay, M. O. Tadé, H. Sun and S. Wang, Submicron sized water-stable metal organic framework (bio-MOF-11) for catalytic degradation of pharmaceuticals and personal care products, 105-114, Copyright (2019), with permission from Elsevier.

Bio-MOF-11-Co was synthesised from cobalt acetate and adenine, an amino acid (Scheme 5) [69]. Water stability was a crucial property in this study. Previous studies have reported that copper is active in catalytic reactions in wastewater treatment processes; therefore, two other MOFs, copper benzene tricarboxylate (CuBTC) and bio-MOF-11-Cu, were tested. However, degradation in water was observed for both materials, marking them unsuitable for wastewater purification. Bio-MOF-11-Co was stable in water and efficiently activated PMS to enable degradation of organic molecules, and the metal leaching observed was negligible. Furthermore, the catalyst was reused three times with only water washing and drying at 130 °C in between. Cobalt, a constituent of the vitamin B_{12}, is essential for life in minute amounts [70]. However, in larger amounts cobalt can be poisonous, with an 'immediate danger to life and health' (IDLH) value of 20 mg m^{-3} [71], making the true biocompatibility of this MOF questionable.

Significant attention was focused on confirming the significance of the Lewis basic sites of the adenine ligands within the MOF. Previous studies have highlighted the effect of strongly basic N atoms in catalytic processes [72]. This, combined with the variable oxidation states of cobalt, allowed the catalytic degradation to occur. It was proposed that the sulfate and hydroxyl radicals were generated by electron transfer. Bio-MOF-11-Co was shown to be mesoporous via a characteristic type IV nitrogen adsorption-desorption isotherm (Figure 9).

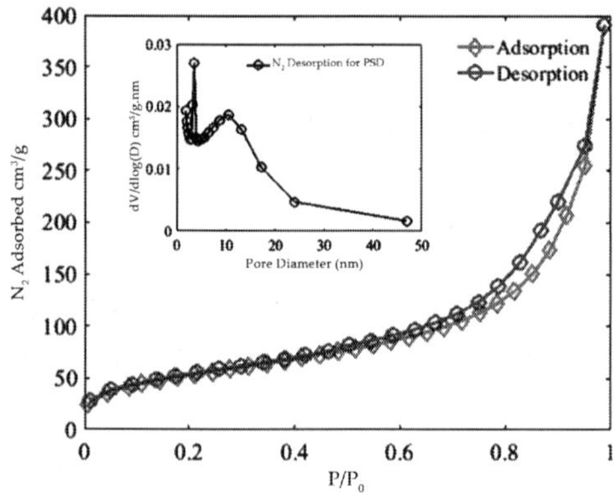

Figure 9. N_2 adsorption-desorption isotherms of bio-MOF-11-Co used to calculate surface area and pore size distribution (PSD) using the Barrett-Joyner-Halenda method [68]· Reprinted from Chemosphere, 196, M. R. Azhar, P. Vijay, M. O. Tadé, H. Sun and S. Wang, Submicron sized water-stable metal organic framework (bio-MOF-11) for catalytic degradation of pharmaceuticals and personal care products, 105-114, Copyright (2019), with permission from Elsevier.

The larger size of mesopores allows less resistance when guest molecules enter the pores, resulting in better adsorption [73]. The researchers noted that in terms of catalyst and PMS loading and reaction time, bio-MOF-11-Co was superior to other catalysts for the degradation of *p*-HBA and SCP. This could be attributed to the aforementioned range of properties of this MOF that make it well suited to the task of catalytic degradation.

BIOINSPIRED MOFS

While there are numerous examples of MOFs being used as drug delivery systems via encapsulation of an API, less prevalent is the use of biologically relevant molecules as linkers. Taking inspiration from biological processes by using biocompatible metals and materials can reduce the risk of toxicity related to release of MOF components once degraded [21]. However, the synthesis of permanently porous and stable MOFs from biomolecules is challenging, in part due to the flexibility of large molecules with high numbers of sp^3 carbons, and a limited number of coordination sites.

APIs as Linkers

Using APIs which have been rigorously tested in clinical trials ensures the maximum dosage and related toxicity is known. Additionally, the prevalence of aromatic motifs in APIs can provide more rigid structures. Incorporation into a MOF provides a novel formulation for the API, potentially improving dosage or solubility issues. This has been exemplified in works by Serre, combining nicotinic acid and iron (2010) [74], and Burrows and Spencer, using zinc or bismuth coordination networks (CN) involving deferiprone, an iron overload drug, as a chelating ligand [75,76]. The latter work observed drug release from the zinc-CN under acidic conditions, while phosphate buffered saline (PBS) released deferiprone from the bismuth analogue, and in Serre's study degraded bioMIL-1 to release nicotinic acid.

APIs as Linkers: Olsalazine and M_2(olz)

Olsalazine, a prodrug of the anti-inflammatory 5-aminosalicylic acid, is widely used in the treatment of gastrointestinal diseases such as ulcerative colitis [77]. Additionally, olsalazine has been shown to inhibit the development of colorectal cancer [78]. Due to the relative scarcity of MOFs constructed wholly from biorelevant materials, without requiring ancillary ligands as in previous olsalazine structures [79,80], in 2016 Long et al. sought to expand a series of isoreticular MOFs of the series M_2(dobdc) (dobdc = 2,5-dioxido-1,4-benzene-dicarboxylate) to create an olsalazine-based MOF with pores of sufficient size to encapsulate additional APIs [81]. M_2(dobdc) structures consist of dicationic metals and dobdc^{4-}, a tetraanionic linker, and have 1D hexagonal pores. Olsalazine has the same coordinating functionalities as the dobpdc^{4-} ligand, but is slightly longer, which the authors theorised could result in expanded pore sizes (Figure 10).

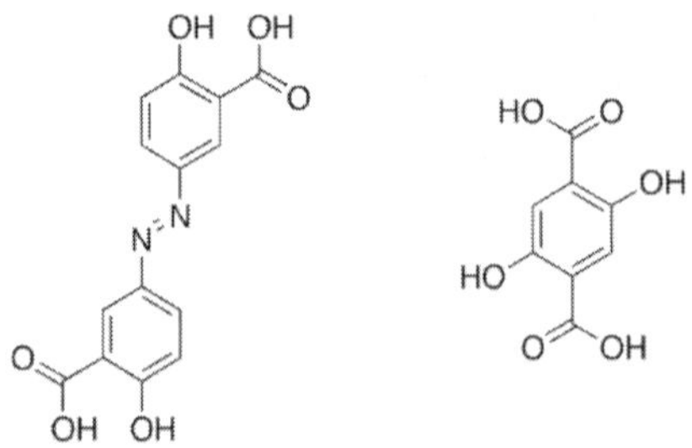

Figure 10. Olsalazine (left) and dobdc.

The MOFs were synthesised using solvothermal techniques (Scheme 6), resulting in isostructural frameworks with mesopores of approximately 2.7 nm and 86 *wt* % API.

As illustrated in Table 2, the surface area of the MOF varied with each metal, with Mg having the largest BET and Langmuir surface area.

Table 2. Brunauer–Emmett–Teller (BET) and Langmuir surface areas of M_2(olz).

M	BET Surface Area ($m^2 g^{-1}$)	Langmuir Surface Area ($m^2 g^{-1}$)
Mg	2545	4593
Fe	1485	2618
Co	2060	3838
Ni	2067	3813
Zn	636	770

In addition to the therapeutic properties of olsalazine, the large pores of Mg_2(olz) allowed the encapsulation of a second therapeutic agent, exemplified here with the model drug phenethylamine (PEA) to form Mg_2(olz)$(PEA)_2$. Degradation of the Mg_2(olz)/Mg_2(olz)$(PEA)_2$ framework to release olsalazine/olsalazine and PEA was observed under physiological conditions, reinforcing the suitability of this structure as a bioactive agent and demonstrating that the MOF could be used as a co-delivery agent.

The use of DMF and methanol in the synthesis of Mg_2(olz) highlights a pertinent question in the field of bioMOFs, as to whether a material can be described as biocompatible if there is the possibility of toxic solvents remaining within the MOF framework. It was reported that Mg_2(olz) was heated under argon and vacuum to remove the residual or coordinated solvent, but it could be alleged a risk remains, and use of biocompatible solvents should be prioritised. No in vivo studies were performed to investigate whether olsalazine in MOF form resulted in improved bioavailability. However, the formulation of olsalazine into a MOF has advantages over free olsalazine as it offers the potential to deliver multiple APIs within one unit.

APIs as Linkers: Curcumin and Medi-MOF-1

A further example utilises the bioactive pharmaceutical ingredient curcumin (Figure 11) [82]. Curcumin has been shown to possess anti-oxidative, anti-inflammatory and anticancer activity [83], and is safe for humans even at high oral doses of 12 g per day [84]. This biocompatibility enables curcumin to be used as a non-toxic building block in MOFs designed for biomedical applications.

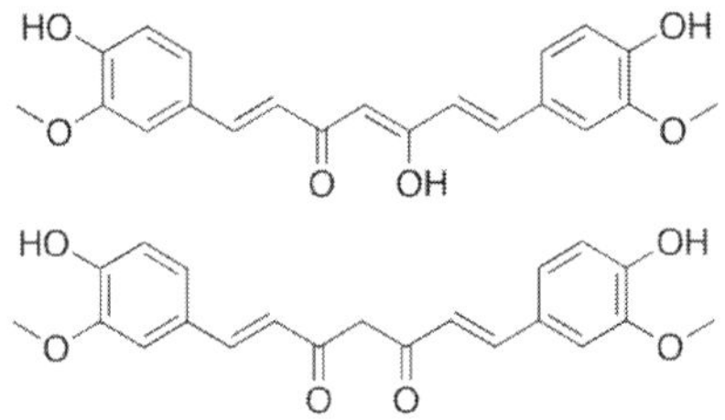

Figure 11. Curcumin in enol (top) and ketone (down) form.

The MOF was synthesised solvothermally using the biocompatible metal Zn with curcumin in the enol form (Scheme 7) [85,86]. Medi-MOF-1 is formed of SBUs consisting of Zn(II) metal-oxo trimers, bridged by hydroxyl O to six other trimers. The O of the 1,3-diketone moiety in its enol form acts as the third coordination site of the linker. Zn-curcumin chains are formed into a 3D distorted primitive cubic (pcu) net by Zn covalently bonding to O atoms of the 1,3-diketone group on other chains. Pores of ~11Å are located in the centre of the pcu lattice cell, with open channels of 6–8 Å. along with specific directions.

Previous studies have demonstrated the ability of curcumin to inhibit the growth of pancreatic cancer cells [87,88]. An in vitro cytotoxicity assay assessed the differences between zinc acetate dehydrate ($Zn(OAc)_2 \cdot 2H_2O$), curcumin and medi-MOF-1, finding similar cytotoxic efficacy between medi-MOF-1 and curcumin. The authors tested the MOF as a co-delivery system using the model drug ibuprofen, resulting in drug loading of about 0.24 g g^{-1}. Under physiological conditions (PBS at 37 °C), 97% of the encapsulated ibuprofen was released within 80 h. This occurred in two distinct stages, with over 50% of the drug, which had formed weaker interactions, released within two hours, and the remaining encapsulated drug, which had made stronger interactions with the MOF framework, slowly releasing in the remaining period. Medi-MOF-1 was unstable in PBS and degraded in the in vitro buffer conditions with 30% of the linker released within the first hour. The authors noted that the synergistic release of curcumin and the loaded drug means medi-MOF-1 acts strictly as a co-delivery system. Medi-MOF-1 exemplifies the potential for MOFs with bioactive linkers to be used in medicine, as they can act as a dual therapy similar to the MOF used for radiotherapy discussed in 2.3.1.

APIs as Linkers: Nalidixic Acid and Mg-MOF

More recently, Mg- and Mn-based MOFs were synthesised using the antibiotic agent nalidixic acid [89]. Nalidixic acid is an antibiotic with high bactericidal and bacteriostatic activity against Gram-negative bacteria, yet its low aqueous solubility leads to poor oral bioavailability [90,91]. Nalidixic acid is particularly relevant to MOF formulation as a synergetic effect of a metal ion to quinolone drugs has been observed, prompting several metal-based nalidixic acid complexes to be synthesised in the past few years [92].

Using mechanochemistry, two bioMOFs based on nalidixic acid and the biocompatible metals Mn and Mg were synthesised (Scheme 8). Water and ethanol were used as solvents during crystal growth by slow evaporation, ensuring full biocompatibility in case of residual solvent in the MOF.

Importantly, the incorporation of nalidixic acid into a MOF enhanced the solubility of the materials in aqueous media. This was tested using high performance liquid chromatography (HPLC) and was attributed to the carboxylate character of the MOFs. While both MOFs were tested on Gram-positive bacteria (Staphylococcus aureus and Enterococcus faecalis), Gram-negative bacteria (*Pseudomonas aeruginosa* and *Escherichia coli*), and yeast strains (*Candida albicans* and *Saccharomyces cerevisiae*), the most significant inhibition of growth was observed against the Gram-negative bacteria E. coli. Mg-MOF demonstrated a bactericidal effect against the Gram-negative bacteria, while Mn-MOF induced a bacteriostatic effect. The antimicrobial activity of both bioMOFs was higher than the nalidixic acid activity alone, demonstrating that an increase in antibiotic efficiency may be achieved by delivery in the form of bioMOFs.

There is clear potential for further bioapplications using biocompatible metals and linkers. MOFs have demonstrated potential as drug delivery agents and use of biocompatible or API linkers can limit potential toxicity issues from the degradation of a MOF within the body. Additionally, co-delivery systems or dual therapeutics using an encapsulated agent in combination with the linker can reduce the number of doses of various APIs, potentially improving patient compliance.

Issues remain, including lack of metal-coordinating functional groups on APIs limiting the ability to form 3D structures, necessitating ancillary linkers. Additionally, harmful solvents, such as DMF, are frequently used in MOF synthesis and could have a toxic effect on the body unless entirely removed from the structure [93,94]. As stated previously, the flexibility of organic ligands can lead to instability in MOFs. However, the number of sp^3 carbons in a drug has been correlated to improved solubility and likelihood of clinical success [95], and so use of APIs in MOF synthesis may remain challenging.

Amino Acids and Endogenous Small Molecules as Linkers

In addition to using APIs as MOF constituents, amino acids have proved useful in the synthesis of medicinally relevant MOFs. These biologically essential compounds are used in the body for the synthesis of proteins or as a source of energy via oxidation to urea and CO_2. Amino acids (except glycine) possess inherent chirality, which has been utilised to create homochiral MOFs for applications, such as separation of chiral structures [96]. 1D and 2D CPs can form due to the single carboxylate coordinating group which can be remedied through the use of ancillary ligands and mixed systems [21].

The Krebs Cycle

The Krebs cycle, or citric acid cycle (CAC) (Figure 12), is a process within the body to release energy through the oxidation of acetyl-coenzyme A (CoA) to adenosine triphosphate (ATP) [97]. Acetyl-CoA is derived from carbohydrates, fats and proteins. Precursors of several amino acids are made during the cycle in addition to the reducing agent nicotinamide adenine dinucleotide (NADH).

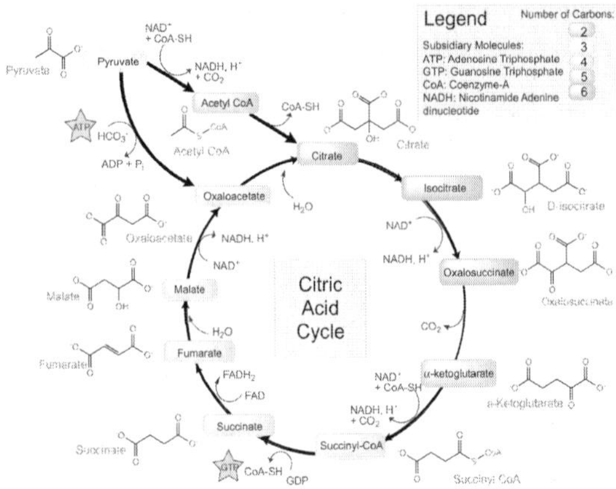

Figure 12. The Krebs cycle, or citric acid cycle.

Taking inspiration from biologically relevant materials such as amino acids and other small molecules involved in the Krebs cycle can ensure biocompatibility in a MOF.

Citric Acid

The starting point of the CAC is the synthesis of citrate from acetyl-CoA. Citric acid has been used as a building block for numerous MOFs due to its three carboxylate groups, which allow trivalent coordination. The first neutral ferrous citrate MOF was reported in 2013, a 1D CP made pseudo-3D by hydrogen bonds between chains [98]. Iron readily forms complexes with oxygen, is biocompatible and is involved in several biological processes [99].

A more recent example uses citrate in the synthesis of a MOF designed for oxygen evolution reaction (OER) electrocatalysis [100]. The OER, $2H_2O \rightarrow O_2 + 4H^+ + 4e^-$, is an oxidative half-reaction of water splitting, to which much research has been directed at the development of efficient catalysts to overcome the slow kinetics and high overpotential which limit this technology [101]. Their porous nature enables MOFs to act as efficient OER electrocatalysts, with the built-in pores and open channels allowing space for the diffusion of reactants and the generated oxygen gas [102]. UTSA-16 was synthesised solvothermally (Scheme 9) and consists of tetranuclear cobalt citrate clusters in which citrate O chelate octahedral cobalt atoms to form the Co_4O_4 cubane motif. This mimics the natural photosystem II and is successful in OER [103].

These clusters act as octahedral linkers and tetrahedral Co(II) atoms as trigonal nodes. The MOF had high activity in alkaline medium, and the authors reported similar efficiency to the benchmark OER catalyst RuO_2. The success of UTSA-16 highlights the value of bioinspired MOFs, not only in medical and bio-related applications, but additionally for more common uses for MOFs, such as catalysis.

Lactic Acid

Recent studies have suggested that lactate, found in the Cori cycle, can be used as a C source in the Krebs cycle [104]. In 2017, Yaghi et al. synthesised two chiral lactic acid-based MOFs, Ca_{14}(L-lactate)$_{20}$(acetate)$_8$(C_2H_5OH)(H_2O) (MOF-1201) and Ca_6(L-lactate)$_3$(acetate)$_9$(H_2O) (MOF-1203) [105]. As with the use of MOFs as drug delivery systems, concerns about the effect that non-biocompatible metals and linkers may have on natural systems inspired the development of fully biocompatible MOFs. Degradation of the MOF is required to release the encapsulated pesticide, and therefore the toxicity of the metal and linker on the environment must be considered; Yaghi's design of a biological acid-based MOF circumvented this issue. Many fumigants, a common form of pesticides, are commonly applied by drip irrigation or shank injection [106]. However, the volatility of conventional pesticides, such as 1,3-dichloropropene and chloropicrin, results in loss and air and groundwater pollution in addition to hazards to users. Porous solids including activated carbon, clay or alumina and adsorption resins have been proposed for use as sorption formulations, allowing adsorbed fumigant to be slowly released [105]. However, limited biodegradability can lead to accumulation of carrier materials, adding to the environmental impact of pesticide use.

Use of ethanol as the solvent ensured full biocompatibility, removing the risk of residual toxic solvents. Acetate was used as an additional ligand, resulting in a three-dimensional arrangement of Ca(−COO, −OH) polyhedra. These contained 1D pores with apertures and internal diameters of 7.8 and 9.6 Å (MOF-1201) and 4.6 and 5.6 Å (MOF-1203). These MOFs represented the first examples of extended porous structures based on Ca(II) and lactate, perhaps reflected by the very low permanent porosity of 430 (MOF-1201) and 160 $m^2\ g^{-1}$ (MOF-1203). Both the high coordination number and geometries of Ca(II) and the flexibility of bio-relevant linkers can result in the formation of dense, nonporous structures, and prior to this work no Ca(II) MOF containing a naturally occurring linker had been reported [105]. MOF-1201 was chosen to be tested as a formulation for the volatile fumigant *cis*-1,3-dichloropropene due to its higher porosity.

Degradation of MOF-1201 to its constituents in water was observed, with 120 ± 10 g dissolved in 1 L of water. The saturated solution had a pH of 7.6, further evidence for its suitability as a pesticide carrier with minimal environmental impact, other than leaving calcium as a nutrient in the soil. Aqueous degradation points to applications not only as a pesticide carrier but medicinally too.

Succinic Acid

In the Krebs cycle, succinate is oxidised to fumarate by succinate dehydrogenase. Succinic acid is a commonly used linker in MOFs, combined with metals including cobalt and terbium [107,108]. As a biocompatible material, materials made from succinic acid have the potential to be used as DDS or adsorbents with reduced environmental impact.

In 2018, a biocompatible MOF consisting of aluminium and succinic acid was reported for the removal of azo dyes in wastewater [109]. Synthetic dyes are common aqueous pollutants; azo dyes, in particular, have highly toxic effects and require complete removal from wastewater in order to mediate any ill effects on aquatic or human life [110,111]. Adsorption techniques hold several advantages over other methods such as electrochemical coagulation or ozonation, mainly in terms of simplicity, selectivity and minimal secondary pollutants [112,113]. Using MOFs as absorbents, due to their porosity, has been an area of significant interest.

The MOF reported here, Al-SA, was synthesised using water as the sole solvent (Scheme 10).

Not only this approach is more environmentally friendly, but the risk of toxic solvents entering wastewater is also removed. Toxic solvents, such as DMA and DMF, are frequently used in the synthesis of MOFs, producing secondary pollutants [93,112].

Al-SA was reported to have a BET surface area of 358.03 $m^2\ g^{-1}$ and pore volume of 0.399 $cm^3\ g^{-1}$. This low BET surface area could be attributed to the flexibility of the succinic acid linker, which, as mentioned, can affect the porosity of MOFs. Al-SA is mesoporous with pores of 4.96 nm. Despite this, it had a similar or higher maximum adsorption capacity to many other adsorbents, including powder activated carbon (PAC) (Table 3, Table 4). Al-SA demonstrated high affinity towards both the synthetic azo dye AB1 and another azo dye, A07, and had a >97.7% removal rate of AB1 across all pH ranges tested.

Table 3. Adsorption capacities of various adsorbents.

Adsorbent	Temperature	Adsorption Capacity (mg g^{-1})
Ammonia-treated CMK-3	25	769.2 [114]
Aluminium ($Al_2(SO_4)_3 \cdot 18H_2O$)-succinic acid MOFs	25	332.5 [115]
Magellanic peat	10	168.9 [116]
Ammonia-tailored CMK-3	25	509.0 [117]
Coal fly ash	60	103.1 [118]
Pumice stone	25	72.5 [119]
Al-SA MOF	25	739.3 [109]

Table 4. Adsorption isotherm parameters for the adsorption of AB1 onto Al-SA MOF and commercial PAC, where Q_{max} = predicted maximum adsorption capacity.

Adsorbents	Langmuir		
	Q_{max}	k_L	R^2
Al-SA MOF	689.38	0.0621	0.9896
Commercial PAC	168.42	0.9326	0.9622

Fumaric Acid

Other bioinspired MOFs have been used for biological and health-related applications. Fumaric acid or fumarate is a common food additive and plays a role in the CAC [120]. Recently a fumaric acid-based MOF was developed for the removal of fluoride from brick tea [121]. Brick tea is consumed in many parts of the world, but its consumption can result in side effects [122,123]. Mature tea plants can absorb and store high levels of fluoride from the air and soil [124], which can pose a danger to regular consumers of brick tea as fluoride is released from the tea during infusion and is highly bioavailable [125]. While small amounts of fluoride are essential for health, large quantities can cause fluorosis. To mitigate this health risk, absorbents can be used to remove F^- from the soil.

Zirconium (MOF-801) and calcium (CaFu) (non-toxic) fumaric acid MOFs were synthesised solvothermally (Scheme 11) and used for this purpose.

The pH of the solution has an impact on the surface charge of the adsorbent; acidic pH resulted in MOF-801 being positively charged, facilitating fluoride adsorption. Brick tea has a pH value of 5.4 and as such experiments were carried out at this pH. Within 5 min, the non-biocompatible Zr MOF was shown to remove almost 80% of fluoride ions in the brick tea infusion. The biocompatible Ca-based MOF was less efficient, with 37% fluoride removed when using 2.0 g L^{-1} adsorbent compared to almost 70% F^- removal under the same conditions for MOF-801. However, for both MOFs, only minor loss of caffeine was observed within 30 min, indicating high selectivity for fluoride over other components in the brick tea infusion.

FTIR, EDX and XPS indicated that the hydroxyl groups in the MOFs were substituted by fluoride. The high selectivity and adsorption rates of CaFu indicate promising applications as an adsorbent.

L-Malic Acid

Malate is one of the final products in the Krebs cycle and is a dicarbonyl compound. L-malic acid has been used to synthesise homochiral MOFs, involving metals such as Mn, Ni, Cu, Co and Ca [126,127,128,129,130]. In 2013 and 2014, Fedin et al. reported multiple Cu and Ni based homochiral MOFs using L-malate and N donor ligands, such as bpy (4,4′-bipyridyl) and bpe (trans-1,2-bis(4-pyridyl)ethylene) (Figure 13) [129,130]. The Cu(II)–malate layers and Ni(II)–malate chains were connected by the N-donor linkers, different lengths of which could be used to control the size of pores. A more recent application of a Cu analogue was as an HPLC separation agent for enantiomers [131].

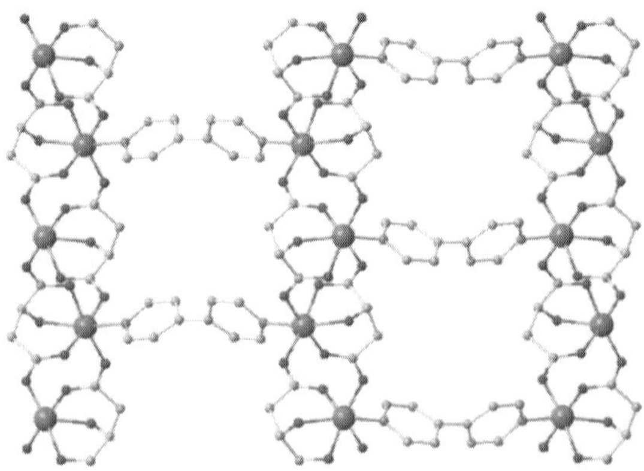

Figure 13. Cu and Ni-based homochiral L-malate MOFs [130].

A Ca based CP was reported in 2013 with the distinction of being a non-mixed system (Figure 14) [128]. The MOF, of the formula $[Ca(HL^1\text{-}MA)]_n$ (H_3L^1-MA = L-malic acid), was synthesised solvothermally and had a highly stable 3D structure, with decomposition observed above 410 °C in the thermogravimetric analysis (TGA). A biocompatible material consisting solely of Ca(II) and L-malic acid, this MOF could have potential bioapplications.

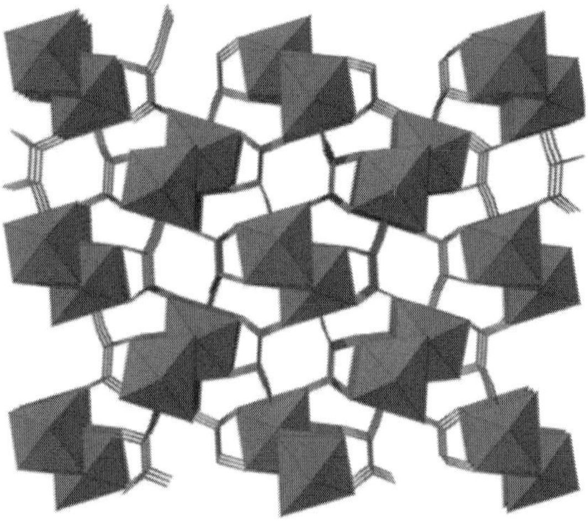

Figure 14. $[Ca(HL^1\text{-}MA)]_n$, a Ca and L-malic acid CP viewed along the a-axis where Ca is represented by green, C by black, O by red and H omitted for clarity. Adapted from [128]. This article is licensed under a Creative Commons Attribution 3.0 International License, which permits use, sharing, adaptation, distribution and reproduction in any medium or format, as long as appropriate credit is given to the original author(s) and the source, a link to the Creative Commons license is provided, and changes made are indicated. https://creativecommons.org/licenses/by/3.0/ This image is a cropped version of the original figure in D. S. Raja, J.-H. Luo, T.-G. Chang, S.-H. Lo, C.-Y. Wu and C.-H. Lin, J. Chem., 2013, 2013, 7.

A 2D porous barium-maleate polymorph was reported in 2014 (Figure 15); the authors suggested it could be useful in optoelectronics or for the typical MOF applications of gas adsorption and storage, however, barium is used as an X-ray contrast medium, potentially highlighting an alternative use for this structure [132].

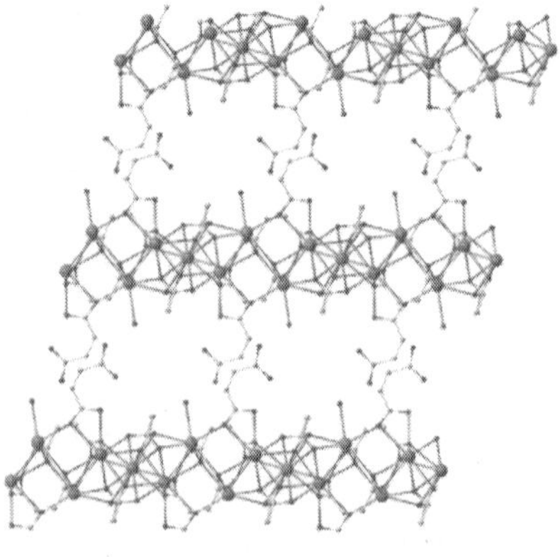

Figure 15. Structure of BM, a barium-maleate 2D material [132].

Peptides as Linkers: Biorelated Structures

Instead of requiring additional, ancillary ligands to create 3D structures, using peptides instead of amino acids can increase the complexity of the ligand and the number of functional groups available for coordination. Peptides can be used as therapeutic agents in their own right, as they naturally act as hormones and transmitters [133].

A Conformationally Active MOF Made of and Mimicking Peptides

Recently, peptides were used to create a flexible MOF, which acted similarly to free peptides which exist in multiple conformations due to rotations about the peptide chain (Figure 16) [134]. Instead of existing in one, stable conformation, nine distinct crystal structures were seen. Change in the peptide conformation could be chemically induced, resulting in a change from a conformation inactive for guest molecule sorption to one that is active.

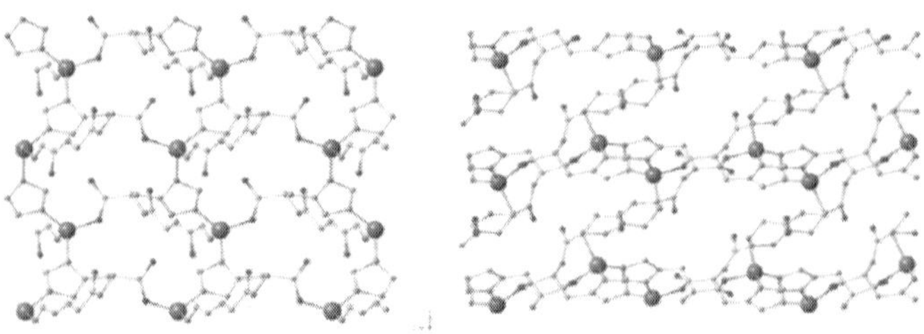

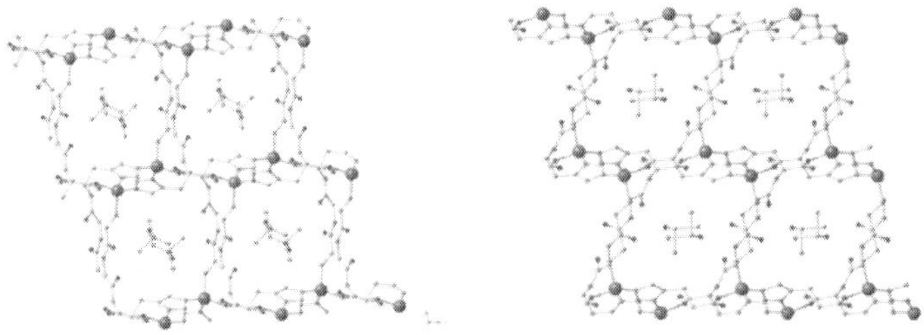

Figure 16. MOFs consisting of the tripeptide glycine–glycine–l-histidine (GGH), Zn and different solvents (clockwise, upper left to lower left: ZnGGH-1, -9, -2 and -3) [134].

The MOF was synthesised solvothermally from the tripeptide glycine–glycine–l-histidine (GGH) and Zn, affording a 3D chiral MOF, ZnGGH-1 (DMF-H_2O) (Scheme 12). Soaking the MOFs in different solvents chemically induced change; DMF and H_2O were able to access the pores of ZnGGH-2 (DMSO) and exchange with DMSO within the pores to form the new structures ZnGGH-3 (DMF) and ZnGGH-4 (H_2O). Crystals of eight solvated ZnGGH structures were obtained by soaking ZnGGH-1 (DMF-H_2O) in different solvents and guests. Using density functional theory (DFT), it was found that, apart from ZnGGH-4 (H_2O), each of the host structures in the absence of molecules within the pores converged to one of three calculated empty structures.

Solvent exchange and subsequent trigger of conformational reorganisation to one of three local minima, involving a twisted or straight linker conformation, created structures active or inactive for guest uptakes, such as dioxane or cyclopentanol. The pore geometry was modified by chemical control of the peptide linker. This study exemplifies the power peptides and their analogues can have in the synthesis of bioinspired structures, but conformational flexibility can lead to unpredictability and instability if not controllable, a potential downside to the use of such flexible molecules in MOF design.

Pseudopeptides as Ligands, for Use in Catalysis

In addition to peptides, pseudopeptides (backbone-modified peptides) can be used as MOF linkers. In 2018, a $ThSi_2$ (ths) CP of the formula $[Cu_3(L^1)_2(H_2O)_8]\cdot 8H_2O$ was reported, which was the first example of a 12-fold interpenetrated ths net [135]. The pseudopeptidic ligand, *N,N,N*-tris(carboxymethyl)-1,3,5-benzenetricarboxamide or trimesoyl-tris-glycine, was based on glycine, the simplest amino acid (Figure 17).

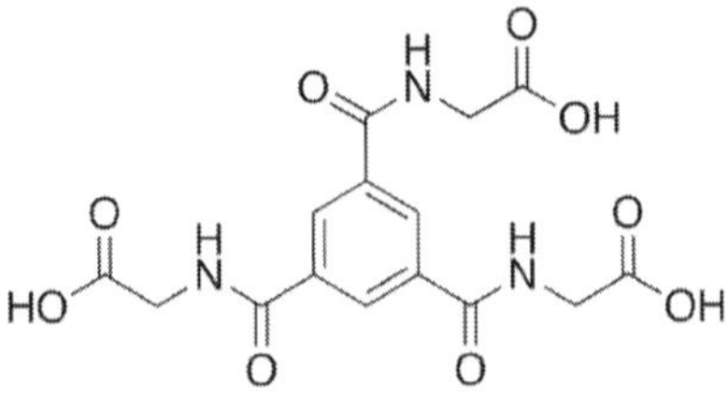

Figure 17. H_3L^1, the ligand used in the study.

The MOF was synthesised solvothermally followed by liquid diffusion in acetone (Scheme 13).

The MOF was tested for use as a catalyst in the A^3 (aldehyde-alkyne-amine) coupling reaction for the synthesis of propargylamines, key intermediates in the synthesis of several N-containing biologically active compounds [136]. The compound demonstrated high activity as a heterogeneous catalyst, particularly with cyclic secondary amines and saturated aliphatic aldehydes, with yields up to 99% and a catalyst loading of 3 mol % (Table 5).

Table 5. Selected results from the A^3 coupling reaction (iPrOH, 90 °C, 24 h, air atmosphere, 3 mol % catalyst).

Entry	**Aldehyde**	**Amine**	**Alkyne**	**Yield [a] (%)**
1	Cyclohexane carboxaldehyde	Pyrrolidine	Phenylacetylene	99
2	Cyclohexane carboxaldehyde	Piperidine	Phenylacetylene	99
3	Cyclohexane carboxaldehyde	Diethylamine	Phenylacetylene	77
4	Benzaldehyde	Pyrrolidine	Phenylacetylene	67
5	Cyclohexane carboxaldehyde	Pyrrolidine	1-hexyne	95
[a] NMR yields based on aldehyde				

The CP lacked porous channels, and as such, it was theorised that the catalysis took place on the CP surface. This work demonstrates that pseudopeptidic ligands can be used to synthesise CPs for the development of organic catalysts.

CONCLUSIONS

Increasingly, researchers are looking beyond applications such as gas separation and fuel storage into areas including medicine [11,137]. It is clear that the possible applications of MOFs are vast, as highlighted in Table 6. These examples demonstrate the unique advantages MOFs can provide in a biomedical context. The synthesis of bioMOFs from biocompatible metals and APIs, and the resultant degradation of the material within a physiological environment offers the potential for slow release drug formulations [74], as does the more traditional encapsulation of an API within the pores of a MOF [33]. These strategies can be combined, to allow the co-delivery of drugs both as part of a framework and within the pores [81,82]. This synergistic effect can also be realised by the simultaneous action of a bioactive cation and linker, such as utilising the antimicrobial properties of zinc [34]. In addition to encapsulating traditional small molecule inhibitors, MOFs can be used to transport and deliver larger, complex molecules, such as hormones [31]. Post-synthetic modification can be a valuable tool to alter properties such as dispersion and chelation abilities [54]. Hydrophilicity and hydrophobicity are important parameters in drug design [138], and the tunable nature of MOFs, allow these to be modified as needed.

Table 6. A summary of the MOFs discussed, with bioapplications and with bioinspired design.

Entry	Compound Name	Formula or Formulation	Applications
MOFs with bioapplications			
1	MIL-100 (Fe)	$[Fe_3O(H_2O)_2OH(C_9H_3O_6)_2]\cdot nH_2O$	Drug delivery (aspirin and ibuprofen) [33]
2	MIL-127 (Fe)	$[Fe_3O(OH)_{0.88}Cl_{0.12}(C_{16}N_2O_8H_6)_{1.5}(H_2O)_3]\cdot nH_2O$	
3	UiO-66 (Zr)	$[Zr_6O_4(OH)_4(C_8O_4H_4)_6]\cdot nH_2O$	
4	CD-MOF-1	$[(C_{48}H_{80}O_{40})(KOH)_2]_n$	Drug delivery (ibuprofen) [40]
5	Hf_6-DBA	$Hf_6(\mu_3\text{-}O)_4(DBA)_6$	Radioenhancers [48]
6	Hf_{12}-DBA	$Hf_{12}(\mu_3\text{-}O)_8(\mu_3\text{-}OH)_8(\mu_2\text{-}O)_6(DBA)_9$	
7	MIL-100 (Fe)	MCM@PEG-CO-DOX	CO therapy, drug delivery (Doxorubicin) [54]
8	MIL-127 (Fe)	$Fe_3(OH)_{0.66}Cl_{0.33}(C_{16}N_2O_8H_6)_{1.5}(H_2O)_3\cdot nH_2O$	Detoxifying agent (aspirin) [56]
9	bio-MOF-11-Co	$Co_2(adenine)_2(CO_2CH_3)_2\cdot 2DMF\cdot 0.5H_2O$	Paraben degradation catalyst [68]
Bioinspired MOFs			

10	Mg_2(olz)	$Mg_2(C_{14}H_{10}N_2O_6)$	Drug delivery/co-delivery (PEA) [81]
11	medi-MOF-1	$[Zn_3(C_{21}H_{20}O_6)_2 \cdot 7(DMA) \cdot 3(EtOH)]$	Drug delivery/co-delivery (ibuprofen) [82]
12	Mg-nalidixic acid	$C_{12}H_{17}N_3O_9Mg$	Antibacterial agents [89]
13	UTSA-16	$[KCo_3(C_6H_4O_7)(C_6H_5O_7)(H_2O)_2] \cdot 8H_2O$	Oxygen evolution reaction (OER) electrocatalyst [100]
14	MOF-1201	Ca_{14}(L-lactate)$_{20}$(acetate)$_8$(C_2H_5OH)(H_2O)	Fumigant sorption formulation [105]
15	MOF-1203	Ca_6(L-lactate)$_3$(acetate)$_9$(H_2O)	
16	Al-SA	-	Azo dye adsorption material [109]
17	MOF-801	$Zr_6O_4(OH)_4$(fumarate)$_6$	Fluoride adsorption material to treat brick tea [121]
18	CaFu	-	
19	BM	$C_{24}H_{14}O_{24}Ba_5$□$7H_2O$	Gas storage and separation, optoelectronics [132]
20	ZnGGH-1-9	Zn(glycine–glycine–L-histidine)-n•(solvent)	Guest molecule sorption [134]
21	$ThSi_2$ CP	$[Cu_3(L^1)_2(H_2O)_8] \cdot 8H_2O$	A^3 coupling catalyst [135]

The process of developing and bringing to market a successful drug or formulation is long and expensive [139], and while significant progress has been demonstrated in terms of proof-of-concept medicinal MOFs, further studies are needed before widespread clinical use becomes a reality. While some MOF studies have demonstrated in vivo compatibility, many only demonstrate activity in vitro, and as such further, comprehensive assessments into their activity and toxicity need to be carried out. In addition, long-term biocompatibility studies will be required before more extensive clinical trials take place. This involves assessing potential degradation pathways, metabolites, target specificity and side effects, in addition to factors such as release profiles and accumulation. The multiple components of a MOF increase the complexity of these assessments; a typical small molecule inhibitor drug is one component under 500 daltons, whereas a MOF DDS could incorporate this in addition to other organic ligands and metal ions. Yet, recent research has focused on novel pharmaceutical agents with more complex behaviours, such as PROTACs and siRNA therapeutics [140,141,142]. As discussed, the use of biocompatible materials can address some of these challenges. The first clinical trial involving a MOF began only in April 2018 [20], and it is clear that while using MOFs for biomedical applications has gained popularity since Férey's seminal work in 2006 [16], further research is required before they are fully understood. While medicinal MOFs may receive less attention than those destined for fuel storage, these porous materials have the potential to improve treatments and human health in a unique manner.

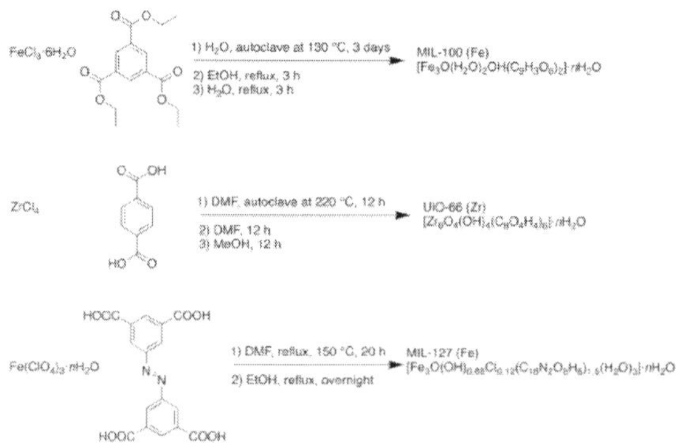

Scheme 1. Synthesis of MIL-100 (Fe), UiO-66 (Zr) and MIL-127 (Fe) using solvothermal and reflux conditions.

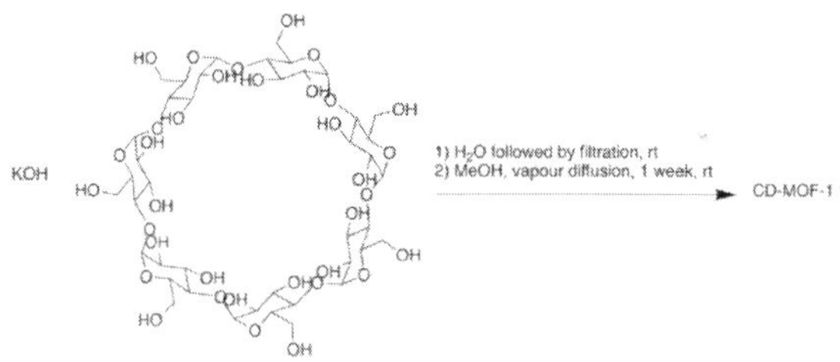

Scheme 2. Synthesis of cyclodextrin (CD)-MOF-1.

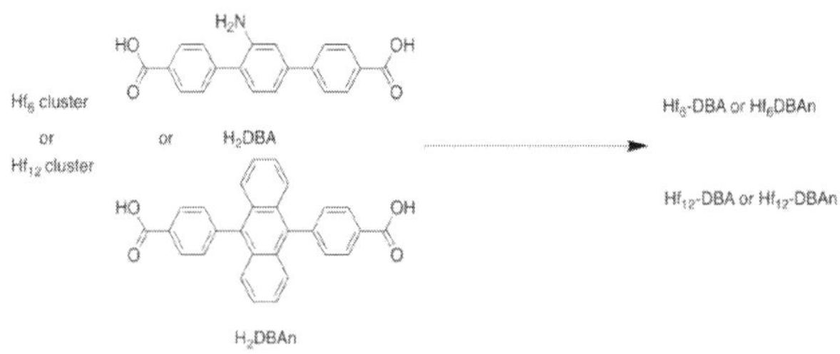

Scheme 3. Synthesis of Hf_6-DBA (DMF, 60 °C, 72 h), Hf_6-DBAn (DMF, trifluoroacetic acid, 60 °C, 72 h), Hf_{12}-DBA (DMF, 80 °C, 72 h) and Hf_{12}-DBAn (DMF, acetic acid, water, 80 °C, 72 h).

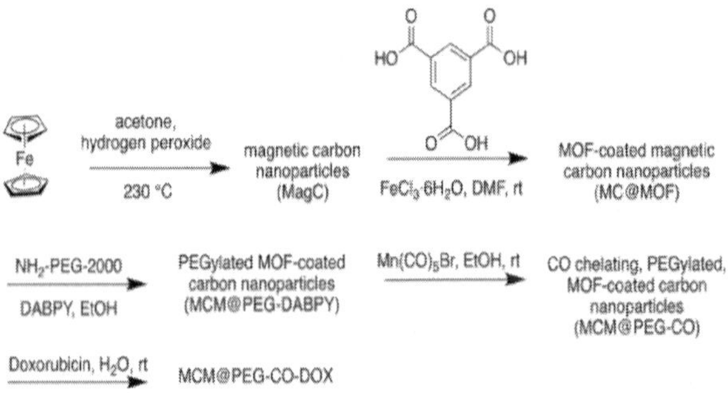

Scheme 4. Synthesis of MCM@PEG-CO-DOX for near-infrared light-responded CO-DOX combination therapy and tumour imaging.

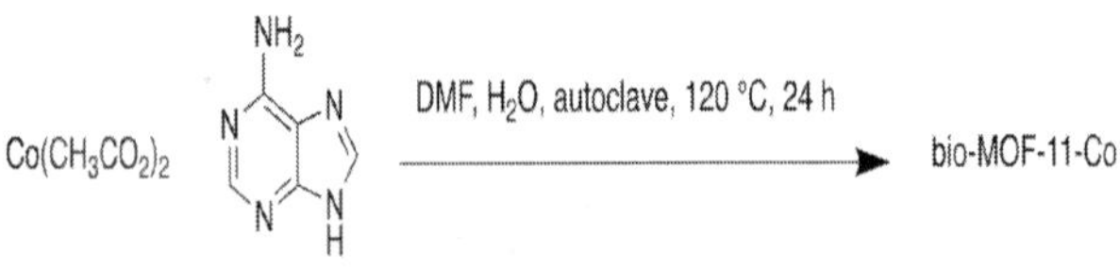

Scheme 5. Synthesis of bio-MOF-11-Co.

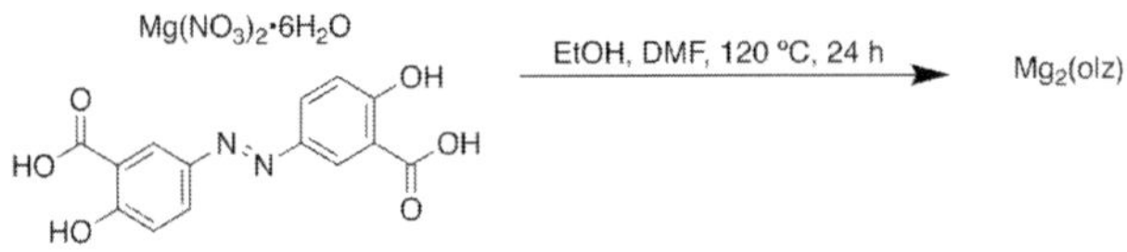

Scheme 6. Synthesis of Mg_2(olz).

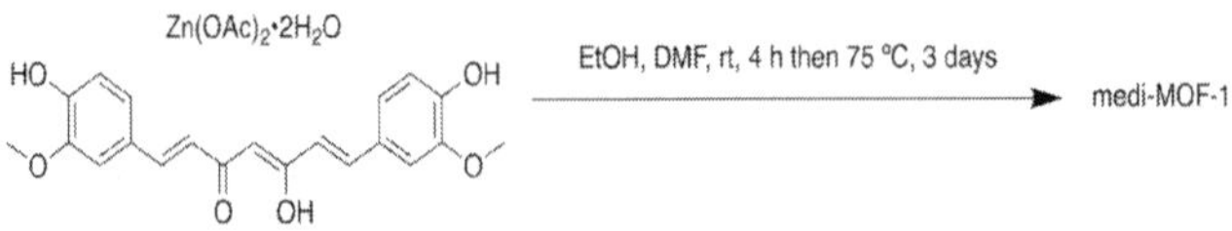

Scheme 7. Synthesis of medi-MOF-1.

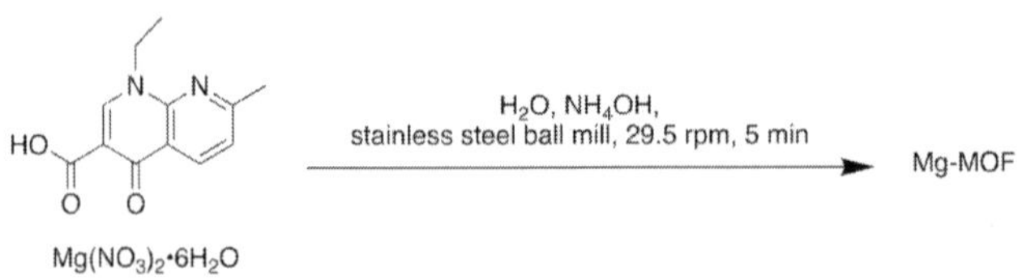

Scheme 8. Synthesis of Mg-MOF.

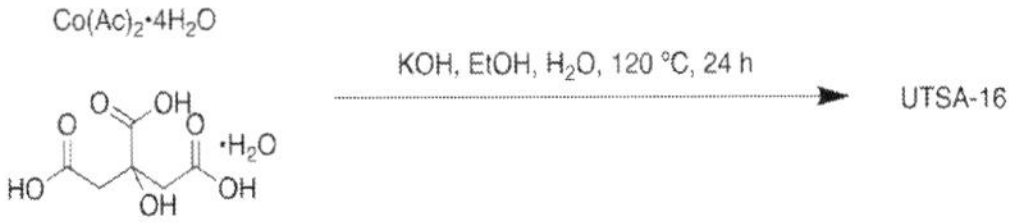

Scheme 9. Synthesis of UTSA-16.

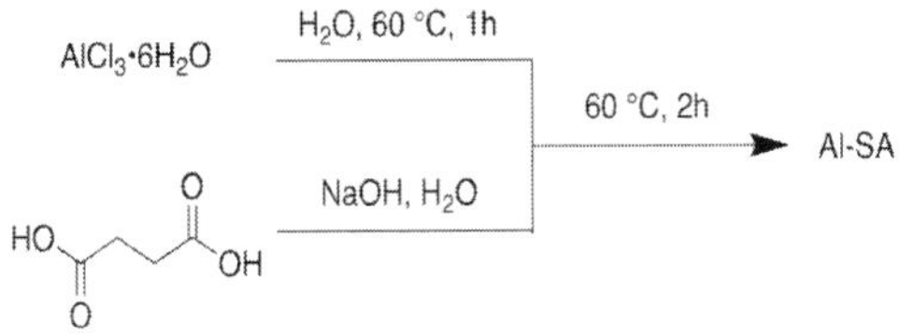

Scheme 10. Synthesis of Al-SA MOF.

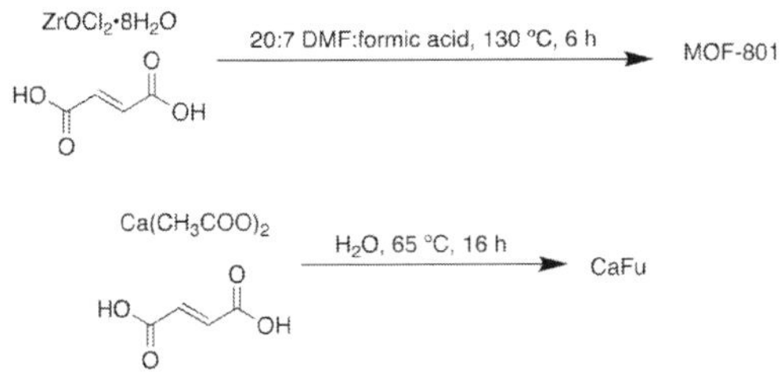

Scheme 11. Synthesis of MOF-801 and CaFu.

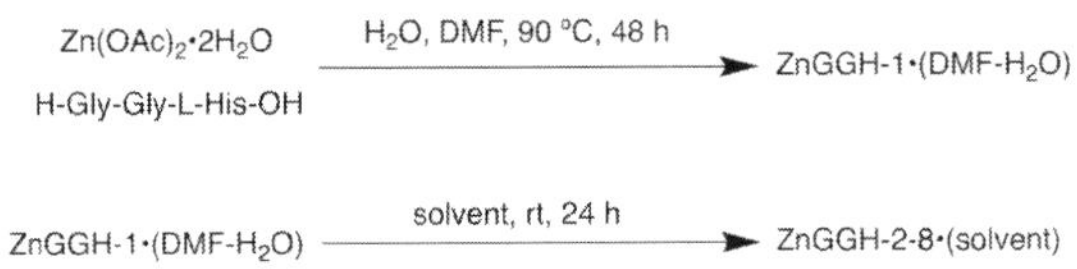

Scheme 12. Synthesis of ZnGGH-1-8•(solvent), including DMF-H_2O, DMSO, DMF, H_2O, H_2O-DMF, furfural and furfuryl alcohol. The term (solvent) in the MOF's name denotes the solvent used in the synthesis and contained in the pores.

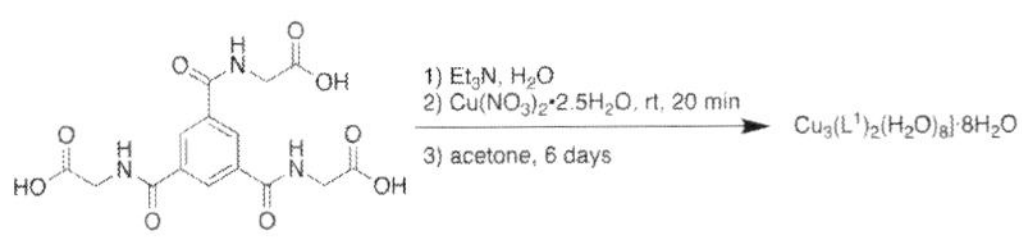

Scheme 13. Synthesis of $Cu_3(L^1)_2(H_2O)_8]\cdot 8H_2O$.

REFERENCES

1. García H., Navalón S. In: Metal-Organic Frameworks: Applications in Separations and Catalysis. Garcia H., Navalón S., editors. Wiley-VCH; Weinheim, Germany: 2018.
2. Hoskins B.F., Robson R. Infinite Polymeric Frameworks Consisting of Three Dimensionally Linked Rod-Like Segments. J. Am. Chem. Soc. 1989;111:5962–5964. doi: 10.1021/ja00197a079.
3. Yaghi O.M., Li H. Hydrothermal Synthesis of a Metal-Organic Framework Containing Large Rectangular Channels. J. Am. Chem. Soc. 1995;117:10401–10402. doi: 10.1021/ja00146a033.
4. Chui S.S.-Y., Lo S.M.-F., Charmant J.P.H., Orpen A.G., Williams I.D. A chemically functionalizable nanoporous material. Science. 1999;283:1148–1150. doi: 10.1126/science.283.5405.1148.
5. Li H., Eddaoudi M., O'Keeffe M., Yaghi O.M. Design and synthesis of an exceptionally stable and highly porous metal-organic framework. Nature. 1999;402:276–279. doi: 10.1038/46248.
6. McGuire C.V., Forgan R.S. The surface chemistry of metal–organic frameworks. Chem. Commun. 2015;51:5199–5217. doi: 10.1039/C4CC04458D.
7. Furukawa H., Ko N., Go Y.B., Aratani N., Choi S.B., Choi E., Yazaydin A.Ö., Snurr R.Q., O'Keeffe M., Kim J., et al. Ultrahigh porosity in metal-organic frameworks. Science. 2010;329:424–428. doi: 10.1126/science.1192160.
8. Klimakow M., Klobes P., Thünemann A.F., Rademann K., Emmerling F. Mechanochemical Synthesis of Metal−Organic Frameworks: A Fast and Facile Approach toward Quantitative Yields and High Specific Surface Areas. Chem. Mater. 2010;22:5216–5221. doi: 10.1021/cm1012119.
9. Klinowski J., Paz F.A.A., Silva P., Rocha J. Microwave-Assisted Synthesis of Metal–Organic Frameworks. Dalton Trans. 2011;40:321–330. doi: 10.1039/C0DT00708K.
10. Batten M.P., Rubio-Martinez M., Hadley T., Carey K.-C., Lim K.-S., Polyzos A., Hill M.R. Continuous flow production of metal-organic frameworks. Curr. Opin. Chem. Eng. 2015;8:55–59. doi: 10.1016/j.coche.2015.02.001.
11. Tansell A.J., Jones C.L., Easun T.L. MOF the beaten track:

Unusual structures and uncommon applications of metal-organic frameworks. Chem. Cent. J. 2017;11:100. doi: 10.1186/s13065-017-0330-0.

12. Sholl D.S., Lively R.P. Defects in Metal-Organic Frameworks: Challenge or Opportunity? J. Phys. Chem. Lett. 2015;6:3437–3444. doi: 10.1021/acs.jpclett.5b01135.
13. Dhakshinamoorthy A., Asiri A.M., Garcia H. Tuneable nature of metal organic frameworks as heterogeneous solid catalysts for alcohol oxidation. Chem. Commun. 2017;53:10851–10869. doi: 10.1039/C7CC05927B.
14. Sajid M. Toxicity of nanoscale metal organic frameworks: A perspective. Environ. Sci. Pollut. R. 2016;23:14805–14807. doi: 10.1007/s11356-016-7053-y.
15. Furukawa H., Cordova K.E., O'Keeffe M., Yaghi O.M. The chemistry and applications of metal-organic frameworks. Science. 2013;341:1230444. doi: 10.1126/science.1230444.
16. Horcajada P., Serre C., Vallet-Regí M., Sebban M., Taulelle F., Férey G. Metal-organic frameworks as efficient materials for drug delivery. Angew. Chem. Int. Ed. 2006;45:5974–5978. doi: 10.1002/anie.200601878.
17. Wang H., Lashkari E., Lim H., Zheng C., Emge T.J., Gong Q., Yam K., Li J. The moisture-triggered controlled release of a natural food preservative from a microporous metal-organic framework. Chem. Commun. 2016;52:2129–2132. doi: 10.1039/C5CC09634K.
18. MOFgen: Biomedical Innovation. [(accessed on 17 January 2019)]; Available online: http://www.mofgen.com/,
19. Hinks N.J., McKinlay A.C., Xiao B., Wheatley P.S., Morris R.E. Metal organic frameworks as NO delivery materials for biological applications. Micropor. Mesopor. Mater. 2010;129:330–334. doi: 10.1016/j.micromeso.2009.04.031.
20. U.S. National Library of Medicine, Clinical Trials Phase I Study of RiMO-301 with Radiation in Advanced Tumours. Primary ID RiMO-CL17-001. ClinicalTrials.gov Identifier: NCT03444714. [(accessed on 18 March 2019)]; Available online: www.clinicaltrials.gov/ct2/show/NCT03444714,
21. Rojas S., Devic T., Horcajada P. Metal organic frameworks based on bioactive components. J. Mater. Chem. B. 2017;5:2560–2573. doi:

10.1039/C6TB03217F.

22. Yang J., Yang Y.-W. Metal–Organic Frameworks for Biomedical Applications. Small. 2020 doi: 10.1002/smll.201906846.
23. Chedid G., Yassin A. Recent Trends in Covalent and Metal Organic Frameworks for Biomedical Applications. Nanomaterials. 2018;8:916. doi: 10.3390/nano8110916.
24. Zhan X., Chen Z., Zhang Q. Recent progress in two-dimensional COFs for energy-related applications. J. Mater. Chem. A. 2017;5:14463–14479. doi: 10.1039/C7TA02105D.
25. Sharma R.K., Yadav P., Yadav M., Gupta R., Rana P., Srivastava A., Zbořil R., Varma R.S., Antonietti M., Gawande M.B. Recent development of covalent organic frameworks (COFs): Synthesis and catalytic (organic-electro-photo) applications. Mater. Horiz. 2020;7:411–454. doi: 10.1039/C9MH00856J.
26. Zhao F., Liu H., Mathe S.D.R., Dong A., Zhang J. Covalent Organic Frameworks: From Materials Design to Biomedical Application. Nanomaterials. 2017;8:15. doi: 10.3390/nano8010015.
27. Latroche M., Surblé S., Serre C., Mellot-Draznieks C., Llewellyn P.L., Lee J.-H., Chang J.-S., Jhung S.H., Férey G. Hydrogen storage in the giant-pore metal-organic frameworks MIL-100 and MIL-101. Angew. Chem. Int. Ed. 2006;45:8227–8231. doi: 10.1002/anie.200600105.
28. Wuttke S., Braig S., Preiß T., Zimpel A., Sicklinger J., Bellomo C., Rädler J.O., Vollmar A.M., Bein T. MOF nanoparticles coated by lipid bilayers and their uptake by cancer cells. Chem. Commun. 2015;51:15752–15755. doi: 10.1039/C5CC06767G.
29. Márquez A.G., Hidalgo T., Lana H., Cunha D., Blanco-Prieto M.J., Álvarez-Lorenzo C., Boissière C., Sánchez C., Serre C., Horcajada P. Biocompatible polymer–metal–organic framework composite patches for cutaneous administration of cosmetic molecules. J. Mater. Chem. B. 2016;4:7031–7040. doi: 10.1039/C6TB01652A.
30. Zhong G., Liu D., Zhang J. Applications of Porous Metal–Organic Framework MIL-100(M) (M = Cr, Fe, Sc, Al, V) Cryst. Growth Des. 2018;18:7730–7744. doi: 10.1021/acs.cgd.8b01353.
31. Chen Y., Li P., Modica J.A., Drout R.J., Farha O.K. Acid-Resistant Mesoporous Metal-Organic Framework toward Oral Insulin Delivery: Protein Encapsulation, Protection, and Release. J. Am. Chem. Soc. 2018;140:5678–5681. doi: 10.1021/jacs.8b02089.

32. Xu J., Wu L., Guo T., Zhang G., Wang C., Li H., Li X., Singh V., Chen W., Gref R., et al. A "Ship-in-a-Bottle" strategy to create folic acid nanoclusters inside the nanocages of γ-cyclodextrin metal-organic frameworks. Int. J. Pharm. 2019;556:89–96. doi: 10.1016/j.ijpharm.2018.11.074.
33. Rojas S., Colinet I., Cunha D., Hidalgo T., Salles F., Serre C., Guillou N., Horcajada P. Toward Understanding Drug Incorporation and Delivery from Biocompatible Metal-Organic Frameworks in View of Cutaneous Administration. ACS Omega. 2018;3:2994–3003. doi: 10.1021/acsomega.8b00185.
34. Tamames-Tabar C., Imbuluzqueta E., Guillou N., Serre C., Miller S.R., Elkaïm E., Horcajada P., Blanco-Prieto M.J. A Zn azelate MOF: Combining antibacterial effect. CrystEngComm. 2015;17:456–462. doi: 10.1039/C4CE00885E.
35. Cunha D., Yahia M.B., Hall S., Miller S.R., Chevreau H., Elkaïm E., Maurin G., Horcajada P., Serre C. Rationale of Drug Encapsulation and Release from Biocompatible Porous Metal–Organic Frameworks. Chem. Mater. 2013;25:2767–2776. doi: 10.1021/cm400798p.
36. Gaudin C., Cunha D., Ivanoff E., Horcajada P., Chevé G., Yasri A., Loget O., Serre C., Maurin G. A quantitative structure activity relationship approach to probe the influence of the functionalization on the drug encapsulation of porous metal-organic frameworks. Micropor. Mesopor. Mater. 2012;157:124–130. doi: 10.1016/j.micromeso.2011.06.011.
37. Cunha D., Gaudin C., Colinet I., Horcajada P., Maurin G., Serre C. Rationalization of the entrapping of bioactive molecules into a series of functionalized porous zirconium terephthalate MOFs. J. Mater. Chem. B. 2013;1:1101–1108. doi: 10.1039/c2tb00366j.
38. Smaldone R.A., Forgan R.S., Furukawa H., Gassensmith J.J., Slawin A.M.Z., Yaghi O.M., Stoddart J.F. Metal-organic frameworks from edible natural products. Angew. Chem. Int. Ed. 2010;49:8630–8634. doi: 10.1002/anie.201002343.
39. Hartlieb K.J., Holcroft J.M., Moghadam P.Z., Vermeulen N.A., Algaradah M.M., Nassar M.S., Botros Y.Y., Snurr R.Q., Stoddart J.F. CD-MOF: A Versatile Separation Medium. J. Am. Chem. Soc. 2016;138:2292–2301. doi: 10.1021/jacs.5b12860.
40. Hartlieb K.J., Ferris D.P., Holcroft J.M., Kandela I., Stern C.L., Nassar M.S., Botros Y.Y., Stoddart J.F. Encapsulation of Ibuprofen in CD-MOF

and Related Bioavailability Studies. Mol. Pharma. 2017;14:1831–1839. doi: 10.1021/acs.molpharmaceut.7b00168.

41. Tiwari G., Tiwari R., Rai A.K. Cyclodextrins in delivery systems: Applications. J. Pharm. Bioall. Sci. 2010;2:72–79. doi: 10.4103/0975-7406.67003.

42. del Valle E.M.M. Cyclodextrins and their uses: A review. Process Biochem. 2004;39:1033–1046. doi: 10.1016/S0032-9592(03)00258-9.

43. Connell P.P., Hellman S. Advances in radiotherapy and implications for the next century: A historical perspective. Cancer Res. 2009;69:383–392. doi: 10.1158/0008-5472.CAN-07-6871.

44. Gianfaldoni S., Gianfaldoni R., Wollina U., Lotti J., Tchernev G., Lotti T. An Overview on Radiotherapy: From Its History to Its Current Applications in Dermatology. Open Access Maced. J. Med. Sci. 2017;5:521–525. doi: 10.3889/oamjms.2017.122.

45. Liu Y., Zhang P., Li F., Jin X., Li J., Chen W., Li Q. Metal-based NanoEnhancers for Future Radiotherapy: Radiosensitizing and Synergistic Effects on Tumor Cells. Theranostics. 2018;8:1824–1849. doi: 10.7150/thno.22172.

46. Grzmil M., Meisel A., Behé M., Schibli R. In: Radiopharmaceutical Chemistry. Lewis J.S., Windhorst A.D., Zeglis B.M., editors. Springer; Berlin/Heidelberg, Germany: 2019.

47. Pottier A., Borghi E., Levy L. New Use of Metals as Nanosized Radioenhancers. Anticancer Res. 2014;34:443–453.

48. Ni K., Lan G., Chan C., Quigley B., Lu K., Aung T., Guo N., la Riviere P., Weichselbaum R.R., Lin W. Nanoscale metal-organic frameworks enhance radiotherapy to potentiate checkpoint blockade immunotherapy. Nat. Comms. 2018;9:2351. doi: 10.1038/s41467-018-04703-w.

49. Notman N. Core Concept: Holey synthetic materials open their pores to medical applications. Proc. Natl. Acad. Sci. USA. 2019;116:6513–6516. doi: 10.1073/pnas.1901321116.

50. Penney D.G. In: Carbon Monoxide Toxicity. Penney D.G., editor. CRC Press; Boca Raton, FL, USA: 2000.

51. Wegiel B., Gallo D., Csizmadia E., Harris C., Belcher J., Vercellotti G.M., Penacho N., Seth P., Sukhatme V., Ahmed A., et al. Carbon monoxide expedites metabolic exhaustion to inhibit tumor growth. Cancer Res. 2013;73:7009–7021. doi: 10.1158/0008-5472.CAN-13-1075.

52. He Q., Kiesewetter D.O., Qu Y., Fu X., Fan J., Huang P., Liu Y., Zhu G., Liu Y., Qian Z., et al. NIR-Responsive On-Demand Release of CO from Metal Carbonyl-Caged Graphene Oxide Nanomedicine. Adv. Mater. 2015;27:6741–6746. doi: 10.1002/adma.201502762.
53. Li W.-P., Su C.-H., Tsao L.-C., Chang C.-T., Hsu Y.-P., Yeh C.-S. Controllable CO Release Following Near-Infrared Light-Induced Cleavage of Iron Carbonyl Derivatized Prussian Blue Nanoparticles for CO-Assisted Synergistic Treatment. ACS Nano. 2016;10:11027–11036. doi: 10.1021/acsnano.6b05858.
54. Yao J., Liu Y., Wang J., Jiang Q., She D., Guo H., Sun N., Pang Z., Deng C., Yang W., et al. On-demand CO release for amplification of chemotherapy by MOF functionalized magnetic carbon nanoparticles with NIR irradiation. Biomaterials. 2019;195:51–62. doi: 10.1016/j.biomaterials.2018.12.029.
55. Horcajada P., Surblé S., Serre C., Hong D.-Y., Seo Y.-K., Chang J.-S., Grenèche J.-M., Margiolaki I., Férey G. Synthesis and catalytic properties of MIL-100(Fe), an iron(III) carboxylate with large pores. Chem. Commun. 2007;27:2820–2822. doi: 10.1039/B704325B.
56. Rojas S., Baati T., Njim L., Manchego L., Neffati F., Abdeljelil N., Saguem S., Serre C., Najjar M.F., Zakhama A., et al. Metal-Organic Frameworks as Efficient Oral Detoxifying Agents. J. Am. Chem. Soc. 2018;140:9581–9586. doi: 10.1021/jacs.8b04435.
57. Vale J.A., Kulig K. American Academy of Clinical Toxicology and European Association of Poisons Centres and Clinical Toxicologists. Position paper: Gastric lavage. J. Toxicol. Clin. Toxicol. 2004;42:933–943.
58. Chyka P., Seger D., Krenzelok E., Vale J. American Academy of Clinical Toxicology and European Association of Poisons Centres and Clinical Toxicologists. Position paper: Single-dose activated charcoal. Clin. Toxicol. 2005;43:61–87.
59. Betten D.P., Vohra R.B., Cook M.D., Matteucci M.J., Clark R.F. Antidote use in the critically ill poisoned patient. J. Intensive Care Med. 2006;21:255–277. doi: 10.1177/0885066606290386.
60. Graham L.M., Nguyen T.M., Lee S.B. Nanodetoxification: Emerging role of nanomaterials in drug intoxication treatment. Nanomedicine. 2011;6:921–928. doi: 10.2217/nnm.11.75.
61. Liu Y., Eubank J.F., Cairns A.J., Eckert J., Kravtsov V.C., Luebke R.,

Eddaoudi M. Assembly of metal-organic frameworks (MOFs) based on indium-trimer building blocks: A porous MOF with soc topology and high hydrogen storage. Angew. Chem. Int. Ed. 2007;46:3278–3283. doi: 10.1002/anie.200604306.

62. Eubank J.F., Wheatley P.S., Lebars G., McKinlay A.C., Leclerc H., Horcajada P., Daturi M., Vimont A., Morris R.E., Serre C. Porous, rigid-metal(III)-carboxylate metal-organic frameworks for the delivery of nitric oxide. APL Mater. 2014;2:124112. doi: 10.1063/1.4904069.
63. Gummin D.D., Mowry J.B., Spyker D.A., Brooks D.E., Osterthaler K.M., Banner W. 2017 Annual Report of the American Association of Poison Control Centers' National Poison Data System (NPDS): 35th Annual Report. Clin. Toxicol. 2018;56:1213–1415. doi: 10.1080/15563650.2018.1533727.
64. Chevreau H., Permyakova A., Nouar F., Fabry P., Livage C., Ragon F., Garcia-Marquez A., Devic T., Steunou N., Serre C., et al. Synthesis of the biocompatible and highly stable MIL-127(Fe): From large scale synthesis to particle size control. CrystEngComm. 2016;18:4094–4101. doi: 10.1039/C5CE01864A.
65. Decker W., Corby D., Ibanez J. Aspirin adsorption with activated charcoal. Lancet. 1968;291:754–755. doi: 10.1016/S0140-6736(68)92204-6.
66. Bu Q., Wang B., Huang J., Deng S., Yu G. Pharmaceuticals and personal care products in the aquatic environment in China: A review. J. Haz. Mater. 2013;262:189–211. doi: 10.1016/j.jhazmat.2013.08.040.
67. Wang W., Kannan K. Fate of Parabens and Their Metabolites in Two Wastewater Treatment Plants in New York State, United States. Environ. Sci. Technol. 2016;50:1174–1181. doi: 10.1021/acs.est.5b05516.
68. Azhar M.R., Vijay P., Tadé M.O., Sun H., Wang S. Submicron sized water-stable metal organic framework (bio-MOF-11) for catalytic degradation of pharmaceuticals and personal care products. Chemosphere. 2018;196:105–114. doi: 10.1016/j.chemosphere.2017.12.164.
69. An J., Geib S.J., Rosi N.L. High and selective CO2 uptake in a cobalt adeninate metal-organic framework exhibiting pyrimidine- and amino-decorated pores. J. Am. Chem. Soc. 2010;132:38–39. doi: 10.1021/ja909169x.
70. Yamada K. In: Interrelations between Essential Metal Ions and Human

Diseases. Sigel A., Sigel H., Sigel R.K.O., editors. Springer; Berlin/Heidelberg, Germany: 2013.

71. 7The National Institute for Occupational Safety and Health (NIOSH), CAS No. 7440-48-4. [(accessed on 16 October 2019)]; Available online: https://www.cdc.gov/niosh/npg/npgd0146.html,
72. Kryszak D., Stawicka K., Trejda M., Calvino-Casilda V., Martin-Aranda R., Ziolek M. Development of basicity in mesoporous silicas and metallosilicates. Catal. Sci. Technol. 2017;7:5236–5248. doi: 10.1039/C7CY00927E.
73. Corma A. From Microporous to Mesoporous Molecular Sieve Materials and Their Use in Catalysis. Chem. Rev. 1997;97:2373–2420. doi: 10.1021/cr960406n.
74. Miller S.R., Heurtaux D., Baati T., Horcajada P., Grenèche J.-M., Serre C. Biodegradable therapeutic MOFs for the delivery of bioactive molecules. Chem. Commun. 2010;46:4526–4528. doi: 10.1039/c001181a.
75. Burrows A.D., Jurcic M., Keenan L.L., Lane R.A., Mahon M.F., Warren M.R., Nowell H., Paradowski M., Spencer J. Incorporation by coordination and release of the iron chelator drug deferiprone from zinc-based metal–organic frameworks. Chem. Commun. 2013;49:11260–11262. doi: 10.1039/c3cc45689g.
76. Burrows A.D., Jurcic M., Mahon M.F., Pierrat S., Roffe G.W., Windle H.J., Spencer J. Bismuth coordination networks containing deferiprone: Synthesis, characterisation, stability and antibacterial activity. Dalton Trans. 2015;44:13814–13817. doi: 10.1039/C5DT02458G.
77. Selby W.S., Barr G.D., Ireland A., Mason C.H., Jewell D.P. Olsalazine in active ulcerative colitis. Brit. Med. J. 1985;291:1373–1375. doi: 10.1136/bmj.291.6506.1373.
78. Brown W.A., Farmer K.C., Skinner S.A., Malcontenti-Wilson C., Misajon A., O'Brien P.E. 5-aminosalicyclic acid and olsalazine inhibit tumor growth in a rodent model of colorectal cancer. Dig. Dis. Sci. 2000;45:1578–1584. doi: 10.1023/A:1005517112039.
79. Xiao D.-R., Sun D.-Z., Liu J.-L., Zhang G.-J., Chen H.-Y., He J.-H., Yan S.-W., Yuan R., Wang E.-B. Two Unprecedented Entangled Metal–Olsalazine Complexes with Coexistence of 2D → 3D Polycatenation and meso-Helix. Eur. J. Inorg. Chem. 2011;2011:4656–4663. doi: 10.1002/ejic.201100366.

80. Chen H.-Y., Xiao D.-R., Yan S.-W., He J.-H., Yang J., Wang X., Yuan R., Wang E.-B. Two three-dimensional pillared metal–olsalazine complexes based on infinite rod-shaped secondary building units. Inorg. Chim. Acta. 2012;387:283–288. doi: 10.1016/j.ica.2012.01.035.
81. Levine D.J., Runčevski T., Kapelewski M.T., Keitz B.K., Oktawiec J., Reed D.A., Mason J.A., Jiang H.Z.H., Colwell K.A., Legendre C.M., et al. Olsalazine-Based Metal-Organic Frameworks as Biocompatible Platforms for H2 Adsorption and Drug Delivery. J. Am. Chem. Soc. 2016;138:10143–10150. doi: 10.1021/jacs.6b03523.
82. Su H., Sun F., Jia J., He H., Wang A., Zhu G. A highly porous medical metal–organic framework constructed from bioactive curcumin. Chem. Commun. 2015;51:5774–5777. doi: 10.1039/C4CC10159F.
83. Maheshwari R.K., Singh A.K., Gaddipati J., Srimal R.C. Multiple biological activities of curcumin: A short review. Life Sci. 2006;78:2081–2087. doi: 10.1016/j.lfs.2005.12.007.
84. Lao C.D., Ruffin M.T., Normolle D., Heath D.D., Murray S.I., Bailey J.M., Boggs M.E., Crowell J., Rock C.L., Brenner D.E. Dose escalation of a curcuminoid formulation. BMC Complement. Altern. Med. 2006;6:10. doi: 10.1186/1472-6882-6-10.
85. Bowen P.K., Guillory R.J., Shearier E.R., Seitz J.-M., Drelich J., Bocks M., Zhao F., Goldman J. Metallic zinc exibits optimal biocompatibility for bioabsorbable endovascular stents. Mater. Sci. Eng. C. 2015;56:467–472. doi: 10.1016/j.msec.2015.07.022.
86. Chen Y., Huang P., Chen H., Wang S., Wang H., Guo J., Zhang X., Zhang S., Yan J., Xia J., et al. Assessment of the Biocompatibility and Biological Effects of Biodegradable Pure Zinc Material in the Colorectum. ACS Biomater. Sci. Eng. 2018;4:4095–4103. doi: 10.1021/acsbiomaterials.8b00957.
87. Kunnumakkara A.B., Guha S., Krishnan S., Diagaradjane P., Gelovani J., Aggarwal B.B. Curcumin potentiates antitumor activity of gemcitabine in an orthotopic model of pancreatic cancer through suppression of proliferation, angiogenesis, and inhibition of nuclear factor-kappaB-regulated gene products. Cancer Res. 2007;67:3853–3861. doi: 10.1158/0008-5472.CAN-06-4257.
88. Dandawate P.R., Vyas A., Ahmad A., Banerjee S., Deshpande J., Swamy K.V., Jamadar A., Dumhe-Klaire A.C., Padhye S., Sarkar F.H. Inclusion complex of novel curcumin analogue CDF and β-cyclodextrin (1:2) and its enhanced in vivo anticancer activity against pancreatic

cancer. Pharm. Res. 2012;29:1775–1786. doi: 10.1007/s11095-012-0700-1.

89. André V., da Silva A.R.F., Fernandes A., Frade R., Garcia C., Rijo P., Antunes A.M.M., Rocha J., Duarte M.T. Mg- and Mn-MOFs Boost the Antibiotic Activity of Nalidixic Acid. ACS Appl. Bio Mater. 2019;2:2347–2354. doi: 10.1021/acsabm.9b00046.

90. Barry A.L., Jones R.N., Thornsberry C., Ayers L.W., Gerlach E.H., Sommers H.M. Antibacterial activities of ciprofloxacin, norfloxacin, oxolinic acid, cinoxacin, and nalidixic acid. Antimicrob. Agents Chemother. 1984;25:633–637. doi: 10.1128/AAC.25.5.633.

91. Mallick S., Sahu A., Pal K. Dissolution behaviour of nalidixic acid solid dispersions using water soluble dispersion carriers. Acta Pol. Pharm. 2004;61:21–30.

92. Psomas G., Kessissoglou D.P. Quinolones and non-steroidal anti-inflammatory drugs interacting with copper(ii), nickel(ii), cobalt(ii) and zinc(ii): Structural features, biological evaluation and perspectives. Dalton Trans. 2013;42:6252–6276. doi: 10.1039/c3dt50268f.

93. Seetharaj R., Vandana P.V., Arya P., Mathew S. Dependence of solvents, pH, molar ratio and temperature in tuning metal organic framework architecture. Arab. J. Chem. 2019;12:295–315. doi: 10.1016/j.arabjc.2016.01.003.

94. Sigma Aldrich: Dorset, UK, 28.06.2019, N,N-Dimethylformamide; SDS No. D4551 [Online] [(accessed on 8 January 2019)]; Available online: https://www.sigmaaldrich.com/MSDS/MSDS/DisplayMSDSPage.do?country=GB&language=en&productNumber=D4551&brand=SIGMA&PageToGoToURL=https%3A%2F%2Fwww.sigmaaldrich.com%2Fcatalog%2Fproduct%2Fsigma%2Fd4551%3Flang%3Den.

95. Lovering F., Bikker J., Humblet C. Escape from Flatland: Increasing Saturation as an Approach to Improving Clinical Success. J. Med. Chem. 2009;52:6752–6756. doi: 10.1021/jm901241e.

96. Zhang J.-H., Nong R.-Y., Xie S.-M., Wang B.-J., Ai P., Yuan L.-M. Homochiral metal-organic frameworks based on amino acid ligands for HPLC separation of enantiomers. Electrophoresis. 2017;38:2513–2520. doi: 10.1002/elps.201700122.

97. Mailloux R.J., Bériault R., Lemire J., Singh R., Chénier D.R., Hamel

R.D., Appanna V.D. The tricarboxylic acid cycle, an ancient metabolic network with a novel twist. PLoS ONE. 2007;2:e690. doi: 10.1371/journal.pone.0000690.

98. Čelič T.B., Jagličič Z., Lazar K., Logar N.Z. Structure and magnetic properties of a new iron(II) citrate coordination polymer. Acta Cryst. B. 2013;69:490–495. doi: 10.1107/S2052519213023713.

99. Abbaspour N., Hurrell R., Kelishadi R. Review on iron and its importance for human health. J. Res. Med. Sci. 2014;19:164–174.

100. Jiang J., Huang L., Liu X., Ai L. Bioinspired Cobalt–Citrate Metal–Organic Framework as an Efficient Electrocatalyst for Water Oxidation. ACS Appl. Mater. Interfaces. 2017;9:7193–7201. doi: 10.1021/acsami.6b16534.

101. Suen N.-T., Hung S.-F., Quan Q., Zhang N., Xu Y.-J., Chen H.M. Electrocatalysis for the oxygen evolution reaction: Recent development and future perspectives. Chem. Soc. Rev. 2017;46:337–365. doi: 10.1039/C6CS00328A.

102. Farid S., Ren S., Hao C. MOF-derived metal/carbon materials as oxygen evolution reaction catalysts. Inorg. Chem. Commun. 2018;94:57–74. doi: 10.1016/j.inoche.2018.06.008.

103. Evangelisti F., Güttinger R., Moré R., Luber S., Patzke G.R. 3d–4f {CoII3Ln(OR)4} Cubanes as Bio-Inspired Water Oxidation Catalysts. J. Am. Chem. Soc. 2013;135:18734–18737. doi: 10.1021/ja4098302.

104. Hui S., Ghergurovich J.M., Morscher R.J., Jang C., Teng X., Lu W., Esparza L.A., Reya T., Zhan L.E., Guo J.Y., et al. Glucose feeds the TCA cycle via circulating lactate. Nature. 2017;551:115–118. doi: 10.1038/nature24057.

105. Yang J., Trickett C.A., Alahmadi S.B., Alshammari A.S., Yaghi O.M. Calcium l-Lactate Frameworks as Naturally Degradable Carriers for Pesticides. J. Am. Chem. Soc. 2017;139:8118–8121. doi: 10.1021/jacs.7b04542.

106. Ajwa H.A., Trout T., Mueller J., Wilhelm S., Nelson S.D., Soppe R., Shatley D. Application of alternative fumigants through drip irrigation systems. Phytopathology. 2002;92:1349–1355. doi: 10.1094/PHYTO.2002.92.12.1349.

107. Livage C., Egger C., Férey G. Hydrothermal versus Nonhydrothermal Synthesis for the Preparation of Organic−Inorganic Solids: The

Example of Cobalt(II) Succinate. Chem. Mater. 2001;13:410–414. doi: 10.1021/cm001148k.

108. Zhang B., Chen L., Yang X., Xu T., Sun T., Wang L., Zhang Q. A luminescent Terbium-Succinate MOF fabricated by co-precipitation for sensing of Fe3+ in aqueous environment. J. Mater. Sci. Mater. Electron. 2017;28:7326–7332. doi: 10.1007/s10854-017-6419-3.

109. Jung K.-W., Choi B.H., Lee S.Y., Ahn K.-H., Lee Y.J. Green synthesis of aluminum-based metal organic framework for the removal of azo dye Acid Black 1 from aqueous media. J. Ind. Eng. Chem. 2018;67:316–325. doi: 10.1016/j.jiec.2018.07.003.

110. Yaseen D.A., Scholz M. Textile dye wastewater characteristics and constituents of synthetic effluents: A critical review. Int. J. Environ. Sci. Technol. 2019;16:1193–1226. doi: 10.1007/s13762-018-2130-z.

111. Puvaneswari N., Muthukrishnan J., Gunasekaran P. Toxicity assessment and microbial degradation of azo dyes. Indian, J. Exp. Biol. 2006;44:618–626.

112. Bu R., Chen F., Li J., Li W., Yang F. Adsorption capability for anionic dyes on 2-hydroxyethylammonium acetate-intercalated layered double hydroxide. Colloids Surf. A. 2016;511:312–319. doi: 10.1016/j.colsurfa.2016.10.015.

113. Blanco S.P.D.M., Scheufele F.B., Módenes A.N., Espinoza-Quiñones F.R., Marin P., Kroumov A.D., Borba C.E. Kinetic, equilibrium and thermodynamic phenomenological modeling of reactive dye adsorption onto polymeric adsorbent. Chem. Eng. J. 2017;307:466–475. doi: 10.1016/j.cej.2016.08.104.

114. Peng X., Hu X., Fu D., Lam F.L.Y. Adsorption removal of acid black 1 from aqueous solution using ordered mesoporous carbon. Appl. Surf. Sci. 2014;294:71–80. doi: 10.1016/j.apsusc.2013.11.157.

115. Jung K.-W., Choi B.H., Dao C.M., Lee Y.J., Choi J.-W., Ahn K.-H., Lee S.-H. Aluminum carboxylate-based metal organic frameworks for effective adsorption of anionic azo dyes from aqueous media. J. Ind. Eng. Chem. 2018;59:149–159. doi: 10.1016/j.jiec.2017.10.019.

116. Sepulveda L.A., Santana C.C. Effect of solution temperature, pH and ionic strengthon dye adsorption onto Magellanic peat. Environ. Technol. 2013;34:967–977. doi: 10.1080/09593330.2012.724251.

117. He C., Hu X. Anionic Dye Adsorption on Chemically Modified Ordered Mesoporous Carbons. Ind. Eng. Chem. Res. 2011;50:14070–14083.

doi: 10.1021/ie201469p.

118. HsuT.-C.Adsorptionofanaciddyeontocoalflyash.Fuel.2008;87:3040–3045. doi: 10.1016/j.fuel.2008.03.026.

119. Samarghandi M.R., Zarrabi M., Amrane A., Soori M.M., Sepehr M.N. Removal of Acid Black Dye by Pumice Stone as a low cost adsorbent; kinetic, thermodynamic and equilibrium studies. Environ. Eng. Manag. J. 2012;12:2137–2147.

120. EU Approved Additives and E Numbers. [(accessed on 8 January 2019)]; Available online: https://www.food.gov.uk/business-guidance/eu-approved-additives-and-e-numbers,

121. Ke F., Peng C., Zhang T., Zhang M., Zhou C., Cai H., Zhu J., Wan X. Fumarate-based metal-organic frameworks as a new platform for highly selective removal of fluoride from brick tea. Sci. Rep. 2018;8:939. doi: 10.1038/s41598-018-19277-2.

122. Cao J., Zhao Y., Liu J. Brick tea consumption as the cause of dental fluorosis among children from Mongol, Kazak and Yugu populations in China. Food Chem. Toxicol. 1997;35:827–833. doi: 10.1016/S0278-6915(97)00049-5.

123. Zhang R., Cheng L., Zhang T., Xu T., Li M., Yin W., Jiang Q., Yang Y., Hu T. Brick tea consumption is a risk factor for dental caries and dental fluorosis among 12-year-old Tibetan children in Ganzi. Environ. Geochem. Health. 2018;41:1405–1417. doi: 10.1007/s10653-018-0216-7.

124. Zhang X., Gao H., Yang T., Wu H., Wang Y., Wan X. Al3+-promoted fluoride accumulation in tea plants (Camellia sinensis) was inhibited by an anion channel inhibitor DIDS. J. Sci. Food Agric. 2016;96:4224–4230. doi: 10.1002/jsfa.7626.

125. Waugh T.D., Potter W., Limeback H., Godfrey M. Risk Assessment of Fluoride Intake from Tea in the Republic of Ireland and its Implications for Public Health and Water Fluoridation. Int. J. Environ. Res. Public Health. 2016;13:259. doi: 10.3390/ijerph13030259.

126. Zingiryan A., Zhang J., Bu X. Cooperative Self-Assembly of Chiral L-Malate and Achiral Succinate in the Formation of Three-Dimensional Homochiral Framework. Inorg. Chem. 2008;47:8607–8609. doi: 10.1021/ic801404p.

127. Nagaraja C.M., Haldar R., Maji T.K., Rao C.N.R. Chiral Porous Metal–Organic Frameworks of Co(II) and Ni(II): Synthesis,

Structure, Magnetic Properties, and CO2 Uptake. Cryst. Growth Des. 2012;12:975–981. doi: 10.1021/cg201447c.

128. Raja D.S., Luo J.-H., Chang T.-G., Lo S.-H., Wu C.-Y., Lin C.-H. Synthesis, Crystal Structure, and Luminescence Properties of a New Calcium(II) Coordination Polymer Based on L-Malic Acid. J. Chem. 2013;2013:7.

129. Yutkin P., Zavakhina M.S., Samsonenko D.G., Dybtsev D.N., Fedin V.P. Synthesis and characterization of expected and unexpected topologies of homochiral porous metal(II) malate frameworks. Inorganica Chim. Acta. 2013;394:367–372. doi: 10.1016/j.ica.2012.08.012.

130. Zavakhina M.S., Samsonenko D.G., Virovets A.V., Dybtsev D.N., Fedin V.P. Homochiral Cu(II) and Ni(II) malates with tunable structural features. J. Solid State Chem. 2014;210:125–129. doi: 10.1016/j.jssc.2013.11.011.

131. Xie S., Hu C., Li L., Zhang J., Fu N., Wang B., Yuan L. Homochiral metal-organic framework for HPLC separation of enantiomers. Microchem. J. 2018;139:487–491. doi: 10.1016/j.microc.2018.03.035.

132. Nair L.P., Bijini B.R., Prasanna S., Eapen S.M., Nair C.M.K., Deepa M., RajendraBabu K. Growth and characterisation of a new polymorph of barium maleate: A metal organic framework. Spectrochim. Acta A. 2015;137:778–784. doi: 10.1016/j.saa.2014.08.073.

133. Purves D., Augustine G.J., Fitzpatrick D., Katz L.C., LaMantia A.-S., McNamara J.O., Williams S.M. In: Neuroscience. 2nd ed. Purves D., Augustine G.J., Fitzpatrick D., Katz L.C., LaMantia A.-S., McNamara J.O., Williams S.M., editors. Sinauer Associates; Sunderland, MA, USA: 2001.

134. Katsoulidis A.P., Antypov D., Whitehead G.F.S., Carrington E.J., Adams D.J., Berry N.G., Darling G.R., Dyer M.S., Rosseinsky M.J. Chemical control of structure and guest uptake by a conformationally mobile porous material. Nature. 2019;565:213–217. doi: 10.1038/s41586-018-0820-9.

135. Loukopoulos E., Michail A., Kostakis E.G. A 12-fold ths interpenetrated network utilizing a glycine-based pseudopeptidic ligand. Crystals. 2018;8:47. doi: 10.3390/cryst8010047.

136. Lauder K., Toscani A., Scalacci N., Castagnolo D. Synthesis and Reactivity of Propargylamines in Organic Chemistry. Chem. Rev. 2017;117:14091–14200. doi: 10.1021/acs.chemrev.7b00343.

137. Ricco R., Pfeiffer C., Sumida K., Sumby C.J., Falcaro P., Furukawa S., Champness N.R., Doonan C.J. Emerging applications of metal–organic frameworks. CrystEngComm. 2016;18:6532–6542. doi: 10.1039/C6CE01030J.
138. Lipinski C.A., Lombardo F., Dominy B.W., Feeney P.J. Experimental and computational approaches to estimate solubility and permeability in drug discovery and development settings. Adv. Drug Deliv. Rev. 1997;23:3–25. doi: 10.1016/S0169-409X(96)00423-1.
139. DiMasi J.A., Grabowski H.G., Hansen R.W., Cao J., Zhao Y., Liu J. Food Chem. Toxicol. 1997, 35, 827–833. J. Health Econ. 2016;47:20–33. doi: 10.1016/j.jhealeco.2016.01.012.
140. Gu S., Cui D., Chen X., Xiong X., Zhao Y. PROTACs: An Emerging Targeting Technique for Protein Degradation in Drug Discovery. BioEssays. 2018;40:e1700247. doi: 10.1002/bies.201700247.
141. Ryther R.C.C., Flynt A.S., Phillips J.A., Patton J.G. siRNA therapeutics: Big potential from small RNAs. Gene Ther. 2005;12:5–11. doi: 10.1038/sj.gt.3302356.
142. Hu B., Weng Y., Xia X.-H., Liang X., Huang Y. Clinical advances of siRNA therapeutics. J. Gene Med. 2019;21:e3097. doi: 10.1002/jgm.3097.

CHAPTER

11

A Mechano-Responsive Supramolecular Metal–Organic Framework (supraMOF) Gel Material Rich in ZIF-8 Nanoplates

Abhijeet K. Chaudhari [a] **and Jin-Chong Tan** [a]

[a]Multifunctional Materials & Composites (MMC) Laboratory, Department of Engineering Science, University of Oxford, Parks Road, Oxford, OX1 3PJ, UK.

ABSTRACT

We report an unconventional ZIF-8 based supramolecular MOF (supraMOF) material, which is straightforwardly synthesised by leveraging the high concentration reaction (HCR) approach. Akin to traditional low-molecular-weight gels (LMWG), we show that mechano-stimulus responsive behavior can be achieved through a reversible and fast sol–gel conversion mechanism.

Citation: Abhijeet K. Chaudhari and Jin-Chong Tan. "A mechano-responsive supramolecular metal–organic framework (supraMOF) gel material rich in ZIF-8 nanoplates" Chem.Commun.,2017,53,8502—8505 https://doi.org/10.1039/C7CC03478D

Remarkably, the supraMOF gel consists of 2D nanoplates of ZIF-8 with a reduced crystal symmetry. The chemistry beyond that of molecules, known as 'supramolecular chemistry'[1] has witnessed a rapid growth over the past two decades.[2] Notably, the traditional candidates in the field of supramolecular science, featuring the self-assembly of small molecules comprising purely organic compounds, have been superseded by exotic and very diverse families of hybrid (organic–inorganic) molecular assemblies and composite systems.[3] Self-assembly methodologies that yield functional gel materials have attracted significant interest owing to their potential applications in medicine,[4] energy,[5] and environmental sustainability.[6] Within this broad class of soft matters, metal-containing gel materials termed 'metallogels'[7] could offer additional controls for tuning the structural and physico-chemical properties for many targeted applications.[8] Metal-containing metallo-gelators studied to date can basically be divided into two main categories: (a) solvent immobilisation attributed to metal-coordination effects[9] and (b) solvent immobilisation conferred by metal complex species used for constructing metallogels.[7c,3b] Recently, researchers have also focused on metal–organic gels (MOGs),[10] which generally contain continuous chains of alternating metal and organic molecules to enable immobilisation of solvents.[11]

An innovative implementation of the basic principle of continuous chain formation of metal–organic building blocks has opened the door to the recent development of smart supramolecular gels,[12] containing porous 3D scaffolding of metal–organic frameworks (MOFs).[13] We termed these materials 'supraMOF', as they are derived from supramolecular gel self-assemblies coexisting with *in situ* generated MOF entities. Our recent study on the HKUST-1 based supraMOF has revealed promising sol–gel reversibility, affecting its chemical structure useful for tuning a range of physical properties including the electrical conductivity.[12a] The literature on MOF compounds has proliferated to cover a very wide and diverse spectrum of applications.[14] Despite their vast potential utilities, MOF science in combination with self-assembled supramolecular materials is, hitherto, a largely unexplored research topic.

One exciting possibility is to yield 'smart' supramolecular materials[12a,c] by leveraging the tunability of MOF compounds, to attain stimuli-responsive behaviours towards magnetism, light, sound waves, mechanical stresses, temperature, electricity, *etc.* Moreover, sol–gel systems will offer new possibilities for facile processing and affordable deposition of functional films towards engineering of sensors, actuators and microelectronic devices.[15]

In this communication, we report the discovery of a new candidate of supra MOF gel material built from the same chemical constituents as for a topical zeolitic imidazolate framework (ZIF) with sodalite topology: ZIF-8 [Zn(mIm)$_2$; mIm = 2-methylimidazolate].[16] Notably, this unconventional ZIF-8 based supraMOF exhibits exceptional mechano-stimulation response, *via* a rapid sol–gel transformation bestowed upon it by its highly reversible mechanical properties in cyclic stress loading.

The ZIF-8 supraMOF was synthesised by adopting the 'high concentration reaction' (HCR) approach.[12a] This involves high concentrations of mIm and $Zn(NO_3)_2$ (see Table 1, reaction A), but prior to mixing the mIm organic linker solution in *N*,*N*-dimethylformamide (DMF) solvent to the Zn(ii) metal ion solution (also in DMF), the former was first deprotonated by taking an equivalent molar concentration of trimethylamine base (NEt_3). Upon combining the above solution reactants, we observed that a white pre-gel first "material" containing fibre-like morphologies formed immediately, which then quickly converted into a gel phase within just a few seconds. Sonication for a further 5 seconds resulted in the formation of a mechanically stable gel phase. After sonication, the sample container was inverted to test for gelation (by inversion test), and we found that, similar to the conventional low-molecular-weight gels (LMWG), this gel material can resist gravity by refusing to flow downwards.

Table 1. Series of reactions (A–E) to investigate the effects of concentration on the framework structure of ZIF-8. Note: DMF volume was divided 50 : 50 between $Zn(NO_3)_2$ and mIm

Reaction	DMF (total) (mL)	$Zn(NO_3)_2$ (mmol)	mIm (mmol)	NEt_3 (mmol)
A (Gel)	6	3	12	12
B	25	3	12	12
C	100	3	12	12
D	300	3	12	12
E	500	3	12	12

Interestingly, this candidate of ZIF-8 supraMOF exhibited rapid sol–gel phase transition upon application of an external mechanical stimulus (stress). Vigorous shaking of the sealed glass vial containing the gel material quickly transformed it into a sol phase, which could easily revert back to a stable gel phase simply by allowing the sol unperturbed for about 5 seconds. The real-time sol ⇌ gel recovery can be seen in the video clip provided as a part of the ESI,† which demonstrates the fast phase reversibility in a matter of seconds.

To gain detailed insights into gel stability and the sol–gel conversion phenomenon, dynamic rheological measurements were performed using controlled parameters as shown in Fig. 1. It can be seen in Fig. 1(a) that the oscillatory frequency sweep between 0.1 Hz and 100 Hz showed a relatively steady behavior for the storage modulus (G') and loss modulus (G''), without any strong dependence on the frequency until beyond ~30 Hz. After which, the magnitude of G' suddenly fell while G'' began to increase; this behavior signifies the collapse of an integrated network of the gel phase. Likewise, the distinct increase of the loss tangent ($\tan(\delta) = G''/G'$) *versus* frequency data above ~30 Hz is evident of an occurrence of the gel-to-sol phase transformation. We have determined that G' is an order of magnitude higher than G'', indicating that the ZIF-8 supra MOF can be considered as a stable gel material unperturbed. Fig. 1b shows the dynamic strain sweep measurements performed from 0.1 to 100% strain at 0.1 Hz, to characterize the stability of the gel against phase change under an incremental shear strain. Horizontal profiles of G' and G'' between 0.1 and ~2% shear strain were found to start becoming strain dependent with an increase in the strain. The G' and G'' profiles coincided at ~4%, marking the yield point of the gel. Subsequently, the magnitude of G'' became much greater than that of G' between shear strains of ~5 and 100%, clearly this meant that the energy dissipation has exceeded the energy storage because the stable gel phase has collapsed into a sol suspension (see Fig. 1b inset).

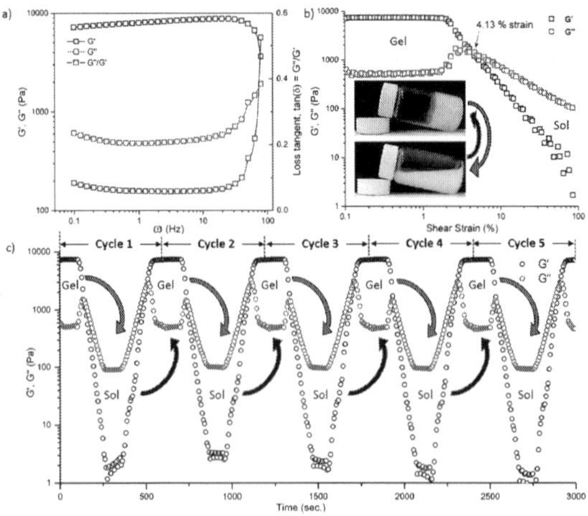

Figure. 1 (a) Oscillatory frequency-sweep rheological experiments for the ZIF-8 based supraMOF gel tested under a constant applied shear strain of 0.1%. G' and G'' are the storage and loss moduli, respectively. Loss tangent, $\tan(\delta) = G''/G'$, data indicating a phase change at above ~30 Hz. (b) Dynamic strain

sweep measurements established the yield point at ~4%, at which the gel converts into sol. Inset: photos showing reversible sol–gel conversion. (c) Mechano-stimulus response of the ZIF-8 supra MOF material tested by performing a cyclic dynamic strain-sweep experiment (consecutive 5 cycles are shown), demonstrating its capability to transform reversibly between gel ⇌ sol with highly reproducible mechanical properties.

By imposing a cyclic (load-unload) mechanical stimulus, we have characterised the phase reversibility of the ZIF-8 supraMOF material as shown in Fig. 1c, so as to establish its thixotropic nature under a cyclic shear strain set between 0.1% ⇌ 100%. We found that the results in Fig. 1b are in fact reversible, that is, upon decreasing the shear strain from 100% to 0.1% the sol recovered to the gel form evidenced by $G' > G''$ (storage > loss). Such gel ⇌ sol reversibility tests were performed for multiple consecutive cycles, for example, shown in Fig. 1c are data representing the profiles obtained by allowing 150 s for the recovery of gel from a collapsed sol phase. Furthermore, the cyclic stress experiments revealed precisely how the complex viscosity can be very reversibly tuned, over a wide range of 10^2–10^4 Pa s as effected by a recoverable sol–gel transformation step.

To study the origin of the mechano-responsive behavior (Fig. 1), we carried out a systematic materials characterisation study to correlate the observed property with its underlying material structure. A gel sample prepared on a piece of glass substrate by layering a thin layer of the material was imaged under a scanning electron microscope (SEM) to characterise its microstructure. The SEM micrographs obtained are shown in Fig. 2a, where the gel material forms what appeared to be a densely packed cellular-like phase, very unlike the typical fibrous network[10d,17] found in conventional gel materials. To further understand the supra MOF gel constitution, we used copious amounts of DMF and methanol to break down (disintegrate) the gel phase, liberating an abundance of 'fine particles' that were then isolated by centrifugation (8000 rpm). Surprisingly, transmission electron microscopy (TEM) revealed that the isolated particles resembled the two-dimensional (2D) morphologies of nanoplates (Fig. 2b). We confirmed the nanoplate geometry *via* atomic force microscopy (AFM) as presented in Fig. 2(c and d), where the averaged nanoplate thickness was found to be of the order of 10s nm and with an aspect ratio (*i.e.* width/thickness) of at least ~5. Intriguingly the mechano-responsive gel phase comprises nanoplates of distorted ZIF-8, whose crystal structure is elucidated below.

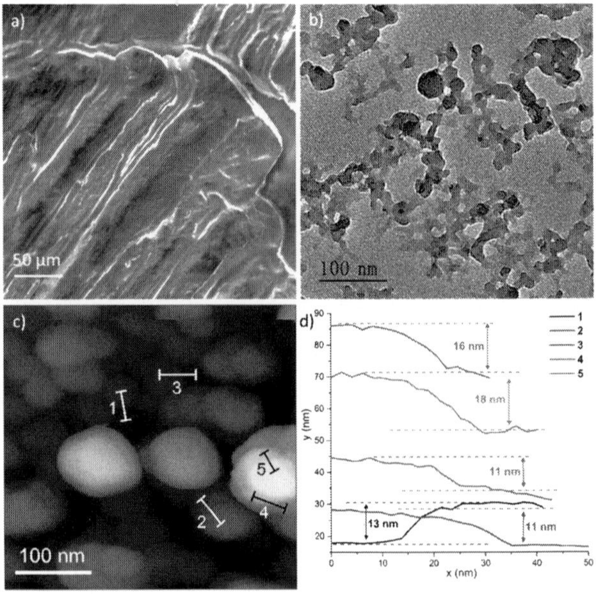

Figure. 2 (a) SEM image of the dried supraMOF gel material. (b) TEM image of the nanoplates obtained after breaking down the gel material into fine powder particles. (c) AFM topography image revealing the 2D morphology of (larger) isolated nanoplates, and (d) their individual thickness profiles of *ca.* 10–20 nm corresponding to the specific regions marked in (c).

The bulk aggregates of nanoplates harvested from the supraMOF gel phase transformed into a monolith after drying at 110 °C. This monolith was carefully ground into a fine powder and then characterised using powder X-ray diffraction (PXRD). Fig. 3a shows that its PXRD pattern has sharp peaks matching the ideal simulated XRD pattern of ZIF-8, but with the appearance of a few broadened extra peaks suggesting the reduction of the crystal symmetry. To investigate this effect, we have carried out a systematic study by varying the reactant concentrations in accordance with Table 1. Specifically, different reactions using mIm and Zn(ii) were performed by maintaining the same synthetic parameters apart from the total volume of the DMF solvent used, which was varied from 6 mL (for highly concentrated reactions) to 500 mL (very diluted reaction conditions). The products obtained from the five reactions were washed and dried prior to PXRD (Fig. 3a), showing a systematic reduction in intensity of the extra peaks with a greater solvent content. This result is supported by the systematic color change of the samples detectable by the naked eye (Fig. 3a inset); we reasoned that additional guest species (*e.g.* NEt_3, DMF) become trapped in the framework from concentrated reactions which decompose upon hearing of the compound producing a yellowish powder (*vs.* a whiter powder of ZIF-8).

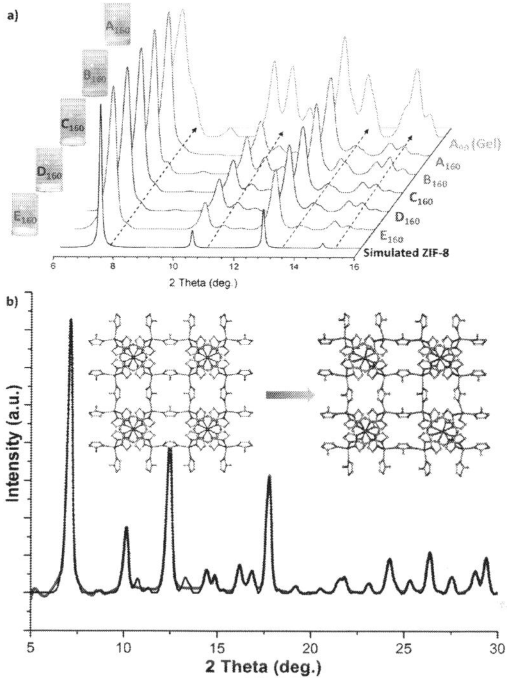

Figure. 3 (a) PXRD patterns of samples derived from the reactions labelled A–E in Table 1, where the subscripts denote the drying temperature of either 90 °C or 160 °C. Sample colors obtained (shown to the left of XRD spectra) revealed the apparent variations between the A to E reaction products. Dotted arrows designate the positions of the extra diffraction peaks due to occluded guests. (b) Pawley refinement method to determine the lattice geometry of the distorted ZIF-8 framework from the experimental PXRD data of the A160 sample. Inset: structural changes from the (standard) cubic cell of ZIF-8 to a triclinic distorted ZIF-8 structure.

To determine the salient changes of the structural parameters of ZIF-8, the PXRD data from the reaction of A160 (gel) were utilized for Pawley structural refinement. The refinement results in Fig. 3(b) show that the high-symmetry (cubic) structure of ZIF-8 has reduced to a low-symmetry (triclinic) crystal structure (R_{wp} = 6.42%) for the nanoplates extracted from the ZIF-8 supraMOF gel. Cell dimensions of the cubic ZIF-8 structure were found to have been modified from a, b, c = 17.086 Å to a = 17.46 Å, b = 17.12 Å, c = 16.95 Å, and the unit cell angles changed from α, β, γ = 90° to α = 89.01°, β = 89.50°, γ = 88.19°. Moreover, thermogravimetric analysis (TGA) of these compounds detected changes in the thermal stability of the framework material when being synthesized under concentrated conditions. Unlike the thermal stability of the conventional ZIF-8 powder till ~350 °C, the nanoplates comprising reduced symmetry ZIF-8 (derived from supra MOF) have a relatively lower thermal stability of up to ~250 °C, which can be ascribed to the decomposition of the entrapped guest species.

To conclude, we have demonstrated the efficacy of the high-concentration reaction (HCR) strategy to derive a previously unreported ZIF-8-based supra MOF gel material. Surprisingly, this supraMOF gel contains an abundance of ZIF-8 nanoplates with only ~10s nm thickness, completely dissimilar to the conventional ZIF-8's rhombic dodecahedral crystals ubiquitous in the literature. Exceptionally, this ZIF-8 supraMOF gel is mechano-stimuli responsive, exhibiting fast sol–gel conversions and perfectly reversible mechanical properties subject to a multi-cyclic strain loading. Our current study not only presents the first discovery of a ZIF-type gel material, but also describes the simple synthetic methodology that may be widened to explore its applicability for other well-studied MOFs to yield supramolecular stimuli-responsive systems and to achieve rare 2D morphologies.

We thank the Samsung GRO and EPSRC RCUK (EP/N014960/1 and EP/K031503/1) for research funding, and the Research Complex at Harwell (RCaH) for the provision of TEM and TGA facilities. We are grateful to Dr Gavin Stenning and Dr Marek Jura for access to PXRD at the R53 Materials Characterisation Laboratory in ISIS RAL, Oxfordshire.

NOTES AND REFERENCES

1. J. M. Lehn, Angew. Chem., Int. Ed., 1990, 29, 1304–1319; (b) F. M. Menger, Proc. Natl. Acad. Sci. U. S. A., 2002, 99, 4818–4822.
2. L. J. Prins, J. Huskens, F. de Jong, P. Timmerman and D. N. Reinhoudt, Nature, 1999, 398, 498–502; (b) O. A. Scherman, Nat. Chem., 2009, 1, 524–525; (c) T. Aida, E. W. Meijer and S. I. Stupp, Science, 2012, 335, 813–817; (d) C. Hansell, Nat. Chem., 2014, 6, 459.
3. L. E. Buerkle and S. J. Rowan, Chem. Soc. Rev., 2012, 41, 6089–6102; (b) A. Y. Tam and V. W. Yam, Chem. Soc. Rev., 2013, 42, 1540–1567; (c) S. Y. Dong, J. Y. Yuan and F. H. Huang, Chem. Sci., 2014, 5, 247–252; (d) Q. Lin, B. Sun, Q. P. Yang, Y. P. Fu, X. Zhu, Y. M. Zhang and T. B. Wei, Chem. Commun., 2014, 50, 10669–10671.
4. M. E. Byrne, K. Park and N. A. Peppas, Adv. Drug Delivery Rev., 2002, 54, 149–161; (b) J. A. Foster, M. O. Piepenbrock, G. O. Lloyd, N. Clarke, J. A. Howard and J. W. Steed, Nat. Chem., 2010, 2, 1037–1043.
5. J. H. Wu, Z. Lan, J. M. Lin, M. L. Huang, S. C. Hao, T. Sato and S. Yin, Adv. Mater., 2007, 19, 4006–4011; (b) Y. Kim, M. Kathaperumal, V. W. Chen, Y. Park, C. Fuentes-Hernandez, M. J. Pan, B. Kippelen and J. W. Perry, Adv. Energy Mater., 2015, 5, 1500767.
6. D. Mohan and C. U. Pittman, Jr., J. Hazard. Mater., 2007, 142, 1–53.
7. K. Murata, M. Aoki, T. Nishi, A. Ikeda and S. Shinkai, Chem. Commun., 1991, 1715–1718; (b) B. G. Xing, M. F. Choi and B. Xu, Chem. – Eur. J., 2002, 8, 5028–5032; (c) A. Y. Tam, K. M. Wong, G. Wang and V. W. Yam, Chem. Commun., 2007, 2028–2030.
8. M. O. M. Piepenbrock, N. Clarke and J. W. Steed, Langmuir, 2009, 25, 8451–8456; (b) M. O. M. Piepenbrock, G. O. Lloyd, N. Clarke and J. W. Steed, Chem. Rev., 2010, 110, 1960–2004; (c) T. Tu, W. Fang, X. Bao, X. Li and K. H. Dotz, Angew. Chem., Int. Ed., 2011, 50, 6601–6605; (d) W. Miao, L. Zhang, X. Wang, H. Cao, Q. Jin and M. Liu, Chem. – Eur. J., 2013, 19, 3029–3036.
9. W. J. Gee and S. R. Batten, Chem. Commun., 2012, 48, 4830–4832.
10. Q. Wei and S. L. James, Chem. Commun., 2005, 1555–1556; (b) S. K. Nune, P. K. Thallapally and B. P. McGrail, J. Mater. Chem., 2010, 20, 7623–7625; (c) F. Yang, Z. Lin, X. He, L. Chen and Y. Zhang, J. Chromatogr. A, 2011, 1218, 9194–9201; (d) S. Samai and K. Biradha, Chem. Mater., 2012, 24, 1165–1173; (e) S. Saha, G. Das, J. Thote and

R. Banerjee, J. Am. Chem. Soc., 2014, 136, 14845–14851.

11. J. Y. Zhang and C. Y. Su, Coord. Chem. Rev., 2013, 257, 1373–1408; (b) Y.-M. Zhang, X.-M. You, H. Yao, Y. Guo, P. Zhang, B.-B. Shi, J. Liu, Q. Lin and T.-B. Wei, Supramol. Chem., 2013, 26, 39–47.
12. A. K. Chaudhari, I. Han and J. C. Tan, Adv. Mater., 2015, 27, 4438–4446; (b) B. Bueken, N. Van Velthoven, T. Willhammar, T. Stassin, I. Stassen, D. A. Keen, G. V. Baron, J. F. M. Denayer, R. Ameloot, S. Bals, D. De Vos and T. D. Bennett, Chem. Sci., 2017, 8, 3939–3948; (c) A. K. Chaudhari, H. J. Kim, I. Han and J. C. Tan, Adv. Mater., 2017, DOI: 10.1002/adma.201701463.
13. B. F. Hoskins and R. Robson, J. Am. Chem. Soc., 1989, 111, 5962–5964; (b) O. M. Yaghi, G. M. Li and H. L. Li, Nature, 1995, 378, 703–706; (c) S. Kitagawa, R. Kitaura and S. Noro, Angew. Chem., Int. Ed., 2004, 43, 2334–2375; (d) J. C. Tan and B. Civalleri, CrystEngComm, 2015, 17, 197–198.
14. G. Ferey, Chem. Soc. Rev., 2008, 37, 191–214; (b) V. Stavila, A. A. Talin and M. D. Allendorf, Chem. Soc. Rev., 2014, 43, 5994–6010; (c) M. G. Campbell, D. Sheberla, S. F. Liu, T. M. Swager and M. Dinca, Angew. Chem., Int. Ed., 2015, 54, 4349–4352; (d) A. K. Chaudhari, M. R. Ryder and J. C. Tan, Nanoscale, 2016, 8, 6851–6859.
15. L. E. Kreno, K. Leong, O. K. Farha, M. Allendorf, R. P. Van Duyne and J. T. Hupp, Chem. Rev., 2012, 112, 1105–1125; (b) M. R. Ryder and J. C. Tan, Mater. Sci. Technol., 2014, 30, 1598–1612; (c) I. Stassen, N. Burtch, A. Talin, P. Falcaro, M. Allendorf and R. Ameloot, Chem. Soc. Rev., 2017, 46, 3185–3241.
16. K. S. Park, Z. Ni, A. P. Cote, J. Y. Choi, R. Huang, F. J. Uribe-Romo, H. K. Chae, M. O'Keeffe and O. M. Yaghi, Proc. Natl. Acad. Sci. U. S. A., 2006, 103, 10186–10191; (b) J. C. Tan, B. Civalleri, C. C. Lin, L. Valenzano, R. Galvelis, P. F. Chen, T. D. Bennett, C. Mellot-Draznieks, C. M. ZicovichWilson and A. K. Cheetham, Phys. Rev. Lett., 2012, 108, 095502.
17. K. Jayaramulu, F. Geyer, M. Petr, R. Zboril, D. Vollmer and R. A. Fischer, Adv. Mater., 2017, 29, 1605307; (b) R. C. T. Howe, A. P. Smalley, A. P. M. Guttenplan, M. W. R. Doggett, M. D. Eddleston, J. C. Tan and G. O. Lloyd, Chem. Commun., 2013, 49, 4268–4270.

CHAPTER 12

Fabrication of a Porous Metal-Organic Framework with Polar Channels for 5-Fu Delivery and Inhibiting Human Osteosarcoma Cells

Li-Chun Zhao,[1] Mei Tang,[1] Qian-Hua Zhang,[1] Zhi-Yi Hu,[1] Hong-Wei Gao,[1] Xia-Yun Liao,[1] Gang Wang,[1] and Jing Leng[1]

[1]College of Pharmacy, Guangxi University of Chinese Medicine, Nanning 530200, China

ABSTRACT

As an emerging kind of crystalline material, the metal-organic framework (MOF) has shown great promise in the biomedical domains such as drug storage and delivery. In this study, a new porous MOF, $[[Dy_2(H_2O)_3(SDBA)_3](DMA)_6]$ (**1**, H_2SDBA=4,4′-sulfonyldibenzoic acid, DMA=*N,N*-dimethylacetamide

Citation: Li-Chun Zhao, Mei Tang, Qian-Hua Zhang, Zhi-Yi Hu, Hong-Wei Gao, Xia-Yun Liao, Gang Wang, and Jing Leng. "Fabrication of a Porous Metal-Organic Framework with Polar Channels for 5-Fu Delivery and Inhibiting Human Osteosarcoma Cells" Journal of Chemistry Volume 2018, Article ID 1523154, 7 pages https://doi.org/10.1155/2018/1523154

(C_4H_9NO)), with uncoordinated O donor sites has been fabricated using a bent polycarboxylic acid organic linker under the solvothermal condition. The structure of the obtained crystalline product has been fully determined by the X-ray single-crystal diffraction, TGA, elemental analysis, XRD, and the gas sorption measurement. Due to the suitable window size and polar atom functionalized 1D channels, the activated **1 (1a)** compound was used for the anticancer drug 5-fluorouracil (5-Fu, $C_4H_3FN_2O_2$) loading by a simple impregnation method. A moderate drug loading and pH-dependent drug-release behavior could be observed for **1a**. Furthermore, as indicated by the MTT assay, this drug/MOF composite shows low toxicity toward the human normal cells and demonstrates obvious anticancer activity against the human osteosarcoma cell line MG63.

INTRODUCTION

Being one of the most life-threatening diseases, cancer has caught the nerves of people all over the world because more and more people are suffering from cancer and die from it [1]. Currently, the most effective treatment method is still chemotherapy which relies on the anticancer drugs [2]. However, chemotherapy has always suffered some unbearable drawbacks such as nonspecific toxicity toward normal cells, poor solubility of the chemotherapy drug, and undesirable side effects [3]. In this case, great notice has been devoted to nanocarriers for anticancer drug delivery for cancer therapy, aiming at a targeted release of anticancer drugs while retarding the drug leaking into the healthy physiological environment [4]. The uncommon microenvironments at tumor sites such as low pH value, high glutathione (GSH) concentration, and different oxygen concentrations can be utilized to trigger drug release from the carriers.

As an emerging kind of crystalline material, the metal-organic framework (MOF) built via the assembly of organic ligands as bridges and metal ions as nodes has underwent a rapid development in the last two decades [5–7]. Recently, porous metal-organic frameworks have been proved to be very successful for drug delivery, possessing advantages such as high drug-loading capacity, suitable pore size, and strong drug-framework interaction [8, 9]. Furthermore, the toxicity of the MOFs toward the human normal cells can be tactfully averted by selecting biocompatible metals and organic ligands. In addition, the pore walls of MOFs could be conveniently gifted polar donor sites via selecting organic ligands with polar donor sites, which could form strong framework-drug interaction to result in the sustainable drug-release behavior. For instance, the anticancer drug-loaded ZIF-8 nanocarriers showing pH-sensitivite drug release have been achieved by Zhao and coworkers [10]; Kotzabasaki et al. revealed that the hydroxyl groups in IRMOF-16 could lead to a much improved drug-loading capacity, which outperforms many other materials targeted for the gemcitabine loading [11]. With the aim in mind of developing novel drug carriers, in this study, a new porous MOF, $[[Dy_2(H_2O)_3(SDBA)_3](DMA)_6]$ (**1**, H_2SDBA=4,4′-sulfonyldibenzoic acid; Figure 1), with uncoordinated O donor sites has been fabricated using a bent polycarboxylic acid organic linker under the solvothermal condition. Although the H_2SDBA ligand has been used in building the Dy-based coordination compounds in the previous literature, none of them are porous structures [12]. The structure of the obtained crystalline product has been fully characterized by the X-ray single-crystal diffraction, elemental analysis, TGA, XRD, and the gas sorption measurement. Due to the suitable window size and polar atom functionalized 1D channels, the activated **1 (1a)** was used for the anticancer drug 5-Fu loading by a simple impregnation method. A moderate drug-loading and sustainable drug-release behavior could be observed for **1a**. Furthermore, as indicated by the MTT assay, this drug/MOF composite shows low toxicity toward the human normal cells and demonstrates obvious anticancer activity against the human osteosarcoma cell line MG63.

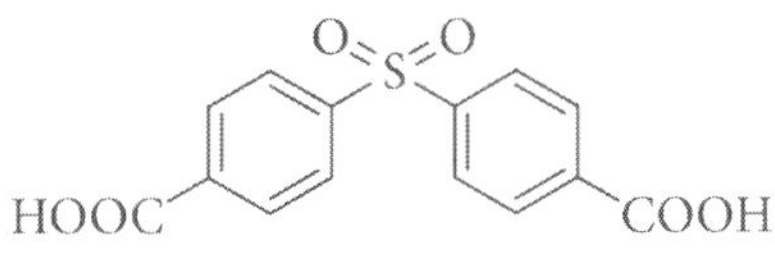

Figure 1. The chemical drawing of 4,4′-sulfonyldibenzoic acid (H_2SDBA) ligand used in the construction of **1**.

EXPERIMENTAL

Chemicals and Instruments

All the chemicals are obtained from the regent companies and used as received. Elemental analyses (C, H, and N) were carried out with a Vario EL III Elemental analyzer. Powder X-ray diffraction (PXRD) curves were collected on a Difrac Plus XRD Commander diffractometer at room temperature. Measurements of the UV-Vis absorption spectra were carried out in a quartz cell using a Camspec M330 UV-Vis spectrometer.

Thermal gravity analysis (TGA) curves were collected using a NETZSCH STA449F3 thermal analyzer. The N_2 sorption data at 77 K were collected on the Autosorb 1 by Quantachrome. The HPLC analysis was conducted using the 1260 Agilent chromatographic system.

Preparation of $[[Dy_2(H_2O)_3(SDBA)_3](DMA)_6]$

In a 50 mL beaker, $Dy(NO_3)\cdot6H_2O$ (46 mg, 0.1 mmol), H_2SDBA (31 mg, 0.1 mmol), DMA (6 mL), and H_2O (1.5 mL) were added. After the addition of 0.2 mL HCl (2 M aq.), the mixture was stirred for 30 min in the air to make a clear solution, and then, the mixture was placed in a 20 mL vial and kept at 90°C for two days. Yellow sheet crystals of **1** were obtained by filtration, washed with water, and left in the air for one day to make them dry. The yield was 46% based on the H_2SDBA ligand. Elemental analysis (%): calcd. for **1** ($C_{66}H_{84}Dy_2N_6O_{27}S_3$): C, 43.69; H, 4.67; and N, 4.63; found: C, 43.19; H, 4.28; and N, 4.66.

Crystal Structure Determination

The X-ray single-crystal diffraction data of **1** were collected on the Oxford diffraction Xcalibur CCD diffractometer with Mo Kα radiation. The structural model of **1** was solved by direct methods using the SHELXT program, and the obtained model was amended using the SHELXL program.

All hydrogen atoms attached to oxygen and carbon were generated geometrically and all non-hydrogen atoms were refined anisotropically. For the disordered nature of the DMA molecules, they could not be figured out from the structural model, so their electrons were removed from the HKL file using the SQUEEZE manipulation in the software PLATON (the total unit cell includes six DMA molecules (48e □ 6=288e), and the SQUEEZE result reveals that 596 electrons were removed from the unit cell). This discrepancy might be owing to the different measurement conditions for the single-crystal X-ray analysis (tested after being sealed in a capillary filled with DMA) and TGA (tested after being dried in the air)). The solvents in the chemical formula of **1** were determined from the TGA curve and elemental analysis results. Relevant data concerning the crystal data and refinement parameters have been provided in Table 1.

Table 1. Crystal data and structure refinements for compound **1**.

Formula	$C_{42}H_{30}Dy_2O_{21}S_3$

Mr	1291.84
T (K)	293 (2)
Crystal system	Monoclinic
Space group	$P2_1/c$
α (°)	90
β (°)	110.892 (3)
γ (°)	90
a (Å)	18.2117 (4)
b (Å)	21.2418 (5)
c (Å)	16.1128 (4)
Volume ($Å^3$)	5823.4 (3)
Z	2
D_{calc} ($g{\cdot}cm^{-3}$)	1.473
μ (Mo Kα) (mm^{-1})	2.719
θ range (°)	2.347 to 24.990
Reflections collected	20165
R_{int}	0.0264
Data number with $I \geq 2\sigma(I)$	7483
R_1	0.0251
R_2 (all data)	0.0558
CCDC	1844814

5-Fu Loading and Release

Twenty milligrams of activated **1** (**1a**) were suspended in 8 ml of 5-Fu solutions, and then, the mixture suspensions were stirred for 72 h at 37°C. Finally, 5-Fu-loaded **1a** was obtained through centrifugation for 5 min, washed with MeOH, and dried at room temperature. **1a** before and after the 5-Fu entrapping were characterized by PXRD. The concentration of 5-Fu in the supernatant was determined by the absorbance of 5-Fu at 265 nm with the help of a calibration curve. Then, the drug-loading efficiency was calculated. Drug encapsulation efficiency = (total amount − without loading amount)/total amount × 100%. The 5-Fu release experiment was compacted in the dialysis bag. The 5-Fu-loaded crystals of **1** were soaked in the PBS solution at different pH values (7.4 and 5.6) at 37°C. In each interval, 0.2 mL of the solution was taken out and fresh PBS with equal amount was added. Then, the 5-Fu content was probed using the HPLC analysis.

GCMC Simulation

The adsorption process of the 5-fluorouracil (5-Fu) molecule to **1a** was investigated using grand canonical Monte Carlo (GCMC) simulations at 300 K. Interactions of 5-Fu with the frameworks were described by a potential composed of van der Waals and Coulomb components. Van der Waals interactions were treated using the Lennard–Jones (LJ) potential and atomic parameters taken from the UFF force field. Coulomb interactions were calculated using atomic partial charges obtained via the charge equilibration (Q_{eq}) method proposed by Snurr and coworkers. Partial charges for 5-Fu molecules were calculated at the B3LYP level using the DMOL3 module. GCMC simulations were carried out with 1×10^7 equilibration steps and 2×10^7 production steps. The configuration-bias Monte Carlo method was used for trial MC moves involving insertion and deletion of 5-Fu molecules. All GCMC simulations were performed using the sorption module embedded in Material Studio 6.1.

In Vitro Toxicity Tests

The cytotoxicity toward HASM (human aortic smooth muscle) cells and MG63 cells was determined using MTT assays. The cells were seeded in a 96-well microplate. After 24 h of cell attachment, the cells were treated and incubated with different concentrations of each complex for 48 h. Then, 25 μl of MTT solution was added into each well. After the cells were incubated for another 4 hours, 150 μl of DMSO was added to each well to dissolve the MTT formazan crystals. At last, the absorbance of each well was monitored through a microplate reader.

RESULTS AND DISCUSSION

Molecular Structure and Physical Characterization of 1

Crystallographic analysis via the X-ray single-crystal diffraction reveals that compound **1** locates in the monoclinic crystal system and $P2_1/c$ space group and shows a three-dimensional channel-type framework. The molecular unit is composed of two independent Dy(III) ions, three $SDBA^{2-}$ ligands, and three coordinated water molecules. As plotted in Figure 2(a), eight O atoms from six different $SDBA^{2-}$ ligands (O1, O5, O6, O7, O7A, O8, and O12) and one coordinated water molecule (O2W) form the DyO8 coordination surrounding, with the Dy1-O bond distances ranging from 2.215 (3) to 2.619

(3) Å; Dy2 is also an eight-atom connected compound with six O atoms from five independent $SDBA^{2-}$ ligands (O2, O11, O13, O14, O17, and O18) and two O atoms of the coordinated water molecules (O1W and O3W), with the Dy2-O bond lengths ranging from 2.249 (2) to 2.531 (3) Å. All the Dy-O bond distances are comparable to those observed in the Dy-based metal-organic frameworks. The Dy1, Dy2, and their symmetry-related atoms are held together via the *syn*-bridging carboxylic groups along the *a*-axis to afford the 1D secondary building unit chains with the Dy1-Dy2 separation of 5.04 Å. As for the $SDBA^{2-}$ ligands, they exhibit three different coordination modes with the two carboxylic groups binding with one or two Dy^{3+} ions forming paddle-wheel units (Figure 2(b)). The SDB ligands are bent with the C-S-C bond angles ranging from 101.7 (2) to 103.6 (2), which connect with the 1D SUB channels to afford the three-dimensional framework with rhombus channels (Figure 2(c)). The channels are filled with uncoordinated O donors site and water-occupied Dy(III) sites, which are potentially favorable for guests binding. The total effective solvent accessible volume is 2172 $Å^3$, corresponding to 37.3% of the cell volume. In the framework of **1**, the three $SDBA^{2-}$ ligands in the molecular unit could be judged as 3-,4-, and 5-connected nodes, and the two Dy(III) ions could be viewed as 5- and 6-connected nodes; so the whole framework of 1 can be abstracted as a 3-, 4-, 4-, 5-, and 6-connected topological network with the Schläfli symbol of {4.6.8}{4^2.6^3.8^5}{4^3.6^2.8}{4^3.6^3}{4^9.6^6}, which has not been included in the TOPOS database (Figure 2(d)).

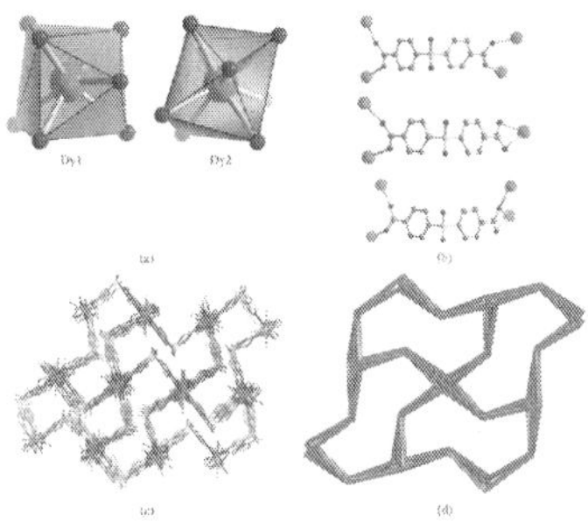

Figure 2. (a) The coordination surroundings for the Dy(III) ions in **1**; (b) view of the binding pattern for the SDBA ligands; (c) view of the 3D network of **1**; (d) 3-, 4-, 4-, 5-, and 6-connected net of **1**.

In order to determine the obtained **1** more fully in terms of phase purity and thermal stability, we carried out the PXRD measurements and TGA experiments. As shown in Figure 3(a), the good match between the simulated and experimental PXRD patterns indicates the high phase purity of the as-synthesized **1**. Considering the following drug delivery experiments, the stability of **1** in water has been checked by soaking the crystalline samples of **1** in water for one day, and then, the corresponding PXRD patterns were collected, which reveal that the water surroundings could not lead to the transformation of the framework of **1**, confirming its good water stability. To probe the lattice solvents in the framework of **1**, the TGA experiments were carried out (Figure 3(b)). A weight loss of 32.3% was observed in the temperature range of 25 to 429°C, which can be attributed to the release of three coordinated water and six lattice DMA molecules. After reaching 429°C, a sharp weight loss could be observed for **1**, which indicates the collapse of the framework. To establish the permanent porosity of the solvent-free samples of **1** (**1a**), the as-prepared crystalline samples of **1** were soaked in the anhydrous MeOH for three days to exchange with the lattice DMA molecules, and then activated at 60°C for 12 hours. The full activation of **1a** has been confirmed via the TGA experiments, which reveals that no obvious weight loss could be found in the temperature range of 25 to 429°C. To determine the surface area of **1a**, N_2 sorption isotherm was collected at 77 K. **1a** shows a type-I sorption behavior with the maximum N_2 uptake capacity of 294 cm^3/g, which indicates complex **1a** is a microporous material (Figure 3(c)). The Brunauer–Emmett–Teller (BET) surface area was calculated to be 1222 m^2/g, indicating the space of **1a** could house considerable amount of guest molecules. Furthermore, the hole diameter and distribution calculated from the 77 K N_2 sorption method reveals a hole size distribution around 0.95 nm, illuminating the potential of accepting guests in the pores (Figure 3(d)).

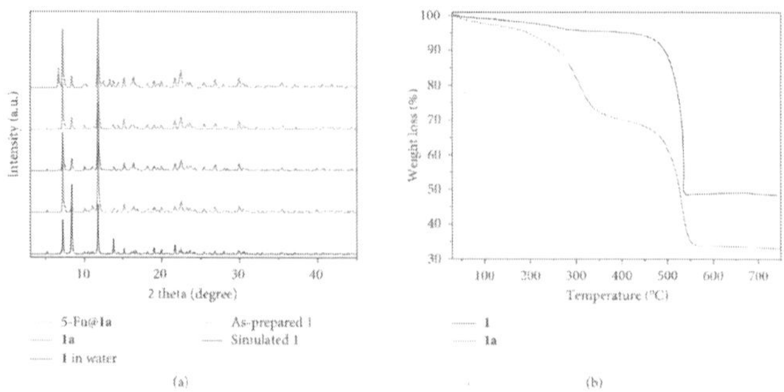

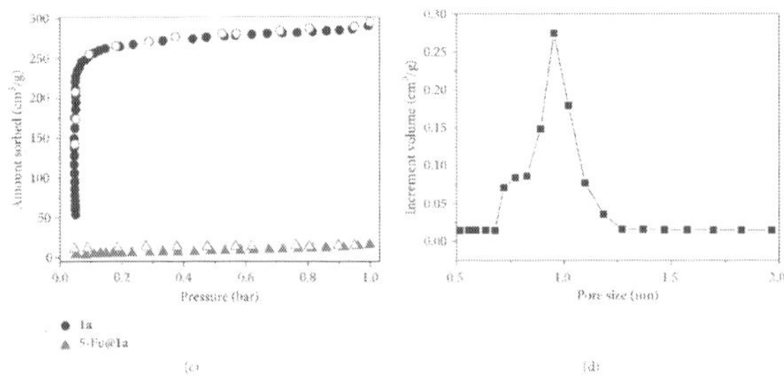

Figure 3. (a) The PXRD patterns of **1** and its different forms; (b) TGA curves of **1** and **1a**; (c) N_2 sorption isotherms of **1** and **1a**; (d) hole size location of **1a** derived from the N_2 sorption data at 77K.

Drug Delivery Experiment

In light of its large BET surface area and suitable pore size, complex 1a has the potential for loading small guest molecules. In this work, 5-fluorouracil (5-Fu) with the molecular size of 5.3×5.0 Å^2 was chosen as the guest molecules because it is a widely used drug molecule for the cancer treatment. The drug-loading experiments were carried out by impregnating of 1a (10mg) in 5mL MeOH solution containing 20mg of 5-Fu with consciously stirring for two hours, and then, the 5-Fu concentration in the MeOH solution was detected via the UV-Vis spectroscopy. As shown in Figure 4(a), there is an obvious decrease in the intensity of the UV-Vis spectrum at 254nm, indicating that the 5-Fu molecules are absorbed into the pores of 1a (Figure 4(a)). The 5-Fu storage capacity is 20.6wt.% (this means every 100g of 5-Fu/1a composite contains 20.6g of 5-Fu) based on the UV-Vis spectrum result. Furthermore, the N_2 sorption experiment of the 5-Fu-loaded 1a indicates that it could hardly adsorb any N_2 under the given conditions, revealing that the pore spaces of 1a are effectively occupied by the drug molecules.

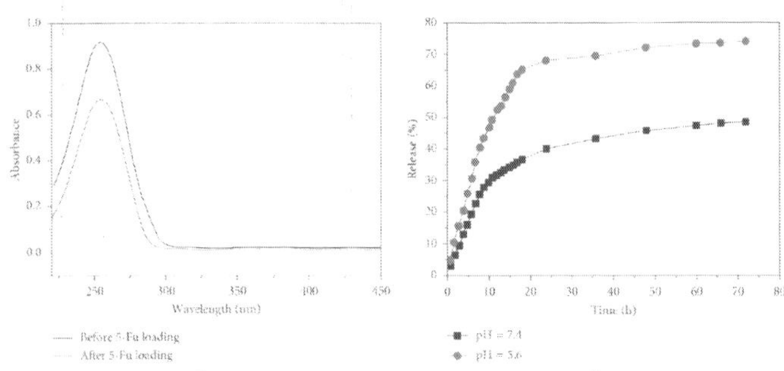

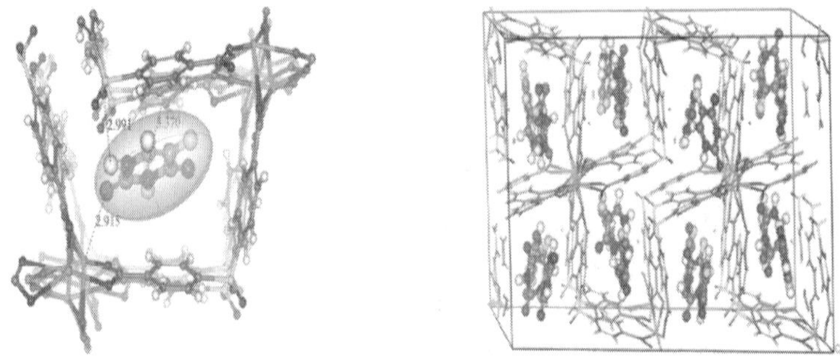

(c) (d)

Figure 4. (a) UV-Vis spectra of the MeOH solution containing 5-Fu before and after loading the drug; (b) drug releasing curves under different pH values; (c) simulated binding site of 5-Fu in the framework of **1** at zero loading; (d) locations of the 5-Fu molecules in the framework of **1** at 37°C and 100 kPa.

The drug sustained-release experiment was studied under different conditions (pH 7.4 at 37°C and pH 5.6 at 37°C), and the 5-Fu released concentration was determined by HPLC. As shown in Figure 4(b), the amount of 5-Fu released from 5-Fu-loaded **1a** is 50% in the first 40 h, and further increasing the time could not lead to more 5-Fu released from the carrier. The present result also indicates that the 5-Fu drug could not be completely released from the carrier under the present conditions. When the pH value of the PBS solution was adjusted to 5.6, it could be observed that the release rate is much faster with nearly 70% of 5-Fu molecules being released into the solution, indicating that the 5-Fu release is pH independent (Figure 4(b)). The slightly large release rate might be due to the partly decomposition of the framework. To understand the interaction between the 5-Fu molecules and the framework, the GCMC simulation was carried out. As shown in Figure 4(c), the simulated favorable binding site of 5-Fu at zero loading indicates that the one 5-Fu molecule prefers to locate in the channel center with short H-bond and Dy-O interactions could be observed (2.991 Å and 2.915 Å). Furthermore, besides the H-bond interactions, there also exists π-π interaction between the 5-Fu molecule and the benzene ring of the ligand with a distance of 3.370 Å. All these interactions contribute to the strong binding energy of 5-Fu with the framework, resulting in the long-lasting drug release. Figure 4(d) shows the distribution of 5-Fu molecules in the framework of **1a** at 37°C and 100 kPa, which reveals that each unit cell of **1** could house 11 5-Fu molecules, corresponding to 22.4 wt.%, which is near to the value calculated from the UV-Vis spectrum.

Anticancer Activity

As a drug delivery system, its cytotoxicity toward the human normal cells should be considered. With this in consideration, the cytotoxicity of the 5-Fu-loaded **1a** and the 5-Fu were studied against the HASM (human aortic smooth muscle) cells via the standard MTT method. The HASM cells were incubated with 5-Fu-loaded **1a** and 5-Fu with various concentrations, and the corresponding cell viabilities were recorded. As shown in Figure 5(a), the cell viabilities are all above 92% under various concentrations of 5-Fu-loaded **1a**, indicating that this drug-MOF composite has little toxicity toward the human normal cells. Furthermore, compared with the drug 5-Fu, the drug-MOF composite shows more cell viability, illuminating that the drug-MOF might possess less side effect compared with the pure drug. To probe the anticancer activity of the 5-Fu-loaded **1a**, the *in vitro* antitumor activity was studied against the human osteosarcoma cell line MG63 under dsifferent concentrations for 24 h and 48 h, respectively. As shown in Figure 5(b), the 5-Fu-loaded **1a** exhibits potent cytotoxicities against the MG63 cells, with about 60% cancer cells killed at 24 h. Interestingly, further prolonging the experimental time could lead to more cell death, indicating the long-lasting drug release.

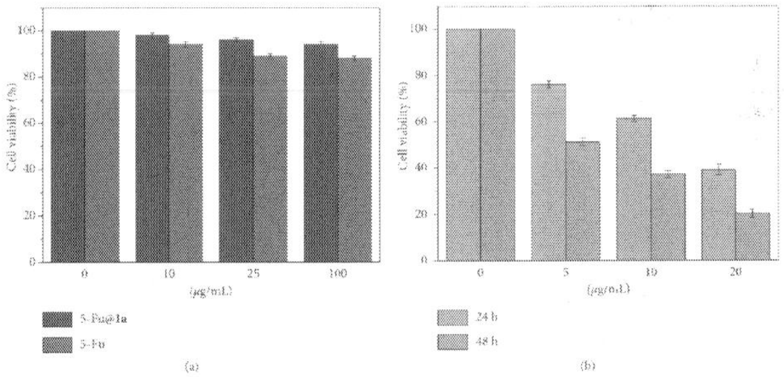

Figure 5. (a) HASM cell viability after exposure to 5-Fu-loaded 1a and 5-Fu at different concentrations after 24 h; (b) viability of MG63 cells after incubation with various concentrations of drug-loaded MOF at 24 h and 48 h.

CONCLUSION

In summary, we have successfully prepared a novel porous MOF using a V-shaped polycarboxylic acid organic linker under the solvothermal condition. The as-prepared MOF is composed of 1D chain-like SBUs that were linked via the ligands to afford the 3D framework with 1D microporous channels running along the *a*-axis. Due to its suitable pore size and high density of O donor sites, it could behave as an effective carrier for the 5-Fu capture and release, and a moderate 5-Fu-loading capacity and pH-dependent drug release behavior could be observed. Furthermore, the MTT assay reveals the low cytotoxicity of the 5-Fu-loaded **1a** and notable anticancer activity against the human osteosarcoma cell line MG63.

DATA AVAILABILITY

The data used to support the findings of this study are included within this article.

CONFLICTS OF INTEREST

The authors declare that there are no conflicts of interest regarding the publication of this paper.

REFERENCES

1. J. Öhlén, J. Cohen, and C. Håkanson, "Determinants in the place of death for people with different cancer types: a national population-based study," *Acta Oncologica*, vol. 56, no. 3, pp. 455–461, 2017.
2. J. Bogach, O. Levine, S. Parpia, M. Valencia, L. Ruo, and P. Serrano, "Does the addition of biologic agents to chemotherapy in patients with unresectable colorectal cancer metastases result in a higher proportion of patients undergoing resection? A systematic review and meta-analysis," *Journal of Gastrointestinal Surgery*, vol. 22, no. 3, pp. 523–528, 2018.
3. J. Zhou, G. Yu, and F. Huang, "Supramolecular chemotherapy based on host–guest molecular recognition: a novel strategy in the battle against cancer with a bright future," *Chemical Society Reviews*, vol. 46, no. 22, pp. 7021–7053, 2017
4. L. Wang, M. Zheng, and Z. Xie, "Nanoscale metal–organic frameworks for drug delivery: a conventional platform with new promise," *Journal of Materials Chemistry B*, vol. 6, no. 5, pp. 707–717, 2018.
5. D. Rambabu, P. Negi, A. Dhir, A. Gupta, and Pooja, "Fe(III) and Cu(I) based metal organic gels for in situ drug loading and drug delivery of 5-fluorouracil," *Inorganic Chemistry Communication*, vol. 93, pp. 6–9, 2018.
6. Y. Georgiou, J. A. Perman, A. B. Bourlinos, and Y. Deligiannakis, "Highly efficient arsenite [As(III)] adsorption by an [MIL-100(Fe)] metal–organic framework: structural and mechanistic insights," *Journal of Physical Chemistry C*, vol. 122, no. 9, pp. 4859–4869, 2018.
7. A. C. Forse, M. I. Gonzalez, R. L. Siegelman et al., "Unexpected diffusion anisotropy of carbon dioxide in the metal–organic framework Zn2 (Dobpdc)," *Journal of American Chemical Society*, vol. 140, no. 5, pp. 1663–1673, 2018.
8. J. Liu, T. Y. Bao, X. Y. Yang et al., "Controllable porosity conversion of metal-organic frameworks composed of natural ingredients for drug delivery," *Chemical Communications*, vol. 53, no. 55, pp. 7804–7807, 2017.
9. X. Du, R. Fan, L. Qiang et al., "Controlled Zn^{2+}-triggered drug release by preferred coordination of open active sites within functionalization indium metal organic frameworks," *ACS Applied Materials and Interfaces*, vol. 9, no. 34, pp. 28939–28948, 2017.

10. Y. Duan, F. Ye, Y. Huang, Y. Qin, C. He, and S. Zhao, "One-pot synthesis of a metal–organic framework-based drug carrier for intelligent glucose-responsive insulin delivery," *Chemical Communications*, vol. 54, no. 42, pp. 5377–5380, 2018.
11. M. Kotzabasaki, E. Tylianakis, E. Klontzas, and G. E. Froudakis, "OH-functionalization strategy in metal-organic frameworks for drug delivery," *Chemical Physics Letters*, vol. 685, pp. 114–118, 2017.
12. R. Łyszczek, A. Ostasz, A. Bartyzel, and A. Lipke, "Thermal, Spectroscopic and Luminescence Investigations of lanthanide(III) coordination polymers based on v-shaped 4,4′-sulfonyldibenzoic acid," *Journal of Analytical Applied Pyrolysis*, vol. 115, pp. 370–378, 2015.

SECTION 3
HYDROGELS AND DYNAMIC COVALENT GELS

CHAPTER

13

History and Applications of Hydrogels

Naziha Chirani[1], L'Hocine Yahia[1], Lukas Gritsch[1,2], Federico Leonardo Motta[1,2], SoumiaChirani[3], Silvia Fare[2]

[1]Laboratoire d'innovation et d'analyse de Bioperformance, Ecole polytechnique, Université de Montréal, Montréal, Quèbec, Canada

[2]Dipartimento di Chimica, Materiali ed Ingegneria Chimica "G. Natta", Politecnico di Milano, Edificio 6, piazza L. da Vinci 32, 20133 Milano (Italy)

[3]Laboratoire de Chimie Organique, Physique et macromeléculaire, Département de pharmacie, Faculté de médecine, Université Djilali Liabés, Sidi Bel-Abbès, Algeria

ABSTRACT

Hydrogels have existed for more than half century, providing one of the earliest records of crosslinked hydroxyethyl methacrylate (HEMA)

Citation: Chirani N, Yahia LH, Gritsch L, et al. "History and Applications of Hydrogels". J Biomedical Sci. 2016, 4:2. http://dx.doi.org/10.4172/2254-609X.100013

hydrogels. Today, hydrogels still fascinate material scientists and biomedical researchers and great strides have been made in terms of their formulations and applications. As a class of material, hydrogels are unique, they consist of a self-supporting, water-swollen three-dimensional (3D) viscoelastic network which permits the diffusion and attachment of molecules and cells. However, hydrogels have recently drawn great attention for use in a wide variety of biomedical applications such as cell therapeutics, wound healing, cartilage/bone regeneration and the sustained release of drugs. This is due to their biocompatibility and the similarity of their physical properties to natural tissue. This review aims to give an overview of the historic and the recent design concept of hydrogels and their several applications based on the old and the most recent publications in this field.

Keywords: Hydrogel; Applications; Proteomics; Wound healing; Tissue engineering; Drug delivery; Vaccine; injectable hydrogel

INTRODUCTION

Hydrogels are polymeric matrixes that swell but don't dissolve (in the short term) in water [1]. The swelling properties are due to the high thermodynamical affinity that this class of materials has for the solvent itself. In the past years this characteristic, coupled with a high versatility and a high tunability of material's properties, lead to deep research and exploitation of hydrogels. These networks establish equilibrium with the liquid and temperature of their surroundings for shape and mechanical strength. Variations in the concentration, structure and/or functionality of the monomer and/or cross-linker used in such gels can change the structure. Indeed, many new gel-form materials, with a plethora of aims were developed and tested in different fields of engineering (e.g. environmental, electronics, biomedical), biotechnology and other disciplines.This phenomenon can be appreciated by analyzing the number of articles on the topic: doing a quick query for the word "hydrogel" on PubMed it is easy to see the clear exponential tendency in the number of articles (**Figure 1**). This is a general trend shared by any kind of scientific publication that is, moreover, underestimated because of the inability of any database to capture the whole scientific literature. This phenomenon is considered not consistent by many academic personalities and, surely, generates concerns about the absence of a real correlation between publishing rates and knowledge [2]. Nevertheless, it is possible to assert that, even if this course of events is not evidence, it is at least a clue of the growing interest of the scientific community on the hydrogel topic.

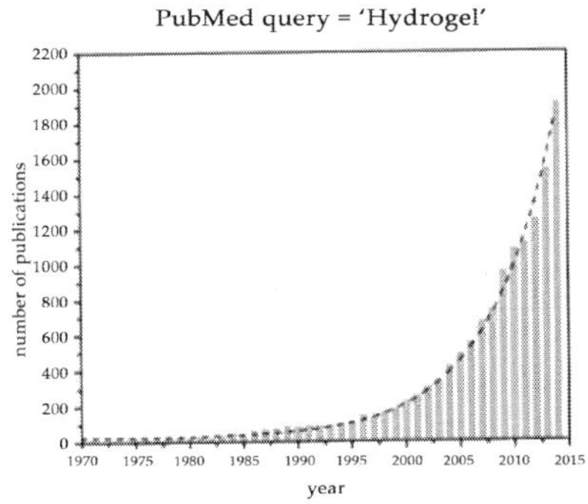

Figure 1: Histogram showing the increase in publications related to the keyword"hydrogel" during the past 45 years. A proper exponential fitting is also detectable.

Smart hydrogel systems with various chemically and structurally responsive moieties exhibit responsiveness to external stimuli including temperature, pH, ionic concentration, light, magnetic fields, electrical fields and chemicals. Polymers with multiple responsive properties have also been developed elegantly combining two or more stimuli-responsive mechanisms. Smart polymer hydrogels change their structural and volume phase transition as a response to external stimuli resulting in an enormous potential for scientific observations and for various advanced technological applications.

Hydrogels can be classified into two distinct categories, the natural and the synthetic hydrogels. Natural hydrogels include collagen, fibrin, hyaluronic acid, matrigel, and derivatives of natural materials such as chitosan, alginate and skill fibers. They remain the most physiological hydrogels as they are components of the extracellular matrix (ECM) *in vivo*. Two main drawbacks of natural hydrogels, however, make their final microstructures and properties difficult to control reproducibly between experiments. First, the fine details of their mechanical properties and their dependence on polymerization or gelation conditions are often poorly understood. Second, due to their natural origin (bovine fibrinogen, rat tail collagen their composition may vary from one batch to another.

In contrast, synthetic hydrogels such as poly (ethylene glycol) diacrylate, poly (acryl amide), poly (vinyl alcohol) are more reproducible, although their final structure can also depend on polymerization conditions in a subtle way, so that a rigorous control of the preparation protocol, including temperature and environment control, may be necessary. Generally speaking, synthetic hydrogels offer more flexibility for tuning chemical composition and mechanical properties; users can, for example vary the concentration or molecular weight of the precursor, or alter the percentage of crosslinkers. They can also be selected or tuned to be hydrolysable or biodegradable over variable periods of time.

Hydrogels may be chemically stable or they may degrade and eventually disintegrate and dissolve. They are called «reversible» or «physical» gels when the networks are held together by molecular entanglements, and/or secondary forces including ionic, H-bonding or hydrophobic forces (Figure 2) [3]. Physical hydrogels are not homogeneous, since clusters of molecular entanglements, or hydrophobically-or ionically-associated domains, can create in homogeneities. Free chain ends or chain loops also represent transient network defects in physical gels.

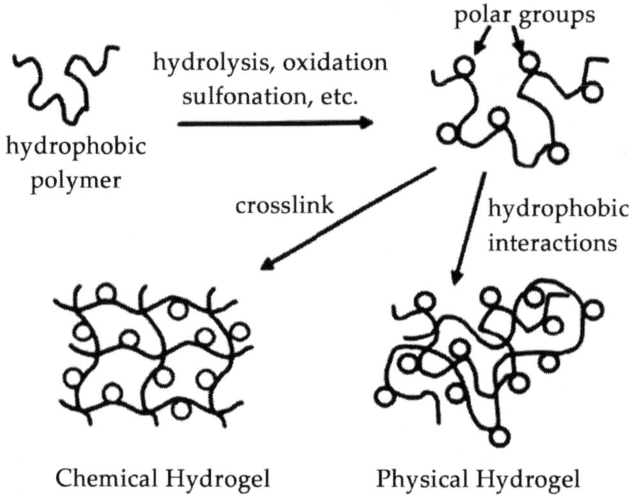

Figure 2: Types of crosslinking, adapted from A.S.Hoffmann, "Advanced Drug Delivery Reviews"[29], on the left of the figure it is possible to see the chemical crosslinked net, on the right the physical one.

When a polyelectrolyte is combined with a multivalent ion of the opposite charge, it may form a physical hydrogel known as an ionotropic hydrogel, calcium alginate is an example of this type of hydrogel. Hydrogels are called "permanent" or "chemical" gels when they are covalently-cross linked networks (Figure 3). The synthetic hydrogels of Wichterle and Lim were based on copolymerization of HEMA with the cross linker EGDMA.

Chemical hydrogels may also be generated by crosslinking of water-soluble polymers, or by conversion of hydrophobic polymers to hydrophilic polymers plus crosslinking is not necessary. In some cases, depending on the solvent composition, temperature and solids concentration during gel formation, phase separation can occur, and water-filled voids or macropores can form.

In chemical gels, free chains ends represent gel network "defects" which do not contribute to the elasticity of the network. Other network defects are chain loops and entanglements, which also do not contribute to the permanent network elasticity.

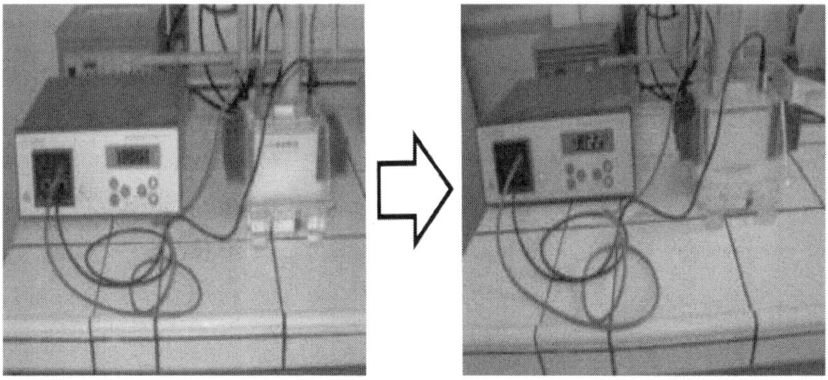

Figure 3: SDS-PAGE migration for separation of proteins mixture.

In many applications, a functional additive is blended into a polymer matrix to enhance its properties. However, when the polymer and functional additive are applied to a surface, the functional molecule may be easily lost. In favorable cases, it may be possible to incorporate the additive directly into the polymer as a comonomer.

In recent study, a functionalized polymer has been obtained through the combination of linking a photodynamic, antimicrobial dye, Rose Bengal, to vinyl benzyl chloride via etherification and then polymerizing this into a water-soluble polymer using chain growth copolymerization [4].

The amount of unincorporated dye was determined using dialysis in buffer solution at pH = 7, 4. When Rose Bengal was first reacted with vinyl benzyl chloride before polymerization, it was found that it was nearly 100% incorporated into the copolymer.

REVIEW

History

The word "hydrogel", according to Lee, Kwon and Park, dates back toan article published in 1894. Anyway, the material described there was not a hydrogel as we describe it today; it was indeed a colloidal gel made with inorganic salts. Is yet remarkable to notice how the history of the term itself is consistently long [5]. Anyhow, the first crosslinked network material that appeared in literature and has been described by its typical hydrogel properties, one for all the high water affinity, was a polyhydroxyethylmethacrylate (pHEMA) hydrogel developed much later, in 1960, with the ambitious goal of using them in permanent contact applications with human tissues, hydrogels are in fact the first materials developed for uses inside the patient [6,7]. Since then the number of studies about hydrogels for biomedical applications began to rise, especially from the decade of 70's [5]. The aims and goals and the number of materials changed and enlarged constantly over the years. As suggested by Buwalda et al. [8], the history of hydrogels can be divided in three main blocks.

A hydrogel's first generation that comprises a wide range of crosslinking procedures involving the chemical modifications of a monomer or polymer with an initiator. The general aim is to develop material with high swelling, good mechanical properties and relatively simple rationale.

Then, starting in the seventies, a different concept of hydrogel grew in importance: a second generation of materials capable of a response to specific stimuli, such as variations in temperature, pH or concentration of specific molecules in solution. These specific stimuli can be exploited to trigger likewise specific events, for example the polymerization of the material, a drug delivery or an *in situ* pore formation [9].

Finally, a third generation of hydrogels focusing on the investigation and development of stereo complexed materials (e.g. PEG-PLA interaction) [10, 11] hydrogels crosslinkedby other physical interactions (e.g. cyclodextrines) [12,13].

This progress in hydrogel's science is quickly leading to an increasing interest in the development of the so called "smart hydrogels", polymeric matrixes with a wide spectrum of tunable properties and trigger stimuli. The topic is theoretically inexhaustible and the possible applications, the engineering and medical devices that can be obtained from it are above any imagination.

Since the pioneering work of Wichterle and Lim in 1960 on crosslinked hydrogels [7], and because of their hydrophilic character and potential to be biocompatible, hydrogels have been of great interest to biomaterial scientists for many years [14-16].

The important and influential work of Lim and Sun in 1980 [17] demonstrated the successful application of calcium alginate microcapsules for cell encapsulation. Later in the 1980s, Yannas and coworkers [18] incorporated natural polymers such as collagen and shark cartilage into hydrogels for use as artificial burn dressings. Hydrogels based on both natural and synthetic polymers have continued to be of interest for encapsulation of cells [19] and most recently such hydrogels have become especially attractive to the new field of "tissue engineering" as matrices for repairing and regenerating a wide variety of tissues and organs [20].

Physical and chemical properties

Despite much progress, a fundamental understanding of gel properties is not yet sufficient for a rational design of novel gel systems. For such designs, it is important to know how solute molecules interact with the gel, in particular, how they partition between the gel phase and the surrounding liquid phase. Partitioning depends on two major effects: size exclusion and molecular attraction/repulsion.

Swelling: Hydrogels are cross-linked polymer networks swollen in a liquid medium. The imbibed liquid serves as a selective filter to allow free diffusion of some solute molecules, while the polymer network serves as a matrix to hold the liquid together. Hydrogels may absorb from 10-20% (an arbitrary lower limit) up to thousands of times their dry weight in water.

The character of the water in a hydrogel can determine the overall permeation of nutrients into and cellular products out of the gel. When a dry hydrogel begins to absorb water, the first water molecules entering the matrix will hydrate the most polar, hydrophilic groups, leading to *primary bound water*. As the polar groups are hydrated, the network swells, and exposes hydrophobic groups, which also interact with water molecules, leading to hydrophobically-bound water, or *secondary bound water*. Primary and secondary bound water are often combined and simply called the *total bound water*. After the polar and hydrophobic sites have interacted with and bound water molecules, the network will imbibe additional water, due to the osmotic driving force of the network chains towards infinite dilution. This additional swelling is opposed by the covalent or physical crosslink, leading to an elastic network retraction force. Thus, the hydrogel will reach an equilibrium swelling level. The additional swelling water that is imbibed after the ionic, polar and hydrophobic groups become saturated with bound water is called free water or bulk water, and is assumed to fill the space between the network chains, and/or the center of larger pores, macropores or voids. As the network swells, if the network chains or crosslink are degradable, the gel will begin to disintegrate and dissolve, at a rate depending on its composition.

There are a number of methods used by researchers to estimate the relative amounts of free and bound water, as fractions of the total water content. All of them are controversial, since there is proton NMR evidence that the interchange of water molecules between the so-called bound and free states is extremely rapid, perhaps as fast as one H_2O molecule every 10^{-9}s. The three major methods used to characterize water in hydrogels are based on the use of small molecular probes, DSC and NMR. When probe molecules are used, the labeled probe solution is equilibrated with the hydrogel, and the concentration of the probe molecule in the gel at equilibrium is measured. Assuming that only the free water in the gel can dissolve the probe solute, one can calculate the free water content from the amount of the imbibed probe molecule and the known (measured) probe molecule concentration in the external solution. Then the bound water is obtained by difference of the measured total water content of the hydrogel and the calculated free water content.

The use of DSC is based on the assumption that only the free water may be frozen, so it is assumed that the endotherm measured when warming the frozen gel represents the melting of the free water, and that value will yield the amount of free water in the HG sample being tested. Then the bound water is obtained by difference of the measured total water content of the HG test specimen, and the calculated free water content.

In another formulation, swelling is the property to absorb water and retain it for a relative long time. It can be evaluated by measuring the dry weight and the swollen-state weight and computing either a ponderal variation (water uptake) or a volume of adsorbed solvent (both the quantities are considered as percentages).

$$W.U. = \frac{swollen\ weight - dry\ weight}{dry\ weight} \times 100 \qquad (1)$$

$$V.A.S. = \frac{swollen\ weight - dry\ weight}{water\ density} \times 100 \qquad (2)$$

As simple as it may seem, the evaluation of swelling is the principal assay to be performed on hydrogel samples, as it can be a measure for many of their properties: crosslinking degree, mechanical properties, degradation rate and so on. For many gels, the evaluation of swelling and swollen state stability is the simplest, cheapest and surest way to discriminate between crosslinked gels and the not crosslinked original polymer [21].

Mechanical properties

The mechanical properties can vary and be tuned depending on the purpose of the material. It is possible to obtain a gel with higher stiffness increasing the crosslinking degree or lowering it by heating the material. The changes in mechanical properties link to a wide range of variables and causes and different analysis must be made according to the material, the conditions and the aim of the study. For example, while gelatin show a noticeable increase in Young Modulus through crosslinking [22], silk fibroin has a very high Young Modulus, but after the revitalization it will decrease [23]. These properties (young modulus, Poisson modulus, storage and loss moduli, tanδ) can be evaluated by a Dynamic Mechanical Analysis (DMA) device or a rheometer, according to the thousands of techniques available on the market that will be no further discussed here [24,25].

It's important to note that in a hydrogel, the Young Modulus is the results of the union between water and gel matrix. If we have to seeds osteoblast cells we will need a more stiff material than if we culture adipocyte, the same rationale is valid for the development of a heterogeneous prosthetic device, for example substitute for the intervertebral disc.

Porosity and permeation: Pores may be formed in hydrogels by phase separation during synthesis, or they may exist as smaller pores within the network. The average pore size, the pore size distribution, and the pore interconnections are important factors of a hydrogel matrix that are often difficult to quantify, and are usually included together in the parameter called « tortuosity ». The effective diffusion path length across a HG film barrier is estimated by the film thickness times the ratio of the pore volume fraction divided by the tortuosity. These factors, in turn, are most influenced by the composition and crosslink density of the hydrogel polymer network.

Labeled molecular probes of a range of molecular weights (MWs) or molecular sizes are used to probe pore sizes in hydrogels [26].

Pore-size distributions of hydrogels are strongly affected by three factors:

- Concentration of the chemical cross-links of the polymer strands. That concentration is determined by the initial ratio of cross-linker to monomer.
- Concentration of the physical entanglements of the polymer strands. That concentration is determined by the initial concentration of all polymerizable monomers in the aqueous solution.
- Net charge of the polyelectrolyte hydrogel. That charge is determined by the initial concentration of the cationic and/or anionic monomer.

These three factors can be quantified using the composition of the hydrogel, that is, by the nominal concentrations of monomer and cross-linker:

$$\%T = \frac{weight\ of\ monomer + weight\ of\ x - crosslinker}{total\ volume} \tag{3}$$

$$\%C = \frac{weight\ of\ x - crosslinker}{weight\ of\ monomer + weight\ of\ x - crosslinker} \tag{4}$$

The porous structure of a hydrogel is also affected by the properties of the surrounding solution, especially by dissolved ionic solutes (Donnan effects) and by dissolved uncharged solutes which partition unevenly between the gel phase and the solution phase (Osmotic effects).

For rational design of hydrogels, it is useful to know the pore-size distribution depends on the hydrogel characterization commonly expressed by %C and %T.

Most techniques used to investigate the porosity of hydrogels are limited because they require the pore solvent and/or temperature to be altered, causing the gel to shrink, swell or require mathematical manipulation and assumption, which may introduce unwanted artifacts.

Porosity is a morphological feature of a material that can be simply described as the presence of void cavity inside the bulk. It is useful to control the porosity in many devices for a wide variety of applications, such as optimal cell migration in hydrogel-based scaffolds or tunable lode/release of macromolecules.

$$Porosity(\%) = \frac{V_{pore}}{V_{bulk} + V_{pore}} \cdot 100 \quad (5)$$

In a sample, pores can show different morphologies: they can be closed, open as a blind end or interconnected, again divided in cavities and throats. These porosities can had been studied and evaluated in papers in the past decades using a various spectrum of techniques. First of all, porosity can be evaluated by theoretical methods, such as unit cube analysis, mass technique, Archimedes method, liquid displacement method. These analysis are commonly coupled with optical and electronic microscopy.

Other interesting techniques are the mercury porosimetry, based on Washburn's equation, with the inconvenience of being a destructive assay, the gas pycnometry, the gas adsorption (that can be issued using different procedures such as small quantity adsorption, monolayer and multilayer adsorption), liquid extrusion porosity, an assay that permits to evaluated sample's permeability too, capillary flow porosity, again a test based on Washburn's equation. Furthermore another important assay is the Micro-CT, also called X-ray Microtomography, a relative new imaging technique, simply described as a non-destructive highresolution radiography, capable of qualitative and quantitative assays on samples and evaluation of their pore interconnections. Between the quantitative assays that can be performed, micro-CT can give information on average pore size, pore size distribution, pore interconnection, struts/walls thickness and anisotropy/ isotropy of the sample (in the sense of presence/absence of preferential orientation of the pores). It is yet, nowadays, an expensive technique both in term of money and time [27,28].

Microscopy techniques can be used in thousands of different assays involving hydrogels. They are both involved in qualitative and quantitative tests, from simple morphological assessing of material's properties to more complex biocompatibility assays.

Briefly, by microscopy techniques topography and surface morphology can be assessed. These techniques can be divided in many classes, by increasing magnification power: optical microscopy (OM), stereo microscopy (SM), electron microscopy (SEM and TEM), tunneling microscopy (STM), atomic force microscopy (AFM) [29].

Crosslinking: Crosslinking is not properly a property of hydrogels, while it is more of a cause of all the other properties of the material itself.

The crosslinking degree can be correlated to basically every characteristic of a hydrogel. The nature of the crosslinking can vary a lot. Indeed, the hydrogel's network can be obtained in many different ways. The processes can be divided into two big categories: first of all the so called physical crosslinking that occurs, for example thanks to hydrophobic interactions between chains, ionic interactions between a polyanionand a polycation (complex coacervation) or ionic interactions between a polyanion and multivalent cations (ionotropic hydrogel). The second category comprises the chemical bound gels. The crosslinking can occur by ultraviolet irradiation, heating or chemical crosslinking via crosslinker with a huge ensemble of reactions, such as Michael's reaction, Michaelis-Arbuzov reaction, nucleophile addition and so on [30].

By controlling the degree of crosslinking is possible to tune the property of the material and optimize it for many different applications getting theoretically, in this way, a wide spectrum of applications starting from the same original polymer [33, 34].

APPLICATIONS: FROM THE LAB TO THE MARKET

Hydrogels are widely used in the area of electrophoresis, bioseparations, proteomic, chromatography, tissue engineering… etc. They are well-known in foods and medicines as absorbents in disposable diapers, as filters for water purification and as separation materials for chromatography and electrophoresis. They are also of interest for controlled drug release and for concentration of dilute solutions of macromolecules [35].

Domestic uses

It's not mandatory for life changing inventions to be high-tech devices with out-of-sight ambitions. Sometimes the simplest expedient can lead to a bigger progress than expensive and futuristic technologies. This is true for hydrogels, too. This class of materials has plenty of domestic usages that exploit their ability of loading, retaining and releasing fluids to significantly improve everyday life.

Diapers: An interesting application of hydrogel's thermodynamically affinity for water, as not fancy as it can be, is the production of super-adsorbent diapers with the property of being dry even after a considerable adsorption of fluids. This is due to, as previously said, the nature of hydrogel's water adsorption: these materials don't act as sponges, unstably trapping liquids into their pores, but instead they retain water (or, sometimes, other solvents) in while if they are carrying considerable quantities of water at the same time. The development of hydrogel-containing diapers, most of them loaded with different formulations of sodium polyacrylate [36-38], in the past two decades cut down on a huge number of dermatological conditions related to a prolonged contact with wet tissues. Nevertheless many health concern aroseon the massive usage of disposable diapers: ≈95% of nappies in western countries are disposable ones and opinions about cloth nappies are still conflicting [39]. Indeed, many chemicals used in the production of such products, like scents, leak-proof materials and super-adsorbent polymers, seem to be key elements for the development of many conditions, from chronic diaper rash and asthma, to more serious problems such as male infertility or even testicular cancer. Furthermore, disposable diapers, since they are used in huge quantity, create a notable environmental issue since it's not easy to dispose of them [40]. However, this is a topic that, as far as interesting and important, goes beyond the purpose of our review.

Watering beads for plants: Another simple application of hydrogels consists in rough powders of polyacrylamide or potassium polyacrylatematrix sold with a huge range of names (Plant-Gel, Super Crystals, Water-Gel Crystals) and used as long term reservoir of water for plant growth in gardening, domestic and sometimes industrial horticulture. On the opposite side as the one of diaper's hydrogel, these materials are optimized for their ability of releasing water, instead of the ability of retaining it. The sustained release of many diverse species is, indeed, one of the main strength of hydrogels on the market, from gardening to genetic engineering. However, even if companies producing such crystals are promoting their practicality and versatility, in the last years the scientific community is questioning about their real utility.

As Chalker-Scott from Washington State University pointed out in her publications on the topic, since the commonly used watering crystals are made out of non-renewable materials, whose monomers can be toxic (e.g. acrylamide), the potential risks of their usage are way higher than the benefits of water storage and controlled release that can, in addition, be obtained in many other ways with lower environmental impact [41].

Perfume delivery: During the nineties patents describing volatile species delivery technologies started to grow in number. In particular, the most significant patented inventions in the field seem to be issued by Procter&Gamble, processing the fragrances into cyclodextrin complexes [42-44].

The general aim was to develop devices capable of slowly dispense fragrances to the surroundings in the long-term and replace the classic salt-based (sodium dodecylbenzenesulphonate) tablets with new, more practical and, let's say it, fancier house care solutions. The role of hydrogels in the process revolves around, once again, their swelling properties that can be exploited in materials "wherein release of a perfume smell is triggered by dynamic swelling force of the polymer when the polymer is wetted" [43]. These devices release volatile particles thanks to osmotic diffusion of the specie from the swollen hydrogel to new water in the environment.

Cosmetics: Cosmetic industry is a market constantly increasing in dimension and product offer. One of the reasons of this behavior is related to the long path to approval necessary for medical devices, medical procedures, drugs or biomolecules, in order to make allowed to be sold on the market. So, while on the other hand, the cosmetic market requires less time and money expenses for the approval of a product, it is common to test new inventions and devices in cosmetics before and in more risky clinical applications.

For a product to be approved in cosmetics, the most important parameter to be assessed is Primary Irritation Index (PII). This index is simple to obtained and exist both for skin and eyes, indeed for each level of PII corresponds a determinate effect [45, 46]. Considering that the majority of hydrogels used in this field are suitable for cells culture and for other biomedical applications, is not surprising that their Irritation Index is among the lowest.

Thus, with a relatively small investment, companies are able tolaunch on the market new cosmetic products based on hydrogels, such as so called "beauty masks".

Usually made with engineered collagen (Masqueology™ by SEPHORA USA Inc., BioCollagen Cosmeceuticals by NOVOSTRATA UK Ltd.), hyaluronic acid (SEPHORA USA Inc.), or polyvinylpyrrolidone (Pecogel®), these masks claim to hydrate the skin, restore its elasticity and promote anti-aging actions [47]. Pecogel by Phoenix Chemicals Inc., is a wide selection of hydrogels, based on polyvinylpyrrolidone, with differences in composition and/or crosslinking method. Pecogels are suitable for cosmetic purposes, such as sunscreen cream or mascara [48]. Furthermore, in some of the commercially available compounds such as Hydro Gel Face Masks by Fruit & Passion Boutiques Inc., the moisturizing action of these organic polymeric gels is coupled with more complex drug-delivery systems developed to release of biomolecules like vitamin C or B3.

The cosmetic industry is on the cutting edge of hydrogels, indeed a pH-Sensitive material P(MAA-co-EGMA) has been developed for release of cosmetics drugs like arbutin, adenosine, and niacinamide, well knowing molecules for wrinkle treatment and for skin-whitening [49]. This hydrogels change is permeability responding to the pH changes: At pH 4.0 it holds the pharmaceuticals inside the matrix, when in contact with skin, at pH 6 and above, the permeability rise and the drugs will delivered. The authors said that this behavior is due by the ionization/deionization of MAA carboxylic groups.

Plastic Surgery

From their first development into the scientific research field, hydrogel where seen as good materials for application in contact with the human body because of their extracellular matrix-like (ECM-like) properties [7,50]. This is the main reason why attempts were made to introduce hydrogels like new materials for plastic reconstruction.

On this path, for many years, Hyaluronic Acid (HA) was thought like the panacea for every pain [51]. Is not surprising, considered this, that HA has been studied to be applied in tissue filling applications. One notable company operating in the field is Macrolane™.

Starting in 2008, Macrolane's treatments and products were specifically studied to enhance breast size and shape and offer a more biocompatible alternative to standard and aggressive silicone prosthesis. Anyhow, soon the scientific community started to point out the controversial behavior of these procedures in latter mammographies: briefly, HA worked as a shielding agent, appearing as denser tissue, and consequently ruining the outcome of the exam. Thus nowadays Macrolane™ is used for diversely situated filling with the exception of breasts. The compound is injected inside the body with a syringe and let it gels restoring the volume. This procedure started to receive criticisms by the scientific community, but yet at the moment there is still not a sufficient amount of publications investigating the properties and possible risks of the gel, concerns started to rise in particular about the long term side effects of this treatment and more specifically about HA's role in cancer control and development [52,53].

Another promising use of hydrogels is bulking agents for treatment of urinary incontinence: smart injectable gels can be involved in clinical procedures where these materials can be used to tighten the urethral channel and reduce patient's incontinence. With such a simple solution it is possible to erase or at least reduce a consistent social handicap and help patients to hold a normal life [54]. An example of product used in this field is Bulkamid® by Contura International, a polyacrylamide hydrogel is a commercial product developed to help women with incontinence narrowing the conduct. There are few motive for the incontinence, and an exhaustive classification is made by Lose et al. where they subdivide each class, Bladder / Urethra incontinent, in overactivity / underactivity, in their work they wrote that bulk injectable systems are promising to improve urethral coaptation. Problematics like infections and body response have been evaluated. In a sample composed by 130 women researchers have found 10 patients with infections, 5 with injection site pain, 2 with incontinence and only 1 with injection site laceration.

Immunotherapy and vaccine

A nanovector composed of peptide-based nanofibrous hydrogel can condense DNA to result in strong immune responses against HIV. HIV infection is presently incurable and has led to more than 30 million deaths [55]. Traditional, vaccines such as live or attenuated viruses are ineffective and pose potential risks [56]. As a consequence, safe alternatives such as DNA vaccines have attracted great attention, they are easily degraded by DNases and lysosomes, and injected naked DNA plasmids are poorly distributed and inefficiently expressed; thus, DNA vaccines can only induce modest humoral and cellular immune reponses [57]. As such, DNA vaccines must be injected with a delivery system to enhance the immune responses. Different synthetic delivery systems have been constructed including polymers [58], liposomes [59] and nano- or microparticles [60]. Some disadvantages limit their further usage, such as toxicity, small amount of antigen loading and decreased biological activity of DNA during complex preparation.

Nanofibers provide an ideal platform [61]. Another platform suited for DNA delivery is based on hydrogel because of its high loading capacity, mild working conditions, and good biocompatibility [62]. Supramolecular hydrogel is a kind of hydrogel composed of nanofibers formed by the self-assembly of small molecules (molecular weight usually <2000) in aqueous solutions [63]. They can quickly respond to various external stimuli as: pH, temperature, ionic concentration or addition of enzymes.

Recently, peptide-based delivery system emerged as an alternative means to efficiently introduce genes or drugs into cells both in-vitro and in-vivo [64]. Some enzyme-triggered peptide based nanofibrous hydrogels are useful for the entrapment of drug molecules without reducing their activity because that they can work in mild conditions [65]. They are also biocompatible and have well-defined structures.

Nap-GFFY (Naphthalene acetic acid-Glycine phenylalanine tyrosine) was demonstrated as a short peptide to construct gelators of nanofibrous hydrogel. This nanovector can strongly activate both humoral and cellular immune responses to a balanced level rarely reported, which is crucial for HIV prevention and therapy. In addition, this hydrogel shows good biosafety in-vitro and in-vivo. Such hydrogel have a critical nanofibrous structure for the dramatically improved immune responses compared to existing materials. The reason of this high efficiency is that this nanovector can condense DNA, promote DNA transfection, and enhance gene expression *in-vitro*. Furthermore, it shows promising biocompatibility and no obvious toxicity. This peptide-based nanofibroushydrogel used as a HIV DNA nanovector can open the doors for effective vaccination based on peptide-based nanofibroushydrogels [66].

Environmental applications

Over past the years, nations gradually started to care about environmental issues and pollution. Many governments decided to opt for greener and safer for the environment policies. Water pollution is one of the biggest issues afflicting especially poor areas of Africa, Asia and South America. Thanks to their affinity for water, hydrogels might be used in two different ways to treat water source.

First the matrix can be used as a holder for purifying microorganism. Many interesting studies, on this particular path, were developed by encapsulating microorganisms inside diverse carrier materials [67]. Chlorella and Spirulina are the most used ones. These microorganisms are already used to remove pollutants chemicals from water resources. The idea is to keep the bacteria inside the network and consequently protect and control the bacteria-colturewhile cleaning the site of depuration. Both synthetic and natural hydrogels were been used. The best working hydrogels in literature appear to be Alginate derived [67] or alternatively carrageenan and agar [68].

A second interesting way to solve the problem of pollutants is to modify the hydrogels to let them seize and keep the pollutant inside the networks. Many authors have tried this way to seize metal ions: the group of Irani has sent to publish a paper in which they discuss a new composite hydrogel for

PbI(II) removal. Briefly, they created a polyethylene-g-poly (acrylic acid)-co-starch/ OMMT (LLDPE-g-PAA-co-starch/OMMT) hydrogel composite, using it like an adsorbent pollutant (Pb(II)) remover. They put the hydrogel in a solution containing lead acetate and then measured the adsorption with an Atomic Absorption Spectrometer (AAS); after that they did a de-absorption phase and repeat for several cycles. The electrostatic attractions, ion exchange and chelation are possible explanation for the metal adsorption happened during the experiments. They reported that the equilibrium adsorption data of the hydrogel was consistent with Langmuir isotherm and the 430mg/g adsorption capacity was in line with other common adsorbents [69,70].

Another feasible way to achieve an interesting water filtering is explained in a paper by Yan et al., where the group performed etherification and consequent functionalization of chitosan beads in order to obtain carboxymethilated chitosan with an enhanced adsorption of metal ions. This has been proven to improve selective adsorption of specific ions like Cu(II), Pu(II) and Mg(II) [70]. It is interesting properties that can be exploited in dye removal application and that has been demonstrate to be possible by magnetical doping of hydrogel microspheres with interpenetrated network (IPN) structures [71].

Hydrogels could be either used like a probe to detect heavy metal ions like in the work by Wang et al. [72]. Moreover, other application of polyacrylamide gels is a flood control device called WATER GEL BAG® and produced by TaiHei Co., Ltd.

One of the biggest environmental problems currently hard to solve is for sure the loss of oil in seas and in other water sources. In the past years the quantity of oil substance dispersed into the hydrosphere around the world is risen dramatically. Food processing, hydrocarbons industry, refining process have increased the risk of pollution. It was discovered that effluents receiving wastewaters from industry have 40.000 mg/L oil concentration [73].

Different kind of depolluting system have been attempted like bentonite organically [74, 75], palygorskite [76], a magnesium aluminum phyllosilicate, and activated carbon. In a study dated 2010 authors attempted to develop a hydrogel to retain water with oil-pollutant molecules [77]. They claim a very promising hydrogel is chitosan thanks to a high presence of amino groups and hydrogen that can react with vinyl monomers. Authors explained use of polyacrylamide grafted on the polysaccharides rises the capability to attract and hold solid particles which remain in suspension in water, called flocculants. The higher crosslinking degree, the lower the retaining capability. This is due to a smaller distance between two nodes in the matrix network. Reducing this distance, the swelling degree will decrease. Furthermore the initial concentration of wasted oil in the environment is another fundamental parameter because of its property of affecting the adsorption kinetic. In fact a higher concentration corresponds to faster swelling kinetics.

Bacterial culture

As already discuss for the environmental applications section, hydrogels can hold inside their matrix a significant number of microorganism for purification of water, for production of biomolecules, or for simple culture of bacteria by themselves.

Indeed, agar is famous as the golden standard substrate for bacterial culture in biotechnological applications [78]. Since it is indigestible by a great number of bacteria and microorganism, it provides a perfect environment for their culture on a solid substrate [79]. Different kinds of agar are been studied, each with a potential use for likewise different kinds of bacterial. Between them *brucella agar, columbi agar, schaedler agar*, or trypicasesoy agar are the most common. None of these gels hassuperior results compared to the others, instead everyone is suitable for different applications in reason of their different pros and their cons [80,81].

Electrophoresis and proteomic

Gel electrophoresis currently represents one of the most standard techniques for protein separation. In addition to the most commonly employed polyacrylamide crosslinked hydrogels, acrylamide agarose copolymers have been proposed as promising systems for separation matrices in two-dimensional (2-D) electrophoresis, because of the good resolution of both high and low molecular mass proteins made possible by careful control and optimization of the hydrogel pore structure. As a matter of fact, a thorough understanding of the nature of the hydrogel pore structure as well as of the parameters by which it is influenced is crucial for the design of hydrogel systems with optimal sieving properties.

What is the proteomics? The term proteomics was first coined in 1995 [82], Proteomics is the study of the proteome, and is the large-scale study of proteins, particularly their structures and functions, in a given type of cell or organism, at a given time, under defined conditions, to make an analogy with genomics, the study of the genome. The proteome is the entire set of proteins, produced or modified by an organism or system. This varies with time and distinct requirements, or stresses, that a cell or organism undergoes. Proteomics is an interdisciplinary domain formed on the basis of the research and development of the Human Genome Project. While proteomics generally refers to the large-scale experimental analysis of proteins, it is often specifically used for protein purification and mass spectrometry.

In the field of proteomics, the ability to detect a large number of proteins in a single analysis represents a key issue to achieve fast and efficient operation [83]. In this context, the combined use of 2-D gel electrophoresis coupled with mass spectrometry has allowed enormous advances during last few decades and has become nowadays one of standard approaches for proteins separation and identification (Figure 3) [84,85].

Among the materials used for 2-D gel electrophoresis, polyacrylamide crosslinked hydrogels have been extensively investigated in the literature [86], because their tunable mesh size porosity appears to be ideal for separating proteins and DNA samples. Typically acrylamide concentrations higher than 5% are used to form the separation matrix, with the acrylamide concentration being selected to maximize resolution of the range of proteins of interest. Lower acrylamide concentrations are necessary when resolution of large high molecular mass (HMM) proteins (>500kDa) is sought, however at the expense of poor mechanical stability of the gel matrix that often yields difficulties in handling these media [74]. Other polymeric systems alternative to polyacrylamide gels have also been proposed as separation matrices for 2-D electrophoresis including agarose, modified polyacrylamide gels and acrylamide-agarose copolymers [87-89]. In particular, the advantages of the acrylamide-agarose system mainly lay in the possibility of improving the resolution of large HMM proteins without compromising the resolution of low molecular mass proteins, partly due to the optimal average pore size of these materials.

Indeed, the electrophoresis migration process through the polymeric gel matrix is driven by the interactions between the protein fragments and the porous network of the gel, causing the quality of protein resolution to be highly dependent on different structural parameters characteristic of the gel matrix [90]. Among these, mean gel pore size, pore size distribution and stiffness of the gel play a crucial role. In order to achieve improved separation performance in 2-D electrophoresis applications, it is therefore essential to understand the specific nature of the pore structure of the gel and the parameters through which this pore structure can be controlled and manipulated [91,92]. In particular, it is of great interest to investigate structure-property relation-ships of these hydrogels in the attempt to optimize their functional performance.

A wide large range of hydrogel chemical compositions was studied and their effect on structural and functional properties of the hydrogel was elucidated.

By employing dynamic rheological tests and creep-recovery tests, a correlation was found between the rheological response of these hydrogels and their sieving properties.

More specifically, the evaluation of the crosslinking density by means of dynamic tests and the use of viscoelastic models for determining the resistance of non-permanent crosslinks to move in the network systems shed light on the pore structure of the hydrogel matrix and helped to clarify its influence on the electrophoretic separation performance. The mechanical stability of the crosslinked hydrogels was also investigated by means of tensile tests and correlated with the crosslinking density of the gel matrix.

Interpenetrating Polymer Networks (IPN)

Interpenetrating polymer networks (IPNs) hydrogels have gained great attention in the last decades, mainly due to their biomedical applications. IPNs are alloys of crosslinked polymers, at least one of them being synthetized and/or cross-linked within the immediate presence of the other, without any covalent bonds between them, which cannot be separated unless chemical bongs are broken [93]. The combination of the polymers must effectively produce an advanced multicomponent polymeric system, with a new profile [94]. According to the chemistry of preparation, IPN hydrogels can be classified in:

i- Simultaneous IPN: When the precursors of both networks are mixed and the two networks are synthesized at the same time by independent, non-interfering routs such as chain and stepwise polymerization [95].

ii- Sequential IPN: Typically performed by swelling of a singlenetwork hydrogel into a solution containing the mixture of monomer, initiator and activator, with or without a crosslinker. If a crosslinker is present, fully IPN result, while in the absence of a crosslinker, a network having linear polymers embedded within the first network is formed (semi-IPN) [96].

When a linear polymer, either synthetic or biopolymer, is entrapped in a matrix, forming thus a semi-IPN hydrogel, fully- IPN can be prepared after that by a selective crosslinking of the linear polymer chains [97,98].

Even if it is still a challenging task, the synthesis of ion imprinted IPN hydrogels constitutes a promising direction in increasing the selectivity of this novel types of sorbents, which is expected to receive much attention in the future.

Tissue Engineering (TE) Applications

Hydrogels are three dimensional polymer scaffolds used in several applications of tissue engineering. A particularly important group of techniques is the so called *in-vivo* tissue regeneration. In this case, a patient's own cells are combined with the polymer, and held *in-vitro* until ready to be implanted. The hydrogel acts as a natural extra-cellular matrix that subsequently promotes cell proliferation and tissue re-growth. The pseudo-extra-cellular matrix, comprised of growth factors, metabolites and other materials, brings cells together and controls tissue structure with the ultimate goal of replacing the natural tissue that was lost or damaged.

When parts of the whole of certain tissues or organs fail, there are several options for treatment, including repair, replacement with a synthetic or natural substitute, or regeneration. The figure below shows how tissue or organ injury, disease or failure has evolved to reach the field of tissue engineering. Tissue repair or replacement with a synthetic substitute is limited to those situations where surgical methods and implants have achieved success. Although implants have been a reasonably successful option, tissue engineering holds out great promise for regeneration of the failed tissue. The first option of the diseased or injured organs is extracorporeal treatment, in which blood is circulated through polymeric membrane exchange devices. These devices are usually passive exchange systems, but more recently experimental systems may contain entrapped or encapsulated cells from other human or animal sources. Those latter systems are called bio-artificial or bio-hybrid organs. Total replacement of the diseased or malfunctioning organ or tissue with a natural substitute requires transplantation of an acceptable, healthy substitute, and there is a limited supply of such organs and tissues. Thus, tissue engineering holds out great promise for regeneration of organs (**Figure 4**). Hydrogels have become increasingly studied as matrices for tissue engineering [85].

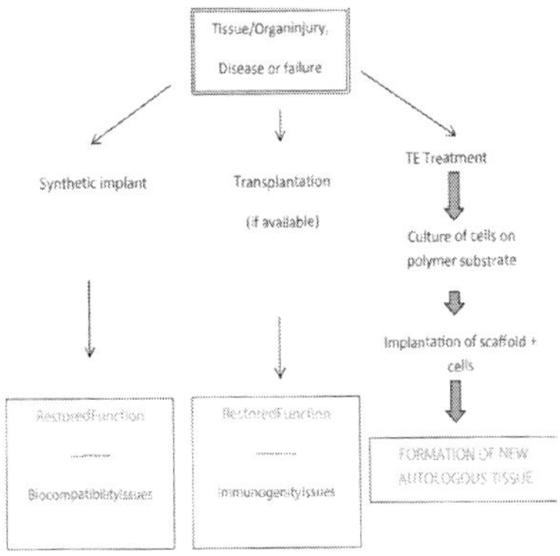

Figure 4: Evolution of various therapeutic methods for treating injured or diseased tissues and organs, to tissue engineering for the repair, regeneration or replacement of such tissues or organs.

Hydrogels designed for use as tissue engineering scaffolds may contain pores large enough to accommodate living cells, or they may be designed to dissolve or degrade away, releasing growth factors and creating pores into which living cells may penetrate and proliferate. **Table 1** lists parameters and properties of hydrogels for these applications.

Table 1: Highlight on important properties and characteristics of hydrogels for tissue engineering.

Type of Hydrogel	**Molecular structures**	**Composition of Hydrogel**	**Important properties**
•Physical •Chemical	•Linearpolymers •Block copolymers •Graftcopolymers •Interpenetrating networks (IPNs) •Polyblends	•Natural polymers and theirderivatives •Syntheticpolymers •Combinations of natural and syntheticpolymers	•Degradability •Injectability •Mechanicalstrength •Ease of handling •Shape and surface/volume ratio (sheets, cylinders, spheres) •Closed/open pores •Water content and character •Chemical modification (e.g. havingattachedcelladhesion ligands) •Addition ofcells and/or drugs •Sterilizability

One significant advantage of hydrogels as tissue engineering matrices vs. more hydrophobic alternatives such as PLGA is the ease with which one may covalently incorporate cell membrane receptor peptide ligands, in order to stimulate adhesion, spreading and growth of cells within the hydrogel matrix. However, a significant disadvantage of hydrogels is their low mechanical strength, posing significant difficulties in handling [99]. Sterilization issues are also very challenging. It is clear that there are both significant advantages and disadvantages to the use of hydrogels in tissue engineering, and the latter will need to be overcome before hydrogels will become practical and useful in this exciting field (**Table 2**).

Table 2: Important advantages and disadvantages of hydrogels as matrices for tissue engineering.

Advantages	Disadvantages
•Aqueous environment can protect cells and fragile drugs (peptides, proteins, oligo nucleotides, DNA) •Good transport of nutrient to cells and products-fromcells •May be easily modified with cell adhesion ligands •Can be injected *in-vivo* as a liquid that gels at body temperature •Usually biocompatible	•Can be hard to handle •Usually mechanically weak •May be difficult to load drugs and cells and then cross link *in-vitro* as a prefabricated matrix •May be difficult to steril-ize

It should be noted that a gel used as a tissue engineering matrix may never be dried, but the total water in the gel is still comprised of bound and free water.

Langer has described tissue engineering as an interdisciplinary field that applies the principles of engineering and the life sciences to the development of biological substitutes that restore, maintain, or improve tissue function [100]. This field is currently advancing rapidly, combining progress in biology and technology, and has raised many hopes in several areas of biology and medicine. First and foremost it has a strong potential as an application in regenerative medicine, for developing lifelike replacement tissues and organs. But it also has an important role to play in fundamental biological research; it allows users to reproduce physiological micro-environments more closely in in-vitro settings than traditional culture methods, offering a way to bridge the gap between *in-vivo* experiments and conventional *in-vitro* studies. On a more operational side, the development of new *in-vitro* models based on human cells has raises the possibility of tackling many of the obstacles that currently hinder pharmaceutical research and drug development. Engineered tissues could reduce the limitations related to the transposition of findings from one organism to another, alleviate ethical problems related to animal testing, increase standardization, and allow more thorough studies of toxicity, metabolism and life cycle of putative drugs before entering clinical testing, this could, in turn, reduce the duration, cost failure rate and risk of clinical trials.

In vivo, the formation of organs and tissues is based on the coordination in time and space of cell differentiation, polarity, shape, division and death. This coordination relies on the cellular integration of signals from microenvironment, mainly consisting of the extracellular matrix (ECM), and intercellular communication [101]. The transduction of a typical combination of these factors coupled to specific cytoplasmic components can induce the three progressive steps in differentiation. First, stem cells are specified towards a certain fate, then they shift from a specified state to a determined state, in which the cell fate cannot be reversed, and finally reach their differentiated state.

The main challenge of tissue engineering is to reconstitute in-vitro an environment that induces the differentiation of cells and their organization in an ordered functional tissue. The cell substrate is of particular importance, since *in-vivo* the extracellular space is occupied by the ECM. The chemical composition of the ECM and the resulting mechanical properties are both important aspects, as the transduction of both chemical and physical signals via cellular adhesion molecules affects cell shape, polarization, migration and differentiation [102]. Furthermore, the ECM topography orients tissue polarity and the morphogenesis of new organs.

Molecular hydrogels hold big potential for cells culture and tissue engineering; peptide-based molecular hydrogels, especially those of long peptides, could provide suitable environments for cell growth, division and differentiation. Zhang group and the Stupp group have demonstrated that peptide-based hydrogels could guide the differentiation of stem cells [103]. The challenge in this field was a method to separate cells from gels post-culture such as using short-peptide-based molecular hydrogels formed by biocompatible methods for 3D cell culture, stem cell controlled differentiation and cell delivery [104]. Responsive molecular hydrogels for the recovery of cells post-culture were also studied [105].

Collagen is the most abundant protein mammals, making up about 30% of the overall body protein content. In order to mimic collagen nanofibers, a serial of short peptides bearing collagen repeating tripeptide of Gly-Xaa-4-Hyp (GXO, X was Lys (K), Glu (E), Ser (S), Ala (A), or Pro (P)) was synthesized [106].

There are two main bottlenecks right now in tissue engineering. The first one is correlated with the possibility of obtaining, *in-vitro*, developed vascular tissue [107,108]. In second instance, once a pseudo-physiological angiogenesis is achieved, the ambitious aim is to create engineered whole organs [109]. First attempts in this direction where made by printing 3D organ-like structures with ECM and cells, or by decellularization and seeding.

Under these circumstances, hydrogels are usually studied as temporary substitutes of the extracellular matrix (ECM) because of comparable physico-chemical properties, such as stiffness and hydrophilicity [51]. Hydrogels used as scaffold in TE must be biocompatible and elicit the smallest response by the body. This is a very complicated problem, indeed, for the culture of eukaryotic cells implies a lot of precautions. Cells morphology, metabolism and overall phenotype are directly correlated with the signals they receive, first of all from the physical and chemicals properties of substrate.

The micro and nanostructure, the porosity and the stiffness of the surfaces are all important signals for the cells (topography and mechanical properties) [109]. A very important phenomenon, for instance, is the *contact guidance*, introduced by Weiss in 1934 [110], for which cells aligns to the shape and the microstructure of a given surface. Especially for some kind of cell culture (e.g. myocytes), this phenomena is crucial for the creation of new working tissue [111,112]. Hydrogels for TE must be developed attempting to fit this necessity.

For the purpose, different materials have been investigated. Actually research is focused mainly on degradable scaffold to allow cell migration while the matrix degenerates. Researches demonstrated the importance of engineered degradation on non-degradable scaffolds by viability test [113,114]. These hydrogels can be divided by the origin of the polymer they are made of: either if it is natural or synthetic. In the present article the attention will be focused on natural derived hydrogels

Dextran: Dextran is natural polysaccharide obtained from the digestion of amylopectin. Low molecular weight dextran is been used like a plasma expander thanks to the relatively inert and nontoxic behavior of this polysaccharide [115]. The most interesting characteristic of dextranareits protein rejection properties, the so called non-fouling [99], coupled with great biocompatibility due to its glycocalyx mimic behavior [100]. This property is useful to create an ECM-like hydrogels for tissue engineering, for instance, Yunxiao and collaborators created a copolymer between methacrylate-aldehyde-bifunctionalized dextran (DEXMA- AD) and gelatin B. This material was obtained by ultraviolet (UV)-crosslinking between a methacrylate groups on Dex-MA-AD and the aldehyde groups, allowing the inclusion of gelatin in the matrix that granting enzymatically degradation and cell adhesive properties. Researchers demonstrate that this kind of hydrogels could promote adhesion of vascular endothelial cells [116,117].

Gelatin: Gelatin is the denatured form of collagen, one of the major component of ECM. Collagen however carries immunogenicity problems due to the presence of antigens from the original tissue. Gelatin is a protein materials with a long α helix with a high content in glycine (≈25%) [118]. Gelatin exist in two different form, processed in acid solution (type A, usually porcine) or in alkali solution (type B, usually bovine) [119]. In its natural form is quickly dissolved in water, but it can be crosslinked to obtain a hydrogel with higher mechanical properties and degradation rate.

The crosslinking can be performed in many different ways, from the physical side we could use UV reticulation or the chain polarity. From the chemical one is very common to polymerize, with enzymes too, the chains with a bound on the side groups of amino acids. Lysine and Glutamic Acid are the most employed for it.

Thanks to his similarity to the natural ECM, many compounds were been developed with gelatin-coating or included element. Chitosan-gelatin, fibroin-gelatin, alginate-gelatin, dextrangelatin are very common. Das and collaborators attempted to create a fibroin-gelatin biomaterial for bioprinting cells-laden in a 3D tissue constructs. They developed two kinds of hydrogels. One crosslinked by sonication, the other crosslinked using tyrosinase enzyme. The results showed that sonication-gelatinfibroin hydrogel shown better osteogenic differentiation while tyrosinased-gelatin-fibroin supported better chondrogenic and adipogenic differentiation [120].

Chitosan: Chitosan is a polysaccharides from chitin of the crustacean skeleton. It is composed by the repetition of N-glucosamine units. A crucial index to assess chitosan's properties is the degree of acetylation, defined as the number of amine in the material. For instance, it is proved that chitosan could decrease the adsorption of protein and the binding of bacteria [121]. Actually it is yet not clear how this materials could been attached by cells. Some studies report a first period in which chitosan repelled cell adhesion, and a second in which cells starts to bind to it. There is no a uniformity in results about this refractory-period, anyhow it can be exploited to seed different kind of cells. For example, Tao Jang et al. published a study in which they used photo polymerization on chitosan in order to deposit cells between the chitosan patterns. They suggest that after the refractory-period, while cells are disposing in nochitosan coated regions, it is possible to seeds another type of cells for the development of a more complex system [122].

In another study chitosan has been coupled with gelatin to create an in-situ gel for cell seeding and/or drug delivery. In particular the group evaluated the difference in crosslinking by two different enzyme [123].

Hyaluronic acid: Hyaluronic Acid – HA is a Glycosaminoglycan GAG enclosed in the natural ECM, core of the material is a polysaccharide with high affinity for water. Usually, to increase the mechanical properties of this biomaterial, a covalent crosslinking between chains is done. Tezel& Fredrickson have reported that a too high degree of modification and cross-linking could influence the biocompatibility property of the material [124].

In vivo HA can have different molecular weights. Low and high molecular weight HA cause an opposite cells behavior [125]. HA macromolecules shown an anti-inflammatory, immunosuppressive properties and blocks angiogenesis, while cleaved small fragments induce the opposite behavior, enabling endothelial cells migration and angiogenesis [126,127]. Indeed, low molecular weight HA have been correlated with some cancers, like prostatic one [125,128]

HA hydrogels were obtained in order to exploit the angiogenic power of the molecule during the degradation of a material. Kisiel's group, for example, crosslinked HA with proteasedegradable peptides and added cell adhesion ligands to improve the cell spreading on the material [129].

On another hand, Shu et Al. tried to mimic the ECM copolymerizing HA and gelatin. They added a thiol group to the Hyaluronan, in this way they were able to crosslinked modified-Hyaluronan with modified-Gelatin by disulfide bonds [130].

Pectin: Another polysaccharides used in tissue engineering hydrogels is pectin. It is obtained from cells walls after a low pH, high temperature processing. Unfortunately, until now, researchers have not reached the goal to standardize this product in an economically sustainable way [131].

Based on its esterification degree pectin is classified from low methoxyl to high methoxyl. Tuning this property changes the mechanical behavior of the material. Reticulation can occur by lowering pH to obtain physical gel, or using divalent or trivalent ions to obtain water-insoluble gel [35,132].

In the tissue engineering field, pectin seems very interesting because of its promotion of nucleation of mineral phase when immersed in a specific biological solution [131,133].

Alginate: Derived from brown algae, alginate is a polysaccharide composed of beta-D-mannuronic acid and alfa-L-gluronic acid. Its reticulation can alsooccurby divalent cations (Ca^{2+}, Fe^{2+}, Ba^{2+}) [134].

In tissue engineering alginate can be used as an immunoisolation barrier [117]. Alginate scaffold for the regeneration of annulus fibrosus are also been developed. This hydrogels have shapememory capability, are cytocompatible and supports proliferation and metabolic activity [135]. Moreover, alginate hydrogels were used to reduce liver cells death [136] and with silk-fibroin as a substrate for stem cells culture [137].

Culture of organs-on-chips

Recent trends towards the development of *in-vitro* multicellular systems with definite architectures, or "organs on chips" are studied. First, the chemical composition and mechanical properties of the scaffold have to be consistent with the anatomical environment *in-vivo.* In this perspective, the flourishing interest in hydrogels as cellular substrates has highlighted the main parameters directing cell differentiation that need to be recapitulated in artificial matrix. Another scaffold requirement is to act as a template to guide tissue morphogenesis. Therefore specific micro-fabrication techniques are required to spatially pattern the environment at microscale. 2D patterning is particularly efficient for organizing planar polarized cell types such as endothelial cells or neurons. However, most organs are characterized by specific sub units organized in three dimensions at the cellular level. The production of such 3D patterns invitro is necessary for cells to fully differentiate, assemble and coordinate to form a coherent micro-tissue. These physiological microstructures are often integrated in micro-fluidic devices whose controlled environments provide the cell culture with more like-life conditions that traditional cell culture method. Such systems have a wide range of applications, for fundamental research, as tools to accelerate drug development and testing, and finally, for regenerative medicine.

The *in-vivo* extra-cellular matrix (ECM) consists mainly of collagen fibers, elastin fibers, glycoproteins and polysaccharides. It acts as a mechanical support to the cells it surrounds and plays an important role in cell shape, cell polarity, cell migration, resistance to external forces, and signal transduction. The primary focus of tissue engineering is to achieve an in-vivo-like cellular environment, by developing *in-vitro* artificial

ECM. In regenerative medicine applications, such scaffolds could either be directly introduced into an injured organ in order to induce *in-vivo* genesis, or first seeded with cells *in-vitro* and then transplanted once those cells have differentiated. For *in-vitro* research and testing applications, they are generally directly used as a substrate. In all cases, the artificial structures need to recapitulate the *in-vivo* environment signals responsible for cell differentiation into the desired tissue.

Biomaterial scaffolds are essentially made of hydrogels. Based on their ability to retain water by swelling, they mimic the high water content of the extracellular matrix [138]. Cell adhesion ligands must be present in hydrogels to allow cells to adhere, spread, migrate and proliferate. There is a large variety of adhesion molecules, such as laminin and its derivatives [138], fibronectin [139] and collagen [140]. It is therefore crucial to select the adhesion molecules for which the seeded cell type has the largest affinity to make adhesion effective. Natural hydrogels are bioactive and usually provide native adhesion sites. Conversely, synthetic hydrogels are inert, since their carbon skeleton presents no adhesion molecules or endogenous factors inducing proliferation and cell differentiation.

To enrich their potential as bioactive materials, synthetic hydrogels are generally supplemented with adhesion molecules [141], either by covalent grafting, adsorption or electrostatic interaction. Adhesion molecules can be grafted after hydrogel polymerization, or added to the pre-polymerized mixture and either physically trapped or chemically incorporated during polymerization. Finally, in the case of photo-activated materials such as PEGDA, adhesion molecules can be chemically modified to covalently attach to the hydrogel backbone. The grafting of PEG polymer has been thoroughly described in the review of Zhu et al. [142].

Culture of cells is very demanding area. It deals with living cells that are highly sensitive to their surrounding environment. It also depends strongly on differentiation, a delicate, complex and still only partly understood phenomenon. It thus requires the development of increasingly sophisticated technologies. Although the use of hydrogels as physiological substrates to replace the *in-vivo* ECM is globally accepted, all parameters must be carefully manipulated to avoid additional unwanted signals. The search for optimal artificial matrices has led to the invention of dynamic hydrogels that can be tuned to reproduce the *in-vivo* micro-environment, which is permanently renewed and crossed by various signals.

Despite these major improvements, challenges remain. Until now, a physiological substrate alone is, in most cases, insufficient to induce cells to assemble into a multicellular entity and coordinate their activity to form a functional tissue. Additional hydrogel structuring at micrometer scale is necessary. In many cases, a unique association of physiological substrate coupled to a 3D structure replicating the physical constraints perceived by the cells *in-vivo* will be needed to induce *in-vitro* the differentiation of cells into a functional tissue.

Nonetheless, the vast majority of these ?organs on chips ?are mostly static, while *in-vivo* cells are often subjected to mechanical deformations associated with the organ's function. Furthermore, the biochemical environment can vary in time depending on the physiology of the entire organism. These physical and chemical stimuli generally affect the differentiation and self-organization of cells into functional tissues and are thus crucial to mimic. Tissue engineering therefore needs to move beyond 3D culture to achieve "4 dimensional structures" that encompass the temporal dimension of organ function as well as it spatial ones. Combining the 3D micro-structured tissues and the dynamic stimuli appears to be the next to reach in micro-organ engineering.

Bone Regeneration

This work gives a general view on the use of hydrogels as scaffold for bone regeneration, with special emphasis on injectable systems, membranes for guided bone regeneration, bio functionalization and biomimetic mineralization.

Alginate hydrogel has been shown to be a useful tool for producing bone and cartilage tissues [142]. It is very biocompatible in humans and peptides can also be covalently coupled to the molecules. Alginate is a naturally occurring polymer found in kelp (seaweed) that is commonly used in gel formation.

The regeneration of large bone defects caused by trauma or disease remains a significant clinical problem. Although osteoinductive growth factors such as bone morphogenetic proteins have entered clinics, transplantation or autologous bone remains the gold standard to treat bone defects. The effective treatment of bone defects by protein therapeutics in humans requires quantities that exceed the physiological doses by several orders of magnitude. This not only results in very high treatment costs but also bears considerable risks for adverse side effects. There issues have motivated the development of biomaterials technologies allowing to better control bimolecular delivery from the solid phase. Importantly, a relatively narrow therapeutic window of growth factor dose is expected to lead to regeneration, whereas too low or too high growth factor doses will most likely lead to severe side effects, as shown in animal models [143]. Since for the generation of constant growth factor levels, multiple injections or even continuous infusion of high doses would be necessary, alternative routes involving biologically inspired growth factor delivery are sought. In an ideal situation, the localized delivery of growth factors should present physiologically relevant doses and preserve its activity for prolonged periods of time. Moreover, carrier materials should facilitate the communication with cells of the host so as to be actively involved in the process of regeneration. Synthetic hydrogels, highly swollen three-dimensional (3D) networks of macromolecules, have emerged as powerful candidate biomaterials to fulfill these requirements [144]. Hydrogels for growth factor delivery applications have either been generated from natural derived biomolecules such as alginate, collagen, and fibrin or from synthetic polymers employing chemical or physical crosslinking reactions [145].

For growth factor administration, the growth factor can either be freely embedded in the hydrogel or bound to it. In the former approach growth factor release is driven by passive diffusion or coupled to material degradation. Release kinetics can be varied by altering material degradation rate or by changing growth factor quantity. To allow covalent tethering to synthetic or biologically derived hydrogels, GF were chemically modified or genetically engineered to contain functional groups such as thiols, acrylates, azides. An example of growth factor delivery from fibrin hydrogels to regenerate bone tissue was reported by Hubbell and co-workers [146].

Recently, a fibrin scaffold possessing integrin-binding sites adjacent to growth factor-binding sites to obtain synergistic effects was used to regenerate bone [147]. For this purpose a trifunctional peptide was created consisting of an N-terminal Gln sequence for covalent incorporation into fibrin matrices, the major integrin binding domain of fibronectin (FN III9-10) and at the C-terminus another domain of fibronectin (FN III12-14) to bind various growth factors. Since FN III12-14 binds growth factors from different families it allows for straightforward delivery of multiple growth factors. By delivering BMP-2 and PDGF-BB, two growth factors known to induce bone formation and previously shown to bind to FN III12-14, the authors tested for improved growth factor efficiency in ectopic positions in nude mice and calvarial critical size defects in rats. By using growth factor concentrations that didn't show any *in-vivo* effects when delivered in empty fibrin matrices, they were able to demonstrate that the FN III9-10/12- 14 functionalized matrices enhanced growth factor-induced bone formation and recruitment of bone forming progenitor cells. Due to the enhanced growth factor signaling, concentrations could be dramatically reduced and bone tissue deposition was observed with much lower doses than elsewhere reported, showing promise for the use of such implants in clinical applications.

In another study a novel photo-cross-linkable chitosan–lactidefibrinogen (CLF) hydrogel was developed and characterized and they also evaluate the efficacy of bone morphogenetic protein-2 (BMP-2) containing a CLF hydrogel for osteogenesis *in-vitro* and *in-vivo.* Radiography, microcomputed tomography and histology confirmed that the BMP-2 containing CLF hydrogels prompted neo-osteogenesis and accelerated healing of the defects in a dose-dependent manner. Thus the CLF hydrogel is a promising delivery system of growth factors for bone regeneration [148].

A photo-cured hyaluronic acid (HA) hydrogels were designed and prepared containing an osteogenesis-inducting growth factor, GDF-5 (growth and differentiation factors 5). Those hydrogels were prepared and confirmed to have controlled GDF-5 release profiles. Cytotoxicity and cell viability suggest that GDF-5loaded HA hydrogel has proper biocompatibility for use as a scaffold which can induce osteogenesis. Moreover, the HA hydrogel showed improved osteogenesis in both *in-vitro* tests, the results from these tests show that the HA hydrogel can be used as a scaffold for bone tissue regeneration [149]. Gelatin hydrogels were also used for bone regeneration at both the critical-sized bone defect.

Cardiac applications

In the last decade, advancements have been made towards developing injectable hydrogels for the purpose of cardiac repair. Hydrogel injections alone have been shown to attenuate the decline in cardiac function and left ventricular remodeling typically seen after myocardial infarction in both large and small animal models. Furthermore, hydrogels have also been shown to improve cell retention when co-injected for cellular cardiomyoplasty and to prolong release of therapeutics when used as a delivery vehicle. This chapter will review the basics of hydrogel properties and discuss recent studies of hydrogels alone and in conjunction with cells or therapeutics for cardiac repair.

Dental applications

Pulp regeneration therapy is important to overcome the limitations of conventional therapy to induce reparative dentinogenesis. Presently, dentists have no choice but to remove the whole dental pulp with an endodontic procedure when a dentin defect with pulp exposure reaches a critical size resulting in an irreversible pulp condition. To overcome this limitation, it is considered important to develop pulp regeneration therapy as well as clarify the mechanisms of pulp wound healing. Pulp wound healing and regeneration have common processes, and results of a number of studies have indicated that pulp wound healing consists of initial inductions of apoptosis of damaged pulp cells [134],followed by reactionary dentinogenesis by surviving odontoblasts and reparative dentinogenesis by odontoblast-like cells [150,151]. Reactionary or reparative dentin is formed toward the residual dental pulp, however, not in the area in which the dentin-pulp complex has been lost. To achieve the regeneration of the dentinpulp complex, induction of appropriate pulp wound healing and formation of new dentin in dentin defects are essential, and a few studies have reported vital pulp therapies to form new dentin in defects [152].

Fibroblast growth factor-2 (FGF-2), which is normally stored in the extracellular matrix and released by enzymatic degradation of extracellular matrix molecules, plays a role in physiologic conditions such as enamel and dentin formation of the tooth germ [153], as well as pathologic conditions [154]. It was previously demonstrated that a gradual and continual release of biologically active FGF-2 was achieved by *in-vivo* biodegradation of gelatin hydrogels that incorporated FGF-2 [155]. Furthermore, a controlled release of FGF-2 from gelatin hydrogels induced neovascularization and regeneration of several tissues, including bone [156], periodontal tissues [157], and others [158].

Wound healing applications

Wound healing is the promise of a new way to heal damaged skin tissue with high biocompatible and bioactive materials. Skin burned, diabetic ulcer, are problems that at the state of the art are very expensive to treat. Prosthetic-tissue engineered skin are been made, unfortunately they are not ready-to-use; they are expensive and have many needs that are not always matched by patients. Theoretically, in wound healing applications a crucial parameter to assess is the wound contraction that can be evaluated in this way, remembering that A_0 is the original burn wound area, and A_t is the burn wound area at the time of biopsy:

$$Wound\ Contraction[\%] = \frac{A_0 - A_t}{A_0} \cdot 100$$

Many systems has been studied, with or without chemicals to aid the skin regeneration. Hyaluronic-acid and gelatin are both two promising materials for the aim because of their natural presence inside human ECM of the skin tissues. Moreover, in literature can be found healing systems made from cellulose [159], alginatechitosan copolymers [160], chitosan-gelatin-honey copolymers [161] and new biphasic gelatin-silk [162].

Most of the products already on the market use a combination of selected materials and proper seeding of cells from various origins (allogenic or autogenic), for instance we cite to applications of HYAFF™ esterified hyaluronic acid [163], both produced by FIDIA Ltd.: LaserskinAutograft®, made of a HA-membrane with keratinocytes, and Hyalo-graft 3D®, made with HA, but with fibroblasts added.

Drug delivery applications

By drug delivery we intend the whole ensemble of procedures, devices and techniques to avoid the problems of standard delivery of medicals, first of all the burst release and quick decay of the effects of the drug overtime, especially for short half-life pharmaceuticals [164].

Hydrogels, and smart hydrogels in particular, can be a very interesting solution in reaching a sustained and targeted release of pharmaceuticals, both increasing the effect of the drug itself and lowering side effects at the same time [165].

Silk Fibroin, protein-based biomaterial with high Young Modulus (1-10GPa) obtained processing the silk from various insects, like spider or silk worm (Bombyx Mori).

The silk fiber is composed by fibroin and sericine. This second protein is correlated with inflammatory reactions when coupled with fibroin. To obtain a pure fibroin wire, silk is boiled in alkaline solution at 120° C for at least 1 hour. Solubilization and regeneration is necessary for use this biomaterial for tissue engineering, like scaffolding. Usually the fibroin is processed in a solution of LiBr [166,167], in a ternary solvent composed by Calcium Chloride / Ethanol / Water to break fiber to fiber bond [168] or by SDS [168,169]. This allows the casting of the polymer and the consequent creation of different structure for geometrical complexity.

In 2011 Chinmoy et Al. used silk fibroin from Bombyx Mori and from Natherea Mylitta to create a 3D scaffold for cardiac tissue engineering. Results showed that metabolic cell response, cardiomyocytes growth and the number of junction between cells for Antheraea Mylitta fibroin are very similar to fibronectin ones. They concluded that silk fibroin from AM enables an efficient attachment of cells without affecting their response to extracellular stimuli.

Hydrogels as soft contact lenses

Contact lenses are one of the reasons why hydrogels where developed in the first place. They come with many different features and procedures of usage. In particular, soft contact lenses with a short turnover are the ones for which hydrogel are mostly useful. The most used polymer exploited to produce soft-contact lenses is silicone, under hydrogel form, thanks to its permeability for oxygen.

Despite the high microbial keratitis risk for a prolonged wear time, which subsists for most of ophthalmic applications in this field, extended wear time of silicone hydrogel lenses are five times safer than other conventional hydrogels for what concern infections [170,171]. Only about 5% of drugs administrated by eye drops are bioavailable, and currently eye drops account for more than 90% of all ophthalmic formulations. The bioavailability of ophthalmic drugs can be improved by a soft contact lens-based ophthalmic drug delivery system. Several polymeric hydrogels have been investigated for soft contact lens-based ophthalmic drug delivery systems: (i) Polymeric hydrogels for conventional contact lens to absorb and release ophthalmic drugs; (ii) Polymeric hydrogels for piggyback contact lens combining with a drug plate or drug or drug solution; (iii) Surface-modified polymeric hydrogels to immobilize drugs on the surface of contact lenses; (iv)Polymeric hydrogels for inclusion of drugs in a colloidal structure dispersed in the lens; (v) Ion ligand-containing polymeric hydrogels; (vi) molecularly imprinted polymeric hydrogels which provide the contact lens with a high affinity and selectivity for a given drug.

The success of the therapy of eye ailments with ophthalmic drugs strongly depends on achieving sufficient drug concentration on the cornea for a sufficient period of time, but the typical delivery of drugs by eye drops which currently account for more than 90% of all ophthalmic formulations is very inefficient and in some instances leads to serious side effects [172]. Upon instillation in the eye, the drug mixes with the fluid present in the tear film and has a short residence time of approximately 2 minutes in the film. Only about 5% of the drug is absorbed into the cornea and the remaining either gets absorbed in the conjunctiva or flows through the upper and the lower canaliculi into the lacrimal sac [173]. The drug containing tear fluid is carried from the lacrimal sac into the nasolacrimal duct. The nasolacrimal duct empties into the nasal cavity, where the drug is then absorbed into the bloodstream. This absorption leads to drug wastage, and more importantly, the presence of certain drugs in the bloodstream leads to undesirable side effects. Furthermore, the application of ophthalmic drugs by eye drops results in a rapid variation in the drug delivery rates to the cornea and this limits the efficacy of the therapeutic system [174]. In addition, dosage through eye drops is inconsistent and difficult to regulate, as most of the drug is released in an initial burst of concentration [175]. To increase the residence time of the drug in the eye, thereby reducing wastage and minimizing side effects, a number of researchers have proposed using contact lenses for ophthalmic drug delivery. It has been shown that in the presence of a lens, ophthalmic drugs have a much longer residence time in the post-lens tear film, compared with 2 minutes in the case of topical application as eye drops [176]. The longer residence time will result in higher drug flux through the cornea and reduce the drug inflow into the nasolacrimal sac, thus reducing drug absorption into the bloodstream. Several methods have been proposed to make ophthalmic drugs deliverable by soft contact lens.

The silicone hydrogel lenses can be designed to release ophthalmic drugs for an extended period varying from 10 days to a few months. The transport of timolol and DX in the silicone gels is diffusion limited, but the release profiles are complex particularly for dexamethasone which is the evidence of complex microstructure of the gels. The silicone hydrogels may also be suitable for other drug delivery applications such as puncta plugs, ophthacoils, retinal implants, transdermal patches, wound healing patches, etc.

Injectable hydrogels

The desire and need to minimize traditional open surgeries and in light of the challenges along with the traditional intravenous administration of chemotherapeutics for example [177], injectable hydrogel-drug system emerges as a powerful tool for noninvasive and in-situ controlled-release of drugs. Such application could also reduce the healthcare expenses and improve the recovery time for the patients. Minimal invasive procedures using endoscopes, catheters and needles have been developed considerably in the last few decades. In the field of tissue engineering and regenerative medicine, there is a need for advancement over the conventional scaffolds and pre-formed hydrogels. In this scenario, injectable hydrogels have gained wider appreciation among the researches, as they can be used in minimally invasive surgical procedures. Injectable gels with their ease of handling, complete filling the defect area and good permeability have emerged as promising biomaterials (**Figure 5**).

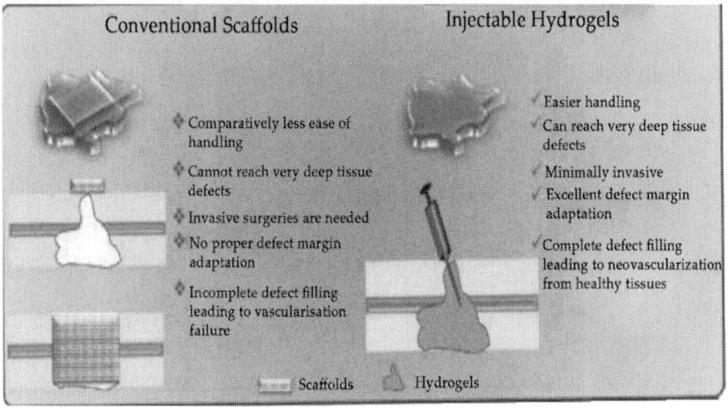

Figure 5: Advantages of injectable hydrogels over conventional scaffolds [179].

Basing on their origin, two kinds of biodegradable polymers are used for preparation of injectable hydrogels: naturally derived and synthetic polymers. Naturally derived polymers such as polysaccharides (cellulose, chitosan, hyaluronic acid, chondroitin sulfate, dextran sulfate, alginate, pectin, etc.), proteins (collagen, gelatin, heparin, fibrin, etc.) and synthetic polymers like polypeptides, polyesters, polyphosphazenes, etc. are widely used [178,179].

In comparison with synthetic polymers, most but not all naturally derived polymers are expected to have better interaction with cells along with increased cell proliferation and differentiation [180]. On the other hand, synthetic polymers possess tunable mechanical properties and degradation profile.

Mutually, exclusive advantages of these polymers have motivated the researchers to investigate combinatorial systems of synthetic and naturally derived polymers, thereby improvising the properties of injectable hydrogels [181]. Injectable hydrogels are prepared using various physical and chemical crosslinking methods as shown in **Figure 6** [182].

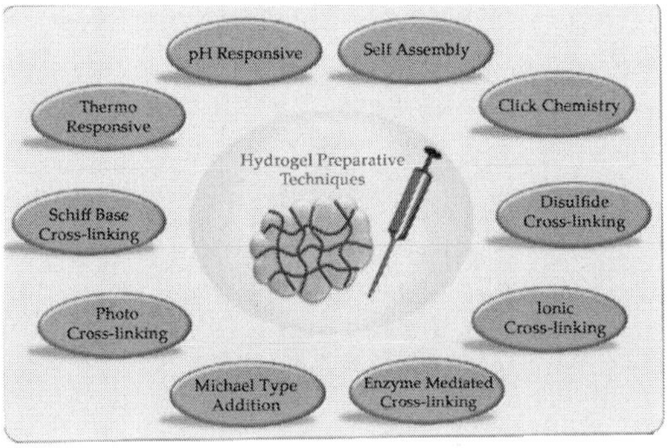

Figure 6: Schematic showing different techniques to prepare hydrogels.

Particle migration, poor shape definition and/or rapid resorption limit the success of current urethral bulking agents. The use of injectable agents to augment urethral tissue function is now common practice because they facilitate minimally invasive delivery of the bulking agent and provide the advantage of low cost and low patient morbidity. The procedure can be performed in an outpatient setting with the patient under local anesthesia. Current therapies using cross-linked collagen, autologous fat or carbon coated beads have provided acceptable cure rates [183]. However, complications or limitations of the treatments include the need for repeat injections, allergic reactions, material resorption and particle migration. Several newly applied materials, including calcium hydroxyapatite and materials consisting of natural or synthetic polymers appear to be efficacious as current therapies but extensive long-term data are still unavailable [184].

Recent clinical trials as well as previous animal studies demonstrated that transplantation of autologous chondrocytes suspended in a polymeric alginate hydrogel produced encouraging results and suggested that injectable alginate hydrogels may have potential use as bulking agents [185]. Covalently cross-linked alginate hydrogels provide a versatile system with potential as an injectable bulking agent. The shape memory characteristics of these materials facilitated the minimally invasive delivery of a bulking construct in a predefined three-dimensional macroporous form, which minimized migration or leakage of the material from the puncture site and ensured integration with host tissue. The hydrogels are easy to prepare and can be stored dehydrated at room temperature, required no specialized injection equipment and can be delivered by a single injection. Such hydrogels have the potential to fulfill many requirements for an optimal urethral or other soft tissue bulking agent.

Injectable hyaluronicacid MRI provides the possibility to monitor the hydrogel and shape of the materials correctly by only H-MRI. There are many very promising studies involving hydrogels and imaging techniques. In one of them, by Yang X. and collaborators, an injectable hyaluronic acid hydrogel has been modified with fluorine-19 and subsequently tested. The authors claim its high potential in many clinical applications since, besides it being biocompatible and easily injectable, it can moreover be tracked with minimal invasive manners [186].

Applications in electronics

The use of hydrogels as matrixes in electronics is very promising considering the high tunability and precision of capacitors with hydrogel dielectrics. By changing the polymer and the solution carried it is possible to build high-performant cheap capacitors.

These Hydrogel electrolytes can be composed by organic polymers. According to Choudhury et al. [187], poly(ethylene oxide), potassium poly(acrylate), poly(vinyl alcohol) and gelatin are among the most promising materials for the purpose. However, organic polymer seem to show lower properties compared to the ones of inorganic hydrogels. Many authors affirm the more interesting characteristic of silica hydrogel compared to standard polymer hydrogels, nevertheless, opinions are still divided on the subject.

CONCLUSIONS

In this review we provide a historical overview of the developments in hydrogel research from simple networks to smart materials. Hydrogels are widely present in everyday products though their potential has not been fully explored yet. These materials already have a well-established role in contact lenses, hygiene products and wound dressing markets but commercial hydrogel products in tissue engineering and drug delivery are still limited. Many hydrogel-based drug delivery devices and scaffold shave been designed, studied and in some cases even patented, however not many have reached the market. More progress is expected in these two areas. Limited commercial products with hydrogels in drug delivery and tissue engineering are related to some extent to their high production costs.

There are numerous patents on hydrogels but only a few reached the market, hydrogels are used for producing contact lenses, hygiene products and wound dressings. Other commercial uses of hydrogels are in drug delivery and tissue engineering.

More developments are expected in drug delivery and tissue engineering, high production costs of hydrogels are limiting their further commercialization.

Over the past decades, significant progress has been made in the field of hydrogels as functional biomaterials. Biomedical application of hydrogels was initially hindered by the toxicity of crosslinking agents and limitations of hydrogel formation under physiological conditions. Emerging knowledge in polymer chemistry and increased understanding of biological processes resulted in the design of versatile materials and minimally invasive therapies. Hydrogel matrices comprise a wide range of natural and synthetic polymers held together by a variety of physical or chemical crosslinks. With their capacity to embed pharmaceutical agents in their hydrophilic crosslinked network, hydrogels form promising materials for controlled drug release and tissue engineering. Despite all their beneficial properties, there are still several challenges to overcome for clinical translation.

ACKNOWLEDGMENTS

The authors gratefully acknowledge MITACS program of Québec and Sarko Holdings for financial support of this work.

REFERENCES

1. Kopecek J (2002) Polymer chemistry: swell gels.Nature 417: 388-389, 391.
2. ISSN International Centre: The ISSN register. https://www.issn.org (2006). Accessed 20 Feb 2007.
3. Van Noorden R (2014) Global scientific output doubles every nine years. Nature News Blog.
4. Campoccia D, Doherty P, Radice M, Brun P, Abatangelo G, et al. (1998) Semisynthetic resorbable materials from hyaluronan esterification. Biomaterials 19: 2101-2127.
5. Wei Yang Seow, Charlotte A.E. Hauser(2014)Short to ultrashort peptide hydrogels for biomedical uses. Materials Today 17: 381-388.
6. Lee SC, Kwon IK, Park K (2013) Hydrogels for delivery of bioactive agents: a historical perspective.Adv Drug Deliv Rev 65: 17-20.
7. Kopecek J (2007) Hydrogel biomaterials: a smart future? Biomaterials 28: 5185-5192.
8. Wichterle O, Lìm D (1960) Hydrophilic Gels for Biological Use. Nature 185: 117-118.
9. Buwalda SJ, Boere KW, Dijkstra PJ, Feijen J, Vermonden T1, et al. (2014) Hydrogels in a historical perspective: from simple networks to smart materials. J Control Release 190: 254-273.
10. Yom-Tov O, Neufeld L, Seliktar D, Bianco-Peled H (2014) A novel design of injectable porous hydrogels with in situ pore formation. ActaBiomater 10: 4236-4246.
11. Abebe DG, Fujiwara T (2012) Controlled thermoresponsive hydrogels by stereocomplexed PLA-PEG-PLA prepared via hybrid micelles of pre-mixed copolymers with different PEG lengths. Biomacromolecules 13: 1828-1836.
12. Chung HJ, Lee Y, Park TG (2008) Thermo-sensitive and biodegradable hydrogels based on stereocomplexedPluronic multi-block copolymers for controlled protein delivery. J Control Release 127: 22-30.
13. Kirakci K, Å Ãcha V, Holub J, KubÃ P, Lang K (2014) Luminescent hydrogel particles prepared by self-assembly of Î²-cyclodextrin polymer and octahedral molybdenum cluster complexes.InorgChem 53: 13012-13018.
14. Kono H, Teshirogi T (2015) Cyclodextrin-grafted chitosan hydrogels

for controlled drug delivery. Int J BiolMacromol 72: 299-308.

15. Peppas NA(1987) Hydrogels in Medicine and Pharmacy. CRC Press 21: 184
16. Park K, Shalaby WSW, Park H (1993). Biodegradable Hydrogels for Drug Delivery. Technomic.
17. RS. Harland, RK. Prud'homme (1992) Polyelectrolyte Gels: Properties, Preparation and Application. American Chemical Society.
18. Lim F, Sun AM (1980) Microencapsulated islets as bioartificial endocrine pancreas.Science 210: 908-910.
19. Yannas IV, Lee E, Orgill DP, Skrabut EM, Murphy GF (1989) Synthesis and characterization of a model extracellular matrix that induces partial regeneration of adult mammalian skin.ProcNatlAcadSci U S A 86: 933-937.
20. Sefton MV, May MH, Lahooti S, Babensee JE (2000) Making microencapsulation work: conformal coating, immobilization gels and in vivo performance.J Control Release 65: 173-186.
21. Griffith LG(2000) Polymeric biomaterials. Acta Mater 48: 263-277.
22. O Okay (2010) General Properties of Hydrogels. In: Gerlach G, Arndt KF. Hydrogel Sensors and Actuators, Engineering and Technology. Springer Series on Chemical Sensors and Biosensors. editor: Springer
23. Byju AG, Kulkarni A, Gundiah N (2013) Mechanics of Gelatin and Elastin based hydrogels as Tissue Engineered Constructs. International Conference on Fracture
24. MondalM, Trivedy K, Nirmal Kumar S (2007)The silk proteins, serin and fibroin in silkworm Bombyxmori. Caspian J EnvSci5: 63-76.
25. Menard KP (2008) Dynamic Mechanical Analysis: A Practical Introduction. 2nd ed. CRC Press
26. Morrison FA. Understandin Rheology. Oxford University Press.
27. Dong LC, Hoffman AS, Yan Q (1994) Dextran permeation through poly(N-isopropylacrylamide) hydrogels.J BiomaterSciPolym Ed 5: 473-484.
28. ASTM International. F2540. Standard Practice Standard Guide for Assessing Microstructure of Polymeric Scaffolds for Use in Tissue Engineered Medical Products. 2004
29. ASTM International. F2603 standard practice Standard Guide for Interpreting Images of Polymeric Tissue Scaffolds. 2006

30. Gulrez SKH, Al-Assaf S, Phillips GO (2011) Hydrogels: Methods of Preparation, Characterisation and Applications, Progress in Molecular and Environmental Bioengineering. In: Carpi A. Analysis and Modeling to Technology Applications
31. Hoffman AS (2001) Advanced Drug Delivery Reviews 3: 3-12
32. Bian L, Hou C, Tous E, Rai R, Mauck RL, et al. (2013) The influence of hyaluronic acid hydrogel crosslinking density and macromolecular diffusivity on human MSC chondrogenesis and hypertrophy. Biomaterials 34: 413-421.
33. Sung HW, Huang DM, Chang WH, Huang RN, Hsu JC (1999) Evaluation of gelatin hydrogel crosslinked with various crosslinking agents as bioadhesives: in vitro study.J Biomed Mater Res 46: 520-530.
34. Weber LM, Lopez CG, Anseth KS (2009) Effects of PEG hydrogel crosslinking density on protein diffusion and encapsulated islet survival and function.J Biomed Mater Res A 90: 720-729.
35. Wellner N, Kacuráková M, Malovi´ková A, Wilson RH, Belton PS (1998) FT-IR study of pectate and pectinate gels formed by divalent cations. Carbohydr Res308: 123–131.
36. Tanaka T (1986) In: Encylopedia of Polymer Science and Engineering. 2nd ed. Vol 6: 514-536. Wiley.
37. Brandt KA, Goldman SA, Inglin TA. US Patent RE32649 E. Hydrogel-forming polymer compositions for use in absorbent structures. 1988
38. Assarsson PG, King PA, Yen SN (1975) US Patent 3901236 A. Disposable absorbent articles containing hydrogel composites having improved fluid absorption efficiencies and processes for preparation
39. Gross JR, Mcfadden RT (1975) US Patent 3926891 A Method for making a crosslinkable aqueous solution which is useful to form soft, water-swellablepolyacrylate articles
40. Statista, The Statistics Portal. https://www.statista.com. Accessed Feb 2015.
41. Umachitra G, Bhaarathidhurai (2012) Disposable baby diaper--a threat to the health and environment.J Environ SciEng 54: 447-452.
42. Chalker-Scott L (2015)The Myth of Polyacrylamide Hydrogels
43. Trinh T, Gardlik JM (1990) European Patent EP 0392608 A2. Solid consumer product compositions containing small particle cyclodextrin complexes

44. Subkowski T, Bollschweiler C, Wittenberg J, Siegel W, Pelzer R (2013) International Patent WO 1999004830 A1. Low molecular weight modulators of the cold-menthol receptor trpm8 and use thereof.
45. Peppas NA, Peppas LB (1997) Controlled release of fragrances from polymers. J ApplPolym Sci. 66: 509–513.
46. NFPA Guidelines & Definitions. Oklahoma State University (2008). Accessed Feb 2015.
47. An SM1, Ham H, Choi EJ, Shin MK, et al. (2014) Primary irritation index and safety zone of cosmetics: retrospective analysis of skin patch tests in 7440 Korean women during 12 years.Int J CosmetSci 36: 62-67.
48. https://phoenix-chem.com/ESW/Files/Pecogel_Brochure.pdf
49. Lorenz DH (1994) US patent 5306504 A Skin adhesive hydrogel, its preparation and uses
50. E. Lee and B. Kim (2011) Smart delivery system for cosmetic ingredients using pH-sensitive polymer hydrogel particles. Korean J ChemEng 28: 1347-1350
51. Slaughter BV, Khurshid SS, Fisher OZ, Khademhosseini A, Peppas NA (2009) Hydrogels in regenerative medicine.Adv Mater 21: 3307-3329.
52. Rhee SM, You HJ, Han SK (2014) Injectable tissue-engineered soft tissue for tissue augmentation.J Korean Med Sci 29 Suppl 3: S170-175.
53. Pienaar WE, McWilliams S, Wilding LJ, Perera IT (2011) The imaging features of MACROLANEâ„¢ in breast augmentation.ClinRadiol 66: 977-983.
54. Yamaguchi S, Nagumo Y, Niwa K (2013) Efficacy and safety of Macrolane(â„¢) for breast enhancement: a 12-month follow-up study in Asian women.J PlastSurg Hand Surg 47: 191-195.
55. Joint United Nations Programme on HIV/AIDS. UNAIDS report on the global AIDS epidemic;https://www.unaids.org/globalreport,2010
56. Wen Y, Collier JH2 (2015) Supramolecular peptide vaccines: tuning adaptive immunity.CurrOpinImmunol 35: 73-79.
57. Wijesundara DK, Jackson RJ, Ramshaw IA, Ranasinghe C (2011) Human immunodeficiency virus-1 vaccine design: where do we go now?Immunol Cell Biol 89: 367-374.
58. Qiao Y, Huang Y, Qiu C, Yue X, Deng L, et al. (2010) The use of PEGylated poly [2-(N,N-dimethylamino) ethyl methacrylate] as a

mucosal DNA delivery vector and the activation of innate immunity and improvement of HIV-1-specific immune responses.Biomaterials 31: 115-123.

59. Greenland JR, Letvin NL (2007) Chemical adjuvants for plasmid DNA vaccines.Vaccine 25: 3731-3741.
60. Jiang L, Qian F, He X, Wang F, Ren D, et al. (2007) Novel chitosan derivative nanoparticles enhance the immunogenicity of a DNA vaccine encoding hepatitis B virus core antigen in mice.J Gene Med 9: 253-264.
61. Liu Y, Sun Y, Yan H, Liu X, Zhang W, et al. (2012) Electrospun fiber template for replica molding of microtopographical neural growth guidance.Small 8: 676-681.
62. Mu X, Zheng W, Xiao L, Zhang W, Jiang X (2013) Engineering a 3D vascular network in hydrogel for mimicking a nephron.Lab Chip 13: 1612-1618.
63. Aida T, Meijer EW, Stupp SI (2012) Functional supramolecular polymers.Science 335: 813-817.
64. Schwartz JJ, Zhang S (2000) Peptide-mediated cellular delivery. CurrOpinMolTher 2: 162-167.
65. Williams RJ, Smith AM, Collins R, Hodson N, Das AK, et al. (2009) Enzyme-assisted self-assembly under thermodynamic control. Nat Nanotechnol 4: 19-24.
66. Zhang X, Chu X, Wang L, Wang H, Liang G, et al. (2012) Rational design of a tetrameric protein to enhance interactions between self-assembled fibers gives molecular hydrogels.AngewChemInt Ed Engl 51: 4388-4392.
67. Reynolds WS, Dmochowski RR (2012) Urethral bulking: a urology perspective.UrolClin North Am 39: 279-287.
68. Moreno-Garrido I (2008) Microalgae immobilization: current techniques and uses.BioresourTechnol 99: 3949-3964.
69. Joint I, Mühling M, Querellou J (2010) Culturing marine bacteria - an essential prerequisite for biodiscovery.MicrobBiotechnol 3: 564-575.
70. IraniM, Ismail H, Ahmad Z, Fan M (2015) Synthesis of linear low-density polyethylene-g-poly (acrylic acid)-co-starch/organo-montmorillonite hydrogel composite as an adsorbent for removal of Pb(??) from aqueous solutions. J. Environ. Sci. 27: 9-20.
71. Yan H, Dai J, Yang Z, Yang H, and Cheng R (2011) Enhanced and

selective adsorption of copper(II) ions on surface carboxymethylated chitosan hydrogel beads. ChemEng J174: 586–594.

72. Ahmad H, Nurunnabi M, Rahman MM, Kumar K, et al. (2014) Magnetically doped multi stimuli-responsive hydrogel microspheres with IPN structure and application in dye removal. Colloids Surf PhysicochemEng Asp. 459: 39-47.
73. Wang X, Ye G, Wang X (2014) Hydrogel diffraction gratings functionalized with crown ether for heavy metal ion detection. Sens Actuators B Chem 193: 413–419
74. Arcadio PS, Gregoria AS (2003) Physical–Chemical Treatment of Water and Wastewater. IWA Publishing, CRC Press
75. Panpanit S, Visvanathan C (2001)The role of bentonite in UF flux enhancement mechanisms for oil/water emulsion. J MembrSci 184: 59-68.
76. Hadi M, Viraraghavan T(1999) Removal of oil from water by bentoniteorgano-clay. In: Hazardous & Industrial Wastes - Proceedings of the Mid-Atlantic Industrial & Hazardous Waste 187–196.
77. Qiu Z, Zhang Y, Fang Y (1995) Removal of oil from concentrated wastewater by attapulgite and coagulant. Water Qual Res J 30: 89-99.
78. Sokker HH, El-Sawy NM, Hassan MA, El-Anadouli BE (2011) Adsorption of crude oil from aqueous solution by hydrogel of chitosan based polyacrylamide prepared by radiation induced graft polymerization.J Hazard Mater 190: 359-365.
79. Heginbothom M, Fitzgerald TC, Wade WG (1990) Comparison of solid media for cultivation of anaerobes.J ClinPathol 43: 253-256.
80. Smith A(2012) History of the Agar Plate. Laboratory News
81. Cover TL (2012) Perspectives on methodology for in vitro culture of Helicobacter pylori.Methods MolBiol 921: 11-15.
82. Murray PR (1978) Growth of clinical isolates of anaerobic bacteria on agar media: effects of media composition, storage conditions, and reduction under anaerobic conditions. J ClinMicrobiol8: 708–714.
83. Wasinger VC, Cordwell SJ, Cerpa-Poljak A, Yan JX, Gooley AA, et al, (1995) Progress with gene-product mapping of the Mollicutes: Mycoplasma genitalium. Electrophoresis 16: 1090-1094.
84. Rabilloud T, Vaezzadeh AR, Potier N, Lelong C, Leize-Wagner E, et al. (2009) Power and limitations of electrophoretic separations in proteomics strategies.Mass Spectrom Rev 28: 816-843.

85. Rabilloud T, Chevallet M, Luche S, Lelong C (2010) Two-dimensional gel electrophoresis in proteomics: Past, present and future.J Proteomics 73: 2064-2077.

86. Rogowska-Wrzesinska A, Le Bihan MC, Thaysen-Andersen M, Roepstorff P (2013) 2D gels still have a niche in proteomics.J Proteomics 88: 4-13.

87. Chirani NA, Dembahri Z, Tokarski C, Rolando C, Benmouna M (2011) Newlydesigned polyacrylamide/dextran gels for electrophoresis protein separation: synthesis and characterization. Polymer International60: 1024-1029.

88. Suh KS, Mutoh M, Gerdes M, Yuspa SH (2005) CLIC4, an intracellular chloride channel protein, is a novel molecular target for cancer therapy.J InvestigDermatolSympProc 10: 105-109.

89. Greaser ML, Warren CM (2012) Protein electrophoresis in agarose gels for separating high molecular weight proteins.Methods MolBiol 869: 111-118.

90. Greppi GF, Roncada P (2005)La componenteproteicadel latte caprino. In: Pulina G. L'alimentazionedellacapra da latte. Avenue Media Publisher 71-99

91. Wang J, Ugaz VM (2006) Using in situ rheology to characterize the microstructure in photopolymerized polyacrylamide gels for DNA electrophoresis.Electrophoresis 27: 3349-3358.

92. Righetti PG, Gelfi C (1997) Electrophoresis gel media: the state of the art.J Chromatogr B Biomed SciAppl 699: 63-75.

93. Anseth KS, Bowman CN, Brannon-Peppas L (1996) Mechanical properties of hydrogels and their experimental determination. Biomaterials 17: 1647-1657.

94. Myung D, Waters D, Wiseman M, Duhamel PE, Noolandi J, et al. (2008) Progress in the development of interpenetrating polymer network hydrogels.PolymAdvTechnol 19: 647-657.

95. Kim SJ, Yoon SG, Kim SI (2004) Synthesis and characterization of interpenetration polymer network hydrogels composed of alginate and poly(diallyldimethylammonium chloride). J ApplPolymSci91: 3705-3709.

96. Wang JJ, Liu F (2013) Enhanced adsorption of heavy metal ions onto simultaneous interpenetrating polymer network hydrogels synthesized by UV irradiation. Polym Bull 70: 1415-1430.

97. Hoare TR, Kohane DS (2008) Hydrogels in drug delivery: progress and challenges. Polymer 49: 1993-2007.
98. Dragan ES, Lazar MM, Dinu MV, Doroftei F. Macroporous (2012) composite IPN hydrogels based on poly(acrylamide) and chitosan with tuned swelling and sorption of cationic dyes. ChemEng J204-206: 198-209
99. Lee KY1, Mooney DJ (2001) Hydrogels for tissue engineering.Chem Rev 101: 1869-1879.
100. Hutmacher DW1 (2001) Scaffold design and fabrication technologies for engineering tissues--state of the art and future perspectives.J BiomaterSciPolym Ed 12: 107-124.
101. Langer R (2000) Tissue engineering. MolTher 1: 12-15.
102. Bryant DM1, Mostov KE (2008) From cells to organs: building polarized tissue.Nat Rev Mol Cell Biol 9: 887-901.
103. Zhou M, Smith AM, Das AK, Hodson NW, Collins RF, et al. (2009) Self-assembled peptide-based hydrogels as scaffolds for anchorage-dependent cells.Biomaterials 30: 2523-2530.
104. Liu H J, Hu YH, H. M. Wang, J. Y. Wang, D. L. Kong, et al.(2011)Soft Matter7: 5430–5436
105. Yang C, Li D, Liu Z, Hong G, Zhang J, et al. (2012) Responsive small molecular hydrogels based on adamantane-peptides for cell culture.J PhysChem B 116: 633-638.
106. Wang H, Yang Z (2012) Short-peptide-based molecular hydrogels: novel gelation strategies and applications for tissue engineering and drug delivery.Nanoscale 4: 5259-5267.
107. Kaivosoja E, Barreto G, Levón K, Virtanen S, Ainola M, et al. (2012) Chemical and physical properties of regenerative medicine materials controlling stem cell fate.Ann Med 44: 635-650.
108. Khademhosseini A, Langer R (2007) Microengineered hydrogels for tissue engineering.Biomaterials 28: 5087-5092.
109. Mironov V, Boland T, Trusk T, Forgacs G, Markwald RR (2003) Organ printing: computer-aided jet-based 3D tissue engineering.Trends Biotechnol 21: 157-161.
110. Tanzi MC, Bianchi A, Farè S, Mantero S, Raimondi MT, et al. (2013) GruppoNazionale di Bioingegneria n°32: ApproccioIntegrato per la MedicinaRigenerativa. PàtronEditore.

111. Weiss P (1934) In Vitro Experiments on the Factors Determining the Course of the Outgrowing Nerve Fiber. Journal of Experimental Zoology 69: 393–448.
112. Clark P, Connolly P, Curtis AS, Dow JA, Wilkinson CD (1991) Cell guidance by ultrafine topography in vitro.J Cell Sci99 : 73-77.
113. Saito AC, Matsui TS, Ohishi T, Sato M, Deguchi S (2014) Contact guidance of smooth muscle cells is associated with tension-mediated adhesion maturation.Exp Cell Res 327: 1-11.
114. Elisseeff J, Anseth K, Sims D, McIntosh W, Randolph M, et al. (1999) Transdermal photopolymerization for minimally invasive implantation.ProcNatlAcadSci U S A 96: 3104-3107.
115. Elisseeff J, McIntosh W, Anseth K, Riley S, Ragan P, et al. (2000) Photoencapsulation of chondrocytes in poly(ethylene oxide)-based semi-interpenetrating networks.J Biomed Mater Res 51: 164-171.
116. Massia SP, Stark J, Letbetter DS (2000) Surface-immobilized dextran limits cell adhesion and spreading.Biomaterials 21: 2253-2261.
117. Holland NB, Qiu Y, Ruegsegger M, Marchant RE (1998) Biomimetic engineering of non-adhesive glycocalyx-like surfaces using oligosaccharide surfactant polymers.Nature 392: 799-801.
118. Liu Y, Chan-Park MB (2010) A biomimetic hydrogel based on methacrylated dextran-graft-lysine and gelatin for 3D smooth muscle cell culture.Biomaterials 31: 1158-1170.
119. Ward AG, Courts A (1997) The Science and Technology of Gelatin. Academic Press 66: 373-374
120. Type A & B Process Definition. Vyse Gelatin Company. https://www.vyse.com/ (2009). Accessed Feb 2015.
121. Das S, Pati F, Choi YJ, Rijal G, Shim JH, et al. (2015) Bioprintable, cell-laden silk fibroin-gelatin hydrogel supporting multilineage differentiation of stem cells for fabrication of three-dimensional tissue constructs.ActaBiomater 11: 233-246.
122. Ignatova M1, Manolova N, Rashkov I (2013) Electrospun antibacterial chitosan-based fibers.MacromolBiosci 13: 860-872.
123. Jiang T, Deng M, James R, Nair LS, Laurencin CT (2014) Micro- and nanofabrication of chitosan structures for regenerative engineering. ActaBiomater 10: 1632-1645.
124. Chen T, Embree HD, Brown EM, Taylor MM, Payne GF (2003) Enzyme-

catalyzed gel formation of gelatin and chitosan: potential for in situ applications.Biomaterials 24: 2831-2841.

125. Tezel A1, Fredrickson GH (2008) The science of hyaluronic acid dermal fillers.J Cosmet Laser Ther 10: 35-42.
126. Lam J, Truong NF, Segura T (2014) Design of cell-matrix interactions in hyaluronic acid hydrogel scaffolds.ActaBiomater 10: 1571-1580.
127. West DC, Kumar S (1989) Hyaluronan and angiogenesis.Ciba Found Symp 143: 187-201.
128. Tempel C, Gilead A, Neeman M (2000) Hyaluronic acid as an anti-angiogenic shield in the preovulatory rat follicle.BiolReprod 63: 134-140.
129. Slevin M, West D, Kumar P, Rooney P, Kumar S (2004) Hyaluronan, angiogenesis and malignant disease.Int J Cancer 109: 793-794.
130. Kisiel M, Martino MM, Ventura M, Hubbell JA, Hilborn J, et al. (2013) Improving the osteogenic potential of BMP-2 with hyaluronic acid hydrogel modified with integrin-specific fibronectin fragment. Biomaterials 34: 704-712.
131. Shu XZ, Liu Y, Palumbo F, Prestwich GD (2003) Disulfide-crosslinkedhyaluronan-gelatin hydrogel films: a covalent mimic of the extracellular matrix for in vitro cell growth.Biomaterials 24: 3825-3834.
132. Munarin F, Tanzi MC, Petrini P (2012) Advances in biomedical applications of pectin gels.Int J BiolMacromol 51: 681-689.
133. McKenna BA, Nicholson TM, Wehr JB, Menzies NW (2010) Effects of Ca, Cu, Al and La on pectin gel strength: implications for plant cell walls.Carbohydr Res 345: 1174-1179.
134. Munarin F, Guerreiro SG, Grellier MA, Tanzi MC, Barbosa MA, et al. (2011) Pectin-based injectable biomaterials for bone tissue engineering. Biomacromolecules 12: 568-577.
135. Jang J, Seol YJ, Kim HJ, Kundu J, Kim SW, et al. (2014) Effects of alginate hydrogel cross-linking density on mechanical and biological behaviors for tissue engineering.J MechBehav Biomed Mater 37: 69-77.
136. Guillaume O, Daly A, Lennon K, Gansau J, Buckley SF, et al. (2014) Shape-memory porous alginate scaffolds for regeneration of the annulus fibrosus: effect of TGF-Î²3 supplementation and oxygen culture conditions.ActaBiomater 10: 1985-1995.

137. Shteyer E, Ben Ya'acov A, Zolotaryova L, Sinai A, Lichtenstein Y et al. (2014) Reduced liver cell death using an alginate scaffold bandage: A novel approach for liver reconstruction after extended partial hepatectomy. ActaBiomater. 10: 3209-3216.
138. Ziv K, Nuhn H, Ben-HaimY, Sasportas LS, Kempen PJ, et al. (2014) A tunable silk-alginate hydrogel scaffold for stem cell culture and transplantation.Biomaterials 35: 3736-3743.
139. Burdick JA, Murphy WL (2012) Moving from static to dynamic complexity in hydrogel design.Nat Commun 3: 1269.
140. Douezan S, Dumond J, Brochard-Wyart F (2012) Wetting trasitions of cellular aggregates induced by substrate rigidity. Soft Matter 8: 4578-4583.
141. Jiang LY, Luo Y (2013) Guided assembly of endothelial cells on hydrogel matrices patterned with microgrooves: a basic model for micro vessel engineering. Soft Matter. 9: 1113-1121.
142. Tse JR, Engler AJ (2010) Preparation of hydrogel substrates with tunable mechanical properties.CurrProtoc Cell Biol Chapter 10: Unit 10.
143. Zhu J (2010) Bioactive modification of poly(ethylene glycol) hydrogels for tissue engineering.Biomaterials 31: 4639-4656.
144. Carmeliet P (2000) Mechanisms of angiogenesis and arteriogenesis. Nat Med 6: 389-395.
145. Silva AK, Richard C, Bessodes M, Scherman D, Merten OW (2009) Growth factor delivery approaches in hydrogels. Biomacromolecules 10: 9-18.
146. Hubbell JA (2003) Materials as morphogenetic guides in tissue engineering.CurrOpinBiotechnol 14: 551-558.
147. Schmoekel HG, Weber FE, Schense JC, Grätz KW, Schawalder P, et al. (2005) Bone repair with a form of BMP-2 engineered for incorporation into fibrin cell ingrowth matrices.BiotechnolBioeng 89: 253-262.
148. Martino MM, Tortelli F, Mochizuki M, Traub S, Ben-David D, et al. (2011) Engineering the growth factor microenvironment with fibronectin do mains to promote wound and bone tissue healing. SciTransl Med 3: 100ra89.
149. Kim S, Bedigrew K, Guda T, Maloney WJ, Park S, et al. (2014) Novel osteoinductive photo-cross-linkable chitosan-lactide-fibrinogen hydrogels enhance bone regeneration in critical size segmental bone

defects.ActaBiomater 10: 5021-5033.

150. Bae MS, Ohe JY, Lee JB, Heo DN, Byun W, et al. (2014) Photo-cured hyaluronic acid-based hydrogels containing growth and differentiation factor 5 (GDF-5) for bone tissue regeneration.Bone 59: 189-198.
151. Ueno Y, Kitamura C, Terashita M, Nishihara T (2006) Re-oxygenation improves hypoxia-induced pulp cell arrest.J Dent Res 85: 824-828.
152. Kawagishi E, Nakakura-Ohshima K, Nomura S, Ohshima H (2006) Pulpal responses to cavity preparation in aged rat molars.Cell Tissue Res 326: 111-122.
153. Nakashima M, Iohara K, Zheng L (2006) Gene therapy for dentin regeneration with bone morphogenetic proteins.Curr Gene Ther 6: 551-560.
154. Madan AK, Kramer B (2005) Immunolocalization of fibroblast growth factor-2 (FGF-2) in the developing root and supporting structures of the murine tooth.J MolHistol 36: 171-178.
155. Annabi B, Naud E, Lee YT, Eliopoulos N, Galipeau J (2004) Vascular progenitors derived from murine bone marrow stromal cells are regulated by fibroblast growth factor and are avidly recruited by vascularizing tumors.J Cell Biochem 91: 1146-1158.
156. Yamamoto M, Ikada Y, Tabata Y (2001) Controlled release of growth factors based on biodegradation of gelatin hydrogel.J BiomaterSciPolym Ed 12: 77-88.
157. Nakayama J, Fujioka H, Nagura I, Kokubu T, Makino T, et al. (2009) The effect of fibroblast growth factor-2 on autologous osteochondral transplantation.IntOrthop 33: 275-280.
158. Nakahara T, Nakamura T, Kobayashi E, Inoue M, Shigeno K(2003) Novel approach to regeneration of periodontal tissues based on in-situ tissue engineering: effects of controlled release of basic fibroblast growth factor from a sandwich membrane. Tissue Eng. 9: 153-162.
159. Yasuda Y, Koyama H, Tabata Y, Fujihara Y, Oba M, et al. (2008) Controlled delivery of bFGF remodeled vascular network in muscle flap and increased perfusion capacity via minor pedicle.J Surg Res 147: 132-137.
160. Laçin NT (2014) Development of biodegradable antibacterial cellulose based hydrogel membranes for wound healing.Int J BiolMacromol 67: 22-27.

161. Lee YH, Chang JJ, Yanga MC, Chienc CT, Lai WF (2012) Acceleration of wound healing in diabetic rats by layered hydrogel dressing. Carbohydrate Polymers. 88: 809-819.

162. Tao Wang, Xiao-Kang Zhu, Xu-Ting Xue, Da-Yang Wu (2012) Hydrogel sheets of chitosan, honey and gelatin as burn wound dressing. Carbohydrate Polymers. 88: 75–83.

163. Kanokpanont S, Damrongsakkul S, Ratanavaraporn J, Aramwit P (2012) An innovative bi-layered wound dressing made of silk and gelatin for accelerated wound healing.Int J Pharm 436: 141-153.

164. Gurny R, Kaufmann B, Moeller M(2012) International Patent WO 2012013670 A1. Process for the esterification of hyaluronic acid with hydrophobic organic compounds

165. Bertrand N, Leroux JC (2012) The journey of a drug-carrier in the body: an anatomo-physiological perspective.J Control Release 161: 152-163.

166. Ashley GW, Henise J, Reid R, Santi DV (2013) Hydrogel drug delivery system with predictable and tunable drug release and degradation rates. ProcNatlAcadSci U S A 110: 2318-2323.

167. Marin MA, Mallepally RR, McHugh MA(2014) Silk fibroin aerogels for drug delivery applications. J Supercrit Fluids 91: 84-89.

168. Patra C, Talukdar S, Novoyatleva T, Velagala SR, Mühlfeld C, et al. (2012) Silk protein fibroin from Antheraeamylitta for cardiac tissue engineering.Biomaterials 33: 2673-2680.

169. Park CH, Jeong L, Cho D, Kwon OH, Park WH (2013) Effect of methylcellulose on the formation and drug release behavior of silk fibroin hydrogel.CarbohydrPolym 98: 1179-1185.

170. Mandal BB, Kundu SC (2008) A novel method for dissolution and stabilization of non-mulberry silk gland protein fibroin using anionic surfactant sodium dodecyl sulfate.BiotechnolBioeng 99: 1482-1489.

171. Yesilirmak N, AltÄnors DD (2013) A silicone hydrogel contact lens after 7 years of continuous wear.Cont Lens Anterior Eye 36: 204-206.

172. Phan CM, Subbaraman L, Jones L (2014) Contact lenses for antifungal ocular drug delivery: a review.Expert Opin Drug Deliv 11: 537-546.

173. Wilson CG (2004) Topical drug delivery in the eye.Exp Eye Res 78: 737-743.

174. Cohen RA, Gebhardt BM, Insler MS(1996) Comparison of contact lens delivery and topical instillation of 4% lidocaine in the normal cornea

and limbal conjunctiva. Invest Ophthal Vis Sci. 37: 332.

175. Eugen B, Liliana V, Thomas GN, John T (2006) Polymeric materials for ophthalmic drug delivery: trends and perspectives. J Mater Chem16: 3439-3443.
176. Lin HR, Sung KC (2000) Carbopol/pluronic phase change solutions for ophthalmic drug delivery.J Control Release 69: 379-388.
177. Jun Chen, Xiaojian Li, LiqianGao, Yi Hu, Wen Zhong, et al. (2014) Nano LIFE.4: 1441004
178. Lowe CP1, Dreischor MW (2005) The size of a polymer in a symmetric solvent.J ChemPhys 122: 84905.
179. Stevens MM1, George JH (2005) Exploring and engineering the cell surface interface. Science 310: 1135-1138.
180. A. Sionkowska (2011) Current research on the blends of natural and synthetic polymers as new biomaterials: review. ProgPolymSci 36: 1254-1276
181. A. Sivashanmugam (2015)An overview of injectable polymeric hydrogels for tissue engineering, Eur.Polym. J.
182. HehlEM1, BeckR, LuthardK, GuthoffR, DrewelowB (1999) Improved penetration of aminoglycosides and fluorozuinolones into the aqueous humour of patients by means of Acuvue contact lenses. Eur J ClinPharmacol 55: 317-323.
183. Dmochowski RR, Appell RA (2000) Injectable agents in the treatment of stress urinary incontinence in women: where are we now? Urology 56: 32-40.
184. Lightner DJ1 (2002) Review of the available urethral bulking agents. CurrOpinUrol 12: 333-338.
185. Bent AE, Tutrone RT, McLennan MT, Lloyd LK, Kennelly MJ, et al. (2001) Treatment of intrinsic sphincter deficiency using autologous ear chondrocytes as a bulking agent. NeurourolUrodyn 20: 157-165.
186. YangX, SunY, KootalaS, HilbornJ, HeerschapA, et al. (2014) Injectable hyaluronic acid hydrogel for 19F magnetic resonance imaging. CarbohydrPolym 110: 95-99.
187. Choudhury NA, Sampath S, Shukla AK (2008) Hydrogel-polymer electrolytes for electrochemical capacitors: an overview. Energy Environ Sci 2: 55-67.

CHAPTER

15

Hydrogels Based on Dynamic Covalent and Non Covalent Bonds: A Chemistry Perspective

Francesco Picchioni [1,] **and Henky Muljana** [1,2]

[1]Department of Chemical Engineering, Engineering and Technology Institute Groningen (ENTEG), University of Groningen, Nijenborgh 4, 9747 AG Groningen, The Netherlands

[2]Department of Chemical Engineering, Parahyangan Catholic University, Ciumbuleuit 94, Bandung 40141, West Java, Indonesia

ABSTRACT

Hydrogels based on reversible covalent bonds represent an attractive topic for research at both academic and industrial level. While the concept of reversible covalent bonds dates back a few decades, novel developments continue to appear in the general research area of gels and especially

Citation: Picchioni, F.; Muljana, H. "Hydrogels Based on Dynamic Covalent and Non Covalent Bonds: A Chemistry Perspective". Gels 2018, 4, 21. https://doi.org/10.3390/gels4010021

hydrogels. The reversible character of the bonds, when translated at the general level of the polymeric network, allows reversible interaction with substrates as well as responsiveness to variety of external stimuli (e.g., self-healing). These represent crucial characteristics in applications such as drug delivery and, more generally, in the biomedical world. Furthermore, the several possible choices that can be made in terms of reversible interactions generate an almost endless number of possibilities in terms of final product structure and properties. In the present work, we aim at reviewing the latest developments in this field (i.e., the last five years) by focusing on the chemistry of the systems at hand. As such, this should allow molecular designers to develop a toolbox for the synthesis of new systems with tailored properties for a given application.

Keywords: Hydrogels; Dynamic Covalent Bonds; Reversible Polymeric Network

INTRODUCTION

In the last few years, a renewed interest in hydrogels has arisen due to an extended range of applications [1]. This stems from the characteristics of these materials such as biocompatibility and responsiveness to a variety of external stimuli [2]. While "classical" applications such as their use as an adsorbent in waste water treatment [3, 4] are still heavily investigated, other areas are gaining significant attention at the moment. In particular, biomedical applications are very popular and include cell culture [5], wound dressing and healing [2, 6], drug delivery [2, 7, 8], tissue engineering scaffolds [9], bone repair [10], and cartilage regeneration [11]. Such wide variety of biomedical applications stems from the peculiar characteristic of hydrogels, namely the fact that they can retain a large amount of water in their structure [1]. This makes it, in turn, also possible to incorporate water-soluble moieties in the final product such as proteins and DNA [5], but also dispersible (nano) particles [12, 13] and (nano) emulsions [14].

Particularly important is the adhesion at the interface between the polymeric chains and several different substrates to be then released (where and when needed) upon action of an external stimulus. On a molecular level, this translates in the presence of reversible bonds between the polymeric network and such "loading." For example, the catechol chemistry (Figure 1) can actually be emplSoyed to ensure adhesion of the polymeric chains to several different moieties while crosslinking the network in an irreversible way [15].

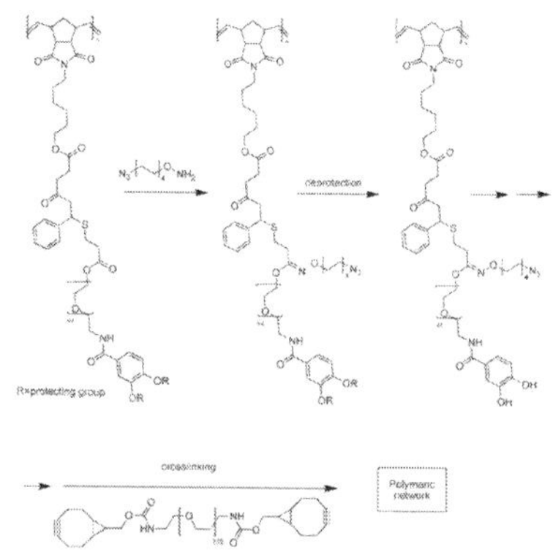

Figure 1. Incorporation of catechol groups in a polymeric network. Adapted and re-drawn from [**15**].

In this particular case, the crosslinking via metal free click chemistry is practically quantitative and does not result in the formation of any by-product, and this rendering it ideal for biomedical applications [**16**,**17**,**18**].

The catechol chemistry is not the only example of a reversible yet covalent bond that can be formed and broken on command upon an external stimulus. Indeed, along the same strategy, the imine formation chemistry can be used to tune the adhesion [**18**].

The molecular reversibility can be actually achieved in two different ways: either by making use of equilibrium reactions (e.g., the Diels-Alder one) or through dynamic exchange reactions (e.g., reaction of an excess amino groups with epoxide ones). Both approaches have been widely used for tuning the adhesion of different substrates on different polymeric networks and for the network formations (and disruption) itself. These are discussed in the following paragraph.

DYNAMIC HYDROGELS BASED ON REVERSIBLE (COVALENT) INTERACTIONS

The same concept of covalent reversible bonds can be used for the hydrogel formation and disruption. Examples of interactions used in this case are (Figure 2) electrostatic ones [19,20], cycloadditions [8,21], redox reactions [22,23,24], and other ones such as imine [25,26,27,28] and enamine formation [29], acylhydrazone [30,31,32,33], and borax acid reaction with hydroxyls [13,34,35,36].

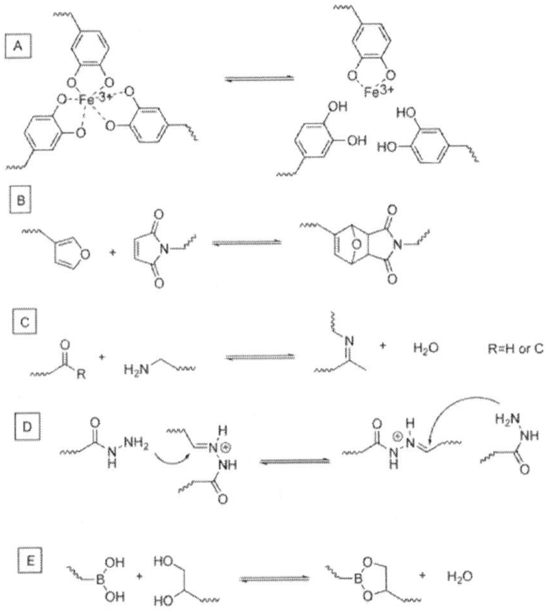

Figure 2. Schematic representation of reversible covalent bonds. (A): catechol-Fe chemistry; (B): furan-maleimide as example of cycloaddition reactions; (C): imine/enamine formation; (D): dynamic acylhydrazone exchange reaction; (E): boric ester formation and hydrolysis.

The general idea is that the use of dynamic covalent bonds allows the polymeric network to adjust itself as a result of an external stimulus. This can be achieved in principle through other weaker interactions, e.g., hydrogen bonding. In particular, a clear trend is detected in the last year, according to which self-assembly driven processes [10, 37] can be conveniently used for hydrogel preparation. However, the use of covalent bonds displays two distinct and clear advantages [25]. In first instance, the network is still covalently linked, which renders it quite robust against small random fluctuations in environmental conditions such as temperature. Furthermore, exchange reactions such as the one of an amine with an imine are often kinetically controlled by the use of catalysts. In turn, this allows the possibility to freeze the network conformation (by slowing the kinetics) when desired.

The general concept behind the use of reversible interactions for the hydrogel polymeric chains is the (reversible) network disruption with immediate release of any loading (Figure 3).

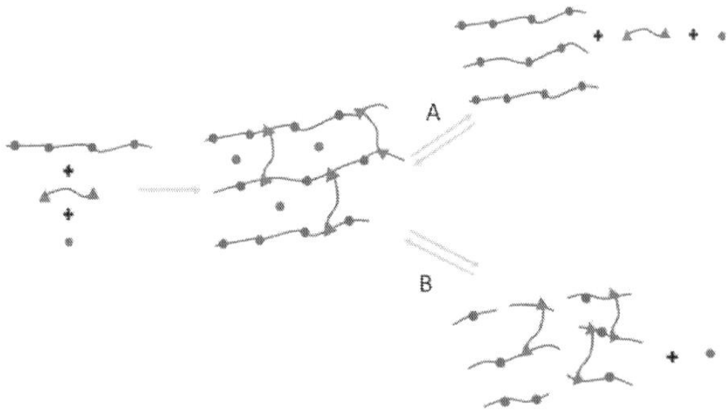

Figure 3. Schematic diagram for the strategy behind the use of reversible covalent bonds in hydrogels. (●) = reversible bonds along the backbone; (Δ) = reversible bonds as crosslinking points; (●) = loading of the network (substrate).

Reversible bonds can be incorporated along the backbone (red circles) or at the crosslinking point (green triangles). The network, when subjected to an appropriate external stimulus, can then break at the crosslinking point (route A) or along the backbone (route B). This generates network fragments that can be quite different in terms of chemical structure even if in both cases the loading (blue circles) will be released. As a result of the network disruption, the load is released as the polymeric chains become soluble and not able anymore to entrap the load. In the specific case of drug delivery, this mechanism entails a controlled release rate depending on the kinetics of the network disruption, which in turn can be linked, at the molecular level, to the kinetics of the reversible bonds.

An interesting example is based, once more, on the catechol chemistry [**38**] for which a hydrogel containing di-methylacrylamide units (DMA) can be reversibly crosslinked by reversible complexation with Fe^{3+} ions (see **Figure 4**).

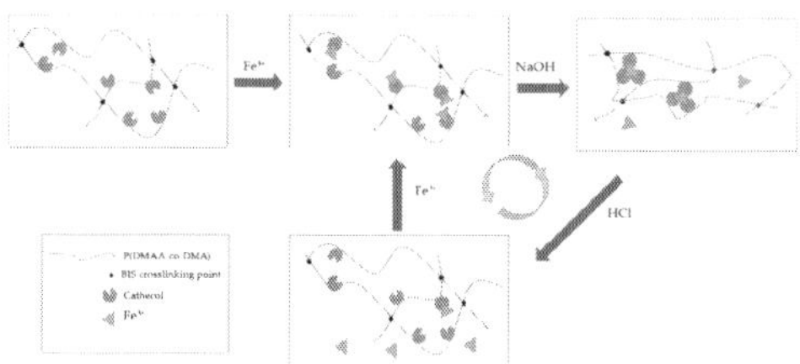

Figure 4. Schematic drawing of reversible complexation of DMA with Fe^{3+} ions. Adapted and re-drawn from [**38**] with permission.

In this particular case, the reversibility is cleverly achieved by switching between a tri- and a mono-chelate complex based on the pH of the solution. The corresponding final product displays a shape memory effect triggered by the pH value.

In order to achieve a sol-gel transition and, possibly, self-healing properties in response to multiple stimuli, different functional groups might be embedded along the backbone, e.g., di-sulfide and acylhydrazone ones, as shown in Figure 5 [39].

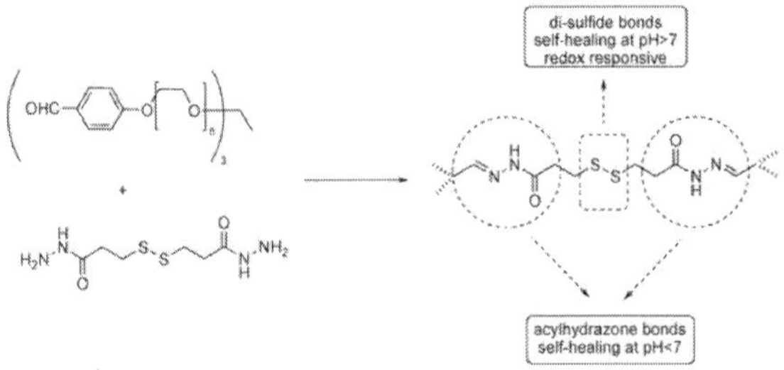

Figure 5. Dynamic hydrogels based on di-sulfide and acylhydrazone chemistry. Adapted and re-drawn from [**39**].

The advantage of such approach is that different functional groups are factually responsible for the response under different environmental conditions. Similar to this, hydrogels have been prepared for which both Diels-Alder and acylhydrazone groups are present along the backbone (Figure 6) [9].

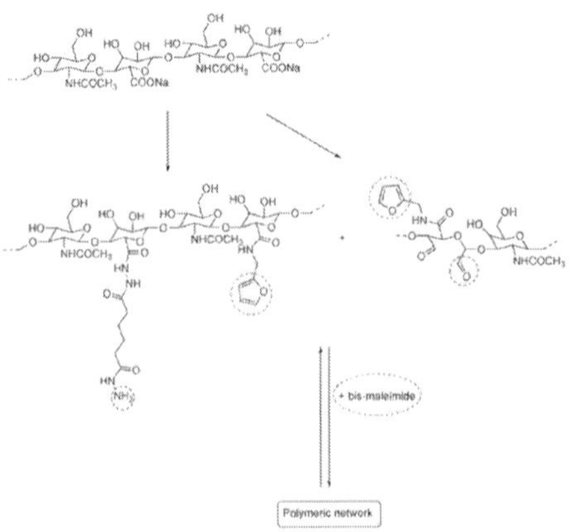

Figure 6. Double cross link network based on Diels-Alder and acylhydrazone reversible chemistry. Adapted and re-drawn from [**9**].

In this case, it has been proposed that the Diels-Alder adducts preserve the gel integrity and endow it with good mechanical properties while the acylhydrazone reversible chemistry is conveniently employed to fine-tune the crosslinking density. The latter is in turn responsible for the self-healing behavior. The same hydrazone reversible chemistry can be conveniently combined with self-assembly of block copolymers to yield (see Figure 7) hydrogels with improved mechanical properties [33].

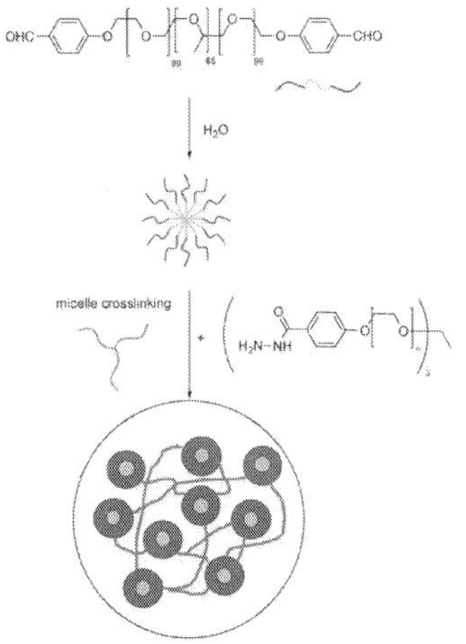

Figure 7. Dynamic covalent hydrogels with triblock copolymer micellization. Adapted and re-drawn from [33]. Yellow: hydrophobic block and/or domain. Blue: hydrophilic block and/or domain. Green: crosslinker.

The resulting hydrogels still display self-healing properties and elongations up to 10.000%, the latter being a function of the pH. This stems from the dependency of the crosslinking density (acylhydrazone bonds) on the pH values. Such outstanding mechanical behavior (i.e., strain values) can also be attributed to the presence of the micelles, which actually act as physical crosslinking points for the all system.

This approach, i.e., the use of two different reversible covalent bonding, represents a very popular choice in the last five years [40,41]. This stems from the fact that, besides the synergy in terms of reversible behavior as result of external stimuli and of combination of different properties, an additional one can be pursued in terms of the synthetic approach. A paradigmatic example is represented by the synthesis of hydrogels based on poly(ethylene-glycol) (PEG) with the use of thiol-ene addition as well as borax-diol chemistry (Figure 8) [35].

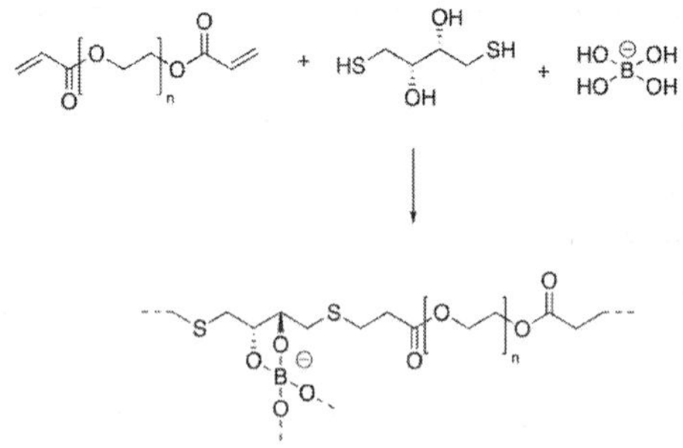

Figure 8. Hydrogels based on poly (ethylene-glycol) (PEG) with the use of thiol-ene addition and a borax-diol chemistry. Adapted and re-drawn from [**35**].

Borax acts in this case as catalyst for the thiol-ene reaction, thus factually helping in building the polymeric backbone, but also as crosslinking agent via the reversible borax-diol chemistry. The reversibility of the crosslinking was demonstrated by self-healing experiments and confirmed by rheology measurements. As typical for these systems, i.e., covalently and reversibly crosslinked gels, a cross over point between the elastic (G′) and loss (G″) modulus is observed for relatively moderate frequency values ($1 < \omega < 100$ rad/s). This is in fact a very general characterization technique and observation [**35**] that can be used to characterize the response of the hydrogel to shear force.

Very recently host-guest chemistry (molecular recognition) has been also reported for the synthesis of hydrogels [**42,43**]. In one example the proposed approach is rather versatile as it relies on fixed molecular recognition interaction even if in different matrixes prepared by in situ polymerization, as illustrated in **Figure 9** [**43**].

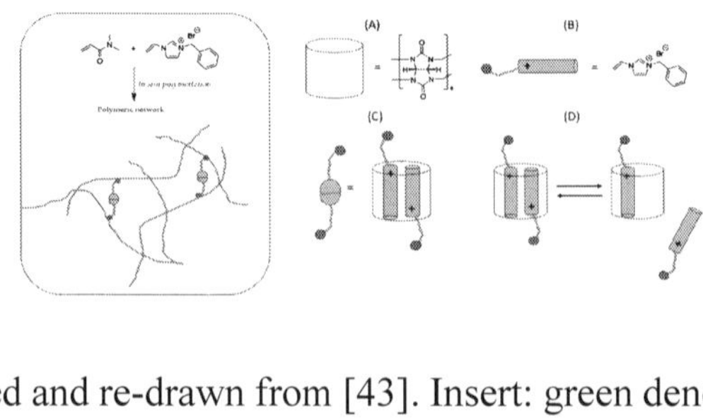

Figure 9. Adapted and re-drawn from [43]. Insert: green denotes the polymeric network. Yellow/orange structure: see (B,C). (A) chemical structure of host (B) chemical structure (C) host guest coupling (D) dynamic equilibrium for reversible crosslinking. Di-methylacrylamide is used here as an example of a monomer that can be used.

The versatility of this method and its compatibility with several different monomeric precursors and polymeric end-products render it particularly attractive.

An interesting consequence of the dynamic character of these bonds is the fact that this significantly favors the adhesion between hydrogels and organo-gels (when in contact with each other), provided that such reaction can take place at the interface between these two gels [44].

A shown above, the choice of covalent reversible bonds for the network formation endows the final product with a multifaceted portfolio of properties (i.e., multiple responses to multiple stimuli). This can be achieved on the basis of several different substrates. Indeed, synthetic polymers are still widely used for the preparation of hydrogels, with polyacrylamide representing the most popular choice [45]. On the other hand, hydrogels based on natural products (e.g., chitosan [4,17], proteins [11,31,36], modified alginate [46], peptides [28,47]) represent a very convenient choice in view of clear advantages related to biological applications, but also to general sustainability principles. Indeed, attention is being paid also to novel synthetic pathways in agreement with green chemistry principles [48].

The choice of PEG as the main constituent of the polymeric backbone is an obvious one [49,50] when making allowances for the commercial availability of many (functionally modified) PEG varieties as well as their well-known thickening effect even in the presence of salts. On the other hand, polysaccharides also represent a popular choice mainly in view of the easiness of the modification and their availability in nature. An elegant example is the combination of modified cellulose with chitosan (Figure 10) [29].

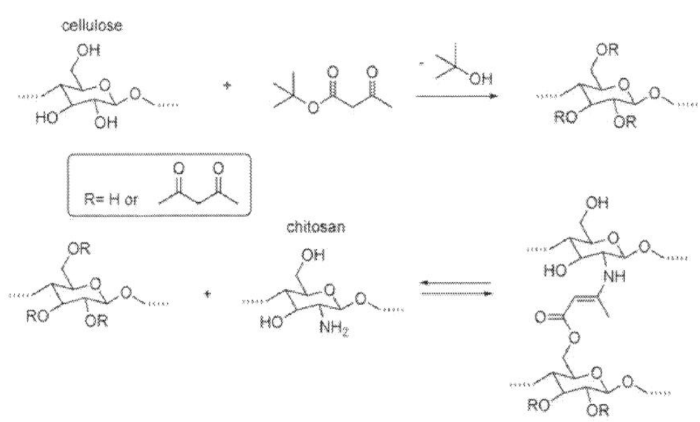

Figure 10. Polysaccharide hydrogels based on covalent enamine bond. Adapted and re-drawn from [29].

The presence of an amino group along the chitosan chain is exploited here in order to achieve high reactivity (even at room temperature) with modified cellulose through enamine formation, in this case or via imine formation in other reported examples [51,52]. A combination of a natural polysaccharide with PEG combines the best of both worlds and has been recently reported [53].

CONCLUSIONS AND FUTURE PERSPECTIVES

As clearly seen form the example discussed above, the use of reversible covalent bonds in hydrogels endows the final product with very peculiar chemical structures, which in turns translate into a kaleidoscopic ensemble of possibilities in terms final properties. As also observed in other chemistry-related research fields, the study of synergistic phenomena seems to represent a very popular trend. In the specific case of hydrogels, this translates into the combination of multiple reversible interactions (and more specifically, reversible covalent bonds) in the same final product. This allows control of the network structure in processes such as self-healing in response to several different external stimuli. This is particularly attractive for biomedical applications where physiological parameters (e.g., pH, temperature, shear stress, etc.) might change in a simultaneous manner. On a molecular level, this generates the need for synthetic strategies allowing the incorporation of such reversible covalent bonds on the polymeric network. The above review of the most recent trends clearly show on one side the high sophistication level of the synthetic strategies employed and the multifaceted properties toolbox achievable in this way.

It must be stressed here that such research trends will probably continue in the coming years as the number of possibilities in terms of chemical bonds is certainly not yet exhausted. On the other hand, it is also conceivable that, as the topic will reach more scientific maturity, more attention will be paid to industrially feasible preparation routes. In this context, the choice of suitable substrates (i.e., polymeric materials) as well as synthetic strategies will probably constitute a focal point of future research projects.

CONFLICTS OF INTEREST

The authors declare no conflict of interest.

REFERENCES

1. Ahmed, E.M. Hydrogel: Preparation, characterization, and applications: A review. *J. Adv. Res.* 2015, *6*, 105–121.
2. Chen, L.; Yin, Y.; Liu, Y.; Lin, L.; Liu, M. Design and fabrication of functional hydrogels through interfacial engineering. *Chin. J. Polym. Sci.* 2017, *35*, 1181–1193.
3. Liu, J.; Zhu, K.; Jiao, T.; Xing, R.; Hong, W.; Zhang, L.; Zhang, Q.; Peng, Q. Preparation of graphene oxide-polymer composite hydrogels via thiol-ene photopolymerization as efficient dye adsorbents for wastewater treatment. *Colloids Surf. A Physicochem. Eng. Asp.* 2017, *529*, 668–676.
4. Piatkowski, M.; Janus, L.; Radwan-Praglowska, J.; Raclavsky, K. Microwave-enhanced synthesis of biodegradable multifunctional chitosan hydrogels for wastewater treatment. *Express Polym. Lett.* 2017, *11*, 809–819.
5. Zhang, Z.; Du, J.; Li, Y.; Wu, J.; Yu, F.; Chen, Y. An aptamer-patterned hydrogel for the controlled capture and release of proteins via biorthogonal click chemistry and DNA hybridization. *J. Mater. Chem. B* **2017**, *5*, 5974–5982.
6. Yang, D.H.; Seo, D.I.; Lee, D.; Bhang, S.H.; Park, K.; Jang, G.; Kim, C.H.; Chun, H.J. Preparation and evaluation of visible-light cured glycol chitosan hydrogel dressing containing dual growth factors for accelerated wound healing. *J. Ind. Eng. Chem.* **2017**, *53*, 360–370.
7. Xu, L.; Cooper, R.C.; Wang, J.; Yeudall, W.A.; Yang, H. Synthesis and Application of Injectable Bioorthogonal Dendrimer Hydrogels for Local Drug Delivery. *ACS Biomater. Sci. Eng.* **2017**, *3*, 1641–1653.
8. Gregoritza, M.; Messmann, V.; Abstiens, K.; Brandl, F.P.; Goepferich, A.M. Controlled Antibody Release from Degradable Thermoresponsive Hydrogels Cross-Linked by Diets-Alder Chemistry. *Biomacromolecules* **2017**, *18*, 2410–2418.
9. Yu, F.; Cao, X.; Du, J.; Wang, G.; Chen, X. Multifunctional Hydrogel with Good Structure Integrity, Self-Healing, and Tissue-Adhesive Property Formed by Combining Die Is Alder Click Reaction and Acylhydrazone Bond. *ACS Appl. Mater. Interfaces* **2015**, *7*, 24023–24031.
10. Lu, S.; Bai, X.; Liu, H.; Ning, P.; Wang, Z.; Gao, C.; Ni, B.; Liu, M. An injectable and self-healing hydrogel with covalent cross-linking in

vivo for cranial bone repair. *J. Mater. Chem. B* **2017**, *5*, 3739–3748.

11. Zhu, D.; Wang, H.; Trinh, P.; Heilshorn, S.C.; Yang, F. Elastin-like protein-hyaluronic acid (ELP-HA) hydrogels with decoupled mechanical and biochemical cues for cartilage regeneration. *Biomaterials* **2017**, *127*, 132–140.
12. Ono, R.J.; Lee, A.L.Z.; Voo, Z.X.; Venkataraman, S.; Koh, B.W.; Yang, Y.Y.; Hedrick, J.L. Biodegradable Strain-Promoted Click Hydrogels for Encapsulation of Drug-Loaded Nanoparticles and Sustained Release of Therapeutics. *Biomacromolecules* **2017**, *18*, 2277–2285.
13. Deng, R.; Derry, M.J.; Mable, C.J.; Ning, Y.; Armes, S.P. Using Dynamic Covalent Chemistry to Drive Morphological Transitions: Controlled Release of Encapsulated Nanoparticles from Block Copolymer Vesicles. *J. Am. Chem. Soc.* **2017**, *139*, 7616–7623.
14. Lei, L.; Zhang, Y.; He, L.; Wu, S.; Li, B.; Li, Y. Fabrication of nanoemulsion-filled alginate hydrogel to control the digestion behavior of hydrophobic nobiletin. *LWT-Food Sci. Technol.* **2017**, *82*, 260–267.
15. Slegeris, R.; Ondrusek, B.A.; Chung, H. Catechol- and ketone-containing multifunctional bottlebrush polymers for oxime ligation and hydrogel formation. *Polym. Chem.* **2017**, *8*, 4707–4715.
16. Fu, S.; Dong, H.; Deng, X.; Zhuo, R.; Zhong, Z. Injectable hyaluronic acid/poly(ethylene glycol) hydrogels crosslinked via strain-promoted azide-alkyne cycloaddition click reaction. *Carbohydr. Polym.* **2017**, *169*, 332–340.
17. Guaresti, O.; Garcia-Astrain, C.; Palomares, T.; Alonso-Varona, A.; Eceiza, A.; Gabilondo, N. Synthesis and characterization of a biocompatible chitosan-based hydrogel cross-linked via 'click' chemistry for controlled drug release. *Int. J. Biol. Macromol.* **2017**, *102*, 1–9.
18. Marin, L.; Ailincai, D.; Morariu, S.; Tartau-Mititelu, L. Development of biocompatible glycodynameric hydrogels joining two natural motifs by dynamic constitutional chemistry. *Carbohydr. Polym.* **2017**, *170*, 60–71.
19. Selegard, R.; Aronsson, C.; Brommesson, C.; Danmark, S.; Aili, D. Folding driven self-assembly of a stimuli-responsive peptide-hyaluronan hybrid hydrogel. *Sci. Rep.* **2017**, *7*, 7013.
20. Shi, L.; Carstensen, H.; Hoelzl, K.; Lunzer, M.; Li, H.; Hilborn, J.; Ovsianikov, A.; Ossipov, D.A. Dynamic Coordination Chemistry

Enables Free Directional Printing of Biopolymer Hydrogel. *Chem. Mater.* **2017**, *29*, 5816–5823.

21. Tsurkan, M.V.; Jungnickel, C.; Schlierf, M.; Werner, C. Forbidden Chemistry: Two-Photon Pathway in [2 + 2] Cycloaddition of Maleimides. *J. Am. Chem. Soc.* **2017**, *139*, 10184–10187.
22. Casuso, P.; Odriozola, I.; Perez-San Vicente, A.; Loinaz, I.; Cabanero, G.; Grande, H.; Dupin, D. Injectable and Self-Healing Dynamic Hydrogels Based on Metal(I)-Thiolate/Disulfide Exchange as Biomaterials with Tunable Mechanical Properties. *Biomacromolecules* **2015**, *16*, 3552–3561.
23. Barcan, G.A.; Zhang, X.; Waymouth, R.M. Structurally Dynamic Hydrogels Derived from 1,2-Dithiolanes. *J. Am. Chem. Soc.* **2015**, *137*, 5650–5653.
24. Gong, C.; Shan, M.; Li, B.; Wu, G. Injectable dual redox responsive diselenide-containing poly(ethylene glycol) hydrogel. *J. Biomed. Mater. Res. Part A* **2017**, *105*, 2451–2460.
25. Whitaker, D.E.; Mahon, C.S.; Fulton, D.A. Thermoresponsive Dynamic Covalent Single-Chain Polymer Nanoparticles Reversibly Transform into a Hydrogel. *Angew. Chem. Int. Ed.* **2013**, *52*, 956–959.
26. Ding, F.; Wu, S.; Wang, S.; Xiong, Y.; Li, Y.; Li, B.; Deng, H.; Du, Y.; Xiao, L.; Shi, X. A dynamic and self-crosslinked polysaccharide hydrogel with autonomous self-healing ability. *Soft Matter* **2015**, *11*, 3971–3976.
27. Mukherjee, S.; Hill, M.R.; Sumerlin, B.S. Self-healing hydrogels containing reversible oxime crosslinks. *Soft Matter* **2015**, *11*, 6152–6161.
28. Popescu, M.; Liontos, G.; Avgeropoulos, A.; Voulgari, E.; Avgoustakis, K.; Tsitsilianis, C. Injectable Hydrogel: Amplifying the pH Sensitivity of a Triblock Copolypeptide by Conjugating the N-Termini via Dynamic Covalent Bonding. *ACS Appl. Mater. Interfaces* **2016**, *8*, 17539–17548.
29. Liu, H.; Sui, X.; Xu, H.; Zhang, L.; Zhong, Y.; Mao, Z. Self-Healing Polysaccharide Hydrogel Based on Dynamic Covalent Enamine Bonds. *Macromol. Mater. Eng.* **2016**, *301*, 725–732.
30. Chang, R.; Wang, X.; Li, X.; An, H.; Qin, J. Self-Activated Healable Hydrogels with Reversible Temperature Responsiveness. *ACS Appl. Mater. Interfaces* **2016**, *8*, 25544–25551.

31. Wang, H.; Zhu, D.; Paul, A.; Cai, L.; Enejder, A.; Yang, F.; Heilshorn, S.C. Covalently Adaptable Elastin-Like Protein-Hyaluronic Acid (ELP-HA) Hybrid Hydrogels with Secondary Thermoresponsive Crosslinking for Injectable Stem Cell Delivery. *Adv. Funct. Mater.* **2017**, *27*, 1605609.
32. Yang, X.; Liu, G.; Peng, L.; Guo, J.; Tao, L.; Yuan, J.; Chang, C.; Wei, Y.; Zhang, L. Highly Efficient Self-Healable and Dual Responsive Cellulose-Based Hydrogels for Controlled Release and 3D Cell Culture. *Adv. Funct. Mater.* **2017**, *27*, 1703174.
33. Wang, P.; Deng, G.; Zhou, L.; Li, Z.; Chen, Y. Ultrastretchable, Self-Healable Hydrogels Based on Dynamic Covalent Bonding and Triblock Copolymer Micellization. *ACS Macro Lett.* **2017**, *6*, 881–886.
34. Yang, T.; Ji, R.; Deng, X.; Du, F.; Li, Z. Glucose-responsive hydrogels based on dynamic covalent chemistry and inclusion complexation. *Soft Matter* **2014**, *10*, 2671–2678.
35. He, L.; Szopinski, D.; Wu, Y.; Luinstra, G.A.; Theato, P. Toward Self-Healing Hydrogels Using One-Pot Thiol-Ene Click and Borax-Diol Chemistry. *ACS Macro Lett.* **2015**, *4*, 673–678.
36. Seidler, C.; Ng, D.Y.W.; Weil, T. Native protein hydrogels by dynamic boronic acid chemistry. *Tetrahedron* **2017**, *73*, 4979–4987.
37. Li, Y.; Liu, Y.; Ma, R.; Xu, Y.; Zhang, Y.; Li, B.; An, Y.; Shi, L. A G-Quadruplex Hydrogel via Multicomponent Self-Assembly: Formation and Zero-Order Controlled Release. *ACS Appl. Mater. Interfaces* **2017**, *9*, 13056–13067.
38. Liao, J.; Huang, J.; Wang, T.; Sun, W.; Tong, Z. Rapid shape memory and pH-modulated spontaneous actuation of dopamine containing hydrogels. *Chin. J. Polym. Sci.* **2017**, *35*, 1297–1306.
39. Deng, G.; Li, F.; Yu, H.; Liu, F.; Liu, C.; Sun, W.; Jiang, H.; Chen, Y. Dynamic Hydrogels with an Environmental Adaptive Self-Healing Ability and Dual Responsive Sol-Gel Transitions. *ACS Macro Lett.* **2012**, *1*, 275–279.
40. Chen, J.; Su, Q.; Guo, R.; Zhang, J.; Dong, A.; Lin, C.; Zhang, J. A Multitasking Hydrogel Based on Double Dynamic Network with Quadruple-Stimuli Sensitiveness, Autonomic Self-Healing Property, and Biomimetic Adhesion Ability. *Macromol. Chem. Phys.* **2017**, *218*, 1700166.
41. Collins, J.; Nadgorny, M.; Xiao, Z.; Connal, L.A. Doubly Dynamic Self-Healing Materials Based on Oxime Click Chemistry and Boronic

Acids. *Macromol. Rapid Commun.* **2017**, *38*, 1600760.

42. Zhou, L.; Li, J.; Luo, Q.; Zhu, J.; Zou, H.; Gao, Y.; Wang, L.; Xu, J.; Dong, Z.; Liu, J. Dual stimuli-responsive supramolecular pseudo-polyrotaxane hydrogels. *Soft Matter* **2013**, *9*, 4635–4641.
43. Liu, J.; Tan, C.S.Y.; Lan, Y.; Scherman, O.A. Toward a Versatile Toolbox for Cucurbit[*n*]uril-Based Supramolecular Hydrogel Networks through In Situ Polymerization. *J. Polym. Sci. Part A Polym. Chem.* **2017**, *55*, 3105–3109.
44. Deng, G.; Ma, Q.; Yu, H.; Zhang, Y.; Yan, Z.; Liu, F.; Liu, C.; Jiang, H.; Chen, Y. Macroscopic Organohydrogel Hybrid from Rapid Adhesion between Dynamic Covalent Hydrogel and Organogel. *ACS Macro Lett.* **2015**, *4*, 467–471.
45. Yan, F.; Zhang, X.; Ren, H.; Meng, X.; Qiu, D. Reinforcement of polyacrylamide hydrogel with patched laponite-polymer composite particles. *Colloids Surf. A Physicochem. Eng. Asp.* **2017**, *529*, 268–273.
46. Inal, M.; Isiklan, N.; Yigitoglu, M. Preparation and characterization of pH-sensitive alginate-g-poly(*N*-vinyl-2-pyrrolidone)/gelatin blend beads. *J. Ind. Eng. Chem.* **2017**, *52*, 128–137.
47. Wang, L.; Hui, X.; Geng, H.; Ye, L.; Zhang, A.; Shao, Z.; Feng, Z. Synthesis and gelation capability of mono- and disubstituted cyclo(*L*-Glu-*L*-Glu) derivatives with tyramine, tyrosine and phenylalanine. *Colloid Polym. Sci.* **2017**, *295*, 1549–1561.
48. Nunes, C.S.; Rufato, K.B.; Souza, P.R.; de Almeida, E.A.M.S.; da Silva, M.J.V.; Scariot, D.B.; Nakamura, C.V.; Rosa, F.A.; Martins, A.F.; Muniz, E.C. Chitosan/chondroitin sulfate hydrogels prepared in [Hmim][HSO_4] ionic liquid. *Carbohydr. Polym.* **2017**, *170*, 99–106.
49. Yesilyurt, V.; Webber, M.J.; Appel, E.A.; Godwin, C.; Langer, R.; Anderson, D.G. Injectable Self-Healing Glucose-Responsive Hydrogels with pH-Regulated Mechanical Properties. *Adv. Mater.* **2016**, *28*, 86–91.
50. Czarnecki, S.; Rossow, T.; Seiffert, S. Hybrid Polymer-Network Hydrogels with Tunable Mechanical Response. *Polymers* **2016**, *8*, 82.
51. Raza-Karimi, A.; Khodadadi, A. Mechanically Robust 3D Nanostructure Chitosan-Based Hydrogels with Autonomic Self-Healing Properties. *ACS Appl. Mater. Interfaces* **2016**, *8*, 27254–27263.
52. Iftime, M.; Morariu, S.; Marin, L. Salicyl-imine-chitosan hydrogels: Supramolecular architecturing as a crosslinking method toward

multifunctional hydrogels. *Carbohydr. Polym.* **2017**, *165*, 39–50.

53. Shao, C.; Wang, M.; Chang, H.; Xu, F.; Yang, J. A Self-Healing Cellulose Nanocrystal-Poly(ethylene glycol) Nanocomposite Hydrogel via Diels-Alder Click Reaction. *ACS Sustain. Chem. Eng.* **2017**, *5*, 6167–6174.

CHAPTER

16

Design of Moldable Hydrogels for Biomedical Applications Using Dynamic Covalent Boronic Esters

B.Marco-DufortM.W.Tibbitt

Macromolecular Engineering Laboratory, Department of Mechanical and Process Engineering, ETH Zürich, Zürich 8092, Switzerland

ABSTRACT

Traditional polymeric materials based on thermosets or thermoplastics are applied broadly as biomedical materials. While attractive for a range of applications, thermosets and thermoplastics can be limited by their relatively static materials properties. Recent efforts in materials design has focused on engineering responsive and adaptive networks based on dynamic covalent chemistries. Installing reversible chemistries within the network backbone enables breaking and reforming of bonds in the network and associated rearrangement of the material on experimental timescales. The complexation

Citation: B. Marco-DufortM.W. Tibbitt. "Design of moldable hydrogels for biomedical applications using dynamic covalent boronic esters" Materials Today Chemistry 12 (2019) 16-33. https://doi.org/10.1016/j.mtchem.2018.12.001

between boronic acids and diols to form reversible boronic esters has emerged as a safe and synthetically tractable dynamic covalent cross-linking motif for the design of stimuli-responsive biomedical materials. Here, we present an instructive review on the design of dynamic covalent networks and gels using boronic ester cross-links. We provide a detailed discussion of boronic ester chemistry with guidelines for tuning the binding based on synthetic modification. We explain how network topology and connectivity influence the macroscale properties of the assembled networks. In addition, we discuss how these design principles have been used in foundational and emerging biomedical applications of boronic ester–based hydrogels. The use of boronic esters as dynamic covalent cross-links will continue to produce materials with emergent dynamic properties, and the design principles presented here will aid in the fabrication of next-generation boronic ester-based biomaterials.

Keywords: Dynamic covalent chemistry, Adaptable polymer networks, Soft matter, Polymer engineering, Biomedical materials

INTRODUCTION

Dynamic covalent chemistries offer an attractive route for soft materials fabrication, where the strength of covalent bonds is combined with the reversibility of non-covalent interactions [1]. Polymer networks are formed through physical or chemical cross-linking of functional monomeric or polymeric precursors. Physically cross-linked networks are held together with non-covalent, reversible interactions, such as hydrogen bonding, hydrophobic interactions, or metal-ligand coordination [2]. These often result in shear-thinning (viscous flow upon increased shear) and self-healing (reformation of gel properties after shear cessation) networks, as the bonds can break and reform in response to external stimuli, including force. However, the structures formed are often unstable to small environmental perturbations and lack robustness. Chemically cross-linked networks, on the other hand, are held together by permanent covalent bonds [3]. These give rise to elastic gels with generally higher mechanical properties as compared with physical networks, but the irreversibility of chemical cross-links limits their use for applications that require shear-thinning and self-healing, such as 3D printing or minimally invasive drug delivery.

Recent research in the design and engineering of polymer networks has introduced a new class of soft matter based on dynamic covalent chemistry, which combines the benefits of both physically and chemically cross-linked materials [4]. In this approach, reversible covalent bonds are installed in the network backbone that can break and reform on experimental timescales. Thus, these dynamic covalent networks can rearrange because of the exchange of bonds, enabling stress relaxation and material flow. Often, the binding affinity of the dynamic covalent chemistry is sensitive to changes in environmental conditions, allowing for stimuli-responsive materials properties. Chemistries that have been used in dynamic covalent network formation include transesterification [5], Diels-Alder (DA) cycloaddition [6], and boronic ester complexation [7]. These have enabled the fabrication of processable and recyclable vitrimers, temperature-induced self-healing networks, and injectable hydrogels, respectively. In each case, the ability to break and reform bonds at the molecular scale led to emergent dynamics at the macroscale.

In this review, we present an overview of how boronic ester bonds can be used to engineer dynamic covalent networks and gels. After introducing dynamic covalent chemistry and boronic ester bonds, we provide a thorough discussion of how chemistry influences the binding kinetics and thermodynamics for boronic ester formation and breakage. We then discuss how network formation and topology influence the macroscale properties of dynamic covalent networks. In addition, we survey foundational and emerging biomedical applications that leverage these chemical and physical insights for materials fabrication using boronic ester cross-links. Finally, we conclude with an outlook on the future of this field. In total, this review is intended for chemists, materials scientists, and bioengineers interested in exploiting the full utility of dynamic covalent boronic ester bonds in the design of responsive polymer networks and gels.

Dynamic covalent chemistry

In dynamic covalent chemistries, the competing chemical reaction pathways leading to different products are under thermodynamic control. The structures produced are thus not only reversible but also strong and tough, since they are held together by covalent bonds. In such systems, the relative stabilities ΔG^0 of the products determine their final distribution; rather than the relative energies ΔE_a required to form them, as in kinetically controlled systems [8]. As a result, slight modifications to the starting materials (constitutional, steric, or electronic) or changes to the external environment (such as pH, temperature, or concentration) can significantly alter the thermodynamic equilibrium [8]. Dynamic covalent systems thus adapt to changing conditions because their molecular components can easily assemble and disassemble into different product distributions in response to shifts in the chemical equilibrium.

Dynamic moieties can therefore be incorporated into polymeric networks, allowing for rearrangement of bond connectivity, as shown in Fig. 1. By continuously breaking and reforming bonds, dynamic covalent networks adapt to physical or chemical cues (such as mechanical loading, pH, or temperature), resulting in highly processable and self-healing materials with frequency-dependent mechanical properties. For example, disulfide cross-links (Fig. 1a) were established early on to be reversible [9], [10]. Later, the formation and cleavage of disulfide bonds in response to light and pH was exploited to fabricate reversible networks [11], [12]. Another common strategy to produce dynamic networks involves the DA cycloaddition reaction. Chen et al. synthesized a highly cross-linked polymer network via the DA reaction of furan and maleimide moieties [6]. Network rearrangement was achieved after heating, which favors the retro-DA reaction. In a different study, Skene and Lehn leveraged acylhydrazone functionalities in dynamic polyamides to reversibly exchange repeating monomers [13]. In each of these reactions, the sequence of bond-breaking and formation involves a transition of decreased network connectivity; a bond must be broken before another one is formed. Different mechanical properties can also be achieved when the bond-breaking sequence occurs without a loss of connectivity, as in reversible exchange reactions [4]. Prominently, Montarnal et al. formed irreversible epoxy networks that can rearrange their topology by undergoing a trans esterification reaction (Fig. 1c), where the shuffling of the bonds occurs without depolymerization [5].

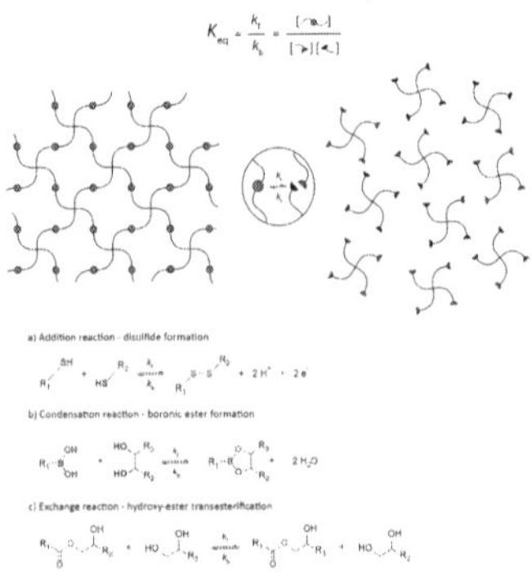

Figure. 1. Dynamic covalent network formation. Moldable polymer networks and gels can be formed by cross-linking macromolecular precursors (for example with multi-arm star polymers as shown here) using dynamic covalent

chemistries. The reversible cross-links can exist in the bound state (orange) or the unbound state (red and blue), and the exchange between bound and unbound is under thermodynamic control, as shown by K_{eq}. Constant breaking and formation of the bonds enables network strand rearrangement and stress relaxation as well as the reversible transition between the gel and liquid state (as shown here). Dynamic covalent chemistries used for moldable network formation can be categorized as **(a)** addition reactions exemplified by disulfide formation, **(b)** condensation reactions including boronic ester formation, which is the focus of this review, and **(c)** exchange reactions such as trans esterification.

Critically, the kinetics associated with dynamic covalent bond formation are usually slow because the lifetime of the bonds must be long enough to produce stable and robust structures. This often requires the use of a catalyst, the application of heat, or extreme changes in pH to equilibrate the thermodynamically stable product on experimental timescales [8]. As a result, a current focus in the field of dynamic covalent networks is to find materials which are suitable for biomedical applications.

Boronic esters as dynamic covalent cross-links

Boronic ester cross-linked networks have emerged as a promising class of dynamic materials. The reversible condensation reaction between boronic acids and *cis*-1,2 or *cis*-1,3 diols to form a cyclic ester (Fig. 1b) occurs in aqueous solutions under mild conditions, at ambient temperature, and requires no catalyst [14]. This enables boronic ester formation under physiological conditions. Furthermore, the reactivity of the boronic acid groups can be tuned using simple neighboring group effects, resulting in materials with highly tuneable mechanical properties that range from mechanically static to extremely malleable [15]. Building from initial studies on the use of boronic acids in medicine, boronic esters have become a common cross-linking motif in the design of biomaterials. The behavior of these materials depends directly on the underlying binding chemistry of the selected boronic ester, and rational design therefore relies on a robust understanding of how to tailor the kinetics and thermodynamics of boronic ester bond formation and breaking.

Biomedical applications of boronic acids

Boronic acids already enjoy wide usage in medicine owing to their low toxicity [16] and biofunctionality [17]. One of the earliest and best-known biomedical uses of boronic acids was as enzyme inhibitors. In

2003, the US Food and Drug Administration (FDA) approved bortezomib as the first proteasome inhibitor for human use [18]. This anticancer drug consists of a dipeptide containing a boronic acid that competitively binds to the hydroxyl groups found in the active site of serine proteases [19]. Upon binding, a reversible tetrahedral intermediate is formed which mimics the normal transition state and inhibits enzymatic activity [20].

Another important biomedical use for boronic acids has been as synthetic chemical sensors for biomolecules. In 1959, Lorand and Edwards calculated the binding affinities of various polyols to the benzeneborate ion by measuring the associated drop in pH upon combining them in aqueous solutions [21].

They showed that the reversible interaction between a boronic acid and a diol shifts when external conditions such as pH and temperature are changed, as it is under thermodynamic control. Conversely, the addition of free diols to an aqueous solution containing boronic acids would also shift the equilibrium, causing an associated change in measurable characteristics such as pH or solubility.

This discovery forms the basis for the real-time molecular sensing of many biologically important diol-containing molecules. Research has focused mainly on the detection of saccharides, although sensing of other molecules such as ATP [22] or even heavy metal ions [23] has been demonstrated. A typical device design is to immobilize boronic acid moieties on a substrate. The selective binding of sugars on the device is accompanied by a shift in thermodynamic equilibrium, which results in changes in the measurable properties of the system [14].

For example, optical sensors rely on the change in UV–Vis absorption when certain boronic acids are bound to saccharides [24] while fluorescence sensors measure the intensity difference of bound and unbound fluorophore-functionalized boronic acids [25]. Several recent reviews have highlighted the extensive progress in the field of molecular sensing by boronic acids of saccharides [26], [27] and of glucose in particular [28].

The broad utility of boronic acids is a direct consequence of their dynamic functionality. Interestingly, the same properties of boronic acids that make them attractive for biomedical applications have also been used to form dynamic networks cross-linked with boronic esters. The unique properties of these materials are related directly to the dynamic covalent bonding between boronic acids and diols, which will now be discussed.

CHEMISTRY OF BORONIC ESTERS

Boronic acid chemistry

Boronic acids are weak organic Lewis acids. The ionic equilibrium in aqueous solution between the neutral and anionic forms of boronic acids is shown in Fig. 2. Under acidic conditions, the neutral form is favored, with a vacant p-orbital [29]. The B-atom is sp^2–hybridized and adopts a trigonal planar configuration with an O-B-O bond angle of 120°. Under more basic conditions, the electron-deficient B-atom is attacked by OH^- ions, forming a hydroxyboronate anion. Upon OH^- ion complexation, the B-atom becomes sp^3-hybridized and adopts a tetrahedral configuration with a bond angle of ~109.5° [21], [29]. These two species have very different electronic properties; the boronic acid is electron-accepting, while the boronate anion is electron-donating [30]. Thus, any chemical reaction involving a boronic acid is highly dependent on the equilibrium between its neutral and anionic forms.

Figure. 2. Ionic equilibrium of boronic acids. In aqueous solution, boronic acids exist in ionic equilibrium between their neutral and anionic forms, which possess different electronic and structural properties. The neutral form is electron-accepting with a trigonal planar configuration, whereas the anion is electron-donating with a tetrahedral configuration. Boronic ester formation occurs through complexation of the boronic acid with *cis*-diols, and the rate depends on the equilibrium between the neutral and anionic forms.

Chemistry of boronic ester complexation

Boronic acids interact with polyols in aqueous solution to form reversible and cyclic esters [31]. Fig. 3 summarizes the reaction landscape of boronic ester formation, highlighting the possible pathways for the reaction of boronic acids with *cis*-diol moieties. As shown in Fig. 3, both cyclic boronic and boronate esters are formed with their B-atoms, sp^2-and sp^3-hybridized, respectively. The boronic and boronate esters exist in ionic equilibrium. Initially, it was believed that the boronate anion was much more reactive than the neutral acid because ester formation is favored at high pH, where the concentration of the anions are high. This was supported by kinetic studies based on the pH-depression method [21], which suggested that the boronate anion was 10^4 times faster than the neutral boronic acid in forming esters with diols [32]. More recently, however, Springsteen and Wang revealed discrepancies with these earlier measurements when investigating boronic ester stability with the fluorescent reporter Alizarin Red S. (ARS) [33]. In fact, the precise kinetics of boronic acid-diol complexation are much more complicated than originally thought. Other factors have to be considered, such as the buffer system. For instance, a medium dependence on the binding affinities between boronic acids and diols was observed because of the formation of binary and ternary complexes with common buffer components [34]. Furthermore, some kinetic paths are indistinguishable from each other due to the effects of 'proton ambiguity' [35]. It is now proposed that the preferred kinetic pathway for esterification is through the addition of diols to the neutral boronic acid, rather than through substitution of the hydroxyl ion in the anionic species [30]. This would imply that the reaction rates of esterification are inversely proportional to pH. Evidence of this can be found in materials cross-linked with boronic esters, as longer gel times are observed with increasing pH of the precursor solutions [36].

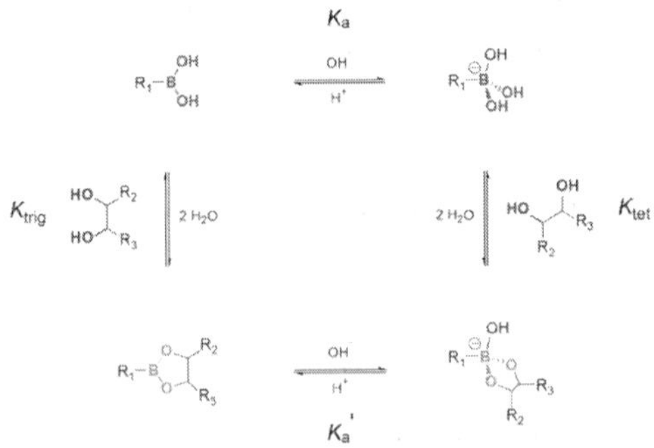

Figure. 3. Reaction landscape of boronic ester formation. Both neutral and anionic forms of the boronic acid exist in thermodynamic equilibrium with *cis*-diols in aqueous solution, forming boronic and boronate esters via condensation reactions. K_{trig} and K_{tet} are the equilibrium reaction constants for the pathways between the trigonal neutral and the tetrahedral anionic forms, respectively. K_a and K_a' are the ionization constants of the boronic acid and boronic ester, respectively.

To account for the multiple ionization states of the acid, ester, and diols in aqueous solution, Van Duin et al. proposed the 'charge rule' to determine the pH-dependent reactivity of the ester. The authors hypothesized that the maximum amount of esters are found at the pH at which the 'sum of the charges of the free esterifying species is equal to the charge of the ester' [37].

Since the pK_a of a molecule defines the pH at which 50% of the neutral groups are converted to their anionic forms, this rule has been interpreted to mean that the optimal pH for ester formation is somewhere in between the pK_a values of the boronic acid and the diol. A simple approximation to this value is to use the mean pK_a of the boronic acid and diol, as shown in Eq. (1):

$$pH_{optimal} = \frac{\mathrm{pK}_{a-acid} + \mathrm{pK}_{a-diol}}{2} \tag{1}$$

The validity of this relationship has been confirmed experimentally by several studies that examined the effects of substituents on the affinity of boronic acids to diols [30], [38]. Recently, a more complete general equation for this relationship was independently derived by Martínez-Aguirre et al. from equilibrium constant measurements and mass balances [39]. Given that diols are almost always less acidic than boronic acids, the charge rule implies that to find an optimal acid-diol pair for physiological use, the pK_a of the boronic acid must be equal to or less than 7.4. To this end, boronic acids have been modified chemically to tune their pK_a, allowing their tailored use for specific conditions and different applications.

Ionization constants (pK_a) of boronic acids

To broaden the utility of boronic acids for ester formation in balanced pH solutions (pH ~7) and for biomedical applications, new synthetic variants of boronic acid have been designed with tailored pK_a. Table 1 highlights several boronic acid derivatives and their associated pK_a values as published in the literature. In general, a few design guidelines have emerged. Electron-withdrawing groups lower the pK_a of the boronic acid, whereas electron-donating groups increase its pK_a. This is because the more acidic boronic acids contain the more electrophilic boron atoms, which improve the formation and stabilization of the boronate anion [29]. Indeed, alkylboronic acids are generally less acidic then arylboronic acids. For example, the pK_a of methylboronic acid (10.4) is much higher than that of phenylboronic acid (PBA) (8.8). From Eq. (1), however, PBA is not ideal for physiological use as its pK_a is greater than 7.4. To improve this, strong electron withdrawing groups have been introduced into the phenyl ring of PBAs in an attempt to lower their pK_a [40]. Table 2 shows how the addition of substituent groups with different electronic properties significantly alters the acidity of PBA derivatives. Electron-rich derivatives of PBA such as 2-methyl and

2-methoxy PBA have higher pK_a values (9.7 and 9.0, respectively) than electron-poor derivatives such as 3-methoxycarbonyl-5-nitro and 2-fluoro-5-nitro PBA (6.9 and 6.0, respectively). Plotting the pK_a of substituted PBAs against the Hammet σ-values of the substituents yields a straight line with a slope of 2.1 [38], indicating the formation of an anionic product as shown in Fig. 2. However, some exceptions to this trend exist. For example, the pK_a of 2-methyl PBA is higher (9.7) than that of 4-methyl PBA (9.3), which is explained by the additional steric hindrance of the 2-methyl group during complexation to the boronate anion [29].

Table 1. Ionization constants for selected boronic acids.

Chemical name	Structure	pK_a	Reference
Boric acid	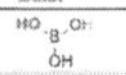	[illegible]	a
Methylboronic acid		10.4	b
Phenylboronic acid	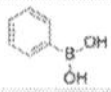	8.8	c
2-Hydroxymethylphenyl boronic acid	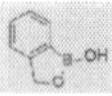	7.2	d
2-Dimethylaminomethyl phenylboronic acid	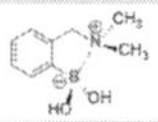	5.2	e
3-Pyridylboronic acid	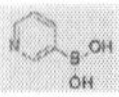	4.0	f

1. The references for the pK_a values taken from literature are (a) [111], (b) [32], (c) [38], (d) [41], (e) [42], and (f) [44].

Table 2. Ionization constants for selected phenylboronic acid derivatives.

Chemical name	Structure	pK_a	Reference 1
2-Methylphenyl boronic acid	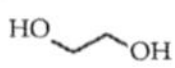	9.7	a
4-Methylphenyl boronic acid	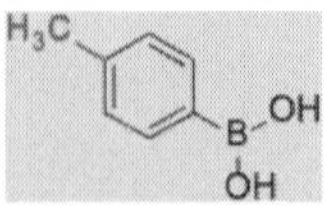	9.3	a
2-Methoxyphenyl boronic acid	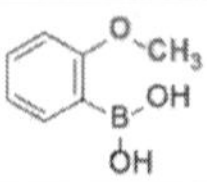	9.0	b
4-Carboxyphenyl boronic acid	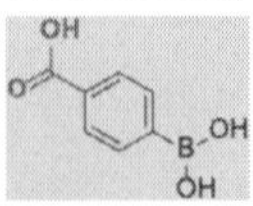	8.4	c
2-Fluorophenyl boronic acid	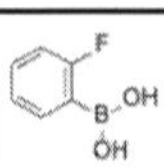	7.9	d

3-Nitrophenyl boronic acid	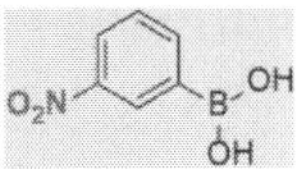	7.0	e
3-Methoxycarbonyl-5-nitro phenylboronic acid	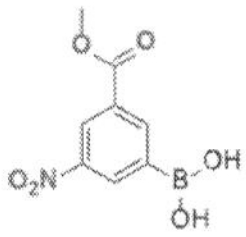	6.9	f
2-Fluoro-5-nitrophenyl boronic acid	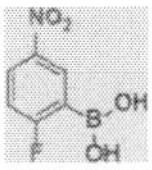	6.0	b

1. The reference for the pKa values taken from literature are (a) [29], (b) [38], (c) [112], (d) [113], (e) [114], and (f) [115].

Recently attention has focused on 2-hydroxymethyl PBA, shown in Table 1. In water, it spontaneously converts to a cyclic benzoboroxole (also known as a benzoxaborole) via intramolecular B-O coordination. This accounts for its relatively low pK_a (7.2), which is largely because of a decrease in ring strain associated with rearrangement from the neutral sp_2–hybridized form to the anionic sp_3–hybridized form. Other key advantages of benzoboroxoles are their high water-solubility and their ability to bind to many different types of diols [41].

'Wulff-type' boronic acids, such as 2-aminomethyl PBA (pK_a = 5.2) shown in Table 1, are commonly used for the sensing of saccharides [41]. In aqueous solution, a dative bond is formed between the N and B atoms in these types of acids, which enhances binding at neutral pH [42]. The exact reasons behind this lower pK_a are still debated, with some citing a shift to sp_3–hybridized boron after the B-N adduct is formed while others believe that there is ionic stabilization of the boronate center by a cationic ammonium center [43]. Finally, zwitterionic and heteroaromatic compounds, such as 3-pyridylboronic acid, shown in Table 1, are known to have pK_a values as low as 4.0 [44].

Therefore, with the appropriate chemical modification, the pK_a of various boronic acids can be varied from ~4.0 to 10. As a result, the equilibrium of esterification can be shifted across a range of pH, dictated by the pK_a of the boronic acid derivative. This enables complexation in acidic and basic environments, including under physiological conditions. It is worth keeping in mind that the acidity of the diol also plays a role in determining the optimal pH, especially in the rare case when the diol is more acidic than the boronic acid. A notable example occurs during the binding between ARS (pK_a = 6.9) and PBA (pK_a = 8.8), where the optimal pH was found to be ~7.0 [33].

Binding affinities of diols to boronic acids

Another critical factor to consider for the stability of boronic ester bonds is the binding affinity of the diol to the boronic acid, which is a measure of the strength of the interaction between the diol and the boronic acid. This can be captured by K_{eq}, the equilibrium dissociation constant for the esterification reaction, which determines the composition of the system at thermodynamic equilibrium, irrespective of the starting conditions. Fig. 1 shows how K_{eq} is related to the quotient of the concentrations of the species involved in dynamic covalent network formation, as well as to the forward and backwards rate constants k_f and k_b of the chemical reaction [8]. In 1959, Lorand and Edwards carried out the first systematic examination of the binding affinities between boronic acids and diols. Their reported formation constants between selected polyols and PBA are listed in Table 3, under Method 1.

The pH depression method used, however, cannot fully capture the complexity of boronic acid-diol binding. Their experimental design was based on the formation constant being directly proportional to the drop in pH associated upon the formation of a boronate ester. This method assumes that only the tetrahedral boronate anions are reacting with the diols; however, it is known that the neutral trigonal species also play a role.

Therefore, recent studies have proposed that the binding constants measured by Lorand and Edwards are values for K_{tet} rather than the overall K_{eq} [33]. In an attempt to measure the overall binding affinities, Springsteen and Wang developed a method based on the competitive displacement of the fluorescent diol–containing ARS. Their findings are summarized in Table 3, under Method 2. Even though there is a lot of variability in the absolute magnitude of the reported binding constants between the two different methods, the observed trends are consistent.

Table 3. Binding affinities for selected polyols. The binding affinities (K_{tet} or K_{eq}) for the complexation of certain polyols with PBA as calculated via pH depression (Method 1) [21] or competitive displacement (Method 2) [33]. PBA, phenylboronic acid.

Chemical name	Structure	Method 1 K_{tet} (M^{-1})	Method 2 K_{eq} (M^{-1})
Ethylene glycol	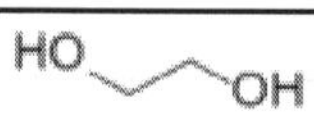	2.76	–

Glycerol	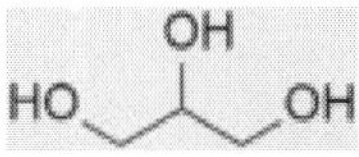	19.7	–
Mannitol	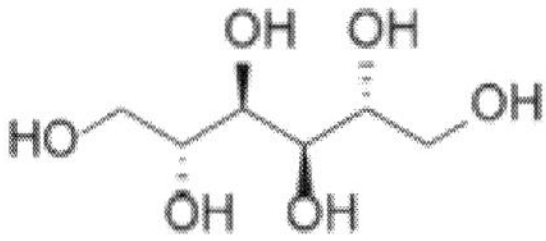	2275	120
Catechol	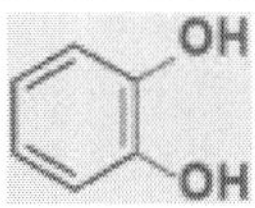	17500	830
Alizarin Red S.	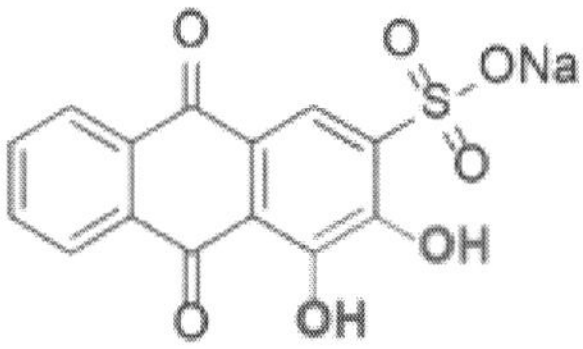	–	1300
D-glucose	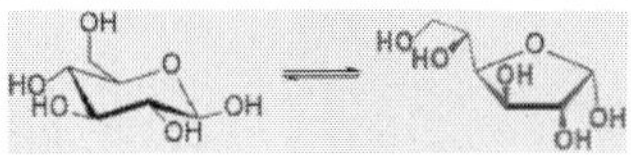	110	4.6
D-galactose		276	15
D-fructose	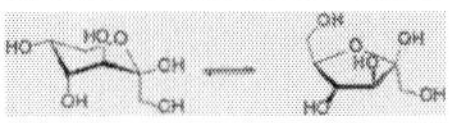	4370	160
Lactose	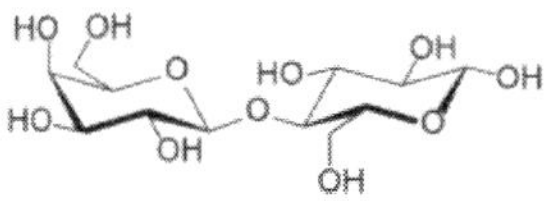	–	1.6

The order of the affinities is determined by the relative position and orientation of the hydroxyl groups in the diol [26]. As the stability of the ester is determined by the amount of strain introduced upon binding of the borate to the diol, the sequence of stabilities should follow the degree of preorganization of the starting diols [30]. Indeed, both studies report that catechol and ARS have the highest binding constants. The rigidity of the coplanar vicinal *cis*-diols in these molecules (highlighted in blue in Table 3) was suggested to account for their enhanced reactivity, as the loss of configurational entropy suffered upon complexation is lower than that of ligands with additional degrees of freedom [41].

A different trend is observed with the acyclic sugar alcohols. Looking at ethylene glycol, glycerol and mannitol in Table 3 reveals how their binding affinity increases with their number of hydroxyl groups, because statistically the possibilities of binding to a boronic acid increase with the number of diols. However, this entropic contribution is not enough to account for the huge increase in stability observed by Lorand and Edwards going from glycerol (19.7 M^{-1}) to mannitol (2275 M^{-1}). Possible explanations for this include a chelating effect in the stepwise hydrolysis of the ester bond and stabilization due to intramolecular hydrogen bonding [45].

The studies also reported large differences in the stability constants of different monosaccharides. For instance, Lorand and Edwards showed that the binding constant of fructose (4370 M^{-1}) is much higher than that of glucose (110 M^{-1}) and galactose (276 M^{-1}). To account for this, the different forms of the saccharides in aqueous solution have to be considered. Table 3 shows the equilibrium between the pyranose (left) and furanose (right) forms of glucose, galactose, and fructose. Studies have shown that the most stable boronic acid-diol complexes are formed with the conformationally locked syn-periplanar hydroxyl groups on furanose rings (highlighted in blue in Table 3), while the more disorganized pyranose diols have much lower binding affinities [46], [47].

Therefore, the saccharides with a larger relative percentage of furanose in their natural speciation should have the higher stability constants. This is exactly what was observed: the composition of furanose in solution is 0.14% for glucose, followed by 2.5% for galactose and 25% for fructose [48]. Furthermore, the disaccharide lactose, which can only exist in the pyranose form, was found to have a lower binding affinity (1.6 M^{-1}) than its components glucose (4.6 M^{-1}) and galactose (15 M^{-1}).

The chemical details of boronic ester formation and breakage are complicated and remain an area of active research. Nonetheless, robust characterization of the pK_a of different boronic acids as well as the binding affinities for common boronic acid-diol pairs provides the user with sufficient guidelines to predict boronic ester behavior as cross-links in polymer networks. While the field will benefit from new chemistries and additional characterization of boronic ester binding kinetics and thermodynamics, the tools already exist to form materials from dynamic covalent boronic ester bonds. Control over materials properties in these dynamic covalent networks requires an understanding of the binding partners as well as how network formation and connectivity dictate macroscale properties. Engineering principles for network formation are discussed in the following section.

ENGINEERING NETWORKS WITH BORONIC ESTER BONDS

As mentioned above, dynamic covalent chemistries, including boronic ester bonds, enable the formation of responsive and moldable materials as the bonds can rearrange by breaking and reforming upon application of external stimuli on experimental timescales [43]. The main considerations in the design of these adaptable polymer networks from dynamic covalent chemistries are the specific chemistry of the binding pair and the mechanism of network formation. The specific boronic acid-diol chemistry defines the thermodynamics and kinetics of boronic ester bond formation, breaking, and reformation, which controls the dynamics of network strand rearrangement. The mechanism of network formation governs the topology of the network or how the network strands are linked together, which in turn influences the mechanical properties (such as gel point, plateau modulus, relaxation time, and swelling) of the resultant viscoelastic material. Simply put, the properties of polymer networks and gels are controlled by how the molecules in the network are connected [49]. Broad use of dynamic covalent networks and gels requires a robust understanding of how these factors influence emergent network properties. In this section, we discuss how network formation and topology relate to the mechanical properties of dynamic covalent networks.

Network formation and gel point

Polymer network formation occurs by cross-linking multifunctional monomer or macromer precursors [50]. Here, a functional group refers to a chemical moiety that can participate in the formation of a single dynamic covalent bond. Boronic ester cross-linking occurs via complementary bonding between boronic acid and diol moieties, as shown in Fig. 4. If each precursor molecule has a functionality of exactly two, the reaction will lead to chain extension and growth of individual polymer chains. On the other hand, if the average functionality is greater than two, branching can occur which bonds polymer chains together. At a critical extent of reaction p_c, the cross-linking of disparate polymer chains through the branch points induces a phase transition from liquid-like behavior to solid-like behavior as a three-dimensional polymer network percolates across the entire system [50]. The critical point p_c is referred to as the gel point. Just below the gel point, the system exists as a solution of polydisperse branched polymers, and beyond the gel point, at least some fraction of the system exists as an 'infinite' molecular weight polymer network that spans the entire volume. The gel point can be calculated based on the functionality and structure of the network forming precursors, using mean-field models of gelation via the Flory [51] or Flory-Stockmayer [52], [53] approaches that calculate p_c as the conversion at which the weight-average molecular weight of the growing, branched polymer chains diverges.

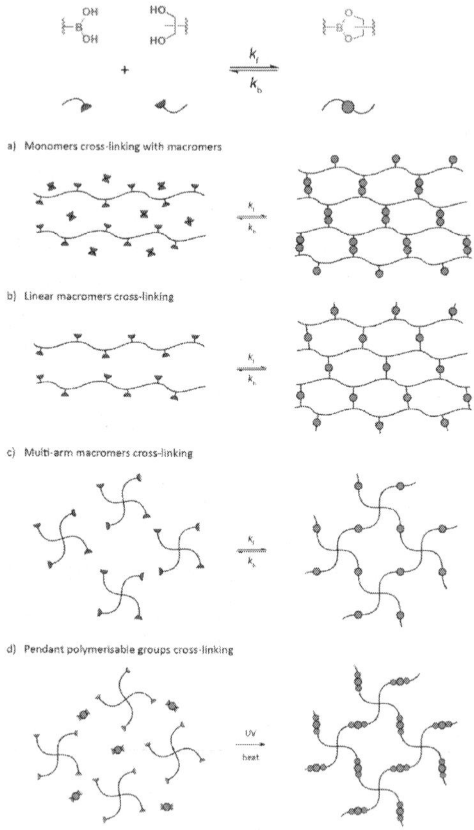

Figure. 4. Formation mechanisms for dynamic covalent networks. The choice of network formation and network connectivity, in cooperation with the selected dynamic covalent chemistry, influence macroscale properties of boronic ester-based networks and gels. In all cases, the reversible complexation between boronic acid (red) and diol (blue) forms dynamic covalent boronic esters (orange). Several strategies exist to form networks and gels using this cross-linking motif. **(a)** Multifunctional small boronic acid molecules can be used to cross-link polyvalent polymer chains, as exemplified by the gel formed by mixing borax and poly(vinyl alcohol). **(b)** Gels can also be formed by mixing polyvalent linear polymers whose backbones have been functionalized with diols and boronic acids. **(c)** In addition, dynamic covalent networks have been formed through the cross-linking of multi-arm star polymers. **(d)** Moldable networks can also be formed by cross-linking monomers that contain a previously formed boronic ester using an orthogonal pendant reactive group (purple). In all cases, the molecular weight of the polymers, the number of functional groups, the polymer concentration, and the stoichiometry between boronic acids and diols can be used to modulate bulk mechanical properties such as plateau modulus G_0' and relaxation time τ_r.

To exploit the reversible nature of the boronic ester bond to form dynamic covalent networks, boronic esters moieties need to be installed into the backbone of the forming polymer network. Two main mechanisms can be employed to introduce boronic ester bonds into the network backbone [4]. First, networks can be formed via direct cross-linking of multifunctional boronic acid and diol monomers or macromers, as shown in Fig. 4a, b, and c. In this manner, the cross-links themselves are composed of reversible and dynamic boronic esters during formation. Alternatively, networks can be formed from monomers or macromers that contain one or multiple boronic esters in their core and pendant polymerizable functional groups (such as (meth) acrylates, thiols and alkenes, or azides and alkynes), as shown in Fig. 4d. Here, the cross-linking chemistry and conditions must maintain the integrity of the boronic ester bond. As most boronic esters are highly dynamic in aqueous solution at standard pH [116], the former mechanism that uses the boronic ester as the cross-linking chemistry itself is more common for the formation of boronic ester–based hydrogels and will be the focus of the rest of this section.

In covalently cross-linked networks and gels, a formed bond is permanent, and calculation of the gel point is relatively straightforward. For example, the critical extent of reaction p_c based on the Flory approach can be calculated for multifunctional precursors as:(2)pc=1f−1where f is the average functionality of the precursors [51], [54]. In the Flory-Stockmayer approach, the critical extent of reaction p_c for a step-growth polymerization between precursor A with functionality f_A and precursor B with functionality f_B can be calculated as:(3)pc=1r(fA−1)(fB−1)where r is the stoichiometric ratio between A and B with $r \in [0,1]$ [53]. Note, the two approaches converge for $r = 1$ and $f_A = f_B$.

For dynamic covalent bonds, such as boronic esters, the reacted bonds are not permanent; the bonds are in flux as they continually rearrange. Therefore, gelation in dynamic covalent networks relates also to the equilibrium constant K_{eq} and the concentration of the functional groups [C]. At sufficiently long times after mixing, one can assume that the system has reached thermodynamic equilibrium and that the number of formed bonds or 'extent of reaction' is determined solely by the equilibrium constant for the dynamic covalent chemistry and the concentration of functional groups. Thus, the number of formed bonds $p \sim f(K_{eq},[C])$ must remain above p_c to maintain a gel. As discussed above, K_{eq} for boronic ester bonds varies with changes in environmental factors, including pH, temperature, or stress. Modulating K_{eq} provides a facile handle to reverse gelation in dynamic covalent networks and enables easy processing of the networks. External stimuli can transform the network from a solid to a liquid and back again repeatedly, allowing for recycling, healing, as well as shaping and reforming through subtle shifts in the external conditions.

A thorough treatment of the gel point for associating polymers was developed by Semenov and Rubinstein [55]. The guidelines here provide descriptive indicators for what systems will or will not form a gel and how external parameters can be used to transition between sol and gel, reversibly. This understanding has enabled the design and synthesis of multifunctional precursors that have been used to engineer responsive materials exploiting the dynamic covalent nature of boronic ester bonds [43]. This has been demonstrated in the design of moldable materials from borax and polysaccharides, glucose-responsive gels, and self-healing polymer networks that will be discussed further below.

Dynamic mechanical analysis

Engineering properties of major interest in the design of polymeric materials, including dynamic covalent networks, are the plateau modulus and the relaxation time. The modulus of elasticity E for a material defines the ratio between the stress applied to a material and the resultant deformation in the linear elastic region. The relaxation time τ_r characterizes the timescale within which strain-induced stresses within the material are dissipated [50]. A standard technique for quantifying the linear mechanical properties of viscoelastic materials is to measure the stress in the material in response to small sinusoidal deformations. The application of cyclic strain, called dynamic mechanical analysis, is employed commonly for soft materials ($E \sim 10^1$–10^8 Pa) characterization using a shear rheometer, which measures the shear modulus G of the material, or the ratio of shear stress to shear strain. G and E are related via the Poisson's ratio [56].

Shear rheometry probes materials properties with an applied oscillatory shear strain. It measures the frequency-dependent dynamic or complex modulus G^* of the material, which can be expressed in terms of the storage modulus $G'(\omega)$—a measure of the stored elastic energy—and the loss modulus $G''(\omega)$—a representation of the viscous energy dissipation [56]. Tuning the frequency of the applied strain enables investigation of different timescales or relaxation modes within the material. Owing to the many hierarchical length scales present in polymeric materials that are inherently coupled to different timescales of molecular motion, mechanical properties are dependent on the frequency of analysis. High frequency measurements probe short times and thus small motions at the atom or monomer level while low frequency measurements probe long times and thus large motions at the chain or material level. As such, the frequency-dependent modulus of all polymeric materials increases at high frequency, indicative of the glassy dynamics associated with relaxation on the smallest length scales [4]. At lower frequencies, polymer segments and chains are able to relax, and the frequency-dependent modulus decreases to a plateau modulus G_0'. Both the plateau modulus and the glassy region are illustrated in Fig. 5a, which plots the frequency-dependent storage modulus $G'(\omega)$ for model polymer networks.

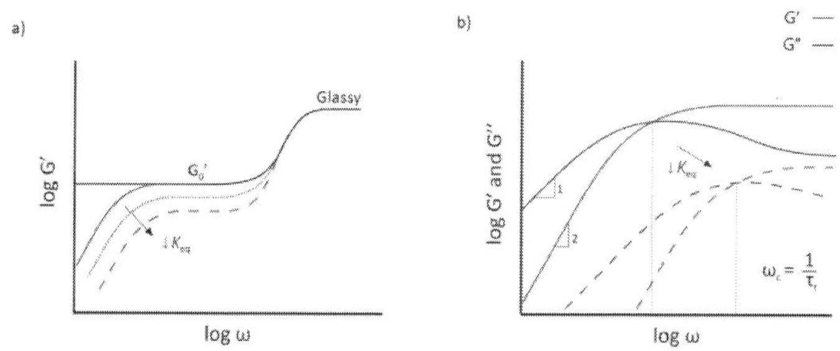

Figure. 5. Rheology of dynamic covalent networks. Dynamic mechanical analysis is used to assess the frequency-dependent properties of soft materials. **(a)** Traditional thermosets (black line) exhibit a plateau modulus G_0' dictated by the cross-linking density with increasing modulus at high frequency associated with glassy dynamics at short length scales. Traditional thermoplastics (red line) exhibit similar behavior except at low frequency where a decrease in modulus is observed associated with network strand rearrangement via curvilinear diffusion or reputation at long timescales. Dynamic covalent networks (red dotted and red dashed lines) exhibit glassy dynamics at high frequency, a plateau modulus at intermediate frequency, and decreasing modulus at low frequency associated with network rearrangement as bonds break and reform on long timescales. In addition, for condensation reactions such as boronic ester formation, the plateau modulus decreases with decreasing K_{eq}. **(b)** Dynamic covalent networks also exhibit network relaxation on experimental timescales with a corresponding relaxation time τ_r. The relaxation time is associated with the crossover frequency ω_c; below ω_c, the material behaves as a viscous fluid, and above ω_c, the material behaves as a viscoelastic solid. The crossover frequency shifts to shorter timescales with decreasing K_{eq}. Furthermore, the crossover point can also be influenced by the exact magnitudes of the forward and backward rate constants k_f and k_b.

Classic polymeric materials are subdivided into thermoplastics—polymer melts that flow with increasing temperature or dissolve upon addition of a good solvent—and thermosets—covalently cross-linked networks that do not change state with temperature prior to combustion and swell without dissolving upon addition of a good solvent. In thermosets, the topology of the network strands is fixed by the permanent covalent cross-links. These can fluctuate in three-dimensional space but prevent flow or dissolution of the network. In thermoplastics, the topology of the polymer strands is constrained by other polymer chains via physical interactions, entanglements, and excluded volume; however, at sufficiently long times, chains can overcome these physical constraints and move or reptate to relax internal stress.

Therefore, as their network topology is fixed, thermosets behave as ideal, elastic materials. They possess a plateau modulus associated with the cross-linking density of the material and a non-finite relaxation time. A representative curve for their ideal behavior is shown in Fig. 5a. Thermoplastics, on the other hand, also demonstrate a plateau modulus, as polymer diffusion is constrained by topological entanglements with other chains that act as effective cross-links at certain timescales. However, with decreasing frequency or longer timescales, the frequency-dependent modulus decays as the polymer chains are able to reptate via curvilinear diffusion, and thus thermoplastics possess a finite relaxation time, as shown in Fig. 5a.

Viscoelastic dynamic covalent networks exhibit rheological behavior that mirrors aspects of both thermosets and thermoplastics. Their behavior is characterized by the Maxwell model of viscoelasticity, which combines in series an elastic element with modulus G_M and a viscous element with viscosity η_M [50]. In the Maxwell model, a critical timescale is defined as $\tau_r = \eta_M/G_M$. The material will behave as a solid at timescales shorter than τ_r and flows like a viscous liquid on timescales longer than τ_r. That is, dynamic covalent networks possess a plateau storage modulus defined by the network cross-link density as well as a finite relaxation time defined by the chemistry of the dynamic covalent cross-links, as illustrated in Fig. 5a. A complete treatment of the polymer physics and scaling laws for the behavior of reversible polymer networks was developed by Rubinstein and Semenov, and we refer the reader to these studies for a more thorough discussion of the theoretical foundations of associating polymer networks [57]. Here, we restrict our discussion to general engineering design criteria for the design of dynamic covalent networks.

Plateau modulus

As previously discussed, the elastic modulus is a critical parameter in the design of soft materials. The theory of rubber elasticity for cross-linked polymer networks relates the molecular details and topology of the network to the modulus of the material [50]. At its core, rubber elasticity states that $E \sim$ T, where is the number density of network strands and T is the temperature of the material. The number density of network strands is directly related to the cross-link density ρ_x and further depends on the molecular properties of the network precursors, e.g., molecular weight, persistence length, and concentration. The recursive method of Miller and Macosko is commonly used to calculate the number density of network strands based on the properties of the precursor molecules [58].

The molecular foundation of rubber elasticity is the entropic penalty caused by deforming individual polymer chains away from their relaxed state, which leads to a restoring force of k_BT per network strand. Here, k_B is the Boltzmann constant and T is the temperature. Thus, the modulus scales linearly with the number density of network strands with each strand contributing k_BT to the modulus. Based on rubber elasticity, the shear modulus can be calculated directly as $G =$ k_BT. Thus, the plateau modulus from shear rheometry G_0' can be related to the number density of network strands in the cross-linked material. Rubber elasticity is based on the underlying assumptions that the network strands are Gaussian, and the cross-links are points in free space [50]. The model stated above, $G = \nu k_B T$, is based on the affine network model, which assumes that each network strand experiences the same deformation as the bulk material. The phantom network model accounts for the fact that the cross-links are not fixed in space but fluctuate around an average location. This leads to an overall decrease of the predicted modulus of the material as given by $G =$ k_BT (1–2/f), where f is the functionality at the cross-link junctions. Recently, more advanced versions of rubber elasticity have been developed that account for more complicated network topology that includes network loops and entanglements including the real elastic network theory or RENT [49]. In all models, the general scaling that $G \sim E \sim$ T holds is inherent to the thermoelastic nature of rubbery materials.As alluded to above, several design elements provide control over the network strand density and thus the elasticity in polymer networks and gels. Consider the step-growth polymerization of 4-arm star-polymers functionalized with boronic acid or diol derivatives, as depicted in Fig. 4c. Here, the network strand density can be tuned by altering the molecular weight of the individual polymer molecules and thereby the molecular weight of the individual arms. Additionally, the overall polymer concentration will affect the density of network strands, as well as the equilibrium bound fraction and the extent of network defects. Selecting backbone polymers with different rigidities or extents of hydration will also influence the modulus for a fixed polymer concentration. The network connectivity, or how many network strands link together at each cross-linking junction, also changes the density of active network strands for a given polymer concentration and can be increased by interchanging the 4-arm star-polymers for 8-arm star-polymers or polyfunctional linear polymers.

Uniquely, dynamic covalent networks also provide control over K_{eq} and thus the effective number of intact network strands. This affects both the gel point as described above but also the plateau storage modulus for dynamic addition and condensation networks such as boronic ester networks and gels. Some fraction of the boronic acid and diol functionalities exist in the non-bound state, and this fraction is dictated by the environmental conditions. As K_{eq} decreases, the network plateau modulus decreases, when all other factors are kept constant, as shown in Fig. 5a. The physical properties of reversible associating networks, which include dynamic covalent networks, have been explored theoretically by several groups. Green and Tobolsky proposed an early molecular theory of the physical properties of associating and viscoelastic polymer media [10]. These initial efforts have been continued by Wang and others to develop extended theories for the behavior of transient networks [59]. Leibler et al. presented a model for the dynamics of networks composed of linear polymers with reversible cross-links [60]. Semenov and Rubinstein provided an understanding of the statics and dynamics in thermoreversible gelation in associative polymers [55], [57]. Recent work from Tang et al. has adapted the phantom network model to predict plateau storage modulus for dynamic covalent networks that accounts for K_{eq} and thus the fraction of formed cross-links [61]. Building upon these directions, additional frameworks are needed to precisely link network topology and binding constants to network mechanics.

Relaxation time

The time-dependent viscoelastic properties of dynamic covalent networks can also be engineered by the cross-linking chemistry and network topology. As discussed above, these materials exhibit frequency-dependent storage $G'(\omega)$ and loss $G'(\omega)$ moduli, owing to the breaking and reformation of boronic ester bonds that occurs on experimental timescales. At high frequencies or short timescales, their properties are analogous to an elastic network; at low frequencies or long timescales, the material behaves like a viscous polymer solution. Oscillatory shear rheometry across a range of frequencies is used to quantify the viscoelastic behavior of dynamic covalent networks. In general, a critical frequency ω_c is observed when $G'(\omega) = G'(\omega)$, which is related to the relaxation time τ_r of the material, τr=2π/ωc. At frequencies below ω_c, the viscoelastic moduli should follow the classic scaling rules for terminal relaxation in a Maxwell model, where $G'(\omega) \sim \omega^2$ and $G'(\omega) \sim \omega^1$ [57]. Fig. 5b illustrates these behaviors and the effect of varying K_{eq} or the magnitude of the forward and reverse rate constants.

The characteristic timescale of the viscoelastic behavior in dynamic covalent networks and gels is controlled by the dynamic covalent chemistry that comprises the reversible cross-links, which ultimately dictates the timescale of relaxation of the network strands. The timescale of the individual bond τ_B is related to the relaxation time τ_r of the network. Tuning the specific chemistry of the boronic acid-diol pair, the environmental conditions and the network topology provide control over the time-dependent properties for this class of materials [57], [62]. For example, bonds with short τ_B timescales produce materials that behave as viscous polymer melts whereas bonds with high τ_B generate materials that can be indistinguishable from covalent networks [63].

In practice, dynamic covalent networks are not characterized by a single relaxation time but display a distribution of relaxation times. This is related both to the variance in dynamic covalent bond properties and cross-link topology in the network. The distribution of relaxation times can be determined by converting the frequency-dependent storage and loss moduli calculated from rheometry into a relaxation spectrum $H(\tau)$, as described by Grindy et al. for metal ligand–based dynamic polymer networks [64]. The relaxation spectrum models the material as an infinite number of Maxwell models in parallel rather than the single Maxwell model mentioned previously. It can be interpreted as how much energy is stored or dissipated by the material during dynamic loading [64]. A more complete understanding of how network topology and binding thermodynamics and kinetics control the relaxation behavior in dynamic covalent networks and gels is a focus of current research and will be necessary to provide a priori design of materials with precise relaxation behaviors for targeted applications.

Materials assembly from boronic ester cross-linking

As described above, the reversible nature of the boronic ester bond can be exploited to build dynamic covalent networks and gels with time-dependent viscoelastic properties that are moldable and self-healing. Excellent reviews of boronic ester-based materials have been published, and we direct the reader to these works for a more comprehensive discussion of boronic ester–based materials design [43], [65]. Here, we restrict our discussion first to some of the foundational work that demonstrates how boronic acid-diol chemistry and network topology can be used to engineer dynamic covalent networks from boronic ester bonds and then highlight some of the emerging biomedical applications of boronic ester–based materials.

The earliest examples of engineering materials with boronic esters involved the combination of borax with polyhydroxy compounds. Deuel and Neukom demonstrated in their seminal paper that borax could be used in aqueous solution to cross-link gels with suitable hydroxyl functional polysaccharides, including poly(vinyl alcohol) (PVA) [66]. The cross-linking of PVA with borax (Fig. 4a) is now commonly used to make moldable and self-healing 'slime' based on these original recipes. In their original article, Deuel and Neukom already demonstrated that materials properties depend on pH, providing evidence for an 'optimal' pH for gel formation and highlighted that the addition of small molecule diols, including fructose, glycerin, and sucrose, influenced materials properties by disrupting the formed cross-links. This work also explored altering the ratio of borax to polyhydroxy, which indicated that the materials properties depend on the extent of cross-linking. As a whole, this manuscript laid the foundation for the myriad of boronic ester–based materials that are now made and how they can be engineered based on their cross-linking chemistry, environmental factors, and network topology.

Building upon these concepts, Pezron et al. investigated the properties of reversible gels formed from borax and galactomannan, guaran, and (hydroxypropyl)guaran [67], [68]. In these studies, ^{11}B NMR was used to probe the extent of free borax as compared to mono-diol and di-diol complexation, providing insights into the molecular structure of the gels and the thermodynamics of binding. This work was extended by Leibler et al. to construct physical models of the properties and dynamics of reversible networks based on the underlying structure and chemistry [60]. Building on these findings, Kesavan and Prud'homme studied the linear viscoelastic properties of guaran and (hydroxypropyl)guaran, corroborating with these data the theory developed by Leibler et al. for the rheology of reversible networks [69].

A major focus for the modern use of boronic ester–based materials has been on the sensing of carbohydrates. Kitano et al. incorporated PBA moieties into poly(*N*-vinyl-2-pyrrolidone) [poly(NVP-co-PBA)], which was combined with PVA to assemble glucose-responsive gels. These responded to glucose by undergoing a gel-sol transition, envisioning applications as glucose-responsive drug delivery platforms [70]. In this work, they demonstrated the influence of polymer molecular weight, stoichiometry, and diol chemistry on materials rheology. The general concept of using boronic ester bonds for the assembly of materials that respond to the presence of carbohydrates has been explored by many other groups. For example, Kataoka et al. engineered a range of glucose-sensing materials from PBA-based materials [71], [72].

An additional emphasis in the community has been on the general design of moldable and self-healing networks and gels using the dynamic covalent boronic ester bond. Roberts et al. synthesized linear polymers of acrylic acid or 2-hydroxypropylmethacrylamide containing 10 mol% of PBA or salicylhydroxamic acid (SHA) [73]. Upon mixing in aqueous solutions, the PBA and SHA containing polymers assembled to form dynamically restructuring hydrogel networks, best illustrated by the scheme in Fig. 4b, whose properties were tuned by the volume fraction of polymer and pH. Similarly, as depicted in Fig. 4c, He et al. assembled reversible hydrogels from catechol-functionalized 4-arm poly(ethylene glycol) (PEG) and 1,3-benzenediboronic acid, exploiting the high affinity binding between PBA and catechol to engineer pH-responsive and self-healing gels [74]. In this work, they demonstrated that higher concentrations of the multi-arm PEG increased the plateau modulus of the formed network. Cash et al. installed dynamic covalent boronic ester bonds directly into polymer networks through the direct thiol-ene photopolymerization of 4-((allyloxy)methyl)-2-(4-vinylphenyl)-1,3,2-di-oxaborolane, a boronic ester containing divinyl monomers, with multifunctional small molecule thiols [75]. The installation of the reversible boronic ester bond into the network backbone during photopolymerization delivered tough and self-healing polymer networks, as illustrated in Fig. 4d. In an analogous approach, Theato et al. built moldable and self-healing gels by incorporating borax into the monomer mixture during network formation via Michael-type addition of PEG diacrylate and dithiothreitol, a 1,2-diol-containing dithiol [76]. Cromwell et al. designed dynamic networks through the chemical synthesis of 1,2-diol-containing polycyclooctene and diphenylboronic ester linkers [15]. In this manner, they fabricated solvent-free polymer networks that were self-healing and reprocessable, with the self-healing rate controlled by the small molecule PBA-diol binding kinetics.

Advancing from these early demonstrations, Yesilyurt et al. fabricated dynamic covalent networks from 4-arm PEG molecules functionalized with 1,2-diols or PBA derivatives (Fig. 4c), which highlighted the ability to tune materials properties with boronic acid-diol chemistry and network topology [7], [77]. In these well-defined networks, they explored the influence of altering the boronic acid chemistry, which modified the optimal pH of network formation and the crossover frequency ω_c [7]. In addition, they showed that tuning the weight percent of polymer controlled the plateau modulus but did not influence significantly the relaxation time of the gel [77]. Network dissolution and release of entrapped molecules was modulated by the presence of free diols, including glucose, and the gels were designed as injectable materials based on their shear-thinning and self-healing properties. In total, the studies presented here on the design of dynamic covalent networks and gels from boronic ester bonds illustrate that control of materials properties is enabled by tailoring both the chemistry of the reversible bond *and* the network topology.

EMERGING BIOMEDICAL APPLICATIONS OF BORONIC ESTER–BASED MATERIALS

Traditional polymeric materials, based on thermosets and thermoplastics, are designed with a specific set of properties for a given application and lifetime. A major force in modern materials science research is the synthesis of dynamic materials with functions beyond those of static polymeric materials [78]. As we transition from static to dynamic materials design, we envision 'intelligent' systems that can respond to external stimuli and interact with the surrounding environment [79], [80]. Self-assembly and supramolecular chemistry are both robust synthetic tools for the fabrication of dynamic soft matter [81], [82]. Within this framework, the use of dynamic covalent bonds, such as boronic ester bonds, provides a versatile approach to assemble responsive networks and gels. The unique ability of dynamic covalent bonds to break and reform in response to external stimuli enables the design of polymeric systems that exhibit macroscale changes to select environmental factors [78]. Responsive polymeric materials composed of boronic ester dynamic bonds have been developed exploiting this approach, with a particular emphasis on self-healing networks and carbohydrate-responsive gels. Traditional applications of boronic ester–based materials, discussed briefly above, have been reviewed in depth elsewhere [43], [65]. Here, we present recent efforts in the emerging areas enabled by insights into the boronic acid-diol binding chemistry, advances in polymer design, and new strategies for network formation. Specifically, we highlight the use of boronic ester-based materials as responsive drug delivery systems (DDSs), dynamic scaffolds for cell culture, and engineered adhesives.

Responsive drug delivery systems

New therapeutic molecules to treat disease are under constant development. While each molecule is developed to modulate the underlying biology and restore homeostasis *in vivo*, the efficacy of a drug is dependent critically on the manner by which it is administered in the body [86]. Complications can arise if the molecule is present at the wrong time, the wrong concentration, or at the wrong site. In addition, many emerging therapeutics are comprised of poorly soluble hydrophobic small molecules or complex biologics, such as nucleic acids, antibodies, and protein conjugates, which require a material carrier for stability and efficient delivery *in vivo* [87]. Therefore, an attractive application of 'intelligent' materials based on dynamic boronic ester bonds is for the design of responsive DDSs that can control the release of emerging and existent therapeutics. In the future, this could include the design of injectable materials that can be applied in a minimally invasive manner at the target site, sense aberrant biological signals, and respond by releasing corrective therapeutics [80].

Initial uses of reversible boronic ester bonds for drug delivery focused on direct modification of therapeutics or delivery vehicles with boronic acid or diol moieties. As an early example, Shiino et al. engineered PBA-functionalized gel beads for the design of diol binding chromatography columns [88]. They then functionalized insulin with gluconic acid (with ~2 gluconic acid residues per insulin molecule) which could be loaded onto the PBA gel bead column. Toward the development of an autoregulated insulin pump, the release of insulin was triggered in response to changes in glucose levels. In a similar approach, Su et al. designed a pH-responsive polymer conjugate of the FDA-approved anticancer drug bortezomib (BTZ) [89].

In this study, a catechol-functionalized PEG molecule was synthesized and bound to BTZ to produce the polymer conjugate. The reversible nature of the boronic ester bond was exploited to release BTZ in acidic environments, e.g., tumor microenvironment or subcellular endosome. Wang et al. extended the conjugation approach to design PBA-functionalized chitosan nanoparticles (CS-PBA-NPs) to bias the biodistribution of doxorubicin-loaded NPs to sialic acid residues present on liver cancer cells [90]. CS-PBA-NPs were synthesized with cell penetrating and adhesive (RGD) peptides for improved uptake. While complex in their design, this system demonstrated the utility of exploiting boronic ester binding to increase the residence time and uptake at a target site.Beyond conjugation approaches, several groups have investigated the use of rationally engineered responsive gels for the controlled release of molecular therapeutics. Matsumoto et al. fabricated hydrogels based on poly (*N*-isopropyl-methylacrylamide) and 4-(2-acrylamidoethylcarbamoyl)-3-fluorophenylboronic acid that were designed to swell in the presence of elevated glucose levels (hyperglycemia) [83]. The principle was based on the creation of phenylboronate anions upon complexation of the PBA with glucose, which induced a volume change and corresponding increase in mesh size of the gel. As shown in Fig. 6a, when the glucose concentration was kept low (normoglycemia), a localized dehydration layer ('skin layer') formed on the outer surface of the gel, enabling control of the release of insulin from the gel. This system demonstrated pulsatile, glucose-responsive release of insulin in vitro using fluctuations between hyperglycemic and normoglycemic conditions. In another approach to design glucose-responsive insulin release, Zhang et al. assembled layer-by-layer films exploiting boronic ester bonding between PVA and poly[acrylamide-co-3-(acrylamide)-phenylboronic acid] [91]. Insulin was incorporated into the PVA layers, and film dissolution, including insulin release, was accelerated by the presence of glucose.

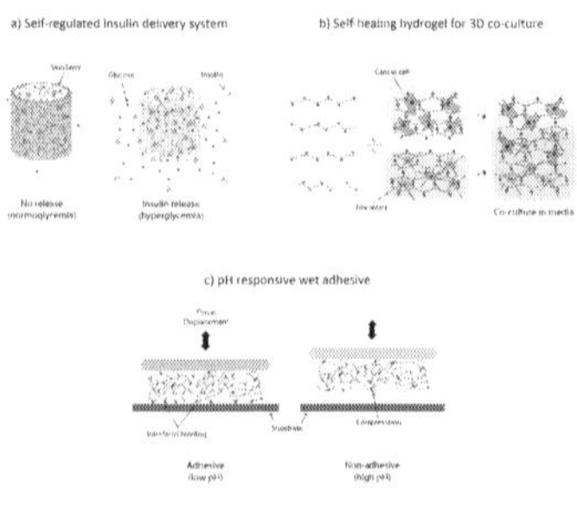

Figure. 6. Emerging biomedical applications of boronic ester–based materials. (a) A self-regulated insulin delivery system based on a cylindrical-shaped PBA-containing hydrogel. Rapid decreases in the glucose concentration induce the formation of an outer dehydrated 'skin layer', whose reduced mesh size controls insulin release as the levels of glucose fluctuate between normo- and hyperglycemia. Figure adapted from Ref. [83]. **(b)**. Two model human cell lines (breast cancer cells and pulmonary fibroblasts) were encapsulated within dynamic hydrogels and cut into 'blocks'. When the 'blocks' are joined, self-healing of the material occurs, enabling 3D co-culture of the two different cell types. Figure adapted from Ref. [84]. **(c)** A pH responsive wet adhesive consisting of copolymerized PBA and dopamine moieties. The adhesive functions at low pH due to the free dopamine groups available for interfacial binding. At high pH, adhesive properties are lost as free dopamine moieties are consumed because of increased boronic ester complexation within the material. Figure adapted from Ref. [85].

Another emerging use of the reversible nature of boronic ester bonds is the design of injectable biomaterials for controlled drug delivery. As the boronic ester bonds can break and reform, gels with shear-thinning and self-healing properties can be assembled using these dynamic covalent bonds as cross-links. Dong et al. assembled injectable hydrogels using an acrylamide copolymer containing boronic acid and glucose functionalities [92]. Owing to the complexation between boronic acid and glucose derivatives on disparate chains, the solution of copolymer formed a gel, and this material exhibited shear-thinning and self-healing properties. In addition, a model therapeutic was released in response to the addition of free glucose. This approach was extended by Yesilyurt et al. to form defined hydrogel networks from PBA and diol functionalized 4-arm PEG molecules [7]. These gels also exploited reversible boronic ester bond formation as a cross-linking chemistry, and the gels exhibited shear-thinning and self-healing properties that enabled injection through standard gauge syringe needles. Again, release of model therapeutics was accelerated in glucose containing aqueous solutions. Recently, Huang et al. presented another use of dynamic boronic ester bonds to fabricate injectable DDS [93]. Here, the bioactive polyphenol, containing two functional diols, was used to form cross-links with boronic acid–functionalized multi-arm PEG molecules. In this manner, the material was injectable owing to the responsive boronic ester bonds and released the bioactive polyphenol as the gel dissolved in aqueous media.

These recent examples using reversible boronic ester bonds to control the release of therapeutic agents highlight the ability of dynamic covalent bonds to engineer 'intelligent' DDS. Special emphasis in the use of boronic ester bonds has been centered on the design of glucose-responsive DDS. It should be noted that as the binding between the PBA and glucose is relatively weak, the release in many of the glucose-responsive materials is not much faster than in aqueous buffer alone. Other biological triggers beyond glucose, such as pH or fructose, may provide more specific release from boronic ester-based responsive gels.

Dynamic scaffolds for cell culture and tissue engineering

As we learn more about the dynamic nature of the native extracellular matrix, increased emphasis has been placed on the design of synthetic mimics that enable temporal modulation of biophysical and biochemical cues in the model cell niche [94]. Engineered hydrogel platforms have advanced 3D cell culture and tissue engineering, and a current focus in biomaterials is to move from static niches to dynamic and user-controlled niches [79], [95]. Increasingly, dynamic niches reveal critical aspects of biology and development that were not observable in static materials. For example, it has been shown that the biophysical properties of the cell niche influence cell function, including differentiation [96], [97], [98]. Recent research with responsive culture scaffolds has highlighted that dynamic physical properties of the cell niche also influence cell fate and function [99], [100], [101].

Initial demonstrations of the utility of boronic ester bonds in the presence of living cells were focused on surface engineering to modulate cell attachment. Aoki et al. generated acrylamide-based polymer surfaces that were functionalized with PBA for the culture of bovine aortic endothelial cells (BAEC) [102]. BAECs adhered to the acrylamide surface without the need for additional protein coating, presumably through interactions between surface-bound PBA moieties and cell surface carbohydrates, and the cells formed model capillary structures. In another demonstration of surface functionalization for control of cell attachment, Liu et al. grafted a poly (acrylamidophenylboronic acid) brush from a silicon nanowire surface array [103]. The engineered, PBA-presenting surface was exploited to selectively capture cancer cells that overexpress sialic acid on the cell surface. Importantly, the ability to break the formed boronic ester between the PBA-containing brush and sialic acid at the cell surface enabled release of captured cells in response to changes in pH and glucose concentration.

The ability to assemble dynamic 3D materials from boronic ester bonds has also been exploited for the direct encapsulation of mammalian cells. Konno et al. synthesized boronic acid containing phospholipid polymers, which formed gels upon mixing with diol containing polymers including

PVA [104]. This approach was used for the reversible encapsulation of murine fibroblast cells, demonstrating a general approach for the 3D encapsulation and release of mammalian cells within dynamic covalent gels. Recently, Chen et al. developed a dynamic 3D culture platform based on dynamic covalent bonds between benzoxaborole and catechol [105]. In this work, zwitterionic polymers based on methacryloyloxyethyl phosphorylcholine were synthesized with pendant catechol or benzoxaborole moieties. The benzoxaborole was chosen on account of its lower pK_a (7.2), which enabled robust gel formation upon mixing of the two polymers at pH 7.4. As expected for dynamic covalent networks, the materials were shear-thinning and self-healing and enabled encapsulation and culture of cells for 24 h, which is similar to the timeframe upon which gels dissolved in pure phosphate buffered saline (PBS). Faster dissolution was observed in PBS with fructose. To extend the length of time for 3D cell culture within viscoelastic boronic ester–based hydrogels, Tang et al. synthesized gels containing dynamic boronic ester cross-links as well as permanent azide-alkyne cross-links [61]. To achieve this, the authors synthesized 8-arm PEG molecules end-functionalized with both 2-fluorophenylboronic acid and azide as well as 8-arm PEG molecules end-functionalized with dibenzylcyclooctyne or nitro dopamine. Upon mixing of all three components, the authors argue that a dual-network was formed with viscoelastic properties from the dynamic covalent chemistry that did not dissolve with cells in culture for up to 7 days.

The ability to reversibly tune the mechanical properties of a dynamic hydrogel scaffold could prove very useful for 3D cell culture, as cellular fate and function are influenced by the dynamic properties of the cell's environment. To this end, Accardo and Kalow tuned the stiffness of a dynamic hydrogel by using photoswitches to control the reactivity of the dynamic covalent cross-links in the material [106]. They functionalized the ends of 4-arm PEG macromers with either azobenzene boronic acids or diols. Upon mixing, viscoelastic and stress-relaxing dynamic hydrogels were produced. Exposure to certain wavelengths of light changed the conformation of azobenzene, which in turn shifted the chemical equilibrium of the boronic acid-diol condensation. Stiffening and softening occurred as the increase/decrease in the equilibrium constant generated a higher/lower cross-linking density in the network.

Beyond 3D cell culture of single-type cell populations, Smithmyer et al. exploited boronic ester–based gels for the combined culture of multiple cells types [84]. This work leveraged the synthesis of a statistical copolymer of 90 mol % *N,N*-dimethylacrylamide and 10 mol % pinacol-protected

2-acrylamidophenylboronic acid to form dynamic and self-healing gels upon mixing with PVA. Uniquely, the authors demonstrated that disparate populations of cells could be encapsulated within separate gels and then assembled into a single microtissue, owing to the self-healing nature of the material. This approach, illustrated in Fig. 6b, will enable future research to investigate cell-cell interactions in 3D culture and to assemble more complicated tissue engineering constructs. In another use of reversible boronic ester gels for tissue engineering applications, Tseng et al. used dynamic covalent gels as sacrificial templates for vascular structures [107]. The gels were synthesized from polyol-containing polymers and borax and deposited in the form of the desired vascular structure. The sacrificial gel was then encapsulated in collagen or fibrin gel and subsequently removed upon addition of glucose-containing media. Engineered vessels were then developed within the resultant tubules.

There is growing interest in the use of responsive, moldable, and viscoelastic hydrogels as 3D culture platforms for cell encapsulation and tissue engineering [100]. Building upon the early studies using boronic ester–based hydrogels discussed here, we anticipate that the boronic ester binding motif will be used to engineer materials for many in vitro and *in vivo* cell culture applications. However, one of the main limitations of boronic ester–based gels is their lack of stability, which limits sustained use in cell culture media. New approaches that combine viscoelastic gels with covalent chemistry or incorporate other features to improve stability are needed. In addition, as the cells can rearrange the network at the cellular and subcellular length scales, characterization techniques to assess local properties (such as modulus or ligand density) are essential, as these may deviate significantly from the values measured for the bulk gel, confounding biological observations [108].

Engineered adhesives with dynamic covalent bonds

Boronic ester–based materials are also being developed for other applications in materials science outside of biomedicine. For instance, next-generation adhesives are being designed to allow materials with arbitrary surface chemistries to hold strongly to disparate surfaces with different chemistries. Traditional adhesives efficiently adhere two materials together; however, few enable stimuli responsive changes to the molecular adhesion state. Ideal adhesives would enable on-demand debonding or release of the formed adhesion through reversible bonding or efficient tuning of relative cohesive and adhesive forces. Therefore, dynamic covalent chemistry is being integrated into engineered adhesives to enable improved bonding to varied surfaces as well as to provide a handle for on-demand release. Within this approach, boronic ester bonds have been incorporated into soft matter adhesives.

In one example, Caretti et al. expanded upon standard PVA-borax gels to engineer sticky adhesives for cleaning soiled art by incorporating co-solvents, such as 1-propanol, into the solvent phase of the material [109]. With the addition of 1-propanol as a co-solvent, the gels remained moldable and could be applied at the surface of sensitive materials, including an oil painting from the 16th-17th century. Further, the addition of co-solvent modulated the polarity of the system and enabled cleaning of oxidized varnish from the painting surface.

To generate self-healing materials with engineered cohesion, Cash et al. fabricated solvent-free polymer networks with boronic ester bonds in the network backbone [75]. This material was able to self-heal upon application of water at the damaged interface through the rearrangement and reformation of boronic ester bonds at the interface. In another study, reversible wet adhesives were developed using dopamine and boronic acid functionalized polymers [85]. These demonstrated pH-responsive adhesion, as illustrated in Fig. 6c.

The wet adhesive functioned at low pH due to the free dopamine groups available for interfacial binding. Adhesive properties were lost at high pH, as free dopamine moieties were consumed due to increased boronic ester complexation within the material. Stimuli-responsive adhesives were also designed using boronic acid–functionalized alginate [110]. Here, boronic ester was formed between the pendant boronic acids and diols inherent to the alginate polymer.

Adhesives are only one example of how the dynamic covalent nature of boronic esters can be exploited in soft materials design. As our understanding of how to engineer dynamic covalent networks with specific moduli, relaxation modes, and surface chemistry increases, boronic ester–based materials will likely find broad application as processable, self-healing, and stimuli-responsive soft matter. Work by Cromwell et al. and Cash et al. have convincingly demonstrated the utility of boronic ester bonds in the design of non-swollen polymer networks [15], [75]. In many cases, the chemistry now exists to develop new polymer architectures which will enable materials with the desired properties for a range of applications.

CONCLUSIONS AND OUTLOOK

The fascinating emergent properties of dynamic covalent networks based on boronic ester cross-links have led to widespread interest in this class of materials, especially within the biomedical materials community. In this review, we have highlighted the current state of the field with an emphasis on how the chemistry and network formation at the molecular scale influence materials properties. The widespread interest and use of dynamic covalent networks based on boronic ester cross-links has been enabled by insight into boronic acid-diol complexation and the polymer physics of non-covalent polymer networks. This has led to applications of these materials as responsive DDSs, dynamic scaffolds for cell culture, and as engineered adhesives. The field of dynamic covalent networks using boronic ester bonds has matured significantly in the recent years; however, we remain limited by the incomplete knowledge of several aspects of these materials.

We do not have a full understanding of boronic ester chemistry, and thus additional efforts to characterize the binding kinetics and thermodynamics of boronic esters are needed. While some aspects of the chemistry are well-characterized, as highlighted in this review, many questions remain. What is the precise reaction landscape during boronic acid-diol complexation? How precisely do substituent affects influence the K_{eq} of the resultant boronic ester? How does this control the optimal pH of binding between boronic acid and diol? Continued interactions between polymer engineers and organic chemists will shed light on these open research areas.

On the other hand, the polymer physics of covalent networks and gels is well-characterized. To a sufficient approximation, one can predict the resultant properties of a network based on the molecular precursors. Continued efforts in the characterization of dynamic covalent networks, building on seminal work by Leibler et al. as well as Semenov and Rubinstein are needed to provide the same set of theoretical tools for boronic ester–based networks and gels. Ideally, one would be able to predict the relaxation modes, plateau modulus, and shear-thinning behavior a priori from the molecular details of the cross-linking chemistry and network forming components. This remains a challenge for dynamic covalent networks; however, new rheological approaches and theoretical frameworks are being developed to provide new insight into these fundamental questions.

As the field matures, we anticipate that dynamic covalent cross-links from boronic esters will continue to play a large role in the fabrication of next-generation, 'intelligent' soft matter. As the distinction between synthetic and living matter dissolves, dynamic covalent networks and gels will provide necessary tools to realize engineered responsive materials [80]. Design of materials that overlay multiple dynamic covalent chemistries or integrate non-covalent and covalent chemistries within a single material will enable new applications of these materials. Improved imaging and modeling will allow us to predict and understand how these responsive materials evolve on different length scales and timescales during use. There is broad potential for this class of materials, and continued interactions between polymer engineers, chemists, and bioengineers will enable the field to develop new materials based on the principles of chemical and network design discussed here.

ACKNOWLEDGMENTS

This work was supported by startup funds from ETH Zürich.

REFERENCES

1. Z. Wei, J.H. Yang, J. Zhou, F. Xu, M. Zrínyi, P.H. Dussault, Y. Osada, Y.M. Chen, Self-healing gels based on constitutional dynamic chemistry and their potential applications, Chem. Soc. Rev. (2014), https://doi.org/10.1039/ c4cs00219a.
2. T.F.A. De Greef, M.M.J. Smulders, M. Wolffs, A.P.H.J. Schenning, R.P. Sijbesma, E.W. Meijer, Supramolecular polymerization, Chem. Rev. (2009), https:// doi.org/10.1021/cr900181u.
3. O. Wichterle, D. Lím, Hydrophilic gels for biological use, Nature (1960), https://doi.org/10.1038/185117a0.
4. C.J. Kloxin, C.N. Bowman, Covalent adaptable networks: Smart, reconfigurable and responsive network systems, Chem. Soc. Rev. (2013), https:// doi.org/10.1039/c3cs60046g.
5. D. Montarnal, M. Capelot, F. Tournilhac, L. Leibler, Silica-like malleable materials from permanent organic networks, Science (80-.) (2011), https:// doi.org/10.1126/science.1212648.
6. X. Chen, M.A. Dam, K. Ono, A. Mal, H. Shen, S.R. Nutt, K. Sheran, F. Wudl, A thermally re-mendable cross-linked polymeric material, Science (80-.) (2002), https://doi.org/10.1126/science.1065879.
7. V. Yesilyurt, M.J. Webber, E.A. Appel, C. Godwin, R. Langer, D.G. Anderson, Injectable self-healing glucose-responsive hydrogels with pH-regulated mechanical properties, Adv. Mater (2016), https://doi. org/10.1002/ adma.201502902.
8. S.J. Rowan, S.J. Cantrill, G.R.L. Cousins, J.K.M. Sanders, J.F. Stoddart, Dynamic covalent chemistry, Angew. Chem. Int. Ed. (2002), https:// doi.org/10.1002/ 1521-3773(20020315)41:63.0.CO;2-E.
9. M.D. Stern, A.V. Tobolsky, Stress-Time-Temperature relations in polysulfide rubbers, J. Chem. Phys. (1946), https://doi. org/10.1063/1.1724110.
10. M.S. Green, A.V. Tobolsky, A new approach to the theory of relaxing polymeric media, J. Chem. Phys. (1946), https://doi. org/10.1063/1.1724109.
11. B.D. Fairbanks, S.P. Singh, C.N. Bowman, K.S. Anseth, Photodegradable, photoadaptable hydrogels via radical-mediated disulfide fragmentation reaction, Macromolecules (2011), https://doi. org/10.1021/ma200202w.
12. K.N. Plunkett, M.L. Kraft, Q. Yu, J.S. Moore, Swelling kinetics of

disulfide cross-linked microgels, Macromolecules (2003), https://doi.org/10.1021/ ma025874f.

13. W.G. Skene, J.-M.P. Lehn, Dynamers: Polyacylhydrazone reversible covalent polymers, component exchange, and constitutional diversity, Proc. Natl. Acad. Sci. (2004), https://doi.org/10.1073/pnas.0401885101.
14. J.N. Cambre, B.S. Sumerlin, Biomedical applications of boronic acid polymers, Polymer (Guildf) (2011), https://doi.org/10.1016/j.polymer.2011.07.057.
15. O.R. Cromwell, J. Chung, Z. Guan, Malleable and self-healing covalent polymer networks through tunable dynamic boronic ester bonds, J. Am. Chem. Soc. (2015), https://doi.org/10.1021/jacs.5b03551.
16. W. Yang, X. Gao, B. Wang, Biological and medicinal applications of boronic acids, Boronic Acids Prep. Appl. Org. Synth. Med. (2006), https://doi.org/ 10.1002/3527606548.ch13.
17. Y. Kubo, R. Nishiyabu, T.D. James, Hierarchical supramolecules and organization using boronic acid building blocks, Chem. Commun. (2015), https:// doi.org/10.1039/c4cc07712a.
18. J. Adams, M. Kauffman, Development of the proteasome inhibitor VelcadeTM (Bortezomib), Cancer Invest. (2004), https://doi.org/10.1081/CNV120030218.
19. J. Adams, M. Behnke, S. Chen, A.A. Cruickshank, L.R. Dick, L. Grenier, J.M. Klunder, Y.T. Ma, L. Plamondon, R.L. Stein, Potent and selective inhibitors of the proteasome: Dipeptidyl boronic acids, Bioorg. Med. Chem. Lett (1998), https://doi.org/10.1016/S0960-894X(98)00029-8.
20. P. D'Arcy, S. Brnjic, M.H. Olofsson, M. Frykn€ as, K. Lindsten, M. De Cesare, P. Perego, B. Sadeghi, M. Hassan, R. Larsson, S. Linder, Inhibition of proteasome deubiquitinating activity as a new cancer therapy, Nat. Med. (2011), https://doi.org/10.1038/nm.2536.
21. J.P. Lorand, J.O. Edwards, Polyol complexes and structure of the benzeneboronate ion, J. Org. Chem. (1959), https://doi.org/10.1021/jo01088a011.
22. Y. Kanekiyo, M. Sano, R. Iguchi, S. Shinkai, Novel nucleotide-responsive hydrogels designed from copolymers of boronic acid and cationic units and their applications as a QCM resonator system to nucleotide sensing, J. Polym. Sci. Part A Polym. Chem. (2000), https://doi.org/10.1002/(SICI)1099- 0518(20000415)38:83.0.CO;2-1.
23. R. Nishiyabu, S. Ushikubo, Y. Kamiya, Y. Kubo, A boronate hydrogel

film containing organized two-component dyes as a multicolor fluorescent sensor for heavy metal ions in water, J. Mater. Chem. A. (2014), https:// doi.org/10.1039/c4ta03268c.

24. K.R.A.S. Sandanayake, S. Shinkai, Novel molecular sensors for saccharides based on the interaction of boronic acid and amines: Saccharide sensing in neutral water, J. Chem. Soc. Chem. Commun. (1994), https://doi.org/10.1039/ C39940001083.
25. J. Yoon, A.W. Czarnik, Fluorescent chemosensors of carbohydrates. a means of chemically communicating the binding of polyols in water based on chelation-enhanced quenching, J. Am. Chem. Soc. (1992), https://doi.org/ 10.1021/ja00040a067.
26. X. Wu, Z. Li, X.X. Chen, J.S. Fossey, T.D. James, Y.B. Jiang, Selective sensing of saccharides using simple boronic acids and their aggregates, Chem. Soc. Rev. (2013), https://doi.org/10.1039/c3cs60148j.
27. S.D. Bull, M.G. Davidson, J.M.H. Van Den Elsen, J.S. Fossey, A.T.A. Jenkins, Y.B. Jiang, Y. Kubo, F. Marken, K. Sakurai, J. Zhao, T.D. James, Exploiting the reversible covalent bonding of boronic acids: Recognition, sensing, and assembly, Acc. Chem. Res. (2013), https:// doi.org/10.1021/ar300130w.
28. X. Sun, T.D. James, Glucose sensing in supramolecular chemistry, Chem. Rev. (2015), https://doi.org/10.1021/cr500562m.
29. D.G. Hall, Structure, Properties, and Preparation of Boronic Acid Derivatives. Overview of Their Reactions and Applications, Boronic Acids Prep. Appl. Org. Synthesis Med. (2006), https://doi. org/10.1002/3527606548.ch1.
30. J.A. Peters, Interactions between boric acid derivatives and saccharides in aqueous media: Structures and stabilities of resulting esters, Coord. Chem. Rev. (2014), https://doi.org/10.1016/j.ccr.2014.01.016.
31. Y. Furikado, T. Nagahata, T. Okamoto, T. Sugaya, S. Iwatsuki, M. Inamo, H.D. Takagi, A. Odani, K. Ishihara, Universal reaction mechanism of boronic acids with diols in aqueous solution: kinetics and the basic concept of a conditional formation constant, Chem. Eur J. (2014), https://doi.org/10.1002/ chem.201403719.
32. L. Babcock, R. Pizer, Dynamics of boron acid complexation reactions. formation of 1:1 boron acid-ligand complexes, Inorg. Chem. (1980), https:// doi.org/10.1021/ic50203a013.
33. G. Springsteen, B. Wang, A detailed examination of boronic acid-diol

complexation, Tetrahedron (2002), https://doi.org/10.1016/S0040-4020(02) 00489-1.

34. L.I. Bosch, T.M. Fyles, T.D. James, Binary and ternary phenylboronic acid complexes with saccharides and Lewis bases, Tetrahedron (2004), https:// doi.org/10.1016/j.tet.2004.08.046.
35. S. Iwatsuki, S. Nakajima, M. Inamo, H.D. Takagi, K. Ishihara, Which is reactive in alkaline solution, boronate ion or boronic acid? Kinetic evidence for reactive trigonal boronic acid in an alkaline solution, Inorg. Chem. (2007), https://doi.org/10.1021/ic0615372.
36. M. Rietjens, P.A. Steenbergen, Crosslinking mechanism of boric acid with diols revisited, Eur. J. Inorg. Chem. (2005), https://doi. org/10.1002/ejic. 200400674.
37. M. Van Duin, J.A. Peters, A.P.G. Kieboom, H. Van Bekkum, Studies on borate esters 1: The ph dependence of the stability of esters of boric acid and borate in aqueous medium as studied by 11B NMR, Tetrahedron (1984), https:// doi.org/10.1016/S0040-4020(01)91300-6.
38. J. Yan, G. Springsteen, S. Deeter, B. Wang, The relationship among pKa, pH, and binding constants in the interactions between boronic acids and diols - It is not as simple as it appears, Tetrahedron (2004), https://doi.org/10.1016/ j.tet.2004.08.051.
39. M.A. Martínez-Aguirre, R. Villamil-Ramos, J.A. Guerrero-Alvarez, A.K. Yatsimirsky, Substituent effects and pH profiles for stability constants of arylboronic acid diol esters, J. Org. Chem. (2013), https:// doi.org/10.1021/ jo400617j.
40. G. Wulff, M. Lauer, H. Bohnke, Rapid proton transfer as cause of an unusually € large neighboring group effect, Angew Chem. Int. Ed. Engl. (1984), https:// doi.org/10.1002/anie.198407411.
41. M. Dowlut, D.G. Hall, An improved class of sugar-binding boronic acids, soluble and capable of complexing glycosides in neutral water, J. Am. Chem. Soc. (2006), https://doi.org/10.1021/ja057798c.
42. G. Wulff, Selective binding of polymers via covalent bonds. the construction of chiral cavities as specific receptor sites, Pure Appl. Chem. (1982), https:// doi.org/10.1351/pac198254112093.
43. W.L.A. Brooks, B.S. Sumerlin, Synthesis and applications of boronic acidcontaining polymers: from materials to medicine, Chem. Rev. (2016), https://doi.org/10.1021/acs.chemrev.5b00300.
44. F.C. Fischer, E. Havinga, Thermal and photoinduced deboronations of

some pyridine- and benzeneboronate anions, Recl. Des Trav. Chim. Des Pays-Bas. (1974), https://doi.org/10.1002/recl.19740930110.

45. M. Van Duin, J.A. Peters, A.P.G. Kieboom, H. Van Bekkum, Studies on borate esters II, Tetrahedron (1985), https://doi.org/10.1016/S0040-4020(01)91300-6.

46. R. van den Berg, J.A. Peters, H. van Bekkum, The structure and (local) stability constants of borate esters of mono- and di-saccharides as studied by11B and13C NMR spectroscopy, Carbohydr. Res. (1994), https://doi.org/10.1016/ 0008-6215(94)80050-2.

47. M. Bielecki, H. Eggert, J.C. Norrild, A fluorescent glucose sensor binding covalently to all five hydroxy groups of a-D-glucofuranose. A reinvestigation, J. Chem. Soc. Perkin Trans. 2 (1999), https://doi.org/10.1039/a808896i.

48. S.J. Angyal, The composition of reducing sugars in solution: current aspects, Adv. Carbohydr. Chem. Biochem. (1991) 19e35, https://doi.org/10.1016/ S0065-2318(08)60180-8.

49. M. Zhong, R. Wang, K. Kawamoto, B.D. Olsen, J.A. Johnson, Quantifying the impact of molecular defects on polymer network elasticity, Science (80-.) (2016), https://doi.org/10.1126/science.aag0184.

50. M. Rubinstein, R.H. Colby, C. Ralph H, Polymer physics, 2003, https://doi.org/ 10.1002/pi.1472.

51. P.J. Flory, Molecular size distribution in three dimensional polymers. I Gelation, J. Am. Chem. Soc. (1941), https://doi.org/10.1021/ja01856a061.

52. W.H. Stockmayer, Theory of molecular size distribution and gel formation in branched-chain polymers, J. Chem. Phys. 11 (1943) 45e55, https://doi.org/ 10.1063/1.1723803.

53. W.H. Stockmayer, Theory of Molecular Size Distribution and Gel Formation in Branched Polymers II. General Cross Linking, J. Chem. Phys. 12 (1944) 125, https://doi.org/10.1063/1.1723922.

54. P.J. Flory, Principles of Polymer Chemistry, 1953.

55. A.N. Semenov, M. Rubinstein, Thermoreversible gelation in solutions of associative polymers. 1. Statics, Macromolecules (1998), https://doi.org/ 10.1021/ma970616h.

56. J.D. Ferry, Viscoelastic Properties of Polymers, third ed., 1980, https://doi.org/ 10.1016/0010-4361(73)90801-X.

57. M. Rubinstein, A.N. Semenov, Thermoreversible gelation in solutions of associating polymers - 2 - linear dynamics, Macromolecules (1998), https:// doi.org/10.1021/ma970617þ.
58. C.W. Macosko, D.R. Miller, A new derivation of average molecular weights of nonlinear polymers, Macromolecules (1976), https://doi.org/10.1021/ ma60050a003.
59. S.Q. Wang, Transient network theory for shear-thickening fluids and physically crosslinked networks, Macromolecules (1992), https://doi.org/ 10.1021/ma00051a043.
60. L. Leibler, M. Rubinstein, R.H. Colby, Dynamics of reversible networks, Macromolecules (1991), https://doi.org/10.1021/ma00016a034.
61. S. Tang, H. Ma, H.C. Tu, H.R. Wang, P.C. Lin, K.S. Anseth, Adaptable fast relaxing boronate-based hydrogels for probing cellematrix interactions, Adv. Sci. (2018), https://doi.org/10.1002/advs.201800638.
62. R.J. Sheridan, C.N. Bowman, A simple relationship relating linear viscoelastic properties and chemical structure in a model Diels-Alder polymer network, Macromolecules (2012), https://doi.org/10.1021/ma301329u.
63. M.J. Webber, E.A. Appel, E.W. Meijer, R. Langer, Supramolecular biomaterials, Nat. Mater. (2015), https://doi.org/10.1038/nmat4474.
64. S.C. Grindy, R. Learsch, D. Mozhdehi, J. Cheng, D.G. Barrett, Z. Guan, P.B. Messersmith, N. Holten-Andersen, Control of hierarchical polymer mechanics with bioinspired metal-coordination dynamics, Nat. Mater. (2015), https://doi.org/10.1038/nmat4401.
65. Y. Guan, Y. Zhang, Boronic acid-containing hydrogels: synthesis and their applications, Chem. Soc. Rev. 42 (2013) 8106e8121, https://doi.org/10.1039/ C3CS60152H.
66. H. Deuel, H. Neukom, Über die reaktion von borsaure und borax mit poly- € sacchariden und anderen hochmolekularen polyoxy-verbindungen, Makromol. Chem. (1949), https://doi.org/10.1002/macp.1949.020030102.
67. E. Pezron, A. Ricard, F. Lafuma, R. Audebert, Reversible gel formation induced by ion complexation. 1. Borax-galactomannan interactions, Macromolecules 21 (1988) 1121e1125, https://doi.org/10.1021/ma00182a045.
68. E. Pezron, L. Leibler, A. Ricard, F. Lafuma, R. Audebert, Complex formation in polymerdion solutions: 1: polymer concentration effects,

Macromolecules (1989), https://doi.org/10.1021/ma00193a030.

69. S. Kesavan, R.K. Prud'homme, Rheology of guar and HPG cross-linked by borate, Macromolecules (1992), https://doi.org/10.1021/ma00033a029.
70. S. Kitano, Y. Koyama, K. Kataoka, T. Okano, Y. Sakurai, A novel drug delivery system utilizing a glucose responsive polymer complex between poly (vinyl alcohol) and poly (N-vinyl-2-pyrrolidone) with a phenylboronic acid moiety, J. Control Release (1992), https://doi.org/10.1016/0168-3659(92)90073-Z.
71. A. Kikuchi, K. Suzuki, O. Okabayashi, H. Hoshino, K. Kataoka, Y. Sakurai, T. Okano, Glucose-sensing electrode coated with polymer complex gel containing phenylboronic acid, Anal. Chem. (1996), https://doi.org/10.1021/ ac950748d.
72. A. Matsumoto, R. Yoshida, K. Kataoka, Glucose-responsive polymer gel bearing phenylborate derivative as a glucose-sensing moiety operating at the physiological pH, Biomacromolecules (2004), https://doi.org/10.1021/ bm0345413.
73. M.C. Roberts, M.C. Hanson, A.P. Massey, E.A. Karren, P.F. Kiser, Dynamically restructuring hydrogel networks formed with reversible covalent crosslinks, Adv. Mater. (2007), https://doi.org/10.1002/adma.200602649.
74. L. He, D.E. Fullenkamp, J.G. Rivera, P.B. Messersmith, PH responsive selfhealing hydrogels formed by boronate-catechol complexation, Chem. Commun. (2011), https://doi.org/10.1039/c1cc11928a.
75. J.J. Cash, T. Kubo, A.P. Bapat, B.S. Sumerlin, Room-temperature self-healing polymers based on dynamic-covalent boronic esters, Macromolecules (2015), https://doi.org/10.1021/acs.macromol.5b00210.
76. L. He, D. Szopinski, Y. Wu, G.A. Luinstra, P. Theato, Toward self-healing hydrogels using one-pot thiol-ene click and borax-diol chemistry, ACS Macro Lett. (2015), https://doi.org/10.1021/acsmacrolett.5b00336.
77. V. Yesilyurt, A.M. Ayoob, E.A. Appel, J.T. Borenstein, R. Langer, D.G. Anderson, Mixed reversible covalent crosslink kinetics enable precise, hierarchical mechanical tuning of hydrogel networks, Adv. Mater. (2017), https://doi.org/ 10.1002/adma.201605947.
78. R.J. Wojtecki, M.A. Meador, S.J. Rowan, Using the dynamic bond to access macroscopically responsive structurally dynamic polymers,

Nat. Mater. (2011), https://doi.org/10.1038/nmat2891.

79. J.A. Burdick, W.L. Murphy, Moving from static to dynamic complexity in hydrogel design, Nat. Commun. (2012), https://doi.org/10.1038/ncomms2271.
80. M.W. Tibbitt, R. Langer, Living biomaterials, Acc. Chem. Res. (2017), https:// doi.org/10.1021/acs.accounts.6b00499.
81. G.M. Whitesides, B. Grzybowski, Self-assembly at all scales, Science (80-.) (2002), https://doi.org/10.1126/science.1070821.
82. J.M. Lehn, Constitutional dynamic chemistry: Bridge from supramolecular chemistry to adaptive chemistry, Top. Curr. Chem. (2012), https://doi.org/ 10.1007/128_2011_256.
83. A. Matsumoto, T. Ishii, J. Nishida, H. Matsumoto, K. Kataoka, Y. Miyahara, A synthetic approach toward a self-regulated insulin delivery system, Angew. Chem. Int. Ed. (2012), https://doi.org/10.1002/anie.201106252.
84. M.E. Smithmyer, C.C. Deng, S.E. Cassel, P.J. Levalley, B.S. Sumerlin, A.M. Kloxin, Self-healing boronic acid-based hydrogels for 3D Co-cultures, ACS Macro Lett. (2018), https://doi.org/10.1021/acsmacrolett.8b00462.
85. A.R. Narkar, B. Barker, M. Clisch, J. Jiang, B.P. Lee, PH responsive and oxidation resistant wet adhesive based on reversible catechol-boronate complexation, Chem. Mater. (2016), https://doi.org/10.1021/acs.chemmater.6b01851.
86. L.L. Brunton, B.C. Knollmann, R. Hilal-Dandan, Goodman and Gilman's The Pharmacological Basis of Therapeutics, 2017, https://doi.org/10.1016/ j.geomorph.2006.10.020.
87. M.W. Tibbitt, J.E. Dahlman, R. Langer, Emerging frontiers in drug delivery, J. Am. Chem. Soc. (2016), https://doi.org/10.1021/jacs.5b09974.
88. D. Shiino, Y. Murata, K. Kataoka, Y. Koyama, M. Yokoyama, T. Okano, Y. Sakurai, Preparation and characterization of a glucose-responsive insulinreleasing polymer device, Biomaterials (1994), https://doi.org/10.1016/0142- 9612(94)90261-5.
89. J. Su, F. Chen, V.L. Cryns, P.B. Messersmith, Catechol polymers for pHresponsive, targeted drug delivery to cancer cells, J. Am. Chem. Soc. (2011), https://doi.org/10.1021/ja203077x.
90. X. Wang, X. Zhen, J. Wang, J. Zhang, W. Wu, X. Jiang, Doxorubicin

delivery to 3D multicellular spheroids and tumors based on boronic acid-rich chitosan nanoparticles, Biomaterials (2013), https://doi.org/10.1016/j.biomaterials. 2013.03.008.

91. X. Zhang, Y. Guan, Y. Zhang, Dynamically bonded layer-by-layer films for self-regulated insulin release, J. Mater. Chem. (2012), https://doi.org/ 10.1039/c2jm33413e.
92. Y. Dong, W. Wang, O. Veiseh, E.A. Appel, K. Xue, M.J. Webber, B.C. Tang, X.W. Yang, G.C. Weir, R. Langer, D.G. Anderson, Injectable and glucoseresponsive hydrogels based on boronic acid-glucose complexation, Langmuir (2016), https://doi.org/10.1021/acs.langmuir.5b04755.
93. Z. Huang, P. Delparastan, P. Burch, J. Cheng, Y. Cao, P.B. Messersmith, Injectable dynamic covalent hydrogels of boronic acid polymers cross-linked by bioactive plant-derived polyphenols, Biomater. Sci. (2018), https:// doi.org/10.1039/c8bm00453f.
94. M.W. Tibbitt, K.S. Anseth, Dynamic microenvironments: The fourth dimension, Sci. Transl. Med. (2012), https://doi.org/10.1126/scitranslmed.3004804.
95. A.M. Rosales, K.S. Anseth, The design of reversible hydrogels to capture extracellular matrix dynamics, Nat. Rev. Mater. (2016), https://doi.org/ 10.1038/natrevmats.2015.12.
96. A.J. Engler, S. Sen, H.L. Sweeney, D.E. Discher, Matrix elasticity directs stem cell lineage specification, Cell (2006), https://doi.org/10.1016/j.cell.2006.06. 044.
97. D.E. Discher, P. Janmey, Y.L. Wang, Tissue cells feel and respond to the stiffness of their substrate, Science (80-.) (2005), https://doi.org/10.1126/ science.1116995.
98. P.M. Gilbert, K.L. Havenstrite, K.E.G. Magnusson, A. Sacco, N.A. Leonardi, P. Kraft, N.K. Nguyen, S. Thrun, M.P. Lutolf, H.M. Blau, Substrate elasticity regulates skeletal muscle stem cell self-renewal in culture, Science (80-.) (2010), https://doi.org/10.1126/science.1191035.
99. C. Yang, M.W. Tibbitt, L. Basta, K.S. Anseth, Mechanical memory and dosing influence stem cell fate, Nat. Mater. (2014), https://doi.org/10.1038/ nmat3889.
100. O. Chaudhuri, L. Gu, D. Klumpers, M. Darnell, S.A. Bencherif, J.C. Weaver, N. Huebsch, H.P. Lee, E. Lippens, G.N. Duda, D.J. Mooney, Hydrogels with tunable stress relaxation regulate stem cell fate and

activity, Nat. Mater. (2016), https://doi.org/10.1038/nmat4489.

101. H. Wang, M.W. Tibbitt, S.J. Langer, L.A. Leinwand, K.S. Anseth, Hydrogels preserve native phenotypes of valvular fibroblasts through an elasticityregulated PI3K/AKT pathway, Proc. Natl. Acad. Sci. (2013), https://doi.org/ 10.1073/pnas.1306369110.
102. T. Aoki, Y. Nagao, E. Terada, K. Sanui, N. Ogata, N. Yamada, Y. Sakurai, K. Kataoka, T. Okano, Endothelial cell differentiation into capillary structures by copolymer surfaces with phenylboronic acid groups, J. Biomater. Sci. Polym. Ed. (1995), https://doi.org/10.1163/156856295X00463.
103. H. Liu, Y. Li, K. Sun, J. Fan, P. Zhang, J. Meng, S. Wang, L. Jiang, Dualresponsive surfaces modified with phenylboronic acid-containing polymer brush to reversibly capture and release cancer cells, J. Am. Chem. Soc. (2013), https://doi.org/10.1021/ja401000m.
104. T. Konno, K. Ishihara, Temporal and spatially controllable cell encapsulation using a water-soluble phospholipid polymer with phenylboronic acid moiety, Biomaterials (2007), https://doi.org/10.1016/j.biomaterials.2006.12.017.
105. Y. Chen, D. Diaz-Dussan, D. Wu, W. Wang, Y.Y. Peng, A.B. Asha, D.G. Hall, K. Ishihara, R. Narain, Bioinspired self-healing hydrogel based on benzoxaborole-catechol dynamic covalent chemistry for 3D cell encapsulation, ACS Macro Lett. (2018), https://doi.org/10.1021/acsmacrolett.8b00434.
106. J.V. Accardo, J.A. Kalow, Reversibly tuning hydrogel stiffness through photocontrolled dynamic covalent crosslinks, Chem. Sci. (2018), https://doi.org/ 10.1039/c8sc02093k.
107. T.C. Tseng, F.Y. Hsieh, P. Theato, Y.Wei, S. hui Hsu, Glucose-sensitive self-healing hydrogel as sacrificial materials to fabricate vascularized constructs, Biomaterials (2017), https://doi.org/10.1016/j.biomaterials.2017.04.008.
108. L. Smith, S. Cho, D.E. Discher, Mechanosensing of matrix by stem cells: From matrix heterogeneity, contractility, and the nucleus in pore-migration to cardiogenesis and muscle stem cells in vivo, Semin. Cell Dev. Biol. (2017), https://doi.org/10.1016/j.semcdb.2017.05.025.
109. E. Carretti, S. Grassi, M. Cossalter, I. Natali, G. Caminati, R.G. Weiss, P. Baglioni, L. Dei, Poly (vinyl alcohol)-borate hydro/cosolvent gels: Viscoelastic properties, solubilizing power, and application to art

conservation, Langmuir (2009), https://doi.org/10.1021/la804306w.

110. S.H. Hong, S. Kim, J.P. Park, M. Shin, K. Kim, J.H. Ryu, H. Lee, Dynamic bonds between boronic acid and alginate: hydrogels with stretchable, self-healing, stimuli-responsive, remoldable, and adhesive properties, Biomacromolecules (2018), https://doi.org/10.1021/acs.biomac.8b00144.
111. F.M. Najib, S. Zewar, A.M. Abdulla, A new sensor for thermometric titrations, Talanta (2007), https://doi.org/10.1016/j.talanta.2006.03.048.
112. S. Soundararajan, M. Badawi, C.M. Kohlrust, J.H. Hageman, Boronic acids for affinity chromatography: Spectral methods for determinations of ionization and diol-binding constants, Anal. Biochem. (1989), https://doi.org/10.1016/ 0003-2697(89)90367-9.
113. C.L. Ricardo, X. Mo, J.A. McCubbin, D.G. Hall, A surprising substituent effect provides a superior boronic acid catalyst for mild and metal-free direct Friedel-Crafts alkylations and prenylations of neutral arenes, Chem. Eur J. (2015), https://doi.org/10.1002/chem.201500020.
114. R. p. Singhal, B. Ramamurhy, N. Govindraj, Y. Sarwar, New ligands for boronte affinity chromatography. Synthesis and properties, J. Chromatogr. A (1991), https://doi.org/10.1016/S0021-9673(01)95752-8.
115. H.R. Mulla, N.J. Agard, A. Basu, 3-Methoxycarbonyl-5-nitrophenyl boronic acid: High affinity diol recognition at neutral pH, Bioorg. Med. Chem. Lett (2004), https://doi.org/10.1016/j.bmcl.2003.10.017.
116. S. Lascano, K. Da Zhang, R. Wehlauch, K. Gademann, N. Sakai, S. Matile, The third orthogonal dynamic covalent bond, Chem. Sci. (2016), https://doi.org/ 10.1039/c6sc01133k.

SECTION 4
POLYMER GELS

CHAPTER

18

Diarylbibenzofuranone-Based Dynamic Covalent Polymer Gels Prepared via Radical Polymerization and Subsequent Polymer Reaction

Keiichi Imato,[1,2] Masamichi Nishihara,[3] Atsushi Irie,[1] Atsushi Takahara,[1,3,] and Hideyuki Otsuka[1,2,3,]

[1]Graduate School of Engineering, Kyushu University, 744 Motooka, Nishi-ku, Fukuoka 819-0395, Japan;

[2]Department of Organic and Polymeric Materials, Tokyo Institute of Technology, 2-12-1 Ookayama, Meguro-ku, Tokyo 152-8550, Japan

[3]Institute for Materials Chemistry and Engineering, Kyushu University, 744 Motooka, Nishi-ku, Fukuoka 819-0395, Japan

ABSTRACT

Diarylbibenzofuranone (DABBF) is a dynamic covalent bonding unit, which is in equilibrium with the corresponding radicals at room temperature,

Citation: Keiichi Imato, Masamichi Nishihara, Atsushi Irie, Atsushi Takahara, and Hideyuki Otsuka. "Diarylbibenzofuranone-Based Dynamic Covalent Polymer Gels Prepared via Radical Polymerization and Subsequent Polymer Reaction" Open Access Gels 2015, 1(1), 58-68; https://doi.org/10.3390/gels1010058

and polymers with DABBF linkages show notable properties such as self-healing. The preparation routes have been strictly limited, however, and no polymer with the linkages has been synthesized via radical polymerization because of the strong antioxidant activity of DABBF. Here we present a new method to prepare DABBF-containing polymers via radical polymerization of the precursor, arylbenzofuranone (ABF), and subsequent polymer reaction, dimerization of ABF units in the linear polymers. Polymer gels cross-linked by DABBF linkages were obtained against the relatively strong antioxidant activity of ABF and showed dynamic network reorganization at room temperature.

Keywords: Self-Healing, Dynamic Covalent Chemistry, Gels, Radical Polymerization, Polymer Reaction

INTRODUCTION

Self-healing is a very attractive ability in the field of polymers because materials with the intrinsic activity can repair damage done to them without human intervention, which is expected to reduce waste and improve the lifetime, durability, and reliability of the materials. Various approaches to self-healing of polymeric materials have been reported [1], one of which is the utilization of dynamic bonding systems [2]. Healing based on these systems originates from preferential cleavage of dynamic bonds when the material is damaged, subsequent recombination and exchange of the bonds, and resulting reorganization of the polymer networks. Such healing is categorized into two types: healing involving supramolecular chemistry [3], and healing based on dynamic covalent chemistry [4]. Generally, supramolecular systems enable the damaged area to heal under ambient conditions, mostly without external stimuli [5,6,7,8,9,10,11,12]. Alternatively, dynamic covalent systems require appropriate stimuli such as heating, light, and catalyst to induce healing, although the systems behave like usual covalent systems in the absence of the stimuli [13,14,15,16,17,18,19,20,21,22,23,24,25,26,27].

Diarylbibenzofuranone (DABBF) is a unique dynamic covalent-bonding unit because it is in equilibrium in air with the corresponding radicals even at room temperature (Figure 1a) [28, 29]. We have found noteworthy properties of DABBF-containing polymers, including self-healing [30, 31], temperature-dependent network reorganization [32, 33], and damage detection [34,35]. However, polymers incorporating DABBF linkages have been synthesized only via step-growth polymerization (polyaddition) or polymer reaction, and no polymer with the linkages has been prepared via radical polymerization, which is a significant technique in the industry because of its versatility. This is because DABBF shows excellent antioxidant activity ahead of 3,5-di-*tert*-butyl-4-hydroxyanisole, a well-known antioxidant [29,36], and thereby is considered to react with the propagating radical chain ends, terminating the polymerization. Thus, despite the fascinating properties of DABBF-containing polymers, their preparation has been strictly limited to one method.

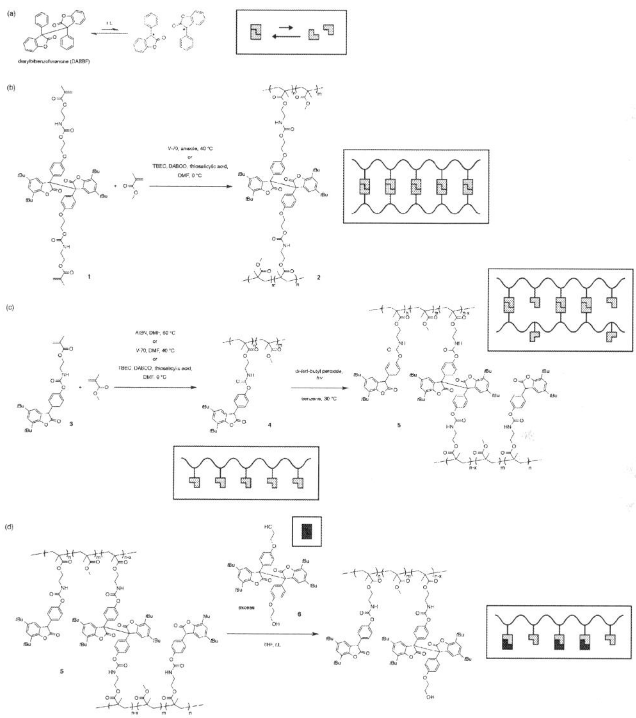

Figure 1. (**a**) Equilibrium between diarylbibenzofuranone (DABBF) and the corresponding arylbenzofuranone (ABF) radicals; (**b**) Synthesis scheme for polymers cross-linked by DABBF linkages via free radical polymerization of DABBF and methyl methacrylate (MMA); (**c**) Synthesis scheme for polymers cross-linked by DABBF linkages via free radical polymerization of ABF-containing methacrylate with MMA and subsequent oxidative coupling reaction; (**d**) De-cross-linking of DABBF-containing cross-linked polymers.

Here we show a new route to synthesize polymers cross-linked by DABBF linkages via free radical polymerization of a DABBF precursor, arylbenzofuranone (ABF), with a methacryloyl group and subsequent dimerization (cross-linking) of the precursor ABF to form DABBF cross-linkages (Figure 1c). ABF is also known to exhibit antioxidant activity, but it is much lower than that of DABBF [29,37,38]; therefore, radical polymerization is expected to proceed in the presence of ABF and ABF-containing polymers can be available. In this study, we also demonstrated free radical polymerization of dimethacryloyl-functionalized DABBF with methyl methacrylate (MMA) to test the hypothesis (Figure 1b), and we investigated the dynamic properties of gels cross-linked by DABBF linkages for self-healing (Figure 1d).

RESULTS AND DISCUSSION

Radical Copolymerization of DABBF-Containing Bifunctional Monomer and MMA

We prepared dimethacrylate-functionalized DABBF monomer **1 by adding** dihydroxy DABBF to two equivalent units of 2-methacryloyloxyethyl isocyanate and allowing them to react. Monomer **1 was employed as a cross-linker for free radical polymerization of MMA, initiated by a low-temperature** azo-type initiator, 2,2′-azobis(4-methoxy-2,4-dimethyl valeronitrile) (V-70), in anisole at 40 °C (Figure 1b). As expected, polymerization did not occur at all because of the radical nature of DABBF, which was confirmed by gel-permeation chromatographic (GPC) measurement of the reaction mixture after 45 h. It has been reported that the antioxidant activity of DABBF depends on temperature because increasing temperature biases the equilibrium to the dissociated radicals and, thus, the increased radicals should more easily react with the radical chain ends [36]. Therefore, we also tried to initiate polymerization at a lower temperature, 0 °C, by using a redox-initiation system, where *tert*-butylperoxy 2-ethylhexyl carbonate (TBEC) was employed as an oxidant, and diazabicyclo[2.2.2] octane (DABCO) and thiosalicylic acid were employed as reductants. The reaction at 0 °C for 120 h formed an insoluble component **2 but the yield was quite small, 0.5%, although the concentration was high enough to form a bulk gel (Figure S1). This result indicates that the lower temperature was effective for the radical polymerization in the presence of DABBF, but DABBF was still reactive with the chain ends at 0 °C.**

Radical Copolymerization of ABF-Containing Monomer and MMA

Methacrylate-functionalized ABF monomer 3 was prepared using a method similar to that used for preparing monomer 1. Monomer 3 and MMA were copolymerized in different ratios at 0, 40, and 60 °C for more than 24 h (Figure 1c). All polymerization was performed in N,N-dimethylformamide (DMF), a polar aprotic solvent, because the antioxidant activity of ABF (hydrogen abstraction), which is strongly dependent on the solvent polarity, is lower in polar solvents [37], while that of DABBF does not have the same dependence [36]. The hydrogen abstraction from ABF depends on the solvent polarity because it is attributed to the enol form, not the lactone form, and the enol fraction decreases with increasing solvent polarity [37]; meanwhile, the antioxidant activity of DABBF originates from only the dissociated radicals, which have little interaction with solvents [36], and thus its hydrogen abstract does not depend on the solvent polarity. As expected, the polymerization proceeded well at all temperatures. Table 1 summarizes the conditions and results of the polymerization. We obtained ABF-containing polymer 4 with different compositions and with molecular weight of more than 10^4; in particular, the polymerization at 0 °C produced higher-molecular-weight polymers. However, the compositions of 4 obtained at 0 °C were largely different from the corresponding feed ratios, and insoluble components were observed in the course of the polymerization at 0 °C, indicating that cross-linking or branching, or both cross-linking and branching occurred. The conversion, yield, and molecular weight decreased with increasing ABF ratio in the reaction mixtures; in contrast, the ratio of the component with molecular weight of approximately 10^3 increased, as shown in Figure 2 and Figure S2. These trends were observed probably because ABF and the generated ABF radicals trapped the propagating radical chain ends and the ABF radicals dimerized to form DABBF, particularly at the beginning of the polymerization.

Table 1. Radical copolymerizations of ABF-containing monomer (**3**) and MMA.

Entry	Initiator [a]	[3]/[MMA] [b]	Temp. (°C)	Time (h)	Conv. (%) [c]	Yield (%)	3/MMA [d]	M_n [e]	M_w/M_n [e]
1	AIBN	1:9	60	24	37	29	1:9.0	20,700	1.83
2	AIBN	1:4	60	24	11	5	1:4.1	24,400	1.44

3	AIBN	1:1	60	72	7	7	1:0.5	15,900	1.27
4	AIBN	1:0	60	120	–	3	1:0	14,400	1.19
5	V-70	1:9	40	61	54	40	1:8.8	22,300	1.81
6	V-70	1:4	40	61	33	23	1:3.6	19,000	1.61
7	V-70	1:1	40	61	13	–	–	–	–
8	V-70	1:0	40	120	5	4	1:0	12,300	1.39
9	redox	1:9	0	72	22	14	1:16.0	47,700	2.13
10	redox	1:4	0	72	10	3	1:5.3	34,300	1.75
11	redox	1:1	0	72	12	1	1:0.6	39,700	1.97

[a] AIBN = 2,2′-azobisisobutyronitrile; V-70 = 2,2′-azobis(4-methoxy-2,4-dimethylvaleronitrile); redox = tert-butylperoxy 2-ethylhexyl carbonate (TBEC), diazabicyclo[2.2.2]octane (DABCO), and thiosalicylic acid. [b] Feed ratio; [c] Calculated by ^{1}H NMR; [d] Composition ratio of 4 calculated by ^{1}H NMR; [e] Determined by GPC in THF.

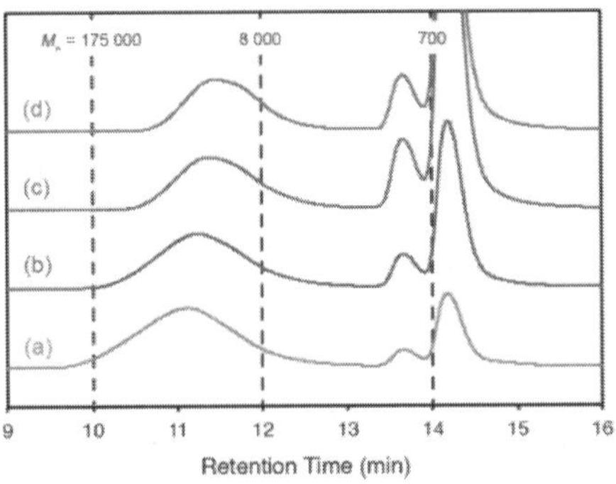

Figure 2. GPC curves of reaction mixtures of entries 5 (**a**), 6 (**b**), 7 (**c**), and 8 (**d**) after polymerization using V-70 as an initiator.

Cross-Linking of ABF-Containing Linear Polymers by Oxidative Coupling

ABF units in the side chains of linear polymers 4 were dimerized to form cross-linked polymers 5 with DABBF linkages (Figure 1c). The linear polymers 4 were irradiated with UV light in the presence of di-tert-butyl peroxide in benzene. The radicals generated by UV excitation of di-tert-butyl peroxide abstracted hydrogen from the ABF units, and the generated ABF radicals dimerized. Linear polymers 4 formed the corresponding insoluble cross-linked polymers 5 (gels) in any organic solvents in a wide range of composition 3/MMA (Figure 3). Increases in the glass-transition temperatures (T_g) after the reaction were observed in differential scanning calorimetric (DSC) analysis of the polymers (Table 2), indicating that the DABBF-linked polymers 5 were successfully obtained and cross-linking suppressed the micro-Brownian motion of the polymer chains between the cross-linking points. T_g of 4 showed a tendency to increase with the ratio of monomer 3, except for the ABF homopolymer, because the degree of polymerization was quite small (approximately 25). We consider that the intramolecular reaction also occurred in the course of the cross-linking of 4. However, there is no way to investigate it because of the dynamic nature of the DABBF linkages at room temperature.

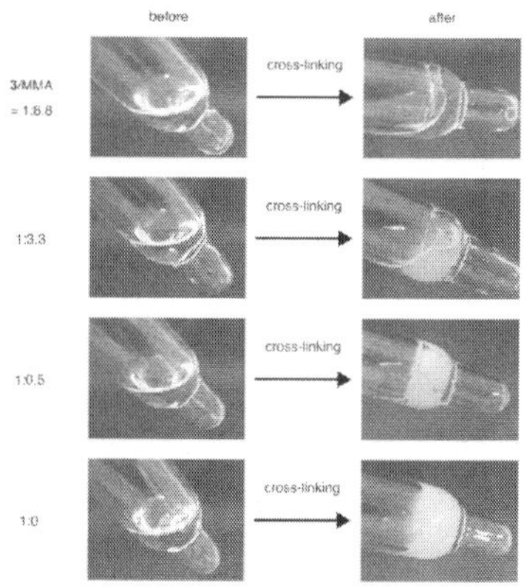

Figure 3. Photographs of linear polymer solutions **4 with different ratios of 3/** MMA (1:8.8, 1:3.3, 1:0.5, and 1:0) before and after cross-linking.

Table 2. Cross-linking of linear polymer **4**.

3/MMA [a]	M_n [b]	T_g before Cross-Linking (°C) [c]	T_g after Cross-Linking (°C) [c]
1:8.8	22,300	108	120
1:3.6	19,000	111	126
1:0.5	15,900	113	131
1:0	12,300	105	130

[a] Composition ratio of 4 calculated by ^{1}H NMR; [b] Determined by GPC in THF; [c] Determined by DSC.

De-Cross-Linking of Polymers with DABBF Cross-Linkages

To investigate the dynamic properties of the cross-linked polymers 5, we performed the de-cross-linking reaction using an excess of low-molecular

DABBF 6 (Figure 1d). DABBF units are known to exchange their bonds at room temperature even when they are incorporated in polymer chains, which is the basis of self-healing by dynamic bonding systems [30, 31, 32, 33]. The cross-linked polymers 5 were swollen with a solution of 6 (10 equiv./ABF units) in tetrahydrofuran (THF), forming gels. After waiting for more than four days in air at room temperature, the cross-linked polymers 5 completely dissolved (Figure 4a) and a THF-soluble high-molecular-weight component was detected by GPC measurement of each reaction mixture (Figure 4b). The products were expected to be linear polymers or cross-linked oligomers, or a mixture of both, all with slightly higher molecular weight than the precursor linear polymers 4 because of the cross-links and capping of the side chains by 6. Therefore, we confirmed that the incorporated DABBF linkages in cross-linked polymers 5 autonomously exchanged their bonds in air at room temperature and the polymer networks reorganized, which would lead to self-healing. Although we tried to perform a detailed analysis by fractionating this component, the cross-linking reaction occurred again after fractionation.

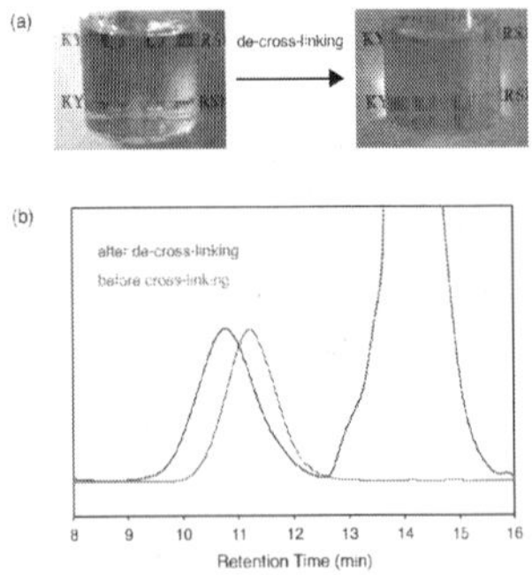

Figure 4. (a) Photographs of cross-linked polymers 5 before and after de-cross-linking; (b) GPC curves of ABF-containing polymer 4 before cross-linking (red line) and the reaction mixture after de-cross-linking (blue line).

Finally, we estimated the conversion ratio resulting from the polymer reaction (cross-linking) from ^{1}H NMR measurements of the reaction mixtures after de-cross-linking (Figure S3). As shown in Table 3, approximately 60% of ABF units in linear polymers **4** dimerized in the course of the reaction, which was not dependent on the composition.

Table 3. Degree of cross-linked DABBF in **5**.

3/MMA [a]	Cross-linked DABBF (%) [b]
1:8.8	53
1:3.6	63
1:0.5	56
1:0	60

[a] Composition ratio of 4 calculated by ^{1}H NMR; [b] Determined by ^{1}H NMR of reaction mixtures after de-cross-linking.

CONCLUSIONS

We successfully developed a method to prepare DABBF-containing polymers via radical polymerization and subsequent polymer reaction. The strategy was based on the difference between the reactivity of DABBF and ABF. As anticipated, the strong antioxidant activity of DABBF completely prevented radical polymerization, but that of ABF did not have much of this effect. The polymer reaction—dimerization of ABF units in the linear polymers—generated insoluble gels, in which approximately 60% of the ABF units were cross-linked. Because the DABBF linkages in the gels were in equilibrium and autonomously exchanged their bonds at room temperature, the gels showed the unique dynamic behavior of network reorganization at room temperature, which is necessary for self-healing. The present route for preparation of DABBF-containing polymers using radical polymerization would expand the availability of their intriguing properties, such as self-healing and damage visualization, from daily necessities to engineering materials.

EXPERIMENTAL SECTION

Radical Copolymerization of DABBF-Containing Bifunctional Monomer and MMA

In a typical run, a solution of 1 (0.50 g, 0.47 mmol) and MMA (0.94 mL, 8.9 mmol) in DMF, a solution of TBEC (8.7 mg, 35 μmol), a solution of DABCO (0.9 mg, 7.9 μmol) in DMF, and a solution of thiosalicylic acid (8.0 mg, 52 μmol) in DMF after freeze–pump–thaw degassing were combined in a test tube to reach a ratio of DMF/MMA = 1:1 (v/v). The tube was sealed and the reaction was allowed to proceed at 0 °C for 120 h. After the mixture was reprecipitated from an excess of methanol, filtered, and dried in vacuo, DABBF-containing cross-linked polymer 2 was obtained as a white solid (6.5 mg, 0.5% yield).

Radical Copolymerization of ABF-Containing Monomer and MMA

In a typical run, 3 (0.13 g, 0.26 mmol), MMA (0.25 mL, 2.4 mmol), V-70 (2.4 mg, 7.8 μmol), and DMF (0.25 mL) were combined in a test tube. After the freeze–pump–thaw degassing cycle, the tube was sealed and the reaction was allowed to proceed at 40 °C for 72 h. After reprecipitation of the mixture from an excess of methanol, filtered, and dried in vacuo, ABF-containing linear polymer 4 was obtained as a white solid (0.14 g, 40% yield). ^{1}H NMR (300 MHz, $CDCl_3$): δ/ pm 0.70–1.22 (br, α-CH_3), 1.23–1.50 (br, CH_3), 1.70–2.10 (br, CH_2), 3.40–3.80 (br, CH_2), 4.13 (br, NH), 4.84 (s, CH), 7.07–7.32 (m, aromatic). FT-IR (KBr, cm^{-1}): 3380, 2990, 2950–2840, 1810–1800, 1740–1720, 1610, 1500–1440, 1270–1240, 1150, 1070, 990, 900, 750.

Cross-Linking of ABF-Containing Linear Polymers

In a typical run, 4 (50 mg, M_n = 22,300, M_W/M_n = 1.81, 3/MMA = 1:8.8), di-tert-butyl peroxide (0.10 mL, 0.55 mmol), and benzene (0.51 mL) were combined in a test tube. The solution was irradiated by UV light (high-pressure mercury lamp) for 120 min at 30 °C. After washing with an excess of benzene and freeze-drying, DABBF-containing cross-linked polymer 5 was obtained as a white solid (52 mg, quantitative).

De-Cross-Linking of Gels with DABBF Cross-Linkages

First, 5 (20 mg) and a THF solution (4.5 mL) of 6 (10 equiv./ABF units) were charged into a sample tube. The mixture was then stirred under air at room temperature. After more than four days, the reaction mixture was evaluated by GPC measurements and ^{1}H NMR measurements.

ACKNOWLEDGMENTS

Hideyuki Otsuka gratefully acknowledges financial support from the Funding Program for Next Generation World-Leading Researchers (No. GR077) and JSPS KAKENHI (No. 26288057 and No. 26620175) from Japan Society of the Promotion of Science (JSPS). A part of this work was also supported by ImPACT Program of Council for Science, Technology and Innovation (Cabinet Office, Government of Japan). Keiichi Imato acknowledges financial support through JSPS Research Fellowships for Young Scientists (No. 24·7074).

CONFLICTS OF INTEREST

The authors declare no conflict of interest.

REFERENCES

1. Burattini S., Greenland B.W., Chappell D., Colquhoun H.M., Hayes W. Healable polymeric materials: A tutorial review. Chem. Soc. Rev. 2010; 39:1973–1985. doi: 10.1039/b904502n. .
2. Wojtecki R.J., Meador M.A., Rowan S.J. Using the dynamic bond to access macroscopically responsive structurally dynamic polymers. Nat. Mater. 2011; 10:14–27. doi: 10.1038/nmat2891. .
3. Brunsveld L., Folmer B.J.B., Meijer E.W., Sijbesma R.P. Supramolecular polymers. Chem. Rev. 2001; 101:4071–4098. doi: 10.1021/cr990125q. .
4. Rowan S.J., Cantrill S.J., Cousins G.R.L., Sanders J.K.M., Stoddart J.F. Dynamic covalent chemistry. Angew. Chem. Int. Ed. 2002; 41:898–952. doi: 10.1002/1521-3773(20020315)41:6<898::AID-ANIE898>3.0.CO;2-E. .
5. Cordier P., Tournilhac F., Soulié-Ziakovic C., Leibler L. Self-healing and thermoreversible rubber from supramolecular assembly. Nature. 2008;451:977–980. doi: 10.1038/nature06669. .
6. Burattini S., Colquhoun H.M., Fox J.D., Friedmann D., Greenland B.W., Harris P.J.F., Hayes W., Mackay M.E., Rowan S.J. A Self-Repairing, Supramolecular polymer system: Healability as a consequence of donor–acceptor π–π stacking interactions. Chem. Commun. 2009:6717–6719. doi: 10.1039/b910648k. .
7. Wang Q., Mynar J.L., Yoshida M., Lee E., Lee M., Okuro K., Kinbara K., Aida T. High-water-content mouldable hydrogels by mixing clay and a dendritic molecular binder. Nature. 2010;463:339–343. doi: 10.1038/nature08693. .
8. Nakahata M., Takashima Y., Yamaguchi H., Harada A. Redox-responsive self-healing materials formed from host-guest polymers. Nat. Commun. 2011;2:511. doi: 10.1038/ncomms1521.
9. Burnworth M., Tang L., Kumpfer J.R., Duncan A.J., Beyer F.L., Fiore G.L., Rowan S.J., Weder C. Optically healable supramolecular polymers. Nature. 2011;472:334–338. doi: 10.1038/nature09963. .
10. Haraguchi K., Uyama K., Tanimoto H. Self-healing in nanocomposite hydrogels. Macromol. Rapid Commun. 2011;32:1253–1258. doi: 10.1002/marc.201100248. .
11. Chen Y., Kushner A.M., Williams G.A., Guan Z. Multiphase design of autonomic self-healing thermoplastic elastomers. Nat.

Chem. 2012;4:467–472. doi: 10.1038/nchem.1314. .

12. Sun T.L., Kurokawa T., Kuroda S., Bin Ihsan A., Akasaki T., Sato K., Haque M.A., Nakajima T., Gong J.P. Physical hydrogels composed of polyampholytes demonstrate high toughness and viscoelasticity. Nat. Mater. 2013;12:932–937. doi: 10.1038/nmat3713. .
13. Chen X., Dam M.A., Ono K., Mal A., Shen H., Nutt S.R., Sheran K., Wudl F. A Thermally Re-mendable cross-linked polymeric material. Science. 2002;295:1698–1702. doi: 10.1126/science.1065879. .
14. Reutenauer P., Buhler E., Boul P.J., Candau S.J., Lehn J.M. Room temperature dynamic polymers based on diels-alder chemistry. Chem. Eur. J. 2009;15:1893–1900. doi: 10.1002/chem.200802145. .
15. Deng G., Tang C., Li F., Jiang H., Chen Y. Covalent Cross-linked polymer gels with reversible sol-gel transition and self-healing properties. Macromolecules. 2010;43:1191–1194. doi: 10.1021/ma9022197.
16. Yoshie N., Watanabe M., Araki H., Ishida K. Thermo-Responsive Mending of polymers crosslinked by thermally reversible covalent bond: polymers from bisfuranic terminated poly(ethylene adipate) and tris-maleimide. Polym. Degrad. Stab. 2010; 95:826–829. doi: 10.1016/j.polymdegradstab.2010.01.032.
17. He L., Fullenkamp D.E., Rivera J.G., Messersmith P.B. pH responsive self-healing hydrogels formed by boronate–catechol complexation. Chem. Commun. 2011; 47:7497–7499. doi: 10.1039/c1cc11928a.
18. Amamoto Y., Kamada J., Otsuka H., Takahara A., Matyjaszewski K. Repeatable photoinduced self-healing of covalently cross-linked polymers through reshuffling of trithiocarbonate units. Angew. Chem. Int. Ed. 2011;50:1660–1663. doi: 10.1002/anie.201003888. .
19. Canadell J., Goossens H., Klumperman B. Self-healing materials based on disulfide links. Macromolecules. 2011; 44:2536–2541. doi: 10.1021/ma2001492.
20. Yuan C., Rong M.Z., Zhang M.Q., Zhang Z.P., Yuan Y.C. Self-healing of polymers via synchronous covalent bond fission/radical recombination. Chem. Mater. 2011; 23:5076–5081. doi: 10.1021/cm202635w.
21. Zheng P., McCarthy T.J. A Surprise from 1954: Siloxane equilibration

is a simple, robust, and obvious polymer self-healing mechanism. J. Am. Chem. Soc. 2012; 134:2024–2027. doi: 10.1021/ja2113257. .

22. Capelot M., Montarnal D., Tournilhac F., Leibler L. Metal-catalyzed transesterification for healing and assembling of thermosets. J. Am. Chem. Soc. 2012; 134:7664–7667. doi: 10.1021/ja302894k. .
23. Amamoto Y., Otsuka H., Takahara A., Matyjaszewski K. Self-healing of covalently cross-linked polymers by reshuffling thiuram disulfide moieties in air under visible light. Adv. Mater. 2012;24:3975–3980. doi: 10.1002/adma.201201928. .
24. Lu Y.-X., Guan Z. Olefin metathesis for effective polymer healing via dynamic exchange of strong carbon–carbon double bonds. J. Am. Chem. Soc. 2012;134:14226–14231. doi: 10.1021/ja306287s. .
25. Ying H., Zhang Y., Cheng J. Dynamic urea bond for the design of reversible and self-healing polymers. Nat. Commun. 2014;5:3218. doi: 10.1038/ncomms4218.
26. Garcia J.M., Jones G.O., Virwani K., McCloskey B.D., Boday D.J., ter Huurne G.M., Horn H.W., Coady D.J., Bintaleb A.M., Alabdulrahman A.M.S., et al. Recyclable, strong thermosets and organogels via paraformaldehyde condensation with diamines. Science. 2014;344:732–735. doi: 10.1126/science.1251484..
27. Gao Y., Luo Q., Qiao S., Wang L., Dong Z., Xu J., Liu J. Enzymetically regulating the self-healing of protein hydrogels with high healing efficiency. Angew. Chem. Int. Ed. 2014;53:9343–9346. doi: 10.1002/anie.201404531. .
28. Frenette M., Aliaga C., Font-Sanchis E., Scaiano J.C. Bond dissociation energies for radical dimers derived from highly stabilized carbon-centered radicals. Org. Lett. 2004;6:2579–2582. doi: 10.1021/ol049111j. .
29. Frenette M., MacLean P.D., Barclay L.R.C., Scaiano J.C. Radically different antioxidants: thermally generated carbon-centered radicals as chain-breaking antioxidants. J. Am. Chem. Soc. 2006;128:16432–16433. doi: 10.1021/ja066870j. .
30. Imato K., Nishihara M., Kanehara T., Amamoto Y., Takahara A., Otsuka H. Self-healing of chemical gels cross-linked by diarylbibenzofuranone-based trigger-free dynamic covalent bonds at room temperature. Angew. Chem. Int. Ed. 2012;51:1138–1142. doi: 10.1002/anie.201104069. .

31. Imato K., Takahara A., Otsuka H. Self-healing of a cross-linked polymer with dynamic covalent linkages at ambient temperature and evaluation at macroscopic and molecular levels. unpublished.
32. Nishihara M., Imato K., Irie A., Kanehara T., Kano A., Maruyama A., Takahara A., Otsuka H. Reversibly crosslinked polymeric micelles formed by autonomously exchangeable dynamic covalent bonds. Chem. Lett. 2013;42:377–379. doi: 10.1246/cl.121289.
33. Imato K., Ohishi T., Nishihara M., Takahara A., Otsuka H. Network reorganization of dynamic covalent polymer gels with exchangeable diarylbibenzofuranone at ambient temperature. J. Am. Chem. Soc. 2014;136:11839–11845. doi: 10.1021/ja5065075. .
34. Imato K., Irie A., Kosuge T., Ohishi T., Nishihara M., Takahara A., Otsuka H. Mechanophores with a reversible radical system and freezing-induced mechanochemistry in polymer solutions and gels. Angew. Chem. Int. Ed. 2015;54:6168–6172. doi: 10.1002/anie.201412413. .
35. Imato K., Kanehara T., Ohishi T., Nishihara M., Yajima H., Ito M., Takahara A., Otsuka H. Mechanochromic dynamic covalent elastomers: quantitative damage evaluation and autonomous recovery. unpublished.
36. Filippenko V., Frenette M., Scaiano J.C. Solvent-independent antioxidant activity from thermally generated carbon-centered radical antioxidants. Org. Lett. 2009;11:3634–3637. doi: 10.1021/ol9014046. .
37. Aliaga C., Stuart D.R., Aspée A., Scaiano J.C. Solvent effects on hydrogen abstraction reactions from lactones with antioxidant properties. Org. Lett. 2005;7:3665–3668. doi: 10.1021/ol050839p. .
38. Zhu X.-Q., Zhou J., Wang C.-H., Li X.-T., Jing S. Actual structure, thermodynamic driving force, and mechanism of benzofuranone-typical compounds as antioxidants in solution. J. Phys. Chem. B. 2011;115:3588–3603. doi: 10.1021/jp200095g.

CHAPTER
19

Assessing Mucoadhesion in Polymer Gels: The Effect of Method Type and Instrument Variables

Jéssica Bassi da Silva,[1] Sabrina Barbosa de Souza Ferreira,[1] Adriano Valim Reis,[1] Michael Thomas Cook,[2] and Marcos Luciano Bruschi[1]

[1]Laboratory of Research and Development of Drug Delivery Systems, Postgraduate Program in Pharmaceutical Sciences, Department of Pharmacy, State University of Maringa, Maringa, Parana CEP 87020-900, Brazil

[2]Research Centre in Topical Drug Delivery and Toxicology, Department of Pharmacy, Pharmacology and Postgraduate Medicine, University of Hertfordshire, Hatfield AL10 9AB, UK

ABSTRACT

The process of mucoadhesion has been widely studied using a wide variety of methods, which are influenced by instrumental variables and experiment

Citation: Bassi da Silva, J.; Ferreira, S.B.S.; Reis, A.V.; Cook, M.T.; Bruschi, M.L. "Assessing Mucoadhesion in Polymer Gels: The Effect of Method Type and Instrument Variables". Polymers 2018, 10, 254. https://doi.org/10.3390/polym10030254

design, making the comparison between the results of different studies difficult. The aim of this work was to standardize the conditions of the detachment test and the rheological methods of mucoadhesion assessment for semisolids, and introduce a texture profile analysis (TPA) method. A factorial design was developed to suggest standard conditions for performing the detachment force method. To evaluate the method, binary polymeric systems were prepared containing poloxamer 407 and Carbopol 971P®, Carbopol 974P®, or Noveon® Polycarbophil. The mucoadhesion of systems was evaluated, and the reproducibility of these measurements investigated. This detachment force method was demonstrated to be reproduceable, and gave different adhesion when mucin disk or ex vivo oral mucosa was used. The factorial design demonstrated that all evaluated parameters had an effect on measurements of mucoadhesive force, but the same was not observed for the work of adhesion. It was suggested that the work of adhesion is a more appropriate metric for evaluating mucoadhesion. Oscillatory rheology was more capable of investigating adhesive interactions than flow rheology. TPA method was demonstrated to be reproducible and can evaluate the adhesiveness interaction parameter. This investigation demonstrates the need for standardized methods to evaluate mucoadhesion and makes suggestions for a standard study design.

Keywords: Pluronic F127, Thermoresponsive Polymers, Thermogelling Polymers, Detachment Force, Rheology, Texture Profile Analysis

INTRODUCTION

Mucosal surfaces cover the nasal, ocular, buccal, rectal, vaginal, and gastrointestinal areas among other parts of the body. Drugs may be administered to these sites for local effect, and their high permeability makes them attractive for systemic drug delivery. However, the natural clearance mechanisms from these sites limit residence time, decreasing drug absorption or duration of local effect [1]. In order to overcome these disadvantages, "mucoadhesive" systems have been developed, which adhere to mucosal membranes through a variety of attractive physicochemical interactions, enhancing retention, and thus the efficacy of medicines [2,3,4,5].

Mucoadhesive polymers are a group of materials employed in different pharmaceutical systems. They are defined as hydrophilic macromolecules, which contain numerous functional organic groups (i.e., carboxylic, hydroxyl, amide, and amine groups) able to establish interactions with mucosal membranes [6,7]. These polymers can be classified according to their interactions with the mucosa (covalent bonds or non-covalent intermolecular interactions). Non-covalent bonds believed to enhance mucoadhesion include hydrogen-bonding, hydrophobic interactions, and electrostatic interactions. Mucoadhesive polymers may be cationic, anionic, or non-ionic [8,9,10]. Anionic polymers, such as poly(acrylic acid) derivates, are believed to form hydrogen bonds below their pKa between their carboxylic groups and the hydroxyl groups of the mucus glycoprotein. It has also been suggested that ion-dipole interactions may occur when in the carboxylate form [9]. Moreover, poly(acrylic acid) derivates may be combined in solution with thermoresponsive polymers, like poloxamer 407 (P407), to enhance retention [11]. Thermoresponsive polymers transition from a liquid to a viscous gel above a critical temperature, allowing for passage through an applicator before thickening upon application to the body [11,12,13].

In vitro or ex vivo techniques are crucial in the performance testing of mucoadhesive drug delivery systems and are cost-effective in selecting efficient systems when compared with in vivo methods. These methods are able to evaluate mucoadhesive formulations, without using animal models, and may offer mechanistic understanding of mucoadhesion [14,15,16,17]. Numerous techniques have been developed to assess and understand the mucoadhesion of drug delivery systems.

The development of new methods should be validated by comparison with a gold standard in vitro technique, or in vivo performance. New methods to investigate the mucoadhesive profile of semisolid polymer systems are typically developed in-house on bespoke equipment, and have not been through validation, which emphasizes the importance of standardized techniques [18,19,20]. Furthermore, each dosage form may require different experimental conditions and comparison may only be possible within dosage form types. The detachment force method (also known as the tensile method) is the most widely employed method to investigate adhesive interactions between a mucosal membrane (or other substrate) and a formulation. This method can be used for solid [19,20,21] and semisolid dosage forms [2,4,5,12,19,21,22,23] and it is known that instrumental parameters and experiment design influence test results. Other techniques, such as the rheological method, can result in different responses and interpretations depending on the analysis type used (flow or oscillatory) [19]. Therefore, it is very important to understand the variables of the method for mucoadhesion testing, considering that standardized methods have been required [1,3].

Therefore, this work aimed to investigate the importance of standardizing the conditions to perform the detachment force and the rheological methods for assessing mucoadhesion of semisolid systems, as well as for assessing mucoadhesive interactions by texture profile analysis.

MATERIALS AND METHODS

Materials

Carbopol 971P® (C971P), Carbopol 974P® (C974P), and Noveon® Polycarbophil (PCB) were kindly donated by Lubrizol (Sao Paulo, Brazil). Triethanolamine, poloxamer 407 (P407), and mucin from porcine stomach (type II) were received from Sigma-Aldrich (Sao Paulo, Brazil). Porcine oral mucosa was sourced from a local slaughterhouse (Maringa, Brazil) and kept frozen at −20 °C. All reagents were used without further purification.

Preparation of Polymeric Systems

Monopolymeric formulations were prepared using P407 (15% or 20%, *w/w*) or C971P, C974P, and PCB (0.10%, 0.15%, 0.20%, 0.25%, or 0.50%, *w/w*). P407 solutions were prepared by dispersing the required amount of the polymer in purified water at 5 °C under mechanical stirring (500 rpm). To prepare the poly(acrylic acid)-containing formulations, the required amount mass of the polymer was dispersed in purified water with mechanical stirring.

Binary polymeric blends were prepared by dispersion of C971P, C974P, or PCB (0.10%, 0.15%, 0.20%, 0.25%, or 0.50%, *w/w*) in purified water under mechanical stirring. The required amount of P407 (15% or 20%, *w/w*) was then added to the preparation. This mixture was maintained for 12 h at the temperature of 4 °C. After this period of time, the preparations were stirred, neutralized using triethanolamine, and refrigerated (4 °C) for at least 24 h before analysis.

Mucoadhesive Analysis by Detachment Force Evaluation

A TA-XTplus texture analyzer (Stable Micro Systems, Surrey, UK) was utilized to investigate the adhesive properties of the formulations [24]. Mucin disks were prepared by the compression of crude porcine mucin (200 mg) using a ring press with a 13-mm diameter die and a compression force of 10 tonnes, applied for 30 s. Moreover, pig buccal mucosa samples were obtained from pigs (white, young, and recent sacrificed) originated from a local slaughterhouse (authorized by the Brazilian Ministry of Agriculture for human consumption). They were cleaned with phosphate saline buffer (PSB), the cheek mucosa was gently removed, and the samples were prepared with the same diameter and area of the mucin disks (132.73 mm^2), using a surgical scalpel. Samples displaying wounds or bruises were not used. The mucosal substrate (disk or tissue) was then horizontally attached to the lower end of the probe (cylindrical, P/6), using double sided adhesive tape. Prior to testing, the disk or the mucosal tissue was hydrated by submersion in a 5% (*w*/*v*) aqueous solution of mucin or in PSB for 30 s, respectively. The excess surface liquid was withdrawn by gentle blotting. Samples of each formulation (5.0 g) were packed into shallow cylindrical vessels with 20 mm diameter, maintained at 37 °C, and placed under the probe, which was lowered at a speed of 1 mm/s until it reached the mucoadhesive hydrogel surface. Immediately, a downward force of 0.03 N was applied and the probe remained on the surface of the sample for 30 s; then the probe was withdrawn at a rate of 10.0 mm/s until complete detachment of the sample from the mucosal substrate. The Texture Exponent 32 software (Stable Micro Systems, Surrey, UK) was used to determine the force required for the detachment (F_{adh}) and the work of adhesion (W_{adh}) (the area under the force/distance curve). All measurements were performed with at least six replicates.

Mucin disks and porcine buccal mucosa were chosen as substrate models, due to the great use of them in the literature. Moreover, pigs have a greater similarity of anatomy, physiology, metabolism and histology than other animals when compared with humans [25,26]. Detachment force, the maximum force necessary to remove the sample from mucosal substrate, and work of adhesion, the area under the force-displacement curve, were evaluated by two-way analysis of variance (ANOVA), using Tukey post hoc test. The p-value < 0.05 was taken to denote significance.

To evaluate the effects of the instrumental parameters on the F_{adh} and W_{adh}, a polymeric system composed of 15% (*w*/*w*) P407 and 0.25% (*w*/*w*) PCB at 37 °C was also studied. A full factorial design $2^4 + 4C$ was created by Statistic 8.0® software (StatSoft Company, Tulsa, OK, USA). The influence of the variables: substrate (X_1), force (X_2), speed of upward probe (X_3) and time of substrate-sample contact (X_4) were investigated. Each factor was set at one of two levels, low (−) or high (+) (Table 1). Four central points were also used to evaluate the curvature and the errors related with isolated effects or the interaction between them.

Table 1. Matrix of factorial design 2^4 + 4C for binary polymeric systems composed of 15% (*w*/*w*) P407 and 0.25% (*w*/*w*) PCB, at 37 °C, and values for the low and high levels of each variable.

Standard Run	Independent Variables			
	X_1	X_2	X_3	X_4
1	A	–	–	–
2	B	–	–	–
3	A	+	–	–
4	B	+	–	–
5	A	–	+	–
6	B	–	+	–
7	A	+	+	–
8	B	+	+	–
9	A	–	–	+
10	B	–	–	+
11	A	+	–	+
12	B	+	–	+
13	A	–	+	+
14	B	–	+	+
15	A	+	+	+
16	B	+	+	+
17	A	0	0	0
18	B	0	0	0
19	A	0	0	0
20	B	0	0	0
Factor		–	0	+
X_1	Substrate	A: Mucin disk		B: Oral pig mucosal
X_2	Force (N)	0.03	0.065	0.10
X_3	Speed (mm/s)	1.00	5.50	10.0
X_4	Time (s)	30.0	75.0	120.0

Mucoadhesive Analysis Using Rheological Methods

Continuous Shear (Flow) Rheology

The increase in consistency index due to a synergism between mucin and polymer systems was measured using a modified version of the method described by Hassan and Gallo [27,28]. The polymeric blends previously prepared, 5% (*w/w*) mucin aqueous solution (prepared just before the measurements), and their mixture were evaluated. The mixtures were composed of the polymeric systems with 5% (*w/w*) mucin added under vigorous stirring for 15 min prior to analysis. Flow rheometry was performed using a controlled stress rheometer (MARS II, Haake Thermo Fisher Scientific Inc., Newington, Germany), equipped with parallel steel cone-plate geometry of 60 mm; separated by a fixed distance of 0.052 mm. Analysis was performed in flow mode at 37 ± 1 °C, over shear rates ranging from 0 to 2000 s^{-1}, increasing over a period of 150 s, and was maintained at the superior limit for 10 s and then decreased during the same period. At least six replicates of each sample were evaluated, and the upward flow curves were fitted using Power Law equation (Oswald-de-Waele) as shown below [2, 27, 29, 30]

$$\tau = k.\, \gamma^{n} \qquad (1)$$

where τ is the shear stress (Pa), k is the consistency index $[(Pa.s)^n]$, γ is the rate of shear (s^{-1}), and n is the flow behavior index (dimensionless).

Oscillatory Rheology

Oscillatory measurements were performed at 37 ± 1 °C, using the same rheometer and geometry previously described, and in oscillation mode. Samples of each polymeric system, mucin aqueous solution 5% (*w/w*) and the polymeric systems with 5% (*w/w*) mucin were placed to the inferior plate, allowing for 1 min equilibration prior to testing. The linear viscoelastic region of each formulation was determined and a frequency sweep was performed from 0.1 to 10.0 Hz.

The elastic modulus (G'), viscous modulus (G''), dynamic viscosity (η'), and the loss tangent (tan δ) were calculated using RheoWin 4.10.0000 (Haake) software (Thermo Fisher Scientific Inc., Newington, Germany). All measurements were performed in at least six replicates [29]. Calculation of the interaction parameter for the polymeric systems (P407/C971P, P407/C974P, and P407/PCB) with mucin was determined by the difference between the storage (elastic) modulus of the mixture, and the theoretical value of the storage modulus obtained by summation of the individual parts [29,31], at an oscillatory frequency of 10.0 Hz as demonstrated in the equation.

$$\Delta G' = G'\text{mixture} - (G'\text{mucin} + G'\text{polymeric system}) \qquad (2)$$

Texture Profile Analysis

The texture profile analysis (TPA) of polymeric blends, 5% (*w*/*w*) mucin aqueous solution and the mixture of both were performed using a TA-XTplus Texture Analyzer (Stable Micro Systems; Surrey, UK) in TPA mode, at 37 °C, to evaluate adhesive interactions. Bottles containing 16 g of formulations were submitted to a double compression by an analytical probe (10 mm diameter).

The two times of compression (at 2 mm/s) were performed on the sample, with a predefined depth (15 mm). A delay period of 15 s was permitted between the end of the first compression and the beginning of the second. From the force-time and force-distance plot, the mechanical properties were obtained, namely: hardness (maximum force obtained during the first compression), compressibility (the work required to deform the sample during the first pass of the probe), adhesiveness (work required to overcome the attractive forces between the surfaces of the probe and the formulation), elasticity (ability to stretch and return to its original size and shape), and cohesiveness (work spent to unite the surface of the sample and the surface of the probe) [2,29]. All analysis was performed in at least five replicates.

The interaction parameter of these variables was calculated by the difference between the values observed for the mixture (polymeric blend with 5% (*w*/*w*) mucin) and the sum of the individual contributions of polymeric system and mucin solution.

Statistical Analysis

The responses obtained were statistically compared using two-way analysis of variance (ANOVA). In all cases, individual differences between means were identified using Tukey's test. Moreover, significant differences were accepted when $p < 0.05$ for all methods [32].

RESULTS AND DISCUSSION

Polymeric Systems

Previously characterized binary polymer blends [23,30,33,34] were selected to evaluate the robustness of the mucoadhesive methods. These systems, composed of P407 and poly(acrylic acid) derivatives possess different rheological and mechanical properties. These polymers contribute in a unique way to the formulation properties. With a temperature increase, P407 increases the viscosity of the system and the other decreases it (C971P, C974P, or PCB). Although other polymers as natural polymers (i.e., gelatin and agar) and synthetic polymers (i.e., poly(vinylpyrrolidone) and poly(vinyl alcohol)) were already widely used to investigate the mucoadhesion process, the selected thermo responsive blends, besides being very well characterized as a complex system, can be a challenge for different mucoadhesion analyses [23,30,33,34].

The carbomers C971P and C974P are mucoadhesive polymers crosslinked with allyl-ethers of pentaerythritol, displaying a concentration ranging from 56% to 68% of carboxylic groups in the chain. However, C974P displays a higher crosslinking degree than C971P. PCB is also a high-molecular-weight polymer of polyacrylic acid, cross-linked with divinylglycol or polyalkenyl ethers, displaying a large number of carboxylic groups (COOH) on the molecular chain, and known by its strong mucoadhesive properties [23,30,34].

The P407 monopolymeric formulation displays higher consistency at body temperatures than the blends containing the same amount of this thermogelling polymer [12,23,35]. Rheological, texture, and mucoadhesion studies have been performed to select the most suitable mucoadhesive semi-solid systems of each blend in previous studies (Table 2).

Table 2. Mucoadhesive profile (using porcine mucin disk), gelation temperature, and rheological interaction parameter of the selected formulations containing poloxamer 407 (P407) and Cabopol 971P® (C971P), Carbopol 974P® (C974P), or Noveon® Polycarbophil (PCB).

Systems	Polymer Amount (%, *w*/*w*)	Mucoadhesive Force (N)	Gelation Temperature (°C)	Interaction Parameter (Pa)
P407/C971P [a]	20/0.20	0.352 ± 0.04	27.88 ± 0.06	1944.73 ± 381.93
P407/C974P [b]	15/0.25	0.231 ± 0.03	36.04 ± 0.06	2509.33 ± 215.85

P407/PCB [c]	15/0.15	0.237 ± 0.01	36.42 ± 0.02	1927.03 ± 93.85

[a] [34]; [b] [30]; [c] [23].

The interaction parameter evaluates the adhesive interaction between polymers and can select optimal formulations for pharmaceutical and biomedical application [29,30]. The gelation temperature and adhesiveness are also important for performance testing of thermogelling systems to ensure that they solidify on application and are retained on the mucosa. The P407/C971P (20/0.20%, *w/w*), P407/C974P (15/0.25%, *w/w*), and P407/PCB (15/0.25%, *w/w*) system were selected based on previously-published information [23,30,34]. A gelation temperature range near the body temperature, between 25 and 37 °C, is considered appropriate for pharmaceutical systems [2,13].

Methods for analyzing the performance (e.g., mucoadhesion) should be investigated for accuracy, precision, specificity, linearity, and range [36]. In this work, some of these characteristics (precision by repeatability and robustness) were studied to evaluate the importance of standardizing the conditions of method. The robustness of an analytical method is defined by ICH guidelines as a measurement of resistance against small variations of the analytical parameter. Thus, it is possible to verify the robustness of the method by demonstrating its reliability during the normal use [36]. In this study, the methods to investigate the mucoadhesive characteristic were kept constant and the semi-solid formulations were changed to consider the formulations as a parameter or independent variable. Even known the temperature is not the same at 37 °C for different routes, and diseases are able to change the body temperature (e.g., fever), 37 °C was kept through the studied methods. Since most mucosal routes demonstrate this mean temperature. In addition, the selected thermoresponsive systems maintain their physicochemical profile with small variation of temperature (30–38 °C) and gel in this temperature range.

Analysis of the Detachment Force

The detachment force is an in vitro method widely used to analyze the mucoadhesive properties of the most dosage forms. By this method, it is possible to measure the W_{adh}, which is based on the area under a force-distance curve obtained during the detachment process, as well as the F_{adh}, defined as the maximum force to separate an adhesive surface from a mucous substrate [1,3,30,35,37]. There is not a standard methodology to this determination [37,38], so not only the choice of the technique, but also the choice of membrane (mucus or other mucosal types) and instrument parameters are critical to investigate the mucoadhesive properties of drug delivery systems. The limited use of the human mucosa means that ex vivo animal models are required, but these require validation and an understanding of intra-species variability. Moreover, the process of mucus/mucosa preparation can modify the physicochemical properties and results in an altered structure when compared to a fresh mucosa [37,38]. To begin this process, a standardization of this method using mucin disks and oral porcine mucosa for the three semi-solid systems was proposed.The results of the detachment force method (Figure 1 and Table S1) demonstrate the reproducibility of the method with a coefficient of variation (or relative standard deviation), lower than 5% for mucin disks and lower than 12% for porcine oral mucosa, with the variation believed to be a result of the mucosa's natural variability [33,39,40]. Previous studies have demonstrated differences among mucin from different sites. Likewise, the variability between individuals and mucosa thickness may also affect results on ex vivo mucosa [16,37,41]. In a previous investigation using mucin disks, polymeric blends containing P407/C971P, P407/C974P, or P407/PCB showed the same values of the mucoadhesion force, when compared by this method, supporting its reliability between laboratories [23,30,34].

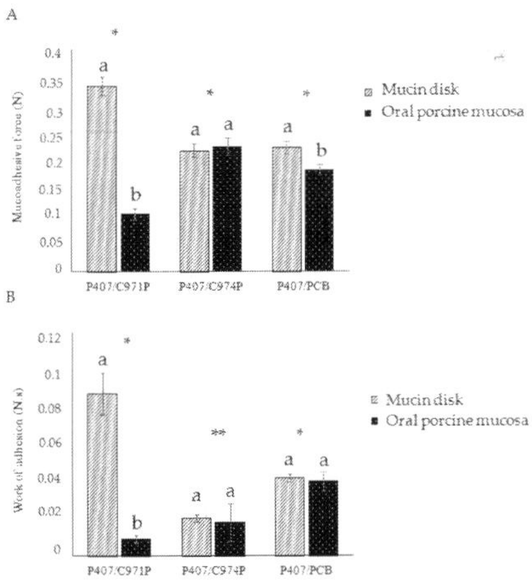

Figure 1. Mucoadhesive force F_{adh} (**A**) and work of adhesion W_{adh} (**B**) of the polymeric blends containing P407 and C971P, C974P, or PCB, obtained by detachment force method using porcine mucin disks or oral porcine mucosa as substrates. The symbols (* and **) and the letters (a and b) represent the significant difference ($p < 0.05$) among the polymeric systems and the substrates, respectively.

The P407/C971P formulation demonstrated the greatest mucoadhesive strength on mucin disks, with the rank order P407/C971P > P407/PCB > P407/C974P. Binary polymeric systems composed of P407 and poly(acrylic acid) derivatives have the availability of the carboxylic groups present in the mucoadhesive polymer decreased. It is believed that interactions between P407 and the mucoadhesive polymer reduces the possibility of interaction between the poly(acrylic acid) derivate and the mucin. However, C971P has higher carboxyl concentration, when compared with C974P and PCB, which allows the P407/C971P system to interact with mucin groups even with the P407 interaction [2,33,34,39,42,43]. Moreover, P407/C971P demonstrated in TPA analysis higher values of hardness, cohesiveness, and compressibility, which contribute with the higher values of mucoadhesion observed with mucin disk. Using porcine buccal mucosa as a substrate gave a different rank-order of mucoadhesion: P407/C974P > P407/PCB > P407/C971P. The PCB with an intermediate amount of free carboxyl groups had in the two cases, also the intermediate mucoadhesive force. Furthermore, C974P, a more cross-linked material with a lower amount of hydroxyl groups, demonstrated similar results using mucin disk and animal mucosa. It is believed that the polymer has greater attractive forces among its polymeric chains, which can be also observed ahead by the high cohesiveness value, in TPA analysis. Creating a more cohesive system, with reduced mucin interaction, macroscopic differences were not observed with the substrate variation [19,23,44].

The discrepant answers observed between the results obtained with the two substrates for the P407/C971P system can suggest this formulation uses hydration to establish adhesive bonds between mucin and polymeric blend [1,40,45]. The C971P polymer exhibits lower cross-linking degree allowing higher swelling in presence of water. Therefore, in a low-hydration surface, such as the mucin disk, this system does not swell and displays a greater response than in high hydrated surface as the animal mucosa [43]. The water concentration also influences the tensile strength method [42], and a super hydration state of some polymers can reduce their mucoadhesive performance [43,46,47,48]. Thus, the selection of the mucosal surface needs to consider the composition and consistency of the formulation analyzed, where the most consistent systems (with high mechanical resistance) have better response using mucin disks.

To compare in vitro and in vivo tests, the same condition of analysis should be used [17]. However, using the same parameters, significant differences were observed between analyses performed with isolated mucin and ex vivo mucosa, particularly in the P407/C971P system. Furthermore, the W_{adh} response, related to elasticity and plasticity, demonstrated a significant difference between the three polymeric systems. On the other hand, for the F_{adh} response, significant difference was not found between the binary blends ($p > 0.05$). Therefore, the W_{adh} is suggested as a better response for the detachment force method and, when calculated as an N·mm unit, it can be converted to Joules, showing the energy dispending at the separation process of the adhesive surfaces.

In order to optimize this method, the responses of the P407/PCB polymeric blend were evaluated (Table 3) under the variation of the instrumental parameters in a factorial design.

Table 3. Response values of the mucoadhesive force (N) and the work adhesion (N·mm) of the binary polymeric system containing 15% (*w/w*) P407 and 0.25% (*w/w*) PCB, at 37 °C, using porcine mucin disks (A) or porcine oral mucosa (B). Each value represents the mean of at least three replicates. In all cases, the coefficient of variation of replicate analyses was less than 12%.

Experiment	Factors				Adhesive Force (N)	Work (N·mm)
	Substrate	Applied Force (N)	Velocity (mm/s)	Time (s)		
1	A	0.03	1.00	30	0.133	0.131
2	B	0.03	1.00	30	0.121	0.081
3	A	0.10	1.00	30	0.189	0.829
4	B	0.10	1.00	30	0.128	0.110
5	A	0.03	10.0	30	0.169	0.313
6	B	0.03	10.0	30	0.255	0.110
7	A	0.10	10.0	30	0.329	0.927
8	B	0.10	10.0	30	0.223	0.405
9	A	0.03	1.00	120	0.138	0.259
10	B	0.03	1.00	120	0.120	0.055
11	A	0.10	1.00	120	0.365	0.740
12	B	0.10	1.00	120	0.110	0.083
13	A	0.03	10.0	120	0.200	0.469
14	B	0.03	10.0	120	0.115	0.245
15	A	0.10	10.0	120	1.247	1.033
16	B	0.10	10.0	120	0.303	0.383

17 (C)	A	0.065	5.50	75	0.191	0.548
18 (C)	B	0.065	5.50	75	0.157	0.125
19 (C)	A	0.065	5.50	75	0.236	0.715
20 (C)	B	0.065	5.50	75	0.169	0.242

The Pareto diagrams (Figure 2) shows the estimated effect of each factor and their interactions. A factor and an interaction are considered to influence the response only if the estimated effect is significant, i.e., $p < 0.05$. It was found that variation of substrate, force, probe velocity rate, and contact time influenced the F_{adh}. Applied force had the highest positive influence, and the substrate had a negative influence on this response of mucoadhesive force. For the W_{adh}, only the contact time and some interactions demonstrated positive influence on the response. Speed and the change of the substrate from mucin disk to animal mucosa had the greatest negative contribution.

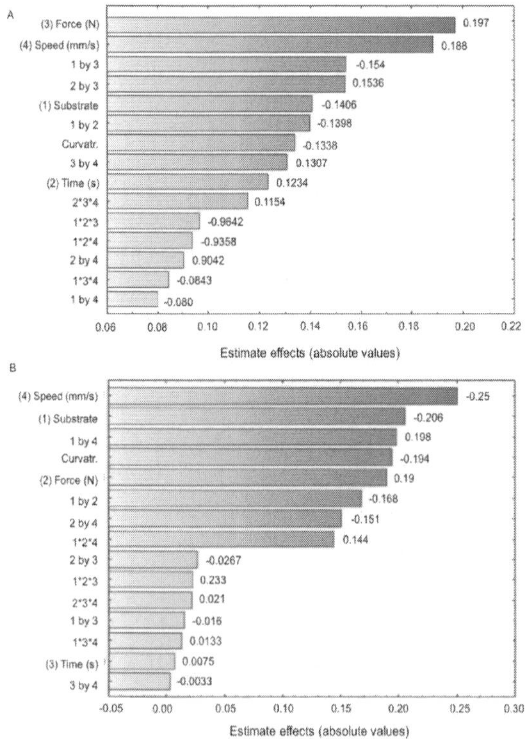

Figure 2. Pareto diagrams with the estimate effect of the parameters about the mucoadhesive force F_{adh} (**A**) and work adhesion W_{adh} (**B**) for the binary polymeric system composed of 15% (*w/w*) P407 and 0.25% (*w/w*) PCB, at 37 °C, and their interactions.

The surface response plots (Figure 3) demonstrate that the largest F_{adh} was obtained using the mucin disk, 0.1 N of contact force, 120 s of contact time, and a speed of 10 mm/s. On the other hand, for the W_{adh} (Figure 4), the highest value was observed using mucin disk, 0.1 N, 1 mm/s, which was time independent. The surface was adjusted to the experimental data by multiple adjusted correlation coefficient values R^2_{adj}. Thus, through the regression analysis, it was observed that 93.98% of the variation in the F_{adh} response is explained with this model (R^2_{adj} = 0.9398). Moreover, 96.08% of the variation in the W_{adh} response is explained with this model (R^2_{adj} = 0.9608). Therefore, the method can estimate the F_{adh} and the W_{adh} according to these parameters.

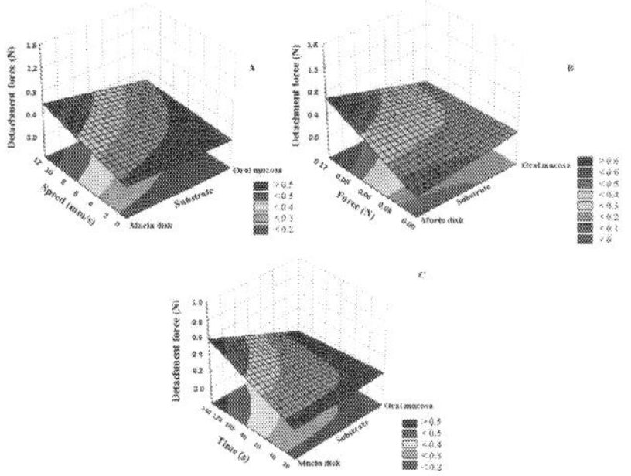

Figure 3. Response surface plots of the detachment force (F_{adh}) using mucin disk or oral mucosa influenced by some factors: (**A**) speed of probe ascent (mm/s); (**B**) applied force (N); (**C**) analysis time (s).

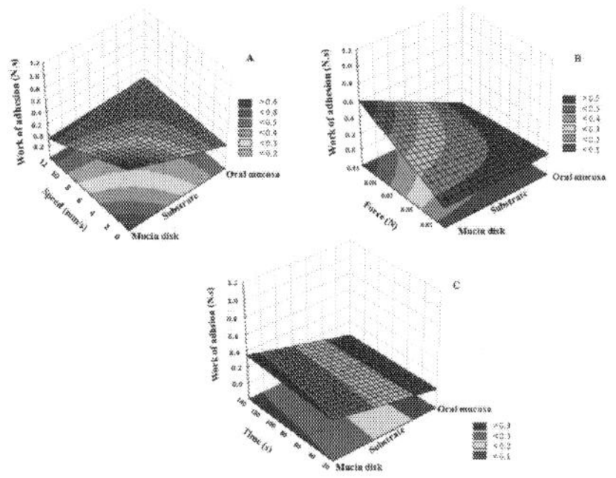

Figure 4. Response surface graphs of the work adhesion (W_{adh}) using mucin disk or oral mucosa influenced by some factors: (**A**) speed of probe ascent (mm/s); (**B**) applied force (N); (**C**) analysis time (s).

Additionally, the desirability of the method was evaluated for both responses, which represents the combination of factors required to obtain a better response [49]. Figure 5 shows the desirability response for the F_{adh} results, which is 85% of the maximum value experimentally obtained (1.147 N).

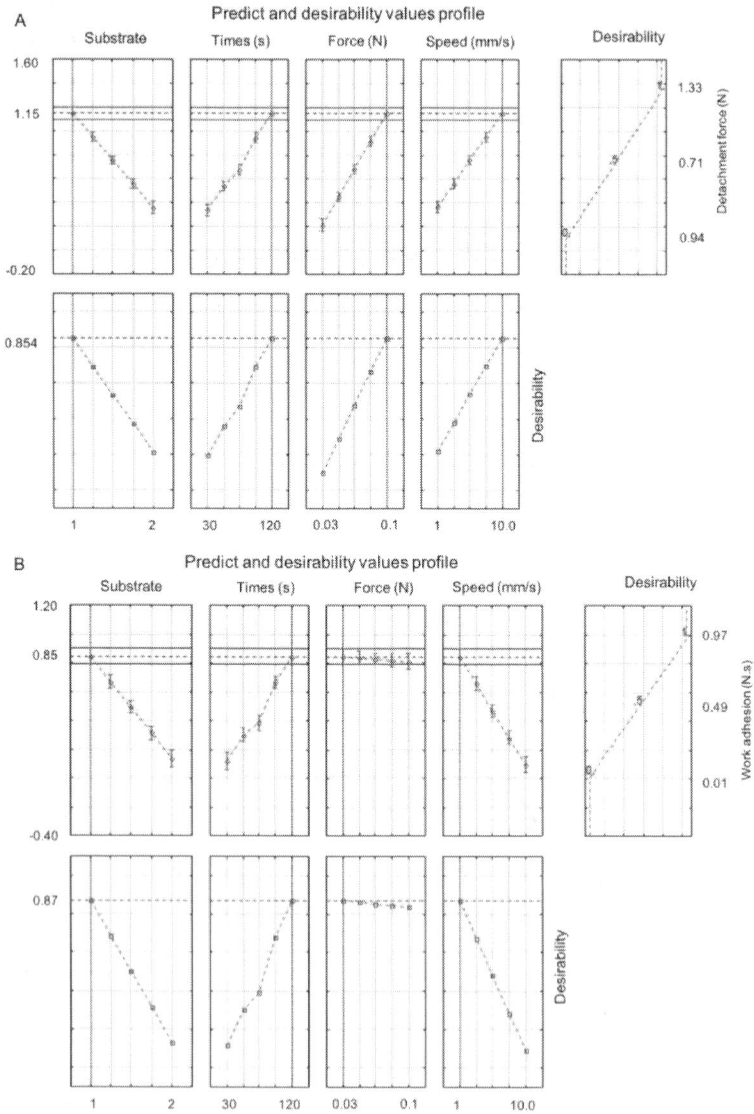

Figure 5. Profile of prediction and desirability values of the detachment force (**A**) and work adhesion (**B**) related to the substrate, time (s), applied force (N), and speed of probe ascent (mm/s).

About the W_{adh} (Figure 5B), the desirability was at 92% of the maximum experimental value (1.019 N.mm). Moreover, the substrate variation from the mucin disk to the animal mucosa reduced both results for the studied semi-solid formulation. It was also observed that the F_{adh} increases when changing the levels from minimum to maximum for time, force, and speed parameters. Prior studies had demonstrated the increase of contact time between formulation and substrate surface increases the interaction between mucin and polymer chains; thus, higher values of adhesion are obtained [43,50].

About the W_{adh}, changing the levels from minimum to maximal for force and speed parameters, the response is greater as well. On the other hand, the time was not relevant to this answer. The statistical analysis suggests that it is not possible to compare results present in the literature, which use different substrates and parameters, for the tensile strength method. Thus, it is clear the necessity of standardization of this method, to have real comparisons among mucoadhesive profiles of different formulations and polymeric systems.

The desirability cannot show the proximity to real values, which characterizes a limitation of the in vitro methods. Nevertheless, this analysis allows a standardization of this method using both responses, since the higher value is obtained using these parameters.

Therefore, for semi-solid formulations containing P407 and poly(acrylic) derivate, it can be suggested that the use of mucin disk, a force of 0.1 N, 120 s of contact time, and 10 mm/s probe speed will obtain better F_{adh} measurement. Using a mucin disk, 0.1 N of force, 30 s of contact time, and a probe speed of 10 mm/s can be used to obtain a better W_{adh} response.

Therefore, with the information obtained, it will be possible in future studies to select and validate the best conditions to perform comparisons.

Rheological Methods

Rheological analysis represents an in vitro model which can predict the in vivo behavior of adhesive formulations and also investigate their structural interactions. The mucoadhesion process is a phenomenon which combines different interaction types and the rheological method is used to evaluate these interactions between mucin and polymeric system [3,16,18,51]. The mucoadhesion process is characterized when the rheological response of a polymer-mucin mixture is higher than the isolated contributions, giving rise to an "interaction parameter" [8,20,28,47].

There are different ways to use the rheological analysis in studying mucoadhesion. It can be performed by consistency measurement over a range of shear rates or by monitoring the viscoelastic properties [27]. Whilst rheology is preferred as a secondary technique, it is often used to measure changes in viscosity or elastic behavior [2,8,23,40,47]. However, a lack of understanding about the meaning of the interaction parameter further confounds the great variability of results in the literature. Therefore, a standardization of the interpretation of results from the rheological analysis was proposed for the three semi-solid systems using the continuous shear analyses to evaluate the mucoadhesive properties (Table 4).

Table 4. Values of the consistency index (*k*) for polymeric blends containing poloxamer 407 (P407) and Carbopol 971P® (C971P), Carbopol 974P® (C974P), or Noveon® Polycarbophil (PCB)[a] at 37 °C. Each value represents the mean (±standard deviation) of at least six replicates.

Formulation	*k* (Pa.s) of Polymeric System	*k* (Pa.s) of Mixture	Interaction Parameter (Pa.s)
P407/C 971P	43,143.33 ± 470.78	227.70 ± 9.74	−42,915.67
P407/C974P	139.40 ± 6.01	107.60 ± 2.84	−31.83
P407/PCB	60.96 ± 3.31	97.63 ± 2.07	36.64
Mucin solution	0.033 ± 0.001		

The addition of mucin to the polymeric system demonstrates reduced consistency of the P407/C971P and P407/C974P systems, with a positive interaction parameter for the PCB-containing system. Consistency is a measure of internal friction and the presence of mucin appears to reduce interaction within the formulation. Comparing with the data in Figure 1A, it can be seen that P407/C971P had the greatest force of adhesion, and a very large, negative, interaction parameter. It is suggested that a negative parameter for the consistency index could be a result of interaction between mucin and mucoadhesive polymer, which reduces bridging of poloxamer micelles in the formulation, lowering the overall consistency. It is also possible that the mucin glycoproteins affect the micellization behavior in an unpredictable manner. On the other hand, the P407/PCB polymeric system mixed with mucin exhibited a consistency index greater than the mucin and polymeric systems isolated. This is evidence of a strong interaction between the polymers and mucin, forming a highly cohesive system.

As already described in the literature, there are disadvantages in the use of flow rheometry as a method to evaluate mucoadhesive properties. The continuous shear analysis is a destructive test, then the mucin–polymer interactions can be disrupted, and it is not often possible to observe the mucoadhesion phenomenon [8,29,40,47]. This study exposes the limitations of the continuous flow rheological method to evaluate the mucoadhesive profile of semi-solid systems, and it is best used to understand mechanistic aspects.

For oscillatory rheology, the synergism between polymer and mucin is employed to evaluate mucoadhesion. This technique has advantageous characteristics, such as being a non-destructive method which simulates the formulation behavior during application. Recently, Jones and co-workers [52] reinforced the correlation between viscoelasticity and mucoadhesion, using linear regression [8]. The adhesive interactions in the polymeric blends were observed by the elastic modulus (G') analysis of the mixture as a function of frequency. When the G' of the mixture is higher than the isolated polymer and mucin contributions, $\Delta G' > 0$, rheological synergism occurs [2,29,53]. The observed and calculated values of G' moduli were obtained at 10.0 Hz, since in this oscillatory frequency the systems are forced at higher oscillatory intensity.

The elastic, or "storage" modulus measures the storage and recovered energy at each deformation cycle, reflecting the solid component of a viscoelastic material [47]. The elastic modulus will demonstrate higher values if a sample is predominantly elastic, i.e., highly structured. In contrast, the loss modulus (G'') demonstrates the lost energy at each cycle, and will be higher when the sample is predominantly viscous. The frequency sweep at the linear viscoelastic region allows the three-dimensional structure of the sample to be preserved throughout analysis. The type of cross-linked structure can be revealed in the oscillations, where a small effort is exerted at each frequency. The oscillatory analysis allows the differentiation of the physical entanglements and secondary bonds, since in low frequencies the polymeric physical entanglements can be separated, while, secondary bonds remain fixed [46,47]. The results are displayed in Table 5.

Table 5. Values of the elastic modulus (*G*′) for polymeric blends containing poloxamer 407 (P407) and Carbopol 971P® (C971P), Carbopol 974P® (C974P), or Noveon® Polycarbophil (PCB)[a] at 37 °C. Each value represents the mean (±standard deviation) of at least six replicates.

Formulation	*G*′ (Pa) Polymeric System	*G*′ (Pa) Mixture	Interaction Parameter (Pa)
P407/C971P	16,646.00 ± 612.97	15,368.17 ± 865.56	−1305.68
P407/C974P	3246.00 ± 227.49	5528.83 ± 320.13	2254.99
P407/PCB	2660.00 ± 85.46	4826.50 ± 279.20	2138.65
Mucin	27.85 ± 2.74		

The viscoelastic response of the systems suggests the entanglement and secondary bonds (hydrogen bonds) between mucoadhesive polymers and mucus glycoproteins for the P407/C974P and P407/PCB binary systems, since the establishment of secondary bonds results in the increase of the elasticity into the formulations. Furthermore, the difference observed on the rheological interaction parameter of the three formulations can be attributed to the structural differences of these polymers [46, 54].

Formulations which show high viscosity have demonstrated to suffer low clearance and, consequently, they remain longer at the action site [55]. However, despite having higher viscosity, the P407/C971P platform displayed rheological antagonism, demonstrating that the addition of mucin did not increase overall interaction in the system. Edsman & Hägerström had already demonstrated positive values (rheological synergism) for low concentrations of cross-linking polymers, while high concentration of them results in negative values. In this sense, the evaluation of the *G*' gives positive values for weakly hydrogels and strongly cross-linked hydrogels will demonstrate a negative interaction parameter [20].

The system composed of P407 and C971P has the higher polymeric concentration when compared with the P407/C974P and P407/PCB formulations. Large concentrations of the polymers into the mixture added mucin, as well as the high viscosity of this system reduces the availability of the solvent, which makes the interpenetration of the polymer and mucin chains difficult, since them flexibility and mobility are reduced [47,56].

Oscillatory rheometry is a valid and sensitive method to evaluate interactions in polymer-mucin mixtures, and although the P407/C974P and P407/PCB formulations had demonstrated similar answers between these methods, a direct comparison with the standard tensile method was not possible. Moreover, it was observed that the results depend more on the polymer concentration than of the polymers chemical structure, since P407/C974P and P407/PCB systems exhibited similar results. In this way, to use the oscillatory rheology as a method to evaluate the mucoadhesive profile of semi-solid formulations, a prior cohesiveness analyses must be done.

Texture Profile Analysis

TPA is a quick and common analytical methodology. It can be used for the mechanical characterization of semi-solid pharmaceutical dosage forms and aid understanding of structure. The results of this analysis allow easy identification of the physicochemical interactions among the components of the formulations and also allow prediction of the behavior of these systems under different analytical conditions, and during use in pharmaceuticals. The values obtained by this technique are: hardness, elasticity, adhesiveness, cohesiveness, and compressibility [57,58].

The development of pharmaceutical dosage forms for topical application requires formation of a target profile. These formulations need to be easily removed from the packing to have good spreadability, bio/mucoadhesion, and adequate viscosity in order to facilitate retention and thus patient compliance with the treatment. Moreover, for mucoadhesive systems the resistance to the natural defense processes needs to be considered [34,40,57,58].

To be easily applied, a semi-solid needs to demonstrate low values for hardness and compressibility, since these values indicate how easy it is to remove the formulation from the packing, as well as the spreadability and removal of the product at the desirable site. However, very low values of them can impair the retention of the formulation. Ideally, semi-solid systems must demonstrate high values of adhesion to ensure retention. In addition, high values of elasticity aid retention, because the systems have a tendency to return to their structure. On the other hand, high values to the cohesiveness parameter are also desirable [2, 23, 29, 57, 58, 59, 60].

In order to obtain pharmaceutical systems with acceptable mechanical characteristics, ensuring in vivo retention and therapeutic efficiency, the texture profile properties of the formulations must be studied. As previously mentioned, the chosen systems have already demonstrated themselves to be favorable pharmaceutical systems, about the TPA properties [23, 30,34]. In addition to studying mechanical properties, it may be used to evaluate the mucoadhesion profile of semi-solid systems by the analysis of the polymer-mucin interaction.

Comparing the results of polymeric systems with and without mucin (Table 6) there was a decrease in the hardness, compressibility and adhesiveness. The cohesiveness had increased values for two of the three systems, and the elasticity was unaltered. The P407/C971P system demonstrated higher values of hardness, compressibility and adhesiveness than P407/C974P and P407/PCB, probably because its polymeric concentration is also greater. In this sense, this formulation displayed also superior hardness, compressibility and adhesiveness responses. However, the decrease of this parameter after mucin addition was more significant, since the C971P presents a lower degree of crosslinking; therefore, it has a greater number of free carboxyl groups for interaction with the mucin, which promotes a more intense disintegration of this blend when compared with those containing other adhesive polymers (C974P and PCB). Similar to rheological analysis, it appears that polymer–mucin interactions may not provide constructive synergism in these systems.

Table 6. Mechanical results obtained by texture profile analysis from the binary polymeric formulations composed of poloxamer 407 (P407) and Carbopol 971P® (C971P), Carbopol 974P® (C974P), or Noveon® Polycarbophil (PCB) with or in the absence of mucin. Each value represents the mean (±standard deviation) of at least six replicates.

System	TPA Results				
	Hardness (N)	Compressibility (N·mm)	Adhesiveness (N·mm)	Elasticity (mm)	Cohesiveness (Dimensionless)
P407/C971P	1.48 ± 0.04	6.74 ± 0.29	7.65 ± 0.45	1.00 ± 0.01	0.87 ± 0.01
P407/C971P + mucin	0.95 ± 0.01	4.06 ± 0.18	3.32 ± 0.19	1.00 ± 0.00	0.92 ± 0.01
P407/C974P	0.64 ± 0.03	2.99 ± 0.12	2.53 ± 0.12	1.00 ± 0.01	0.91 ± 0.01
P407/C974P + mucin	0.34 ± 0.00	1.48 ± 0.08	1.20 ± 0.05	1.00 ± 0.00	1.02 ± 0.01
P407/PCB	0.56 ± 0.02	2.66 ± 0.10	2.46 ± 0.06	1.00 ± 0.00	0.92 ± 0.01
P407/PCB + mucin	0.26 ± 0.01	1.14 ± 0.06	0.93 ± 0.05	1.00 ± 0.01	0.91 ± 0.02

The absence of change in the elasticity of the systems, even with the mucin addition, reflects the results obtained in the oscillatory rheology where the viscoelastic property was retained. Furthermore, the decrease of hardness and compressibility in the added mucin systems agrees with the flow rheology and demonstrates the loss of internal friction in these systems after mucin addition. The cohesiveness of the P407/C974P blend with mucin was superior to the two other blends. Probably, it is due to the C974P chemical structure, since it is a poly (acrylic acid) derivative with a higher degree of cross-linking, providing a more cohesive system, and the increase of the attractive force in the formulation is more dramatic.

The interaction parameters of the responses obtained by TPA were calculated between the polymeric blends and mucin solution (Figure 6); this is an important value to investigate the mucoadhesive profile of the semi-solid preparations. All polymer blends showed a negative interaction parameter, i.e., the mixture with mucin in all the cases obtained lower values when compared with the pure polymeric systems. In this way, the results suggest the interaction between mucoadhesive polymer and mucin, which promotes mechanical changes.

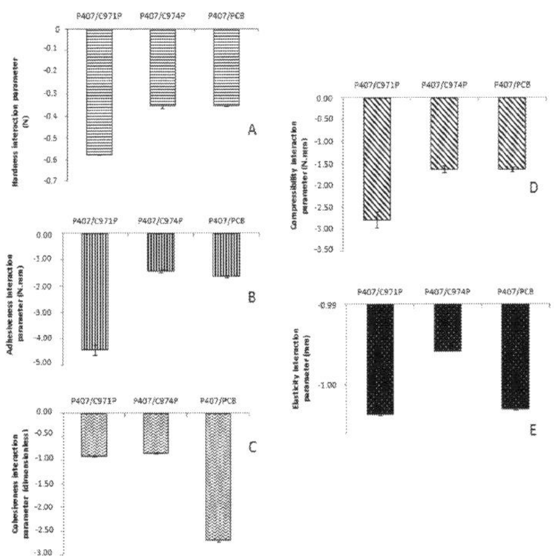

Figure 6. Mechanical properties obtained by texture profile analysis (TPA) of the polymeric systems containing poloxamer 407 (P407), mucin and Carbopol 971P® (C971P), Carbopol 974P® (C974P) or Noveon® Polycarbophil (PCB) and of pure mucin solution: (**A**) hardness, (**B**) adhesiveness, (**C**) cohesiveness, (**D**) compressibility, (**E**) elasticity.

The adhesiveness interaction parameter is a result of interaction between the formulation and polycarbonate probe. Therefore, negative values for mucoadhesive formulations are expected. Considering a higher adhesive polymer-mucin interaction, there are fewer free carboxyl groups to interact with the probe, and the adhesiveness of the mixture on the probe is lower when compared with the mixture with added mucin. Thus, as a new way to measure mucoadhesive properties of semi-solid systems, the analysis of the adhesiveness interaction parameter can be simple and accessible with fast execution. Moreover, a correlation was observed between this method and the tensile strength method using the mucin disks, since the adhesive profile of the three evaluated systems demonstrated a similar ranking evaluation.

CONCLUSIONS

This study utilized the tensile strength method with porcine mucin disks and porcine oral (cheek) mucosa, flow and oscillatory rheometry, as well as texture profile analysis, to evaluate the mucoadhesive performance of the three polymeric systems composed of poloxamer 407 and poly (acrylic acid) derivatives. It was also possible to investigate these methods to understand the parameters which can influence experimental results, highlighting the need for standardization. The reproducibility of the methods for these semi-solid formulations was also shown. The tensile strength method demonstrated differences when comparing the mucin disk and oral ex vivo mucosa. The factorial design displayed that all evaluated parameters have an effect in the F_{adh}; but the same was not observed for W_{adh}, for which most interactions did not influence response. W_{adh} was suggested as a more appropriate metric for evaluating mucoadhesion. The oscillatory rheology was more capable of showing adhesive interactions than continuous flow rheology. However, each rheological analysis needs to be associated with complementary analyses. Furthermore, the texture profile analysis method with the mucin addition was shown to be reproducible by the evaluation of the adhesiveness interaction parameter. In this sense, each one of the methods has an important place within the evaluation of mucoadhesion of semi-solid pharmaceutical systems, but it is very important to understand the importance and influence of the conditions of analysis on experimental results. There is a clear need for standardized methods to evaluate mucoadhesive properties of semisolid drug delivery systems.

ACKNOWLEDGMENTS

The authors would like to thank CAPES (Higher-Education Education Personnel Improvement Coordination), CNPq (National Counsel of

Technological and Scientific Development), and FINEP (Financier of Studies and Projects) of Brazil for financial assistance.

CONFLICTS OF INTEREST

The authors declare no conflict of interest.

REFERENCES

1. 1. Bassi da Silva J., de Souza Ferreira S.B., de Freitas O., Bruschi M.L. A critical review about methodologies for the analysis of mucoadhesive properties of drug delivery systems. Drug Dev. Ind. Pharm. 2017; 9045:1–67. doi: 10.1080/03639045.2017.1294600. .
2. 2. Bruschi M.L., Jones D.S., Panzeri H., Gremião M.P.D., de Freitas O., Lara E.H.G. Semisolid Systems Containing Propolis for the Treatment of Periodontal Disease: In Vitro Realease Kinetics, Syringeability, Rheological, Textural and Mucoadhesive Properties. J. Pharm. Sci. 2007; 99:4215–4227. doi: 10.1002/jps.20843. .
3. 3. Cook M.T., Khutoryanskiy V.V. Mucoadhesion and mucosa-mimetic materials—A mini-review. Int. J. Pharm. 2015; 495:991–998. doi: 10.1016/j.ijpharm.2015.09.064. .
4. 4. Borghi-Pangoni F.B., Junqueira M.V., de Souza Ferreira S.B., Silva L.L., Rabello B.R., Caetano W., Diniz A., Bruschi M.L. Screening and In Vitro Evaluation of Mucoadhesive Thermoresponsive System Containing Methylene Blue for Local Photodynamic Therapy of Colorectal Cancer. Pharm. Res. 2015 doi: 10.1007/s11095-015-1826-8. .
5. 5. Nho Y.-C., Park J.-S., Lim Y.-M. Preparation of Poly (acrylic acid) Hydrogel by Radiation Crosslinking and Its Application for Mucoadhesives. Polymers (Basel) 2014; 6:890–898. doi: 10.3390/polym6030890. .
6. 6. Lejoyeux F., Ponchel G., Wouessidjewe D., Peppas N.A., Duchěne D. Bioadhesive tablets influence of the testing medium composition on bioadhesion. Drug Dev. Ind. Pharm. 1989; 15:2037–2048. doi: 10.3109/03639048909052517. .
7. 7. Almeida H., Amaral M.H., Lobão P., Lobo J.M.S. In situ gelling systems: A strategy to improve the bioavailability of ophthalmic pharmaceutical formulations. Drug Discov. Today. 2014; 19:400–412. doi: 10.1016/j.drudis.2013.10.001. .
8. 8. Mansuri S., Kesharwani P., Jain K., Tekade R.K., Jain N.K. Mucoadhesion: A promising approach in drug delivery system. React. Funct. Polym. 2016; 100:151–172. doi: 10.1016/j.reactfunctpolym.2016.01.011. .
9. 9. Peppas N.A., Huang Y. Nanoscale technology of mucoadhesive interactions. Adv. Drug Deliv. Rev. 2004; 56:1675–1687. doi:

10.1016/j.addr.2004.03.001. .

10. 10. Woertz C., Preis M., Breitkreutz J., Kleinebudde P. Assessment of test methods evaluating mucoadhesive polymers and dosage forms : An overview. Eur. J. Pharm. Biopharm. 2013; 85:843–853. doi: 10.1016/j.ejpb.2013.06.023. .
11. 11. Karolewicz B. A review of polymers as multifunctional excipients in drug dosage form technology. Saudi Pharm. J. 2015 doi: 10.1016/j.jsps.2015.02.025.
12. 12. Junqueira M.V., Borghi-Pangoni F.B., de Souza Ferreira S.B., Bruschi M.L. Evaluation of the methylene blue addition in binary polymeric systems composed by poloxamer 407 and Carbopol 934P using quality by design: Rheological, textural, and mucoadhesive analysis. Drug Dev. Ind. Pharm. 2016;9045:1–41. doi: 10.1080/03639045.2016.1188111. .
13. 13. Yun Chang J., Oh Y.K., Soo Kong H., Jung Kim E., Deuk Jang D., Taek Nam K., Kim C.K. Prolonged antifungal effects of clotrimazole-containing mucoadhesive thermosensitive gels on vaginitis. J. Control. Release. 2002;82:39–50. doi: 10.1016/S0168-3659(02)00086-X. .
14. 14. Accili D., Menghi G., Bonacucina G., Di Martino P., Palmieri G.F. Mucoadhesion dependence of pharmaceutical polymers on mucosa characteristics. Eur. J. Pharm. Sci. 2004; 22:225–234. doi: 10.1016/j.ejps.2003.12.011. .
15. 15. Reineke J., Cho D.Y., Dingle Y.L., Cheifetz P., Laulicht B., Lavin D., Furtado S., Mathiowitz E. Can bioadhesive nanoparticles allow for more effective particle uptake from the small intestine? J. Control. Release. 2013; 170:477–484. doi: 10.1016/j.jconrel.2013.05.043. .
16. 16. Vetchý D., Landová H., Gajdziok J., Dolezel P., Danek Z., Stembírek J. Determination of dependencies among in vitro and in vivo properties of prepared mucoadhesive buccal films using multivariate data analysis. Eur. J. Pharm. Biopharm. 2014; 86:498–506. doi: 10.1016/j.ejpb.2013.12.002. .
17. 17. Laulicht B., Cheifetz P., Tripathi A., Mathiowitz E. Are in vivo gastric bioadhesive forces accurately reflected by in vitro experiments? J. Control. Release. 2009; 134:103–110. doi: 10.1016/j.jconrel.2008.11.012. .
18. 18. Eshel-Green T., Bianco-Peled H. Mucoadhesive acrylated block copolymers micelles for the delivery of hydrophobic drugs. Colloids Surf. B Biointerfaces. 2016; 139:42–51. doi: 10.1016/j.

colsurfb.2015.11.044. .

19. 19. Hägerström H., Edsman K. Interpretation of mucoadhesive properties of polymer. J. Pharm. Pharmacol. 2001; 53:1589–1599. doi: 10.1211/0022357011778197. .
20. 20. Hägerström H., Edsman K. Limitations of the rheological mucoadhesion method: The effect of the choice of conditions and the rheological synergism parameter. Eur. J. Pharm. Sci. 2003; 18:349–357. doi: 10.1016/S0928-0987(03)00037-X. .
21. 21. Pund S., Joshi A., Vasu K., Nivsarkar M., Shishoo C. Gastroretentive delivery of rifampicin: In vitro mucoadhesion and in vivo gamma scintigraphy. Int. J. Pharm. 2011; 411:106–112. doi: 10.1016/j.ijpharm.2011.03.048. .
22. 22. Bromberg L., Temchenko M., Alakhov V., Hatton T.A. Bioadhesive properties and rheology of polyether-modified poly(acrylic acid) hydrogels. Int. J. Pharm. 2004; 282:45–60. doi: 10.1016/j.ijpharm.2004.05.030. .
23. 23. De Souza Ferreira S.B., Bassi da Silva J., Borghi-Pangoni F.B., Junqueira M.V., Bruschi M.L. Linear correlation between rheological, mucoadhesive and textural properties of thermoresponsive polymer blends for biomedical applications. J. Mech. Behav. Biomed. Mater. 2017; 68:265–275. doi: 10.1016/j.jmbbm.2017.02.016. .
24. 24. Otero-Espinar F.J., Delgado-Charro B., Anguiano-Igea S., Blanco-Mendez J. Use of tensile test in the study of the bioadhesive and swelling properties of polymers. In: Muñoz-Ruiz A., Vromans H., editors. Data Adquisition and Measurement Techniques. Interpharm Press Inc.; Buffalo Grove, IL, USA: 1998. pp. 295–342.
25. 25. Aka-Any-Grah A., Bouchemal K., Koffi A., Agnely F., Zhang M., Djabourov M., Ponchel G. Formulation of mucoadhesive vaginal hydrogels insensitive to dilution with vaginal fluids. Eur. J. Pharm. Biopharm. 2010;76:296–303. doi: 10.1016/j.ejpb.2010.07.004. .
26. 26. Varum F.J.O., Veiga F., Sousa J.S., Basit A.W. Mucoadhesive platforms for targeted delivery to the colon. Int. J. Pharm. 2011;420:11–19. doi: 10.1016/j.ijpharm.2011.08.006. .
27. 27. Ivarsson D., Wahlgren M. Comparison of in vitro methods of measuring mucoadhesion: Ellipsometry, tensile strength and rheological measurements. Colloids Surf. B Biointerfaces. 2012;92:353–359. doi: 10.1016/j.colsurfb.2011.12.020. .

28. 28. Hassan E.E., Gallo J.M. A Simple Rheological Method for the in Vitro Assessment of Mucin-Polymer Bioadhesive Bond Strength. Pharm. Res. 1990;7:491–495. doi: 10.1023/A:1015812615635. .

29. 29. Jones D.S., Bruschi M.L., de Freitas O., Gremião M.P.D., Lara E.H.G., Andrews G.P. Rheological, mechanical and mucoadhesive properties of thermoresponsive, bioadhesive binary mixtures composed of poloxamer 407 and carbopol 974P designed as platforms for implantable drug delivery systems for use in the oral cavity. Int. J. Pharm. 2009;372:49–58. doi: 10.1016/j.ijpharm.2009.01.006. .

30. 30. De Souza Ferreira S.B., Bassi da Silva J., Junqueira M.V., Borghi-Pangoni F.B., Guttierres R., Bruschi M.L. The importance of the relationship between mechanical analyses and rheometry of mucoadhesive thermoresponsive polymeric materials for biomedical applications. J. Mech. Behav. Biomed. Mater. 2017;74:142–153. doi: 10.1016/j.jmbbm.2017.05.040. .

31. 31. Hemphill T., Campos W., Pilehvari A. Yield-power law model more accurately predicts mud rheology. Oil Gas J. 1993;91:45–50.

32. 32. Jones D.S. Pharmaceutical Statistics. Pharmaceutical Press; London, UK: 2002.

33. 33. Zhu Z., Zhai Y., Zhang N., Leng D., Ding P. The development of polycarbophil as a bioadhesive material in pharmacy. Asian J. Pharm. Sci. 2013;8:218–227. doi: 10.1016/j.ajps.2013.09.003. .

34. 34. De Souza Ferreira S.B., Moço T.D., Borghi-Pangoni F.B., Junqueira M.V., Bruschi M.L. Rheological, mucoadhesive and textural properties of thermoresponsive polymer blends for biomedical applications. J. Mech. Behav. Biomed. Mater. 2015;55:164–178. doi: 10.1016/j.jmbbm.2015.10.026. .

35. 35. Bassi da Silva J., Khutoryanskiy V.V., Bruschi M.L., Cook M.T. A mucosa-mimetic material for the mucoadhesion testing of thermogelling semi-solids. Int. J. Pharm. 2017;528:586–594. doi: 10.1016/j.ijpharm.2017.06.025. .

36. 36. Ich Topic Q2 (R1). Validation of Analytical Procedures: Text and methodology. [(accessed on 7 January 2018)];Int. Conf. Harmon. 2005 :1–17. Available online: http://www.ich.org/fileadmin/Public_Web_Site/ICH_Products/Guidelines/Quality/Q2_R1/Step4/Q2_R1__Guideline.pdf.

37. 37. Nair A.B., Kumria R., Harsha S., Attimarad M., Al-Dhubiab B.E.,

Alhaider I.A. In vitro techniques to evaluate buccal films. J. Control. Release. 2013;166:10–21. doi: 10.1016/j.jconrel.2012.11.019. .

38. 38. Nep E.I., Conway B.R. Grewia gum 2: Mucoadhesive properties of compacts and gels. Trop. J. Pharm. Res. 2011;10:393–401. doi: 10.4314/tjpr.v10i4.4. .
39. 39. Tang X., Wang C. Study on the physical properties of bioadhesive polymers. Chin. Pharm. J. 2005;40:361–364.
40. 40. Carvalho F.C., Bruschi M.L., Evangelista R.C., Gremião M.P.D. Mucoadhesive drug delivery systems. Braz. J. Pharm. Sci. 2010;46:1–18. doi: 10.1590/S1984-82502010000100002. .
41. 41. Deacon M.P., Davis S.S., White R.J., Nordman H., Carlstedt I., Errington N., Rowe A.J., Harding S.E. Are chitosan–mucin interactions specific to different regions of the stomach? Velocity ultracentrifugation offers a clue. Carbohydr. Polym. 1999;38:235–238. doi: 10.1016/S0144-8617(98)00097-6. .
42. 42. Leung S.S., Robinson J.R. In the last decade, bioadhesive polymers/co- polymers have received considerable attention for controlled drug delivery. J. Control. Release. 1988;5:223–231.
43. 43. Blanco-Fuente H., Vila-Dorrío B., Anguiano-Igea S., Otero-Espinar F.J., Blanco-Méndez J. Tanned leather: A good model for determining hydrogels bioadhesion. Int. J. Pharm. 1996;138:103–112. doi: 10.1016/0378-5173(96)04542-5. .
44. 44. Lejoyeux F., Ponchel G., Duchene D. Influence of some technological parameters on the bioadhesive characteristics of polyacrilic acid matrices. STP Pharma. 1989;5:893–898.
45. 45. Haugstad K.E., Håti A.G., Nordgård C.T., Adl P.S., Maurstad G., Sletmoen M., Draget K.I., Dias R.S., Stokke B.T. Direct determination of chitosan-mucin interactions using a single-molecule strategy: Comparison to alginate-mucin interactions. Polymers (Basel) 2015;7:161–185. doi: 10.3390/polym7020161. .
46. 46. Madsen F., Eberth K., Smart J.D. A rheological examination of the mucoadhesive/mucus interaction: The effect of mucoadhesive type and concentration. J. Control. Release. 1998;50:167–178. doi: 10.1016/S0168-3659(97)00138-7. .
47. 47. Callens C., Ceulemans J., Ludwig A., Foreman P., Remon J.P. Rheological study on mucoadhesivity of some nasal powder formulations. Eur. J. Pharm. Biopharm. 2003;55:323–328. doi:

10.1016/S0939-6411(03)00024-9. .

48. 48. Saiano F., Pitarresi G., Cavallaro G., Licciardi M., Giammona G. Evaluation of mucoadhesive properties of α,β-poly(*N*-hydroxyethyl)-dl-aspartamide and α,β-poly(aspartylhydrazide) using ATR–FTIR spectroscopy. Polymer (Guildf) 2002;43:6281–6286. doi: 10.1016/S0032-3861(02)00504-9. .
49. 49. Yang P., Chen H., Liu Y.-W. Application of response surface methodology and desirability approach to investigate and optimize the jet pump in a thermoacoustic Stirling heat engine. Appl. Therm. Eng. 2017;127:1005–1014. doi: 10.1016/j.applthermaleng.2017.08.077. .
50. 50. Jones D.S., Woolfson A.D., Brown A.F. Textural, viscoelastic and mucoadhesive properties of pharmaceutical gels composed of cellulose polymers. Int. J. Pharm. 1997;151:223–233. doi: 10.1016/S0378-5173(97)04904-1. .
51. 51. Khutoryanskiy V.V. Advances in Mucoadhesion and Mucoadhesive Polymers. Macromol. Biosci. 2011;11:748–764. doi: 10.1002/mabi.201000388. .
52. 52. Jones D.S., Laverty T.P., Morris C., Andrews G.P. Statistical modelling of the rheological and mucoadhesive properties of aqueous poly(methylvinylether-*co*-maleic acid) networks: Redefining biomedical applications and the relationship between viscoelasticity and mucoadhesion. Colloids Surf. B. Biointerfaces. 2016;144:125–134. doi: 10.1016/j.colsurfb.2016.03.008. .
53. 53. Andrews G.P., Jones D.S. Rheological characterization of bioadhesive binary polymeric systems designed as platforms for drug delivery implants. Biomacromolecules. 2006;7:899–906. doi: 10.1021/bm050620y. .
54. 54. Hägerström H. Ph.D. Thesis. Faculty of Pharmacy, Uppsala University; Uppsala, Sweden: 2003. Polymer Gels as Pharmaceutical Dosage Forms: Rheological Performance and Physicochemical Interactions at the Gel-Mucus Interface for Formulations Intended for Mucosal Drug Delivery.
55. 55. Illum L., Farraj N.F., Fisher A.N., Gill I., Miglietta M., Benedetti L.M. Hyaluronic acid ester microspheres as a nasal delivery system for insulin. J. Control. Release. 1994;29:133–141. doi: 10.1016/0168-3659(94)90129-5. .

56. 56. Lee J.W., Park J.H., Robinson J.R. Bioadhesive dosage forms: The next generation. J. Pharm. Sci. 2000;89:850–866. doi: 10.1002/1520-6017(200007)89:7<850::AID-JPS2>3.0.CO;2-G. .
57. 57. Jones D.S., Woolfson A.D., Djokic J. Texture profile analysis of bioadhesive polymeric semisolids: Mechanical characterization and investigation of interactions between formulation components. J. Appl. Polym. Sci. 1996;61:2229–2234. doi: 10.1002/(SICI)1097-4628(19960919)61:12<2229::AID-APP24>3.0.CO;2-0. .
58. 58. Baloglu E., Karavana S.Y., Senyigit Z.A., Guneri T. Rheological and mechanical properties of poloxamer mixtures as a mucoadhesive gel base. Pharm. Dev. Technol. 2011;16:627–636. doi: 10.3109/10837450.2010.508074. .
59. 59. Tugcu-Demiröz F., Acartürk F., Erdo-an D. Development of long-acting bioadhesive vaginal gels of oxybutynin: Formulation, in vitro and in vivo evaluations. Int. J. Pharm. 2013;457:25–39. doi: 10.1016/j.ijpharm.2013.09.003. .
60. 60. Gratieri T., Gelfuso G.M., Rocha E.M., Sarmento V.H., de Freitas O., Lopez R.F.V. A poloxamer/chitosan in situ forming gel with prolonged retention time for ocular delivery. Eur. J. Pharm. Biopharm. 2010;75:186–193. doi: 10.1016/j.ejpb.2010.02.011.

SECTION 5
INORGANIC GELS

CHAPTER

20

Hydrogels as Porogens for Nanoporous Inorganic Materials

Christian Weinberger [1], Dirk Kuckling [2], and Michael Tiemann [1]

[1]Department of Chemistry—Inorganic Functional Materials, Paderborn University, 33098 Paderborn, Germany

[2]Department of Chemistry—Organic and Macromolecular Chemistry, Paderborn University, 33098 Paderborn, Germany

ABSTRACT

Organic polymer-hydrogels are known to be capable of directing the nucleation and growth of inorganic materials, such as silica, metal oxides, apatite or metal chalcogenides. This approach can be exploited in the synthesis of materials that exhibit defined nanoporosity. When the organic polymer-based hydrogel is incorporated in the inorganic product, a

Citation: Weinberger, C.; Kuckling, D.; Tiemann, M. "Hydrogels as Porogens for Nanoporous Inorganic Materials". Gels 2018, 4, 83. https://doi.org/10.3390/gels4040083

composite is formed from which the organic component may be selectively removed, yielding nanopores in the inorganic product. Such porogenic impact resembles the concept of using soft or hard templates for porous materials. This micro-review provides a survey of select examples from the literature.

Keywords: Nanoporous; Mesoporous; Metal Oxide; Silica; Hydrogel; Thin Film; Porogen; Template; Nanocasting

INTRODUCTION

Inorganic materials, with uniform nanopores (pore width ≤ 100 nm [1]), play an important role in a large variety of applications, including catalysis [2,3], energy storage [4,5], sensors [6,7], controlled drug release [8,9], biomaterials [10,11], and separation [12,13]. There are various chemical synthesis methods for porous materials that include distinct strategies for the formation of nanopores in the desired size. Among these, sol-gel chemistry-based methods are the most frequent ones. Standard sol-gel protocols yield materials with, more or less, high degrees of porosity [14], but usually with a broad pore size distribution. Extending the sol-gel approach by introducing porogenic 'soft templates', such as organic molecules or molecular ions (especially for microporous zeolites [15,16]) or supramolecular arrays of self-assembled amphiphilic species (surfactants or block co-polymers for ordered mesoporous materials [17,18,19]) allows for controlling the pore size and size uniformity (By definition, micropores have widths up to 2 nm; mesopore widths range from 2 to 50 nm [1]). The 'soft template' is incorporated in the inorganic solid during its synthesis under sol-gel-chemical conditions, thereby serving as a pore filler and often also as a structure director; it is later removed (by thermal combustion or by ion exchange), yielding uniform nanopores. Utilization of amphiphilic porogens also facilitates the synthesis of mesoporous inorganic materials in form of films by evaporation-induced self-assembly (EISA) [20,21]. However, the 'soft templating' approach cannot be regarded as a uniform synthesis method for mesoporous materials; in many cases, the formation of the respective product (such as most metal oxides) goes along with phase-separation and segregation from the amphiphilic species. For that reason, the concept of 'nanocasting' or 'hard templating' is frequently used for mesoporous materials as an alternative to the 'soft templating' method. Here a solid, porous silica or carbon matrix serves as a structural mold. The desired product is synthesized within the pores of the matrix, followed by selective removal of the latter; the product is obtained as a 'negative replica' of the pore system in the matrix [18,22,23]. The nanocasting approach may be regarded as a more universal synthesis concept than 'soft templating', as the risk of phase separation is avoided.

More recently, organic polymer-based hydrogels have been explored as porogenic matrices for the synthesis of inorganic mesoporous materials. A hydrogel is a network of polymer chains that are hydrophilic and are cross-linked either chemically (by covalent or ionic bonds) or physically (by steric entanglement). Upon exposure to water hydrogels 'swell', i.e., take up large quantities of water in the spaces between the polymer strands [24]. The aqueous phase of the swollen gel can be used for the synthesis of an inorganic solid (by sol-gel-like methods or simply by precipitation). The product will then contain the organic polymer matrix embedded in the inorganic material. When the matrix is later removed (e.g., by combustion), it will leave behind a network of mesopores. This porogenic impact of the hydrogel may be regarded as an intermediate between 'soft' and 'hard templating'. One the one hand, the hydrogel is a pre-formed continuous network, similar to a 'hard' structure matrix; on the other hand, the swollen hydrogel is very flexible, similar to a 'soft' matrix. The general concept of hydrogels serving as matrices for inorganic (though not necessarily porous) phases is also known from bio mineralization [25] and tissue engineering [26]. This micro-review gives some examples of the concept of using hydrogels as porogenic matrices for the synthesis of inorganic porous materials.

PHYSICALLY CROSS-LINKED POROGENIC HYDROGELS

In the literature, there are several reports on the utilization of non-chemically cross-linked organic polymers serving as matrices for inorganic materials. Quite often the polymers are not actually arranged in three-dimensional hydrogel networks, but rather, form fibrous entities that serve as gelators for inorganic nanostructures with a limited degree of porosity. Also, the reaction medium is often an organic solvent rather than water. For example, fibrous SiO_2 [27,28,29,30] and TiO_2 [31,32] nanostructures have been synthesized by using a variety organogelators, including chiral compounds for products with helical morphology [28,29]. Sanchez et al. used 2,3-Bis(n-decyloxy) anthracene (DDOA, a typical low-molecular-mass organogelator) for the synthesis of porous alumina (Al_2O_3) in a water/ethanol mixture [33]. The products showed a fibrous morphology and (after calcination, i.e., combustion of the organic component) bimodal micro-and mesoporosity. Tang et al. also prepared porous Al_2O_3 with high porosity by using an agarose hydrogel [34]. Further examples of low-molecular-mass gelators for inorganic materials with fibrous morphology were reported by Lu et al. who used β-d-Glucopyranoside-functionalized thiosemicarbazides for CdS [35] and N-Lauroyl-l-glutamic acid for CuS [36].

A chitosan/poly (vinyl alcohol) hydrogel was used by Zhu and Xu et al. for the synthesis of mesoporous titania (TiO_2) to be used as photo-catalytically active material [37]. Chitosan and poly (vinyl alcohol) spontaneously form spherical hydrogel beads in water; these served as matrices for millimeter-sized TiO_2 spheres (Figure 1). After drying, but before removal of the matrix, the TiO_2 product exhibited mesopores with uniform sizes of ca. 5 nm. Removal of the matrix by calcination led to a gradual decrease in porosity due to structural collapse.

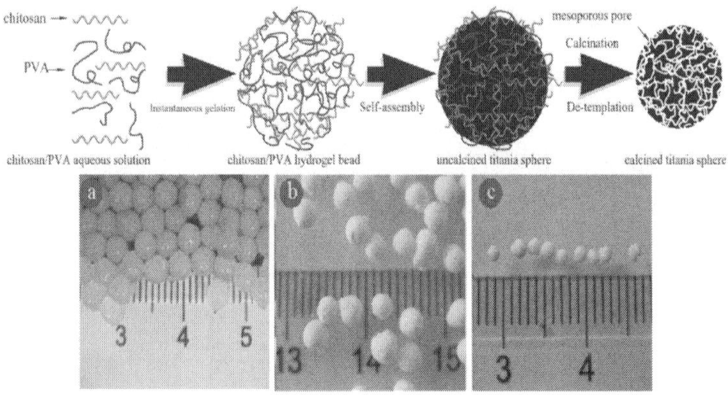

Figure 1. Top: schematic of the synthesis of mesoporous TiO_2 spheres by using a physically cross-linked chitosan/poly(vinyl alcohol) hydrogel as a porogenic matrix. Bottom: optical photographs of (**a**) hydrogel beads, (**b**) hydrogel/TiO_2 composite beads, and (**c**) TiO_2 beads. Reprinted with permission from [**37**], Copyright 2014 Elsevier.

We have recently prepared mesoporous alumina (γ-Al_2O_3) and magnesia (MgO) by using dimethylacrylamide-based hydrogels, both with [see below] and without chemical cross-linking [**38**]. The alumina products had pore sizes in the range between 3.6 to nm and 6.4 nm specific pore volumes up to 0.54 cm^3 g^{-1}; Magnesium oxide exhibited mesopore sizes of 3.6 nm and a lower specific pore volume of ca. 0.37 cm^3 g^{-1}.

CHEMICALLY CROSS-LINKED POROGENIC HYDROGELS

Chemically cross-linked hydrogels may be regarded as the more obvious choice for utilization as porogenic matrices than physically cross-linked hydrogels. Chemical cross-linking may be more stable against potential de-mixing and phase separation during the growth of the inorganic phase. Moreover, the degree of cross-linking can be varied; by changing the density of cross-links it is possible to vary the 'mesh size' in the gel which, in return, will likely affect the porosity of the final product (Figure 2).

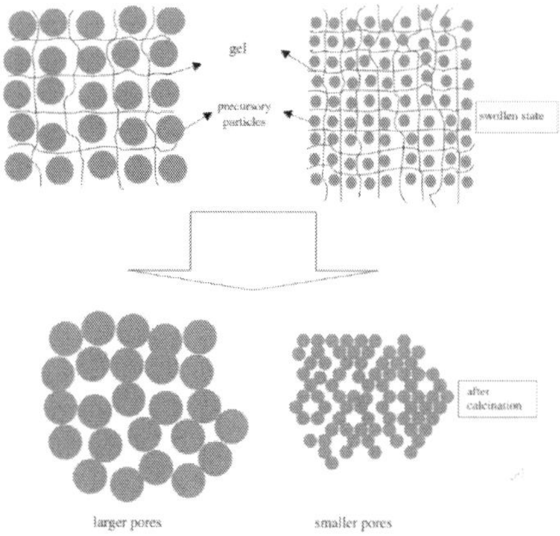

Figure 2. Potential impact of hydrogel mesh size on the porosity of the product. Reprinted with permission from [**42**], Copyright 2003 Elsevier.

Kurumada et al. prepared hydrogels from hydroxyl ethyl methacrylate (HEMA) and ethylene glycol dimethacrylate (EGDMA) as a cross-linker and used them as porogenic matrices for SiO_2, ZnO, and MgO [39,40,41,42]. Aqueous silica sols (from tetraethyl orthosilicate, TEOS) or methanolic solutions of $Zn(NO_3)_2$ or $Mg(NO_3)_2$, respectively, were used as the precursor systems and swelling agents. After conversion to the oxides and subsequent removal of the hydrogel matrices by thermal combustion, the porosity of the inorganic products varied depending on the mesh size, i.e., degree of crosslinking (HEMA/EGDMA ratio) in the hydrogel (Figure 2). Interestingly, contrary trends were observed for amorphous SiO_2 [39] on the one hand, and ZnO [40] and MgO [41] on the other hand; high degrees of cross-linking resulted in smaller pores in case of silica (average pore sizes ranging from 3 to 16 nm; specific pore volumes from 0.3 to 0.65 $cm^3\ g^{-1}$), but in larger pores for ZnO and MgO (pore volumes from 0.1 to 0.5 $cm^3 g^{-1}$). The authors suggested a model for the oxide formation in the hydrogel mesh to explain the different behavior [42]. The same group later used various polyacrylamide hydrogels as matrices for porous SiO_2 [43,44,45]. They mixed two precursor solutions, one containing N-isopropylacrylamide (NIPAM) monomer and N,N'-methylenbisacrylamide (BIS) cross-linker, the other containing the inorganic precursor sol (from TEOS), to carry out the hydrogel formation (polymerization) and sol-gel transition (SiO_2 formation) simultaneously [43,44]. The ratio of monomer to silica was varied and the smallest and most uniform pores (ca. 2 nm) were obtained for the lowest relative amount of monomer. The authors also varied the polarity of the hydrogel mesh by acrylamide monomers with different side chains (N,N'-dimethyl, N,N'-diethyl, N-isopropyl, N-isobutoxy) and found that higher polarity resulted in smaller pores [45].

Poly(2-hydroxyethyl methacrylate-co-acrylic acid) hydrogel films were used by Ford and Yang as matrices for porous amorphous SiO_2 layers on silicon wafers [46]. Photo-induced cross-linking of the hydrogel took place after spreading the polymer at the wafer surface by spin-coating. This procedure allows for micropatterning of the film by masking during UV exposure. Poly(ethylenimine) brushes were grafted to the hydrogel backbone; their amino functions served as initiators for the polycondensation of silica under sol-gel conditions. After removal of the organic component SiO_2 layers with thicknesses around 1 μm were obtained. The authors described the layers as consisting of SiO_2 nanoparticles; the layer thickness varied, depending on the polyethylenimine (PEI) molecular weight.

Bertozzi et al. have used a similar approach for the deposition and growth of hydroxyapatite in a poly(2-hydroxyethyl methacrylate) hydrogel matrix [47]. They used a urea-mediated precipitation technique that leads to hydrolysis of the 2-hydroxyethyl esters; the thus-created carboxylate functions promoted heterogeneous nucleation and 2-dimensional growth of calcium phosphate (CP). This study focused on biomimetic mineralization (for biocompatible bone-like composite materials), rather than attempting to create porosity by post-synthetic removal of the hydrogel matrix. Tong and Hu et al. employed a chemically cross-linked chitosan/glutaraldehyde hydrogel to prepare apatite; after removal of the hydrogel by oxidation with sodium hypochlorite, products with a certain degree of porosity were obtained [48].

We have recently reported the synthesis of mesoporous aluminum oxide (γ-Al_2O_3) by using photo-cross-linked dimethylacrylamide-based hydrogels as porogenic matrices [49,50]. The respective (non-cross-linked) polymers were swelled in a saturated aqueous solution of aluminum nitrate, followed by photo-induced cross-linking. Then the materials were dried and exposed to the vapor of an aqueous ammonia solution to convert aluminum nitrate to aluminum hydroxide. Finally, the materials were calcined to yield aluminum oxide and to simultaneously combust the hydrogel. The γ-Al_2O_3 products exhibited mesopore sizes in the range of 3 to 8 nm, large specific BET surface areas up to 360 $m^2\ g^{-1}$, and specific pore volumes up to 0.50 $cm^3\ g^{-1}$. Variation of the composition of the hydrogel matrix by adding various co-monomers in variable relative quantities turned out to have only little impact on the porosity of the products. The porogenic impact of the hydrogels was attributed to the presence of bundles of polymer strains that are incorporated in the inorganic phase during its precipitation and annealing (Figure 3). We also found that chemical cross-linking is not imperative for this synthesis concept; similar mesoporous Al_2O_3 and MgO materials were obtained by using the same kinds of polymers but without cross-linking [38]. (MgO exhibited lower porosity; its average pore size was 3.6 nm and the specific pore volume was 0.37 $cm^3\ g^{-1}$.).

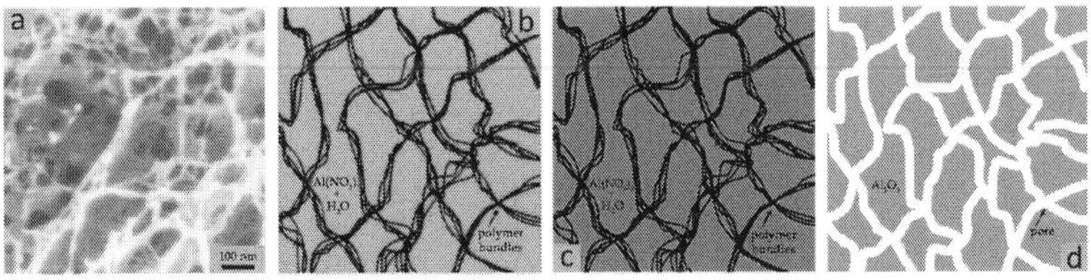

Figure 3. Schematic of the synthesis of mesoporous Al_2O_3 by using a dimethylacrylamide-based hydrogel as a porogenic matrix; (**a**) FESEM image of a hydrogel (shrunken state), (**b**) bundles of polymer strains in $Al(NO_3)_3$ solution, (**c**) after treatment with ammonia, and (**d**) after thermal decomposition of the polymer. Reprinted with permission from [**49**], Copyright 2014 Springer Nature.

Utilization of hydrogel matrices offers the opportunity to prepare films and layers of porous materials by anchoring the hydrogel to the surface of a substrate (Figure 4). We have prepared poly(dimethylacrylamide) hydrogel films with sub-micrometer thicknesses (dry, non-swollen state) on silicon wafer substrates. The films were immobilized by means of a reactive chlorosilyl compound as an adhesion promoter; the promoter acts as a co-monomer in the polymerization and binds to the oxidized surface of the wafer by forming a siloxane bridge. Mesoporous Al_2O_3 and MgO were then prepared in the hydrogel film matrices. After combustion of the organic phase the inorganic products remained attached to the substrate, forming mesoporous layers with thicknesses of a few hundred nanometers (Figure 5) [51].

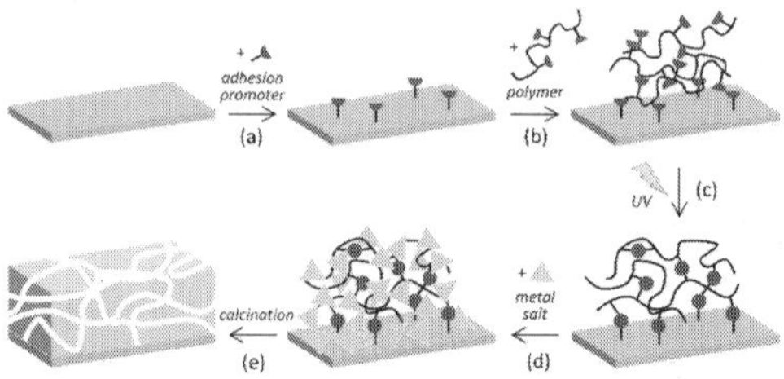

Figure 4. Schematic of the preparation of mesoporous metal oxide: (**a**) anchoring of an adhesion promoter on a substrate; (**b**) spreading of the polymer by spin coating; (**c**) hydrogel formation and immobilization on the substrate by photo-induced cross-linking; (**d**) swelling in metal salt solution; (**e**) formation of the porous metal oxide and combustion of the hydrogel by calcination [**51**].

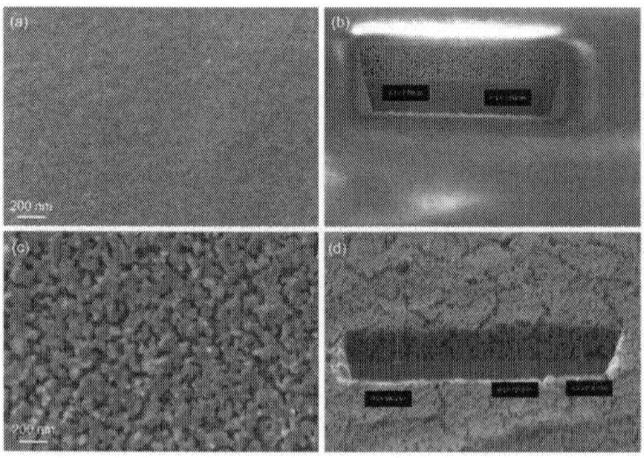

Figure 5. SEM images and focused ion beam (FIB) ablation analysis of mesoporous films of (**a**,**b**) Al_2O_3 (average thickness: 1.77 μm) and (**c**,**d**) MgO (average thickness: 0.59 μm) [**51**].

FURTHER EXAMPLES OF HYDROGEL MATRICES

Some authors have used hydrogels as matrices for the synthesis of individual nanoparticles. In these cases, the matrix does not exactly serve as a porogen, as the inorganic phase is formed only between adjacent meshes of the hydrogel, but does not incorporate the polymer strands. Subsequent removal of the polymer does therefore not yield pores in the inorganic product; rather, the role of the hydrogel network is to provide confined spaces (corresponding to the mesh size) for size-limited growth of the inorganic product. For example, Sahiner prepared various spherical metal (Fe, Co, Ni, Cu) and cadmium sulfide (CdS) particles in the sub-micron size range ('quantum dots') inside an anionic poly(2-acrylamido-2-methyl-1-propansulfonic acid) hydrogel [**52**]. The density of anionic functional groups in the hydrogel (binding the metal cations during the synthesis) had a clear influence on the particle size, whereas the mesh size turned out to have less impact. Hu et al. used a polyacrylamide hydrogel to synthesize cadmium selenide (CdSe) nanoparticles [**53**]. Likewise, Yan et al. prepared zeolite nanocrystals in a physically-cross-linked methylcellulose hydrogel [**54**].

CONCLUSIONS

In summary, physically or chemically cross-linked hydrogels are frequently used as matrices for the synthesis of inorganic materials. The hydrogels may initiate the nucleation of crystalline or amorphous phases and further direct their growth. The coordination abilities of the hydrogel atoms (especially N and O) influence the interactions with the metal ions and might be a possibility for tuning the structure of the resulting composite and the metal oxide, respectively. When the organic polymer strands are incorporated in the inorganic phase, their subsequent removal may leave behind mesopores, in which case the hydrogel acts as a porogen for the inorganic material. The removal of the organic phase (typically by thermal combustion) is a crucial point during the synthesis because the inorganic phase must be formed before the hydrogel starts to depredate. Otherwise, materials with a poor porosity might be obtained. Immobilization of the hydrogel matrix at a substrate surface facilitates the creation of uniform porous films and layers. By a selective chemical cross-linking it might be possible to create patterned structures, such as films, and utilize them in microfluidic devices, in catalysis or as a catalyst support. Since hydrogels can be modified in numerous respects, such as chemical composition, internal and external structure, the main challenges for future work will be to develop structure-property relationships between hydrogel structures, that can be controlled by synthesis protocols, and the final properties of the inorganic target structure.

CONFLICTS OF INTEREST

The authors declare no conflict of interest.

REFERENCES

1. Thommes, M.; Kaneko, K.; Neimark, A.V.; Olivier, J.P.; Rodriguez-Reinoso, F.; Rouquerol, J.; Sing, K.S.W. Physisorption of gases, with special reference to the evaluation of surface area and pore size distribution. (IUPAC technical report). *Pure Appl. Chem.* **2015**, *87*, 1051–1069.
2. Schmidt, W. Solid Catalysts on the Nanoscale: Design of Complex Morphologies and Pore Structures. *ChemCatChem* **2009**, *1*, 53–67.
3. Tüysüz, H.; Schüth, F. Ordered Mesoporous Materials as Catalysts. *Adv. Catal.* **2012**, *55*, 127–239.
4. Bruce, P.G.; Scrosati, B.; Tarascon, J.-M. Nanomaterials for Rechargeable Lithium Batteries. *Angew. Chem. Int. Ed.* **2008**, *47*, 2930–2946.
5. Ren, Y.; Ma, Z.; Bruce, P.G. Ordered mesoporous metal oxides: Synthesis and applications. *Chem. Soc. Rev.* **2012**, *41*, 4909–4927.
6. Tiemann, M. Porous metal oxides as gas sensors. *Chem. Eur. J.* **2007**, *13*, 8376–8388.
7. Wagner, T.; Haffer, S.; Weinberger, C.; Klaus, D.; Tiemann, M. Mesoporous materials as gas sensors. *Chem. Soc. Rev.* **2013**, *42*, 4036–4053.
8. Vallet-Regí, M.; Balas, F.; Arcos, D. Mesoporous Materials for Drug Delivery. *Angew. Chem. Int. Ed.* **2007**, *46*, 7548–7558.
9. Rosenholm, J.M.; Sahlgren, C.; Lindén, M. Towards multifunctional, targeted drug delivery systems using mesoporous silica nanoparticles—Opportunities & challenges. *Nanoscale* **2010**, *2*, 1870–1883.
10. Mamaeva, V.; Sahlgren, C.; Lindén, M. Mesoporous silica nanoparticles in medicine—Recent advances. *Adv. Drug Deliv. Rev.* **2013**, *65*, 689–702.
11. Ehlert, N.; Mueller, P.P.; Stieve, M.; Lenarzd, T.; Behrens, P. Mesoporous silica films as a novel biomaterial: Applications in the middle ear. *Chem. Soc. Rev.* **2013**, *42*, 3847–3861.
12. Sierra, I.; Pérez-Quintanilla, D. Heavy metal complexation on hybrid mesoporous silicas: An approach to analytical applications. *Chem. Soc. Rev.* **2013**, *42*, 3792–3807.
13. Triantafillidis, C.; Elsaesser, M.S.; Hüsing, N. Chemical phase separation strategies towards silica monoliths with hierarchical

porosity. *Chem. Soc. Rev.* **2013**, *42*, 3833–3846.

14. Brinker, C.J.; Scherer, G.W. *Sol-Gel Science: The Physics and Chemistry of Sol-Gel Processing: The Physics and Chemistry of Sol-Gel Processing*; Academic Press: San Diego, CA, USA, 1990; ISBN 0-12-134970-5.
15. Corma, A.; Davis, M.E. Issues in the Synthesis of Crystalline Molecular Sieves: Towards the Crystallization of Low Framework-Density Structures. *ChemPhysChem* **2004**, *5*, 304–313.
16. Masoumifard, N.; Guillet-Nicolas, R.; Kleitz, F. Synthesis of Engineered Zeolitic Materials: From Classical Zeolites to Hierarchical Core-Shell Materials. *Adv. Mater.* **2018**, *30*, 1704439.
17. Meynen, V.; Cool, P.; Vansant, E.F. Synthesis of siliceous materials with micro- and mesoporosity. *Microporous Mesoporous Mater.* **2007**, *104*, 26–38.
18. Gu, D.; Schüth, F. Synthesis of non-siliceous mesoporous oxides. *Chem. Soc. Rev.* **2014**, *43*, 313–344.
19. Wei, J.; Sun, Z.; Luo, W.; Li, Y.; Elzatahry, A.A.; Al-Enizi, A.M.; Deng, Y.; Zhao, D. New Insight into the Synthesis of Large-Pore Ordered Mesoporous Materials. *J. Am. Chem. Soc.* **2017**, *139*, 1706–1713.
20. Brinker, C.J.; Lu, Y.; Sellinger, A.; Fan, H. Evaporation-Induced Self-Assembly: Nanostructures Made Easy. *Adv. Mater.* **1999**, *11*, 579–585.
21. Grosso, D.; Cagnol, F.; Soler-Illia, G.J.A.A.; Crepaldi, E.L.; Amenitsch, H.; Brunet-Bruneau, A.; Bourgeois, A.; Sanchez, C. Fundamentals of mesostructuring through evaporation-induced self-assembly. *Adv. Funct. Mater.* **2004**, *4*, 309–322.
22. Tiemann, M. Repeated templating. *Chem. Mater.* **2007**, *20*, 961–971.
23. Deng, X.; Chen, K.; Tüysüz, H. Protocol for the Nanocasting Method: Preparation of Ordered Mesoporous Metal Oxides. *Chem. Mater.* **2017**, *29*, 40–51.
24. Döring, A.; Birnbaum, W.; Kuckling, D. Responsive hydrogels—Structurally and dimensionally optimized smart frameworks for applications in catalysis, micro-system technology and material science. *Chem. Soc. Rev.* **2013**, *40*, 7391–7420.
25. Addadi, L.; Joester, D.; Nudelman, F.; Weiner, S. Mollusk Shell Formation: A Source of New Concepts for Understanding Biomineralization Processes. *Chem. Eur. J.* **2006**, *12*, 980–987.
26. Gkioni, K.; Leeuwenburgh, S.C.G.; Douglas, T.E.L.; Mikos, A.G.;

Jansen, J.A. Mineralization of Hydrogels for Bone Regeneration. *Tissue Eng. B* **2010**, *16*, 577–585.

27. Jung, J.H.; Ono, Y.; Shinkai, S. Novel preparation method for multi-layered, tubular silica using an azacrown-appended cholesterol as template and metal-deposition into the interlayer space. *J. Chem. Soc. Perkin Trans.* **1999**, *2*, 1289–1291.
28. Jung, J.H.; Ono, Y.; Hanabusa, K.; Shinkai, S. Creation of Both Right-Handed and Left-Handed Silica Structures by Sol-Gel Transcription of Organogel Fibers Comprised of Chiral Diaminocyclohexane Derivatives. *J. Am. Chem. Soc.* **2000**, *122*, 5008–5009.
29. Jung, J.H.; Ono, Y.; Shinkai, S. Sol-Gel Polycondensation of Tetraethoxysilane in a Cholesterol-Based Organogel System Results in Chiral Spiral Silica. *Angew. Chem. Int. Ed.* **2000**, *39*, 1862–1865.
30. Jung, J.H.; Ono, Y.; Shinkai, S. Novel Silica Structures Which Are Prepared by Transcription of Various Superstructures Formed in Organogels. *Langmuir* **2000**, *16*, 1643–1649.
31. Kubo, W.; Murakoshi, K.; Kitamura, T.; Wada, Y.; Hanabusa, K.; Shirai, H.; Yanagida, S. Fabrication of Quasi-Solid-State Dye-Sensitized TiO_2 Solar Cells Using Low Molecular Weight Gelators. *Chem. Lett.* **1998**, *27*, 1241–1242.
32. Kobayashi, S.; Hanabusa, K.; Hamasaki, N.; Kimura, M.; Shirai, H. Preparation of TiO_2 Hollow-Fibers Using Supramolecular Assemblies. *Chem. Mater.* **2000**, *12*, 1523–1525.
33. Llusar, M.; Pidol, L.; Roux, C.; Pozzo, J.L.; Sanchez, C. Templated Growth of Alumina-Based Fibers through the Use of Anthracenic Organogelators. *Chem. Mater.* **2002**, *14*, 5124–5133.
34. Cui, X.; Tang, S.; Zhou, H. Mesoporous alumina materials synthesized in different gel templates. *Mater. Lett.* **2013**, *98*, 116–119.
35. Bao, C.; Lu, R.; Jin, M.; Xue, P.; Tan, C.; Zhao, Y.; Liu, G. Synthesis and Characterization of Nanostructural Hydrogel and Template for CdS Nanofibers. *J. Nanosci. Nanotechnol.* **2004**, *4*, 1045–1051.
36. Tan, C.; Lu, R.; Xue, P.; Bao, C.; Zhao, Y. Synthesis of CuS nanoribbons templated by hydrogel. *Mater. Chem. Phys.* **2008**, *112*, 500–503.
37. Jiang, R.; Zhu, H.Y.; Chen, H.H.; Yao, J.; Fu, Y.Q.; Zhang, Z.Y.; Xu, Y.M. Effect of calcination temperature on physical parameters and photocatalytic activity of mesoporous titania spheres using chitosan/poly(vinyl alcohol) hydrogel beads as a template. *Appl. Surf.*

Sci. **2014**, *319*, 189–196.

38. Chen, Z.; Weinberger, C.; Tiemann, M.; Kuckling, D. Organic Polymers as Porogenic Structure Matrices for Mesoporous Alumina and Magnesia. *Processes* **2017**, *5*, 70.
39. Jiu, J.; Kurumada, K.I.; Tanigaki, M. Preparation of nanoporous silica using copolymer template. *Mater. Chem. Phys.* **2002**, *78*, 177–183.
40. Jiu, J.; Kurumada, K.I.; Tanigaki, M. Preparation of nanoporous ZnO using copolymer gel template. *Mater. Chem. Phys.* **2003**, *81*, 93–98.
41. Jiu, J.; Kurumada, K.I.; Tanigaki, M.; Adachi, M.; Yoshikawa, S. Preparation of nanoporous MgO using gel as structure-direct template. *Mater. Lett.* **2003**, *58*, 44–47.
42. Jiu, J.; Kurumada, K.; Tanigaki, M. Preparation of oxide with nano-scaled pore diameters using gel template. *J. Non-Cryst. Solids* **2003**, *325*, 124–132.
43. Kurumada, K.; Nakamura, T.; Suzuki, A.; Umeda, N.; Kishimoto, N.; Hiro, M. Nanoscopic replication of cross-linked hydrogel in high-porosity nanoporous silica. *J. Non-Cryst. Solids* **2007**, *353*, 4839–4844.
44. Kurumada, K.I.; Nakamura, T.; Suzuki, A.; Umeda, N. Replication of a nano-scale mesh of hydrogel by assembled nanoparticles. *Adv. Powder Technol.* **2007**, *18*, 763–773.
45. Kurumada, K.; Suzuki, A.; Baba, S.; Otsuka, E. Relationship between polarity of template hydrogel and nanoporous structure replicated in sol–gel-derived silica matrix. *J. Appl. Polym. Sci.* **2009**, *114*, 4085–4090.
46. Ford, J.; Yang, S. Directed Synthesis of Silica Nanoparticles on Micropatterned Hydrogel Templates Tethered with Poly(ethyleneimine). *Chem. Mater.* **2007**, *19*, 5570–5575.
47. Song, S.; Saiz, E.; Bertozzi, C.R. A New Approach to Mineralization of Biocompatible Hydrogel Scaffolds: An Efficient Process toward 3-Dimensional Bonelike Composites. *J. Am. Chem. Soc.* **2003**, *125*, 1236–1243.
48. Shen, X.; Tong, H.; Jiang, T.; Zhu, Z.; Wan, P.; Hu, J. Homogeneous chitosan/carbonate apatite/citric acid nanocomposites prepared through a novel in situ precipitation method. *Compos. Sci. Technol.* **2007**, *67*, 2238–2245.
49. Birnbaum, W.; Weinberger, C.; Schill, V.; Haffer, S.; Tiemann, M.; Kuckling, D. Synthesis of mesoporous alumina through photo

cross-linked poly(dimethylacrylamide) hydrogels. *Colloid Polym. Sci.* **2014**, *292*, 3055–3060.

50. Weinberger, C.; Chen, Z.; Birnbaum, W.; Kuckling, D.; Tiemann, M. Photo-cross-linked polydimethylacrylamide hydrogels as porogens for mesoporous alumina. *Eur. J. Inorg. Chem.* **2017**, 1026–1031.
51. Chen, Z.; Kuckling, D.; Tiemann, M. Porous Aluminum Oxide and Magnesium Oxide Films by Using Organic Hydrogels as Structure Matrices. *Nanomaterials* **2018**, *8*, 186.
52. Sahiner, N. In situ metal particle preparation in cross-linked poly(2-acrylamido-2-methyl-1-propansulfonic acid) hydrogel networks. *Colloid Polym. Sci.* **2006**, *285*, 283–292.
53. Hu, Y.; Chen, J.; Chen, W.; Ning, J. Preparation of hollow CdSe nanospheres. *Mater. Lett.* **2004**, *58*, 2911–2913.
54. Wang, H.; Holmberg, B.A.; Yan, Y. Synthesis of Template-Free Zeolite Nanocrystals by Using in Situ Thermoreversible Polymer Hydrogels. *J. Am. Chem. Soc.* **2003**, *125*, 9928–9929.

CHAPTER

21

Development and Evaluation of Organic/Inorganic Combined Gel for Conformance Control in High Temperature and High Salinity Rreservoirs

Hong He[1], Yefei Wang[1], Xiaojie Sun[1], Peng Zhang[2] and Dandan Li[1]

[1]College of Petroleum Engineering, China University of Petroleum (East China), No.66, West Changjiang Road, Qingdao 266580, China

[2]College of Chemistry and Chemical Engineering, Chongqing University of Science and Technology, Chongqing 401331, China

ABSTRACT

Polymer gels have been successfully applied to control undesired water production for many years, however, due to poor long-term thermal stability and salt tolerance, conventional polymer gels present some limitations

---

Citation: He, H., Wang, Y., Sun, X. et al. "Development and evaluation of organic/inorganic combined gel for conformance control in high temperature and high salinity reservoirs". J Petrol Explor Prod Technol 5, 211–217 (2015). https://doi.org/10.1007/s13202-014-0127-9

for application in high temperature and high salinity reservoirs. In this paper, an organic/inorganic combined gel composed of polymer with low degree of hydrolysis, phenol formaldehyde resin and inorganic silicate was proposed and developed. The effects of concentration, temperature on gelation time were studied. Long-term thermal stability, salt tolerance and permeability reduction ability were investigated. Results showed that based on proper gelation time and economical cost, the formulation of the organic/inorganic combined gel was optimized. The gelation time decreased from 54.0 to 15.5 h with increasing temperature from 100 to 130 °C. After aging for 60 days in simulated formation brine with high total dissolved solids (TDS: 20 % NaCl + 2.5 % $CaCl_2$) at 130 °C, compared with polymer gel, the organic/inorganic combined gel had better long-term thermal stability and salt tolerance. Sand pack flow experimental results showed that for the organic/inorganic combined gel composed of 0.3 wt% polymer, 4.0 wt% resin and 3.0 wt% inorganic silicate, after aging for 60 days at 130 °C, the combined gel showed approximately 85 % permeability reduction. Therefore, these results indicate that the combined gel studied here can be a good candidate for controlling water production in high temperature and high salinity reservoirs.

Keywords: Combined gel, Conformance control, High temperature, High salinity

INTRODUCTION

Water flooding is one of the most common methods used to improve oil recovery. However, as more and more oilfields become mature, the breakthrough of water along the high permeability channels often occurs, resulting in poor sweep efficiency and excessive water production. Excessive water production has caused major problems due to decreased oil production and increased costs of lifting, separating and disposing large amounts of produced water, which has attracted more concerns on water control technology. Conformance control that generally refers to using chemical or mechanical methods to reduce undesired water production has been proved as an effective water control technology, which can improve sweep efficiency and thus result in an increase in ultimate oil recovery (Liu et al. 2010). Chemical-based conformance control has been successfully used to improve oil recovery of mature water flooding reservoirs. However, for high temperature and high salinity reservoirs, not many options are available. Polymer gels have been successfully applied to control water production from the hydrocarbon wells (Sydansk and Southwell 2000). However, Due to thermal hydrolysis under high temperature, some of the amide groups (–$CONH_2$) on polymer backbone convert to carboxylate groups that crosslink with divalent cations, resulting in syneresis or precipitation of polymer gel (Kabir 2001). Therefore, due to poor long-term thermal stability and salt tolerance, the conventional polymer gels present some limitations for the application in high temperature and salinity reservoirs. To extend the polymer gel application scope, some thermally stable polymer gels that composed of copolymers and organic crosslinkers have been reported in previous literatures (Moradi-Araghi et al. 1993; Albonico et al. 1995; Bryant et al. 1997; Hardy et al. 1998; Mary et al. 1999; Moradi-Araghi 2000; Vasquez et al. 2005; Eoff et al. 2007; Banerjee et al. 2008; Al-Muntasheri et al. 2009; Sengupta et al. 2012; Yadav and Mahto 2013). The copolymers reported in previous literatures are copolymer of acrylamide and *t*-butylacrylate (PAtBA), and copolymer of acrylamide and acrylamido-2-methylpropane sulfonic acid (AMPS) and the organic crosslinkers are phenol formaldehyde resin and polyethyleneimine (PEI). However, we found little research focused on the gel thermal stability and salt tolerance in the high salinity formation brines, which are crucial to the effectiveness of gel treatments. Sodium silicate based gel has been used to control water production (Nasr-El-Din and Taylor 2005), however, the main problems associated with the sodium silicate based gel are short gelation time and low gel strength in high temperature and high salinity reservoirs. Moreover, the challenge for sodium silicate based gel is the compatibility with the formation brine in the presence of calcium ions or magnesium ions. A white precipitation of calcium silicate or magnesium silicate forms immediately upon mixing with the formation brine.

The single polymer gel or sodium silicate based gel has some disadvantages, to improve the gel compatibility, thermal stability and salt tolerance, organic/inorganic combined gel composed of polymer, organic crosslinker and inorganic material was proposed, which can have synergetic effect and achieve better gel treatment efficiency. Therefore, in this paper, an organic/inorganic combined gel composed of polymer with low degree of hydrolysis, phenol formaldehyde resin and inorganic silicate was developed. The possible gelation mechanism is as follows: (1) Gelation reaction between the amide group (–$CONH_2$) of polymer and hydroxymethyl (–CH_2OH) group of phenol formaldehyde resin can occur to form three-dimensional network structure.

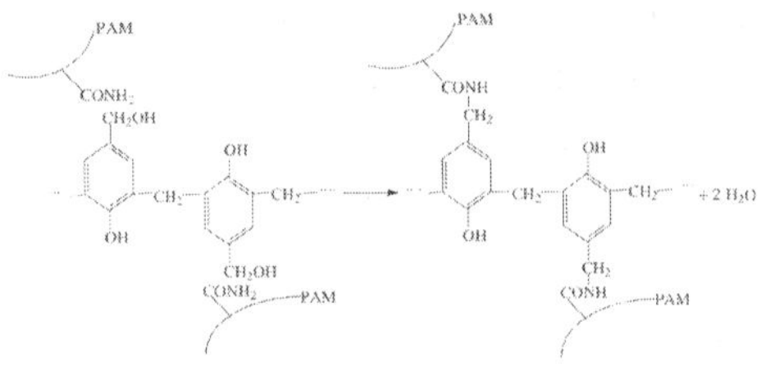

(2) The water-soluble inorganic silicate is converted into silicic acid at high temperature.

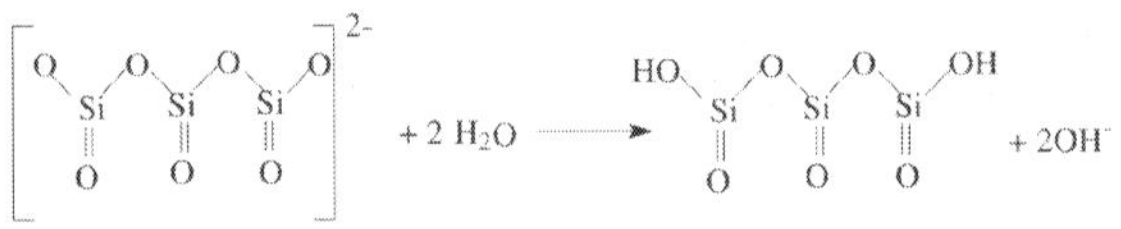

(3) The polycondensation of silicic acid occurs and another three-dimensional network structure is formed.

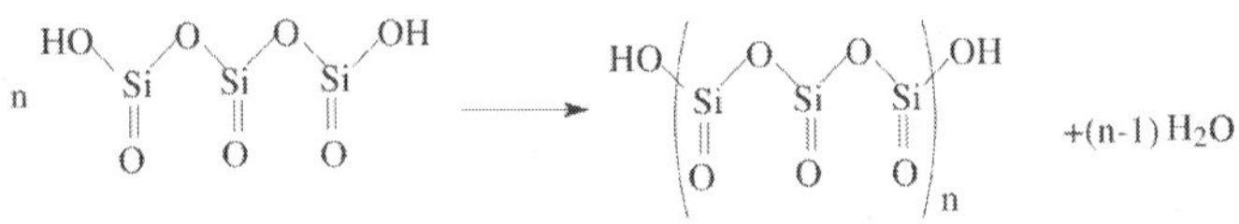

Based on the previous literatures, we found that little research had been done on the use of organic/inorganic combined gel to control water production in high temperature and salinity reservoirs. To ensure that the combined gel can be applied to control water production in high temperature and salinity reservoirs, the objectives of this work are to: (1) optimize the formulation of combined gel based on proper gelation time and economical cost; (2) evaluate the long-term thermal stability and salt tolerance; (3) evaluate permeability reduction to brine as a function of aging time, which can account for the feasible application of water control using the combined gel in high temperature and salinity reservoirs.

EXPERIMENTAL STUDIES

Materials

The polyacrylamide (PAM) used in this study was supplied in powder form. The molecular weight and degree of hydrolysis was 9.6×10^6 g/mol and 3.62 % respectively. The inorganic silicate solution was a viscous fluid with an active content of 35 wt%. Deionized water and the fresh water were obtained in own laboratory. The simulated formation brine contained 20 % NaCl and 2.5 % $CaCl_2$. The gelant solutions with various formulations were prepared with fresh water containing 250 mg L^{-1} NaCl.

Preparation of gelant solution

First, deoxygenated fresh water was used to prepare the polymer solution. Then the predetermined amounts of resin and silicate were added to the polymer solution. The gelant solution was thoroughly mixed by using a magnetic stirrer. Then the gelant solution was injected into an ampoule tube placed inside an oven set at the desired temperature. The ampoule tube was periodically removed from the oven for observation.

Gelation time and gel strength measurement

The gelation time was measured by Sydansk (1988) gel-strength code method. Sydansk gel-strength code method is a semi-quantitative measurement of the gelation time and the gel strength. In this study, the gelation time is defined as the time when the gel strength of gelant solutions reaches grade G (Grade G: the gel is described as moderately deformable nonflowing gel and flows approximately half way down the bottle upon inversion).

Long-term thermal stability and salt tolerance

The long-term thermal stability and salt tolerance were evaluated by measuring the degree of syneresis as a function of aging time. The experimental procedure was as follows: (1) measure the initial mass of formed combined gel m_0; (2) place the formed combined gel inside a steel container containing simulated formation brine with high total dissolved solids (TDS: 20 % NaCl + 2.5 % $CaCl_2$). The steel container was placed inside an oven at 130 °C and the steel container was periodically removed from the oven. Measure the mass of formed combined gel at different aging time $m_{(t)}$; (3) calculate the degree of syneresis (DS) at different aging time according to the following equation:

$$\mathrm{DS} = \frac{m_0 - m_{(t)}}{m_0} \times 100\%. \qquad (1)$$

Sand pack flow experiments

Sand pack flow experiments were conducted on the sand packs with the diameter of 2.5 cm and a length of 20 cm to evaluate the permeability reduction of the combined gel as a function of aging time at 130 °C. The experimental procedure was as follows: (1) The sand packs were filled with clean quartz sands and saturated with simulated formation brine; (2) the sand packs were flooded with simulated formation brine until the pressure drop across the sand packs was stable. Permeability to brine k_{w0} was measured at a flow rate of 1 mL min^{-1}. (3) Then 0.3 PV (pore volume) gelant solution was injected into sand packs at a flow rate of 1 mL min^{-1}. After injection, the sand packs were shut in for 48 h to ensure that the gelant solution can form gel; (4) then the simulated formation brine was injected into the sandpacks to measure the permeability k_{w1} after the gel treatment; (5) calculate the percent permeability reduction (PPR) to evaluate the effectiveness of the gel in reducing permeability according to the following equation:

$$\text{PPR} = (k_{w0} - k_{w1})/k_{w0} \times 100\,\%. \quad (2)$$

RESULTS AND DISCUSSION

Formulation of combined gel optimization

The formulation of combined gel that composed of polymer, phenol formaldehyde resin, and inorganic silicate was optimized by studying the effects of component concentrations on the gelation time and the gel strength. In this study, the gel strength of the combined gel with various formulations can mainly reach grade I (Grade I: the gel is described as rigid gel and there is no gel-surface deformation upon inversion). The appearance of the formed combined gel is shown in Fig. 1.

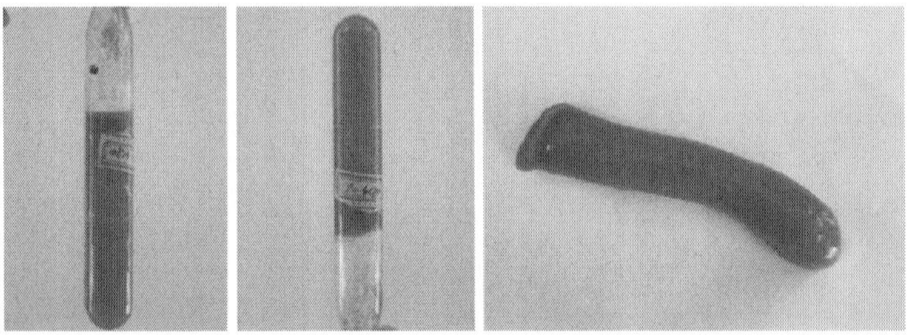

Figure. 1. The appearance of formed combined gel

Effect of polymer concentration

The polymer concentration is critical to the formation of three-dimensional network structure, which can greatly affect the gelation time. The effect of polymer concentration on gelation time was studied by preparing different gelant solutions containing 4.0 wt% resin, 3.0 wt% inorganic silicate and polymer concentrations varied from 0.1 to 0.4 wt% at 130 °C. Figure 2 shows the gelation results of gelant solutions containing various polymer concentrations. The gelation time decreased from 32.0 to 12.0 h with increasing the polymer concentration. As the polymer concentration increased, the cross-linking reaction chances between the amide group ($-CONH_2$) and hydroxymethyl group ($-CH_2OH$) increased, resulting in a decrease in gelation time. Therefore, according to proper gelation time and economical cost, the reasonable polymer concentration should range from 0.2 to 0.3 wt%.

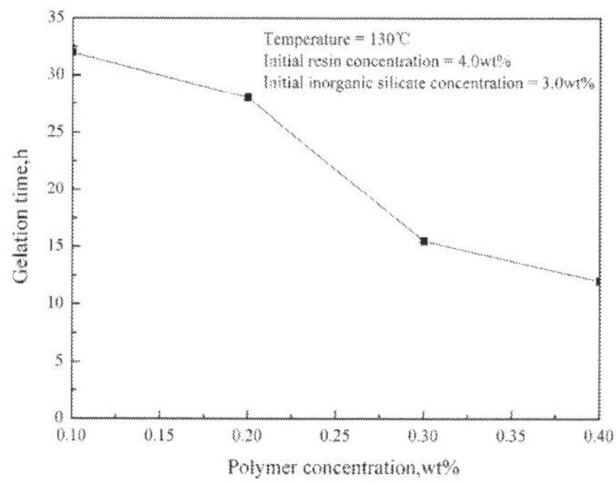

Figure. 2. Effect of polymer concentration on gelation time (130 °C)

Effect of resin concentration

The effect of resin concentration on gelation time was studied by preparing different gelant solutions containing 0.3 wt% polymer, 3.0 wt% inorganic silicate and the resin concentrations varied from 1.5 to 5.0 wt% at 130 °C. Figure 3 shows the gelation results of gelant solutions containing various resin concentrations. The gelation time decreased from 30.0 to 7.0 h with increasing resin concentration. As the resin concentration increased, the cross-linking reaction chances between the amide group ($-CONH_2$) and hydroxymethyl group ($-CH_2OH$) increased, resulting in a decrease in gelation time. Therefore, according to proper gelation time and economical cost, the reasonable resin concentration should range from 2.5 to 4.0 wt%.

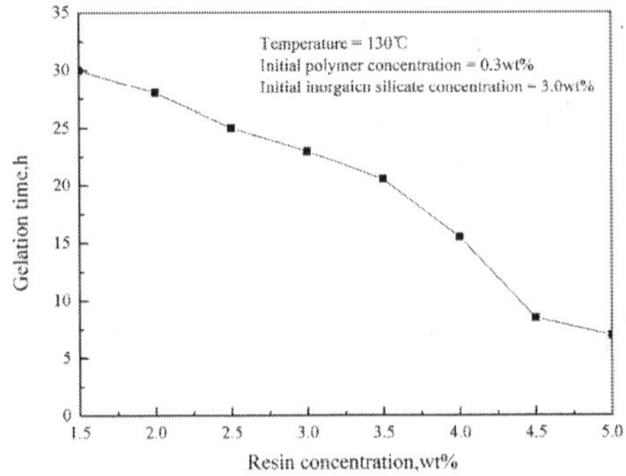

Figure. 3. Effect of resin concentration on gelation time (130 °C)

Effect of inorganic silicate

The inorganic silicate can improve the gelation performance greatly by forming another three-dimensional network structure. The effect of inorganic silicate concentration on gelation time was studied by preparing different gelant solutions containing 0.3 wt% polymer, 4.0 wt% resin, and the inorganic silicate concentrations varied from 1.0 to 8.0 wt% at 130 °C. Figure 4 shows the gelation results of gelant solutions containing various inorganic silicate concentrations. As the inorganic silicate concentration increased from 1.0 to 8.0 wt%, the gelation time decreased from 19 to 4 h. As the inorganic silicate concentration increased, the reaction rate of inorganic silicate forming three-dimensional network structure at high temperature was accelerated, resulting in a decrease in gelation time. Therefore, according to proper gelation time and economical cost, the reasonable inorganic silicate concentration should range from 3.0 to 5.0 wt%.

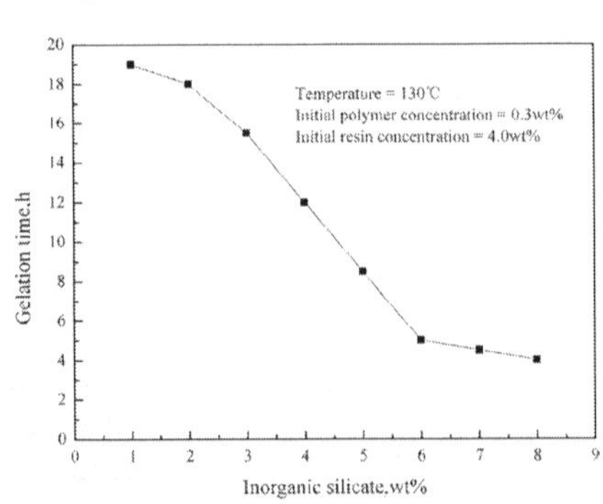

Figure. 4. Effect of inorganic silicate on gelation time (130 °C)

Effect of temperature

The gelation time is dependent on the component concentrations and the temperature. For the gelant solution containing 0.3 wt% polymer, 4.0 wt% resin and 3.0 wt% inorganic silicate, the effect of temperature on gelation time was studied. Figure 5 shows the gelation time of gelant solution containing 0.3 wt% polymer, 4.0 wt% resin and 3.0 wt% inorganic silicate at various temperatures. The gelation time decreased from 54.0 to 15.5 h with increasing temperature from 100 to 130 °C. As the temperature increased, the cross-linking reaction rate increased and resulted in a decrease in gelation time. The gelation time was correlated to the temperature according to the Arrhenius' equation:

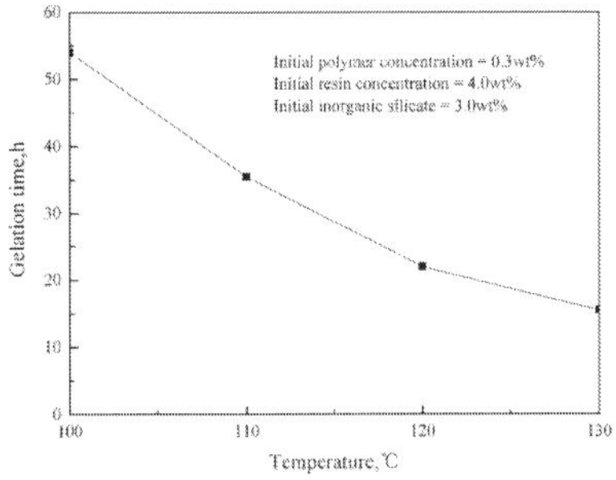

Figure. 5. Effect of temperature on gelation time

$$k = k_0 e^{-E_a/RT} \tag{3}$$

where k is reaction rate constant, k_0 is the frequency factor, E_a is the activation energy(kJ/mol), R is the gas constant (kJ/mol K) and T is absolute temperature (K) (Fig. 6). The Eq. (3) was modified by introducing the gelation time (t_g) instead of the reaction rate constant k. The modified equation was as follows (Nasr-El-Din and Taylor 2005):

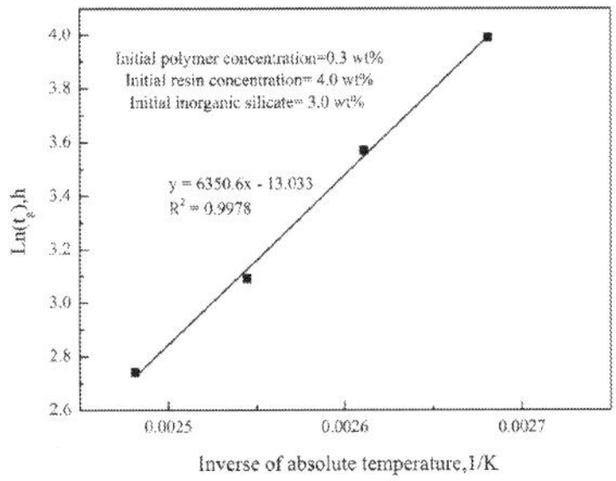

Figure. 6. Arrhenius-type plot of gelation time and temperature

$$t_g = Me^{E_a/RT} \tag{4}$$

where t_g is the gelation time (h), M is the time constant (h), E_a is the activation energy(kJ/mol), R is the universal gas constant (kJ/mol K) and T is absolute temperature (K). Then the Eq. (4) was transformed into the following equation:

$$\ln(t_g) = \frac{E_a}{RT} + \ln M. \tag{5}$$

As we can see from the Eq. (5), there is a linear relationship between the $\ln(t_g)$ and the inverse of the absolute temperature(1/T). Figure 5 shows the plot of ln (t_g) versus the inverse of absolute temperature. Calculated from the slope of Fig. 5, the activation energy E_a of the gelant system was 52.8 kJ/mol. Moreover, it is worth noting that the effect of temperature on the gelation time can be predicted by using Eq. (4).

Long-term thermal stability and salt tolerance

The long-term thermal stability and salt tolerance are crucial to the effectiveness of gel treatments in high temperature and high salinity reservoirs. In this study, the long-term thermal stability and salt tolerance were evaluated by measuring the degree of syneresis as a function of aging time. Figure 7 shows the degree of syneresis observed in bulk gel samples as a function of aging time. After aging for 60 days in simulated formation brine with high total dissolved solids (TDS: 20 % NaCl + 2.5 % $CaCl_2$) at 130 °C, for polymer gel composed of 0.3 wt% polymer and 4.0 wt% resin, about 92 % syneresis was observed, while for the organic/inorganic combined gel that composed of 0.3 wt% polymer, 4.0 wt% resin and 3.0 wt% inorganic silicate, about 40 % syneresis was observed.

The reasons for the phenomenon could be as follows (Liang and Li 2001): Gelation reaction between the amide group (–$CONH_2$) of polymer and hydroxymethyl (–CH_2OH) group of phenol formaldehyde resin could occur to form three-dimensional network structure. Moreover, the inorganic silicate could form another three-dimensional network structure at high temperature. Then compared with single polymer gel network structure, the combined network structure had less free water content and high strength structure, which can improve the gel thermal stability properties.

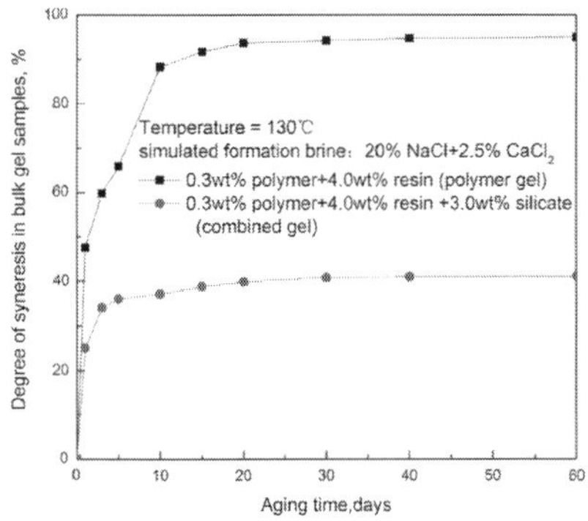

Figure. 7. The degree of syneresis observed in bulk gel samples of polymer gel and combined gel versus aging time

Therefore, compared with polymer gel, the organic/inorganic combined gel had better long-term thermal stability and salt tolerance.

Sand pack flow experiment

Sand pack flow experiments were conducted to evaluate the permeability reduction of the combined gel as a function of aging time at 130 °C. The initial permeability and pore volume of the testing sand pack were 1.47 μm^2 and 33.0 mL respectively. Figure 8 shows the permeability reduction of the combined gel as a function of aging time. For the organic/ inorganic combined gel composed of 0.3 wt% polymer, 4.0 wt% resin and 3.0 wt% inorganic silicate, after aging for 60 days at 130 °C, the combined gel showed approximately 85 % permeability reduction.

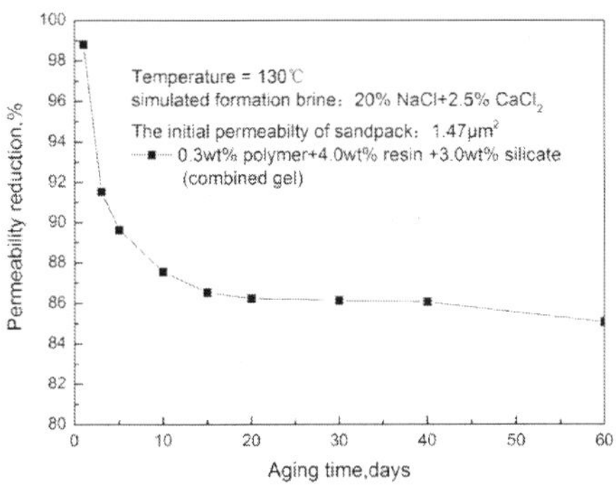

Figure. 8. The permeability reduction of combined gel versus aging time

CONCLUSIONS

(1) According to proper gelation time and economical cost, the formulation of the organic/inorganic combined gel was optimized as follows: the polymer concentration was 0.2–0.3 wt%, the phenol formaldehyde resin concentration was 2.5–4.0 wt% and the inorganic silicate concentration was 3.0–5.0 wt%.

(2) The gelation time decreased from 54.0 to 15.5 h with increasing temperature from 100 to 130 °C. After aging for 60 days in simulated formation brine with high total dissolved solids (TDS: 20 % NaCl + 2.5 % $CaCl_2$) at 130 °C, for the polymer gel composed of 0.3 wt% polymer and 4.0 wt% resin, about 92 % syneresis was observed, while for the organic/inorganic combined gel composed of 0.3 wt% polymer, 4.0 wt% resin and 3.0 wt% inorganic silicate, about 40 % syneresis was observed. The results indicated that the organic/inorganic combined gel had better long-term thermal stability and salt tolerance.

(3) Sandpack flow experimental results showed that for the organic/inorganic combined gel composed of 0.3 wt% polymer, 4.0 wt% resin and 3.0 wt% inorganic silicate, after aging for 60 days at 130 °C, the combined gel showed approximately 85 % permeability reduction.

ACKNOWLEDGMENTS

This work was financially supported by National S&T Major Project (Grant No. 2011ZX05024-004) on chemical flooding in Bohai heavy oil reservoir and the "Taishan Scholars" Construction Engineering (No.ts20070704).

REFERENCES

1. Albonico P, Bartosek M, Malandrino A, Bryant S, Lockhart T (1995) Studies on phenol–formaldehyde crosslinked polymer gels in bulk and in porous media. In: Paper SPE 28983, presented at international symposium on oilfield chemistry, San Antonio, Texas pp 14–17
2. Al-Muntasheri G, Nasr-El-Din H, Al-Noaimi K, Zitha P (2009) A study of polyacrylamide-based gels crosslinked with polyethyleneimine. SPE J 14(2):245–251
3. Banerjee R, Ghosh B, Khilar K, Boukadi F, Bemani A (2008) Field application of phenol formaldehyde gel in oil reservoir matrix for water shut-off purposes. Energy Sour Part A 30(19):1779–1787
4. Bryant SL, Bartosek M, Lockhart TP (1997) Laboratory evaluation of phenol–formaldehyde/polymer gelants for high-temperature applications. J Petrol Sci Eng 17(3):197–209
5. Eoff L, Dalrymple E, Everett D, Vasquez J (2007) Worldwide field applications of a polymeric gel system for conformance applications. SPE Prod Oper 22(2):231–235
6. Kabir A (2001) Chemical water and gas shutoff technology—an overview. In: Paper SPE 72119, presented at SPE Asia Pacific improved oil recovery conference, Kuala Lumpur, Malaysia, pp 6–9
7. Liang B, Li BY (2001) High strength combined gel CY-98 for selective water plugging: structure and properties. Oilfield Chem 18(1):27–30 (In Chinese)
8. Liu Y, Bai B, Wang Y (2010) Applied technologies and prospects of conformance control treatments in China. Oil Gas Sci Technol Revue d'IFP Energies nouvelles 65(6):859–878
9. Mary H, Wouter B, Aly H, Jarl V, John W (1999) The first carbonate field application of a new organically crosslinked water shutoff polymer system. In: Paper SPE 50738, presented at the SPE international symposium on oilfield chemistry, Houston, pp 16–19
10. Moradi-Araghi A (2000) A review of thermally stable gels for fluid diversion in petroleum production. J Petrol Sci Eng 26(1):1–10
11. Moradi-Araghi A, Bjornson G, Doe P (1993) Thermally stable gels for near-wellbore permeability contrast corrections. SPE Adv Technol Ser 1(1):140–145
12. Nasr-El-Din H, Taylor K (2005) Evaluation of sodium silicate/urea gels used for water shut-off treatments. J Petrol Sci Eng 48(3):141–160

13. Sengupta B, Sharma V, Udayabhanu G (2012) Gelation studies of an organically cross-linked polyacrylamide water shut-off gel system at different temperatures and pH. J Petrol Sci Eng 81:145–150
14. Sydansk R (1988) A new conformance-improvement-treatment chromium gel technology. In: Paper SPE 17329, SPE presented at the SPE enhanced oil recovery symposium, Tulsa, Oklahoma pp 16–21
15. Sydansk R, Southwell G (2000) More than 12 years? Experience with a successful conformance-control polymer-gel technology. Old Prod Facil 15(4):270–278
16. Vasquez J, Dalrymple E, Eoff L, Reddy B, Civan F (2005) Development and evaluation of high-temperature conformance polymer systems. In: Paper SPE 93156, presented at SPE international symposium on oilfield chemistry, The Woodlands, Texas, pp 2–4
17. Yadav US, Mahto V (2013) Investigating the effect of several parameters on the gelation behavior of partially hydrolyzed polyacrylamide–hexamine–hydroquinone gels. Ind Eng Chem Res 52(28):9532–9537

INDEX

B

C

D

E

F

G

H

I

O

P

S

X

Z

X

Z